MW01701848

Basic Engineering Thermodynamics

6

Basic Engineering Thermodynamics

SI UNITS

F. J. Wallace

DSc PhD MSc BSc CEng FIMechE

Professor of Engineering
Bath University of Technology

W. A. Linning

BSc CEng MIMechE

Lecturer-in-charge
Thermodynamics Laboratories
University of Birmingham

First published 1968
Revised and first metric edition 1970
First paperback edition 1970

ISBN: 0 273 31476 9
(Cased edition: 0 273 31475 0)

Preface

It is with considerable diffidence that the authors offer yet another text on Engineering Thermodynamics. In recent years there have appeared a number of excellent books, all based on a progressive and rational development of the subject from the first and second laws, an approach pioneered by Keenan in his *Thermodynamics.* As teachers of the subject with many years' experience, we nevertheless feel that this progressive unfolding of the subject, while eminently desirable in itself, is by no means obvious to the student and that the very concise treatment adopted by Keenan himself in its very nature does not commend itself to the elementary student.

While, therefore, not in the least wishing to abandon the rigour and coherence of an approach based firmly on the first and second laws, it is our intention in the present text to make the unfolding of the many new concepts involved in engineering thermodynamics more readily intelligible. In our view a basic course should be largely confined to these concepts, albeit with frequent and illuminating illustrations taken from practice, rather than with the applications themselves.

Power cycles, though not themselves a fundamental topic in the above sense, are included in a last chapter in order to illustrate the application of all the principles discussed in earlier chapters to the most important function of engineering thermodynamics, namely the conversion of latent energy into useful work. On the other hand, details of thermodynamic machinery, and the flow processes occurring within such machinery, e.g. flow through compressor or turbine blading, have been omitted. Likewise, the topic of transfer phenomena (mass, momentum and heat transfer) has not been included.

A first-year course would not necessarily include all topics covered by the present text, which is intended as a reference text on concepts not only for the first year but for all subsequent years up to honours standard. Thus the first twelve chapters and part of the last chapter could probably be covered fairly readily in the space of one session, possibly with certain omissions, while the last two chapters dealing respectively with processes involving chemical change and availability, as well as the mathematically more difficult appendices, may be deferred until the student has consolidated his knowledge.

We do not claim any very great originality in the order adopted for the first twelve chapters. With comparatively minor variations this has been found by many authors to lend itself best to the gradual unfolding of the subject. Chapters 4, 5, 6, 7, and 14 dealing respectively with the first law, reversibility, the second law, entropy and availability, contain separate sub-sections on non-flow and flow systems. While frequent references are made in these chapters to physical and formal similarities between these two types of system, we nevertheless feel very strongly that this insistence on separate treatments is amply justified as, in our experience, students often carry misconceptions concerning these matters right into the final year—and beyond.

The most important subject of thermodynamic irreversibility, which is again a frequent cause of confusion in the student's mind, has been given a good deal of prominence and we feel we may claim some originality in the initial presentation of the subject in Chapter 5 in which the physical causes of both external and internal irreversibilities are described in a rather more lucid manner than the authors have found in other texts. Chapter 14 which deals with availability attempts to give a sufficiently detailed treatment to enable the student to use this important theoretical tool with confidence and with a full understanding of its physical implications.

The appendices are intended primarily for the "advanced" student or for the more gifted and curious first-year student who wishes to probe more deeply into the fundamentals of the subject. Appendix I gives a simplified statistical treatment of entropy for which the authors are indebted largely to Obert. This is felt to have a definite educational value in underlining the true physical nature of entropy as a molecular phenomenon rather than a mathematical abstraction. It is only here and in Chapter 9 dealing with the kinetic theory of gases that the authors have thought it necessary to depart from the continuum treatment of thermodynamics.

Appendix II gives the thermodynamic relations which, because of their essentially mathematical nature, might be found "off-putting" by a first-year student, but which are nevertheless extremely important in later work. Appendix III summarizes the important connections between p–v–T relations and other properties of the pure substance, and includes also a discussion of equations of state.

The authors hope and believe that the present text will provide a coherent and readily intelligible account of the subject and will be found suitable by students throughout their period of University training.

F.J.W. & W.A.L. *January* 1970

Contents

Chapter 1

Introduction

1.1 Scope of Subject

Thermodynamics is the science which deals with changes in STATE as measured by changes in certain observable characteristics or PROPERTIES experienced by substances as a result of energy transfers, primarily in the form of WORK and HEAT. (These and other terms are used rather loosely in this introductory chapter, but will be defined precisely in Chapter 2.)

We shall see in Chapter 2 that the most convenient framework within which thermodynamics can be studied is the SYSTEM, i.e. a quantity of matter, not necessarily constant, contained within a BOUNDARY, which may be

(*a*) physical, e.g. an engine cylinder and piston, with all valves or ports closed,
(*b*) partly physical and partly imaginary, e.g. an engine cylinder with one or more valves or ports open, or
(*c*) wholly imaginary, e.g. a hypothetical closed surface in a fluid field.

The nature of the boundary is generally unimportant, but depending on whether the system experiences only energy transfer, or combined energy and mass transfer at the boundary, we speak of CLOSED or OPEN SYSTEMS.

The practical importance of engineering thermodynamics is evidently immense, since it determines our understanding of all those processes whereby LATENT ENERGY in the following forms is converted into useful work.

(*a*) *Heat*, e.g. solar radiation, and high-pressure high-temperature steam (from natural sources such as geysers in Iceland or New Zealand, or created artificially in steam boilers).
(*b*) *Chemical energy*, i.e. energy released by changes in chemical composition of certain substances, primarily solid, liquid or

gaseous fuels, and which becomes available once again as heat.

(*c*) *Atomic energy*, i.e. energy released by changes in the atomic structure of certain substances such as Uranium 235, resulting in the first instance in the release of heat.

Thus engineering thermodynamics is absolutely basic to our understanding of the mode of operation of all PRIME MOVERS.

It should be noted that such energy conversions as

(*d*) electrical energy to useful work
(*e*) wind power to useful work
(*f*) water power to useful work

do not fall within the scope of engineering thermodynamics since they constitute conversions from one kind of "high-grade" energy to another (see Chapter 14), whereas the first group constitutes conversion from low-grade to high-grade energy and is subject to the peculiar restrictions imposed by the SECOND LAW OF THERMODYNAMICS (see Chapter 6). However, all forms of energy conversion are subject to the FIRST LAW OF THERMODYNAMICS (see Chapter 4) which is merely a particular formulation of the general principles of conservation of energy.

In engineering thermodynamics we are primarily concerned with the conversion process (*a*), i.e. heat into useful work, though in two of the last chapters of the book (Chapters 13 and 14) we shall also consider combustion, i.e. the conversion of chemical energy into useful work. Chemical and physical thermodynamics in a wider sense deal with the behaviour of systems such as electric cells, gas flow through strong magnetic, electric or gravitational fields, and systems subject to capillary action (surface tension). These latter effects, which may be summarized as electricity, magnetism, gravity and capillarity, are excluded from our studies.

1.2 Microstructure and Macrostructure

We are all aware of the fact that ultimately all matter is composed of atoms and molecules, i.e. minute particles with which are associated a whole range of phenomena: for instance, the equivalence of energy and mass as expressed by Einstein's famous formula

$E = mc^2$ where E = energy, m = particle mass, c = particle velocity

and also energy emission in quanta, as expressed by Planck's constant h, to mention only two of these. Whilst we should certainly be aware of this state of affairs, it is only in rare cases that consideration of the behaviour of matter at atomic or molecular level becomes important to the engineer (e.g. in systems containing highly rarefied gas, such as the upper atmosphere, or in plasma flow, i.e. the flow of highly rarefied gas through strong electrical or magnetic fields). However, in the overwhelmingly greater proportion of the processes and systems we consider in engineering thermodynamics, it is sufficient to treat matter as a CONTINUUM, i.e. to adopt a MACROSCOPIC rather than a MICROSCOPIC approach. This means that we define the state of a substance in terms of measurable or observable characteristics associated with vast aggregations of atoms or molecules.

Nevertheless, in some instances our understanding of certain concepts in thermodynamics rather than our success in solving specific problems is enhanced by considering the ultimate microscopic nature of matter. For this reason the KINETIC THEORY OF GASES is discussed in Chapter 9, whilst a simplified statistical approach to the exposition of that very important thermodynamic property ENTROPY (see Chapter 7) is given in Appendix 1.

The situation in thermodynamics is not unlike that in mechanics. We know that under certain conditions the classical or Newtonian concepts concerning, for instance, motion in straight lines no longer apply and that, instead, we have to resort to the theory of relativity. Nevertheless, Newtonian concepts are still so immensely useful (and so much more readily intelligible) that to discard them would be like the proverbial throwing out of the baby with the bath water.

1.3 Scope of the Present Book

In the present book we are concerned with the fundamental laws governing thermodynamic processes, in particular the first and second laws: these are discussed at length in Chapters 4 to 7, the most important basic definitions being given in Chapters 2 and 3. These laws are generalizations from a vast number of experiments rather than from abstract principles. However, they are general to the extent that they can be applied to systems irrespective of the nature of the working fluid. Since, nevertheless, we cannot divorce the analysis of thermodynamic processes from the particular substance used, a detailed discussion of the properties of gases and vapours is included and given in Chapters 8 and 10.

In many cases, the laws of thermodynamics as well as those of mechanics have to be invoked in order to solve a problem. This is particularly so in the case of gas flow with heat transfer or friction. Chapter 11 is therefore devoted to a survey of the techniques of FLUID MECHANICS, and in particular of GAS DYNAMICS.

More complex processes, involving mixtures of two or more different substances in more than one state of aggregation, are discussed in Chapter 12, whilst processes involving chemical change are dealt with in Chapter 13.

Chapter 14 presents a powerful method of analysis of thermodynamic processes, designed to highlight sources of waste or inefficiency, and known as AVAILABILITY ANALYSIS.

Chapter 15 is a survey of vapour and gas power cycles in which the most important thermodynamic characteristics, namely thermal efficiency, work ratio and specific consumption, are analysed and compared. For more detailed treatments the reader is referred to a number of excellent texts written specifically on this topic.

Finally a number of "advanced" topics are dealt with in the Appendices.

The present book does not include any detailed reference to MASS, MOMENTUM AND HEAT TRANSFER or to certain important basic concepts associated with these phenomena, particularly the boundary layer. Since all these transfer phenomena are dependent on GRADIENTS, e.g. temperature gradients for momentum transfer and concentration gradients for mass transfer, and since these gradients in turn determine the rate at which the processes proceed, we may refer to these phenomena also as RATE PHENOMENA. The study of rate phenomena is an essential part of our understanding of the detailed functioning of thermodynamic devices. However, the underlying physical concepts and mathematical treatments are, in the view of the authors, more likely to cause difficulty to students in their first and second year than the topics treated in the present book. It is, therefore, suggested that these topics should be dealt with in later years, preferably as a unified topic under the general heading TRANSFER PHENOMENA. An excellent example of the type of treatment envisaged is W. M. Rohsenow, and H. Y. Choi, *Heat, Mass and Momentum Transfer* (Prentice Hall, 1961).

Chapter 2

Units and Definitions

2.1 Units

All physical quantities encountered in engineering thermodynamics may be expressed in terms of three mutually independent fundamental quantities, of which length and time are two. The choice of the third lies between mass and force, although the former, being the more truly fundamental, is preferred.

Thus force, work, power, heat, internal energy, enthalpy, entropy, to cite only a selection of quantities encountered in the study of thermodynamics, may all be expressed in terms of mass, length and time. This applies even to temperature, which represents a molecular energy level, and has therefore the units appropriate to energy, although in many practical applications it is convenient to introduce temperature as a fourth independent quantity.

2.1.1 Force, work and energy

Since in thermodynamics we are concerned with processes involving ENERGY exchange, it is essential to establish a unit of energy. We shall meet many forms of energy in our study of thermodynamics, namely, potential, kinetic, internal, chemical, etc., but all are covered by the common definition—

ENERGY IS THE CAPACITY, EITHER LATENT OR APPARENT, TO EXERT FORCE THROUGH A DISTANCE.

The capacity to exert force through a distance is, of course, mechanical work, measured as the product of force and distance moved by the point of application in the direction of the force,

$$W = Fx \tag{2.1}$$

Since FORCE is not included in our three fundamental quantities, we have to find a suitable relationship. This is provided by NEWTON'S SECOND LAW, which may be stated in the form

$$\text{Force} = (\text{rate of change of momentum}) \times \text{constant}$$

When any *coherent* system of units is used, such as the SI system, the constant reduces to unity, i.e. when a force F is applied to a body of mass m, the former equals the resultant change of momentum, or

$$F = \frac{d}{dt}(mc) \qquad \text{where } c = \text{velocity } \frac{dx}{dt}$$

$$= m\frac{dc}{dt}$$

$$= m\frac{d}{dt}\left(\frac{dx}{dt}\right)$$

$$= m\frac{d^2x}{dt^2} \qquad (2.2)$$

In the present text in which the SI system of units is employed (see refs. 2.1, 2.2, 2.3, 2.4) the fundamental quantities, mass, length and time, are defined as follows—

Unit of Mass: The unit of mass called the KILOGRAMME is the mass of the international prototype which is in the custody of the Bureau International des Poids et Mesures (BIPM) at Sèvres, near Paris, France (3rd CGPM, 1901).

Unit of Length: The unit of length called the METRE is 1 650 763·73 wavelengths in vacuo of the radiation corresponding to the transition between the energy levels $2p_{10}$ and $5d_5$ of the krypton-86 atom (11th CGPM, 1960).

Unit of Time: The unit of time called the SECOND is defined by the fraction:

1/31 556 925·9747 of the tropical year for 1900 January 0 at 12 h ephemeris time (11th CGPM, 1960).

(The tropical year is the time interval between consecutive passages, in the same direction, of the Sun through the Earth's equatorial plane.)

From these units we may derive the units of all other physical quantities met with in thermodynamics.

From equation (2.2) we define a derived *unit of force* as follows—

$$1 \text{ derived unit of force} = 1 \text{ kg} \times 1 \text{ m/s}^2$$

$$= 1 \text{ kg-m/s}^2$$

$$= 1 \text{ newton (N)} \qquad (2.2a)$$

The derived unit of force is designated the NEWTON (N) and is such that, when a force of 1 newton is applied to a mass of 1 kg, it results in an acceleration of 1 m/s^2.

The MKS system (metre-kilogramme-second system) unlike the SI system is not coherent and employs the kilogramme-force (kgf) as the standard unit of force. This is defined as that force which, when applied to a mass of 1 kg, will produce the standard gravitational acceleration $g = 9{\cdot}80665\ m/s^2$. Since the MKS system is *not* coherent, i.e. the product (or quotient) of any two quantities is *not* the unit of the resultant quantity, a constant has to be introduced into equation (2.2), i.e.

$$F \times C = m\frac{d^2x}{dt^2}$$

i.e. in the MKS system

$$1\ \text{kgf} \times C = 1\ \text{kg} \times 9{\cdot}80665\ \text{m/s}^2 \qquad (2.2b)$$

$$\text{where } C = g_0 = 9{\cdot}80665\ \text{kg-m/kgf-s}^2$$

It is clear that the introduction of the constant $C = g_0$ is an unnecessary and undesirable complication and is likely to lead to errors on the part of the student.

The basic *unit of work* in the SI system is given by equation (2.1), i.e.

$$1\ \text{unit of work} = 1\ \text{N} \times 1\ \text{m} = 1\ \text{N-m}$$

$$= 1\ \text{joule (J)} \qquad (2.3)$$

In the MKS system

$$1\ \text{unit of work} = 1\ \text{kgf} \times 1\ \text{m} = 1\ \text{kgf-m} \qquad (2.3a)$$

The chief weakness of the MKS system represented by equations (2.2b) and (2.3a), apart from its inconvenience, is that it depends on a standard value of the gravitational acceleration. Clearly g varies from point to point on the Earth's surface, depending on altitude, as well as latitude. Altitude determines the magnitude of the Earth's gravitational pull, while latitude influences the centripetal force which has to be provided in order to effect the centripetal acceleration $f_c = \omega^2 R$ (ω = angular velocity of the Earth, R = radius of section through the Earth parallel to equatorial plane) which in turn is given by equation (2.2)

$$F_c = mf_c$$

and which results in a reduction of the effective gravitational pull and hence in the local value of g.

2.1.2 Pressure p

Pressure may be defined as the force per unit area exerted by a fluid on a real or imaginary surface, in a direction normal to that surface. In SI units pressure is measured in newtons per square metre (N/m^2).

In relation to atmospheric pressure, which serves as a useful reference standard, the N/m^2 is an inconveniently small unit and is replaced for convenience by the BAR where

$$1 \text{ bar} = 10^5 \text{ N/m}^2 \tag{2.4}$$

The conversion to British units is as follows:

$$\begin{aligned} 1 \text{ bar} = 10^5 \text{ N/m}^2 &= 14{\cdot}50 \text{ lbf/in}^2 \\ &= 14{\cdot}50 \times 144 \times 32{\cdot}2 \text{ pdl/ft}^2 \end{aligned} \tag{2.4a}$$

In practice we distinguish between ABSOLUTE PRESSURE and GAUGE PRESSURE. The former is measured by the height of a column of suitable fluid (usually mercury) supported by the applied pressure in a manometer designed so that no pressure acts on the top of the fluid column, i.e. having a vacuum above it, while the latter is measured by the height of column of fluid supported by the applied pressure when atmospheric pressure acts on the top of the column.

Such pressure-measuring instruments are known as manometers.

ABSOLUTE PRESSURE p_a

From elementary principles of hydrostatics,

$$\begin{aligned} \text{Absolute pressure} &= \text{weight of liquid column per unit area} \\ &= \text{gravitational force on liquid column per unit area} \\ &= A\rho hg/A = \rho hg \frac{\text{kg.m.m}}{\text{m}^3.\text{s}^2} \\ &= \rho hg \text{ N/m}^2 \end{aligned} \tag{2.5}$$

Since ρ and g may be regarded as constant, the height of the column h is a measure of pressure.

The standard instrument for recording atmospheric pressure (i.e. the weight of the atmosphere per unit area of the Earth's surface) is the barometer, similar in form to the instrument shown in *Fig.* 2.1a. Hence it is usual to measure the atmospheric pressure p_a in metres of mercury (m.Hg). The density of mercury being $\rho = 13616$ kg/m^3, the conversion from m.Hg to N/m^2 is given by

$$\begin{aligned} 1 \text{ m.Hg} = \rho hg &= 13616 \times 1 \times 981 \\ &= 1{\cdot}3366 \times 10^5 \text{ N/m}^2 \\ &= 1{\cdot}3366 \text{ bar} \end{aligned}$$

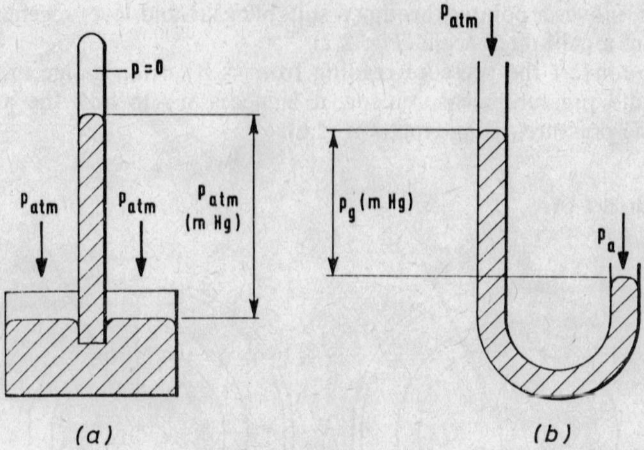

Fig. 2.1
(*a*) *Barometer*
(*b*) *Differential manometer*

Standard atmospheric pressure is defined as

$$p_{atm} = 1 \text{ atmosphere} = 1{\cdot}01325 \text{ bar}$$
$$= 101{\cdot}325 \text{ kN/m}^2$$
$$= 0{\cdot}760 \text{ m.Hg}$$

GAUGE PRESSURE p_g

From the note on the measurement of absolute and gauge pressure it is evident that

$$p_g = p_a - p_{atm} \tag{2.6}$$

where $p_{atm} = 1{\cdot}01325$ bar

Gauge pressure is measured by a differential manometer, in which the pressure to be measured is applied to one limb of a U-tube containing the barometric fluid, and atmospheric pressure to the other (*Fig*. 2.1b). The *difference* in height between the columns is clearly proportional to the pressure difference $p_a - p_{atm}$, i.e. to the gauge pressure p_g.

In practice gauge pressure is measured with the help of so-called Bourdon gauges rather than with manometers. These usually rely on the deformation of a bent hollow tube of suitable material which, when subjected to the pressure to be measured on the inside (and atmospheric pressure on the outside) tends to unbend and, in so doing, moves a pointer through a suitable gear-and-lever mechanism against a calibrated scale (*Fig.* 2.2).

To convert the pressure reading from a Bourdon gauge into an absolute pressure measurement, it is necessary to add the atmospheric pressure, using equation (2.6).

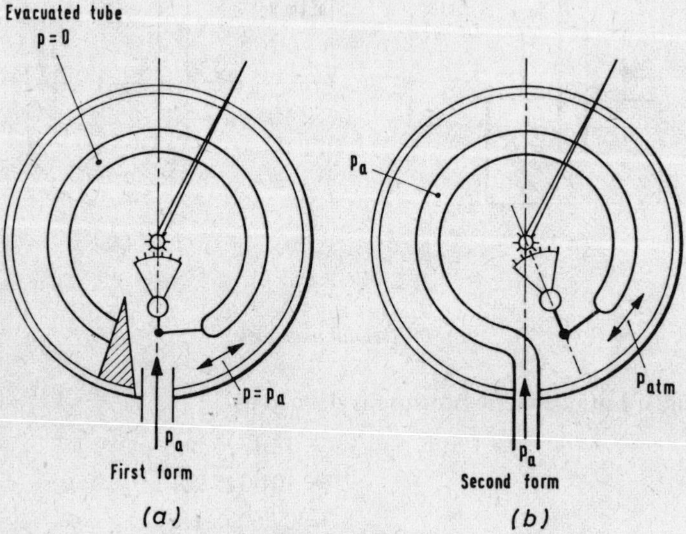

Fig. 2.2 Bourdon pressure gauge

Gauge pressure can be positive,

$$p_g = p_a - p_{atm} > 0 \tag{2.6a}$$

or negative (sub-atmospheric),

$$p_g = p_a - p_{atm} < 0 \quad \text{or} \quad p_{atm} - p_a > 0 \tag{2.6b}$$

Both Bourdon gauges and manometers can be adapted to read either positive or negative gauge pressures (*Fig.* 2.3).

Negative gauge pressures arise particularly in the case of steam plant in which the exhaust pressure from the turbine or engine (see *Fig.* 4.14) is usually well below atmospheric, of the order of 0·02 to 0·1 bar. In such cases it is normal to speak of a "vacuum" of x bar (or m.Hg), this "vacuum" being given by equation (2.6b)

$$\text{"vacuum"} = p_{atm} - p_a \tag{2.6c}$$

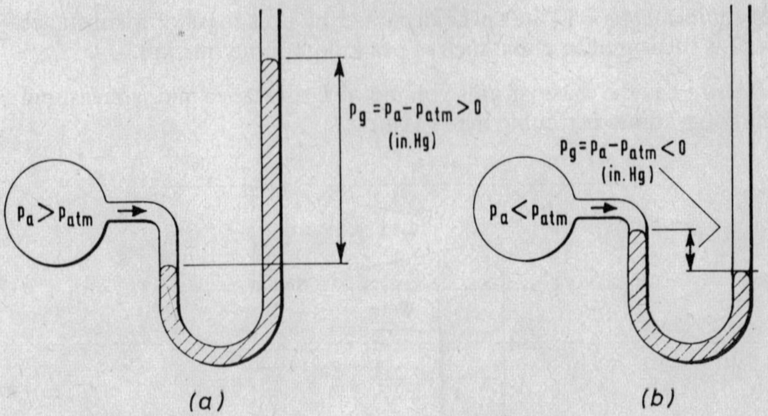

Fig. 2.3 Positive and negative gauge pressure

EXAMPLE 2.1

(*a*) A steam turbine exhausts into a condenser, the gauge on which reads 0·650 m.Hg vacuum. Express this in bar absolute pressure, assuming the atmospheric pressure is 1·013 bar.

SOLUTION

$$\text{Absolute pressure} = 1{\cdot}013 - 0{\cdot}650 \times 1{\cdot}3366$$
$$= 1{\cdot}013 - 0{\cdot}869 = 0{\cdot}144 \text{ bar} \quad \textit{Ans.}$$

(*b*) A venturi meter measures the pressure difference on the up- and downstream sides of a contraction in area of the channel through which the fluid is flowing. The pressure difference is very often measured by means of a U-tube with mercury in the bottom of the tube and the fluid flowing on top of each column of mercury. The flow is proportional to h, the pressure difference, m.Hg. Express this in bar, assuming the fluid to be water. S.G. of mercury 13·6.

SOLUTION

Referring to *Fig*. 2.4 (p. 12),

$$\text{Pressure difference} = [(13{\cdot}6 - 1)/13{\cdot}6]h \text{ m.Hg}$$
$$= (12{\cdot}6/13{\cdot}6) \times 1{\cdot}3366h \text{ bar}$$
$$= 1{\cdot}239h \text{ bar} \quad \textit{Ans.}$$

2.1.3 Volume (V), specific volume (v) and density (ρ)

The above quantities occur very frequently in thermodynamics and may be defined as follows—

Volume V is the space occupied by a substance, and is measured in cubic metres (m^3).

Specific volume v is the space occupied by unit mass of a substance and is measured in cubic metres per kilogramme (m^3/kg).

Density ρ is the mass of unit volume of a substance and is measured in kilogramme per cubic metre (kg/m^3).

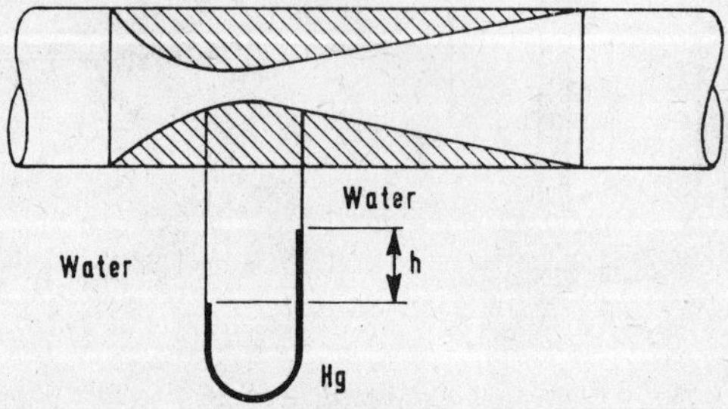

Fig. 2.4

This leads to the simple relationships

$$v = V/m \tag{2.7a}$$

$$\rho = m/V \tag{2.7b}$$

$$v = 1/\rho \tag{2.7c}$$

2.2 Work

We have already seen that the product of force and distance moved by the point of application in the direction of the force is mechanical work (see equation (2.1)). We know from experience that work in thermodynamic processes may assume a number of forms, all equivalent to, and hence measurable in the same units as, displacement work.

(*a*) DISPLACEMENT WORK (*Fig.* 2.5)

Gas contained in a cylinder by a piston can do work on that piston provided its initial pressure is greater than that of the surroundings. The NORMAL force acting on the piston is given by the product of pressure and surface area

$$F = pA$$

so that, for a small displacement δx of the piston, using equation (2.1),

$$p \;\delta W_{disp} = A\delta x \tag{2.8}$$

Furthermore, since $A\delta x = \delta V$, the change in cylinder volume,

$$\delta W_{disp} = p\delta V \tag{2.8a}$$

Equation (2.8) is the basis of work transfer evaluation in all reciprocating engines. Displacement work is discussed in much greater detail in Chapters 4 and 5.

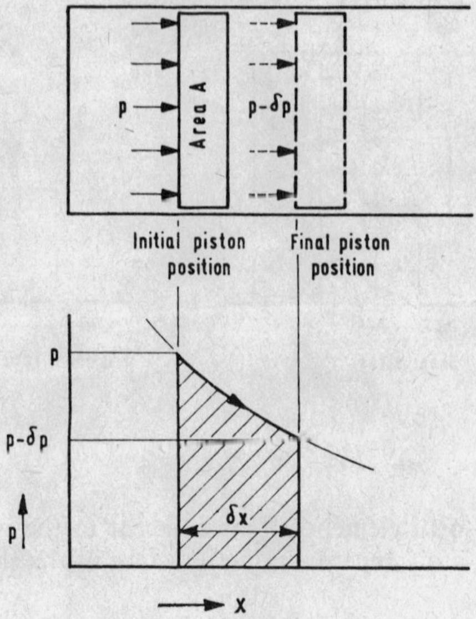

Fig. 2.5 Displacement work

Equations (2.8) may be generalized in the form of the statement

Displacement work = (normal stress or pressure)
× (area of surface element)
× (displacement normal to surface element) (2.8b)

The units of displacement work are given by equation (2.8a) with p in N/m^2, V in m^3, i.e. W_{disp} in N-m or JOULES (J).

(*b*) SHEAR WORK—FLUID FRICTION (*Fig.* 2.6)

If, in the cylinder piston arrangement shown in *Fig.* 2.5, the expansion process takes place rapidly, the *viscosity* of the working fluid will cause shear stresses σ to be set up, generally in the opposite direction to the piston and hence to the gas motion. In particular, at the cylinder wall a wall shear stress σ arises which is proportional to the product of the coefficient of viscosity and the local velocity gradient. (*Fig*. 2.6a.)

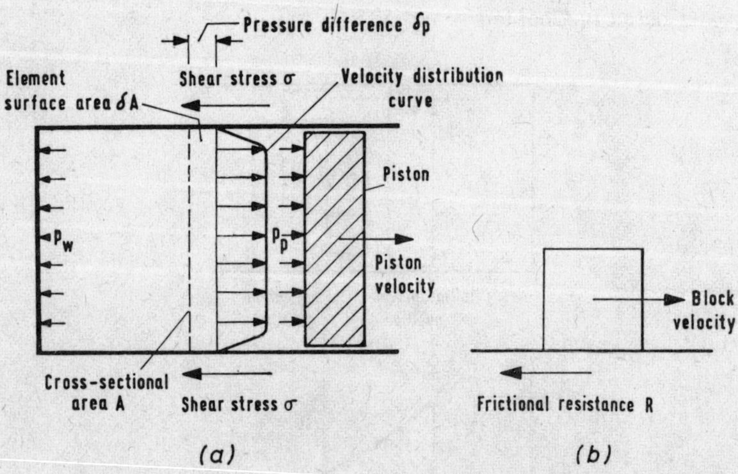

Fig. 2.6 Shear work

Considering a small element of gas adjacent to the wall it is clear that to move it a distance δy in the direction of the shear stress the work becomes

$$\delta W_{shear} = \sigma \delta A \delta y \tag{2.9}$$

where σ = shear stress, N/m^2; δA = area of element adjacent to wall, m^2; δy = distance moved in direction of the shear stress, m,

i.e. δW_{shear} has the unit N-m or joules (J)

Equation (2.9) may be generalized in the form of the following statement:

Shear work = (tangential or shear stress)
× (area of surface element)
× (displacement tangential to surface element) (2.9a)

The situation is analogous to that in mechanics of solids, when a block is caused to slide over a surface in the presence of friction (*Fig*. 2.6*b*), the frictional resistance acting in the plane of contact being R. The work in overcoming friction, for a displacement δy, becomes

$$\delta W_f = R\delta y \tag{2.9b}$$

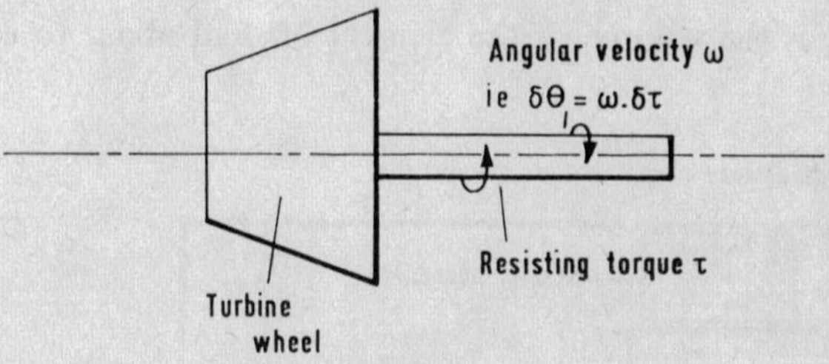

Fig. 2.7 Shaft work

(*c*) SHAFT WORK (*Fig.* 2.7)

Shaft work is associated with rotating members, e.g. an engine crankshaft or output shaft of a gas turbine, and is equal to the product of TORQUE and ANGULAR DISPLACEMENT, i.e.

$$\delta W_{shaft} = \tau\delta\theta \tag{2.10}$$

where τ = torque = force × distance (N-m)

$\delta\theta$ = angular displacement in radians

i.e. δW_{shaft} has the unit N-m or J

Note: The magnitude of an angle in radians is defined as

$$\theta = \frac{\text{arc subtended by angle}}{\text{radius}}$$

and is therefore a dimensionless quantity. It follows that the product $\tau\delta\theta$ has the units of torque, i.e. m-N. However, to differentiate between torque and work, the units of the former are always written as m-N, whilst those of the latter are written as N-m, as with displacement and shear work.

(*d*) FLOW WORK (*Fig.* 2.8)

In processes involving fluid flow, it is useful to consider the work done by an imaginary piston acting on a fluid element about to enter the SYSTEM being analysed (see section 2.5).

This work is therefore analogous to displacement work, the pressure p being a fluid pressure in the plane of the imaginary piston and acting in a direction normal to it. The work done *on* this imaginary piston *by* the external pressure is given by equation (2.8a)

$$\delta W_{flow} = p\delta V \tag{2.10}$$

where δV is the volume of the element of fluid about to enter the system.

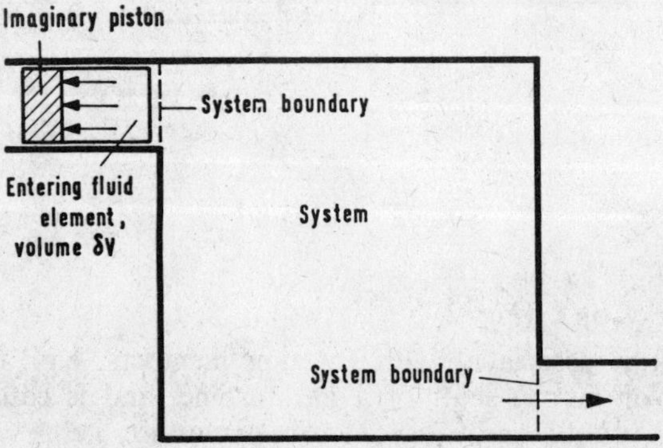

Fig. 2.8 Flow work

Using the specific volume relationship (2.7a) equation (2.10) becomes

$$\delta W_{flow} = pv\delta m \tag{2.10a}$$

which constitutes work done by the imaginary piston *on* the system.

Since in processes involving flow there is usually simultaneous flow of fluid into and out of the system, we may distinguish between

$$(\delta W_{flow})_{in} = pv\delta m_{in} \tag{2.10b}$$

for a fluid element entering the system, and

$$(\delta W_{flow})_{out} = pv\delta m_{out} \tag{2.10c}$$

which constitutes work done by the system *on* a second imaginary piston being moved by a fluid element leaving the system. The quantity

$$w = pv \tag{2.11}$$

is known as the SPECIFIC FLOW WORK.

The units of flow work are clearly those of displacement work, namely N-m (=J), whilst those of specific flow work are N-m/kg (=J/kg).

EXAMPLE 2.2

(*a*) *Displacement work*

A steam engine cylinder has a bore of 0·2032 m and a stroke of 0·254 m. The area of the indicator diagram is 7·0967 cm², its maximum length is 6·125 cm² and the indicator spring number used in the indicator is 2·04 bar/cm.

Calculate the indicated horsepower (ihp) when the engine is running at 200 rev/min.

The atmospheric pressure p_a and back pressure p_b are shown in the indicator diagram (*Fig*. 2.9).

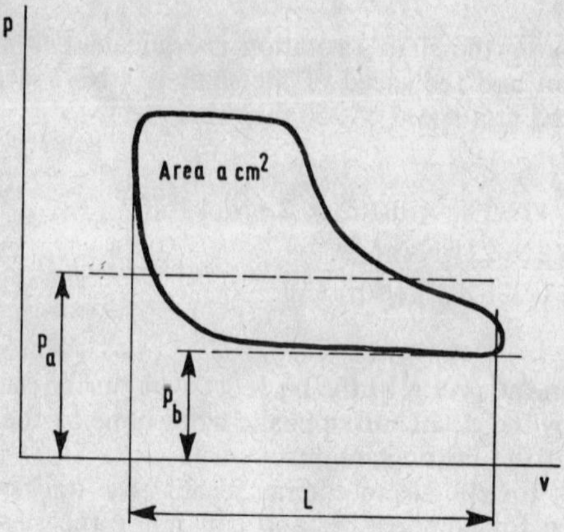

Fig. 2.9

SOLUTION

(*a*) $p_m = \dfrac{aS}{l}$

where a = area of indicator diagram (cm²)

S = spring number (bar/cm)

l = length of indicator diagram (cm)

Therefore

$$p_m = 7{\cdot}0967 \times 2{\cdot}04/6{\cdot}125 = 2{\cdot}298 \text{ bar}$$

$$1 \text{ bar} = 10^5 \text{ N/m}^2$$

$$\text{Power } W = 10^5\, p_m ALN \text{ N-m/s}$$

$$= 100 p_m ALN \text{ kJ/s or kW}$$

where cylinder area $A = \frac{1}{4}\pi \times 0{\cdot}2032^2 = 0{\cdot}03243 \text{ m}^2$

Stroke $L = 0{\cdot}254$ m

Number of revolutions per second $N = 200$ cycles/min $= 3{\cdot}333$ cycles/s
whence $W = 10^5 \times 2{\cdot}298 \times 0{\cdot}03243 \times 0{\cdot}254 \times 3{\cdot}333$

$= 6482$ Nm/s
$= 6{\cdot}482$ kJ/s or kW *Ans.*

(*b*) *Shaft Work*

When a shaft has a ready load resisting rotation, and is running at a steady speed, the torque τ exerted by the shaft is constant. The work transmitted by the shaft is the product of the torque and the radians per unit time through which the shaft is turning:

$$W = \tau\omega$$

If the load resisting the shaft's rotation is equivalent to 11·3398 N at a radius of 0·762 m and the speed of the shaft is 4·1667 rev/s, what is the power transmitted at a speed of 250 rev/min?

SOLUTION

$$W = \tau\omega = 11{\cdot}3398 \times 0{\cdot}762 \times 2\pi \times 4{\cdot}1667$$
$$= 226{\cdot}221 \text{ N-m/s}$$
$$= 0{\cdot}226221 \text{ kJ/s or kW} \quad \textit{Ans.}$$

Notes

1. Work done on the piston by the back pressure during the inward stroke is cancelled by the equal and opposite work done by the back pressure on reversal of the piston motion.
2. Single acting, for the steam engine, means that only one side of the piston is used for compressing and expanding the steam and so the working cycles per minute equal the revolutions per minute for each cylinder.
3. Double acting, which is more usual with steam engines, means that both sides of the piston are used and so, if we neglect the small difference of area involved due to the piston rod, the work done is twice that for single acting working. In other words, the working cycles per minute are twice the revolutions per minute. The ihp in the question then becomes $2 \times 8{\cdot}38 = 16{\cdot}76$.
4. Similarly, when we consider the internal combustion engine there are two types of cycle used, namely the two-stroke and the four-stroke cycles. The numbers denote the strokes required to complete one cycle, so we may say theoretically that, for identical cycles, the two-stroke cycle engine is twice as powerful as the four-stroke engine.

2.3 Power

This is defined as the rate of doing work and is therefore measured in N-m/s or J/s. The standard unit of power is the WATT (W) defined as

$$1 \text{ watt (W)} = 1 \text{ N-m/s} = 1 \text{ J/s} \qquad (2.12a)$$

In many cases a larger unit is preferred, the kilowatt (kW) given by

$$\begin{aligned} 1 \text{ kilowatt (kW)} &= 10^3 \text{ W} \\ &= 10^3 \text{ J/s} \\ &= 1 \text{ kJ/s} \end{aligned}$$

The kilowatt is also the standard unit of ELECTRICAL POWER, but in mechanical engineering we retain the horsepower (hp) where 1 metric hp = 75 kgf-m = 75 × 9·80665 N-m/s. That is,

$$\begin{aligned} 1 \text{ metric hp} &= 0{\cdot}7355 \text{ kW} \\ &= 735{\cdot}5 \text{ W} \end{aligned} \tag{2.12b}$$

To add to the confusion, the British horsepower differs slightly from the metric horsepower:

$$\begin{aligned} 1 \text{ British hp} &= 0{\cdot}7457 \text{ kW} \\ &= 745{\cdot}7 \text{ W} \end{aligned} \tag{2.12c}$$

Since mechanical (and electrical) work are equivalent, in a limited sense (discussed in Chapters 4, 5 and 6), to HEAT, it follows that a common system of units can be adopted. However, historically, thermal units have evolved through calorimetry. Thus, in many works the thermal unit of energy is the CALORIE (cal) or kilocalorie (kcal) related to SI units of work as follows:

$$1 \text{ cal} = 4{\cdot}1868 \text{ J} \tag{2.13a}$$

$$1 \text{ kcal} = 4{\cdot}1868 \text{ kJ} \tag{2.13b}$$

The conversion constant in equations (2.13a) and (2.13b) is known as the MECHANICAL EQUIVALENT OF HEAT or JOULE'S EQUIVALENT (see section 4.1), denoted by J.

$$J = 4{\cdot}1868 \text{ J/cal} \tag{2.14a}$$

$$= 4{\cdot}1868 \text{ kJ/kcal} \tag{2.14b}$$

The calorie itself is defined in section 2.12.

EXAMPLE 2.3

(*a*) Express the following quantities in terms of kJ and kcal.
(*i*) the horsepower-hour (hph)
(*ii*) the kilowatt-hour (kWh)

SOLUTION

(*i*) 1 British hp = 0·7457 kW = 0·7457 kJ/s
1 hph = 0·7457 × 3600 = 2685 kJ *Ans.*

Since 1 kJ = 1/4·1868 kcal,

$$1 \text{ hph} = 2685/4{\cdot}1868 = 640{\cdot}2 \text{ kcal} \quad \textit{Ans.}$$

(*ii*) Since 1 W = 1 J/s = 1 N-m/s = 1 ampere × 1 volt,

$$1 \text{ kWh} = 3600 \text{ kJ} \quad \textit{Ans.}$$

$$= \frac{3600}{4{\cdot}1868} \text{ kcal} = 859{\cdot}85 \text{ kcal} \quad \textit{Ans.}$$

(*b*) An electric motor drives an air compressor. The current and voltage supplied to the motor are 15 A and 250 V. What is the power input W to the motor in kilowatts?

If the drive from the electric motor to the compressor is by means of a V-belt and pulleys, and the speed of the compressor shaft is 6·67 rev/s, estimate the torque on the compressor shaft, assuming the belt transmission is 80 per cent efficient.

SOLUTION

$$W = 15 \times 250 = 3\,750 \text{ W} = 3{\cdot}75 \text{ kW} \quad \textit{Ans.}$$

$$\text{Power to compressor shaft} = 0{\cdot}8 \times 3{\cdot}75 = 3{\cdot}00 \text{ kW}$$

That is,

$$\tau\omega = 3{\cdot}00 \text{ kW (or kJ/s)} = 3000 \text{ N-m/s}$$

$$\omega = 2\pi \times 6{\cdot}67 \text{ rad/s}$$

$$\tau = 3000/2\pi \times 6{\cdot}67 \text{ N-m}$$

$$= 3000/41{\cdot}909 = 71{\cdot}584 \text{ N-m} \quad \textit{Ans.}$$

2.4 Energy

Energy is the capacity, either latent or apparent, to exert a force through a distance.

We have already seen in subsection 2.1.1 that the capacity to exert a force through a distance is mechanical work, given by equation (2.1) and measured in joules.

A number of forms of energy have already been listed, namely, potential, kinetic, internal, chemical, etc. These are defined and discussed in the present section.

POTENTIAL ENERGY Potential energy is energy of position. In practice it is restricted to gravitational energy, i.e. energy arising from the gravitational attraction of the Earth. From Newton's second law, equation (2.2b), the gravitational force on a body of mass m is given by

$$F_g \text{ (N)} = m \text{ (kg)} \times g \text{ (m/s}^2\text{)}$$

where g = gravitational acceleration = 9·80665 m/s^2.

Hence, if a body of mass m is at a height z above an arbitrary datum plane, usually the Earth's surface in the particular locality in question, its potential energy, i.e. its ability to do work due to the gravitational force F_g acting through a distance z, is given by

$$\mathrm{PE} = F_g z = mgz \text{ N-m} \tag{2.15}$$

KINETIC ENERGY Kinetic energy is energy of motion. It may readily be shown from Newton's second law that kinetic energy is proportional to the product of mass m and the square of velocity c^2, the constant of proportionality being $\frac{1}{2}$, i.e.

$$\mathrm{KE} = \tfrac{1}{2}mc^2 \text{ N-m} \tag{2.16}$$

where the mass is measured in kg and the velocity in m/s.

The velocity c is generally measured relative to the Earth's surface in the particular locality in question.

INTERNAL ENERGY Internal energy (U) is energy of mass composition. Fundamentally it may be related to the kinetic energy level of the molecules constituting matter (see Chapter 9). As with potential and kinetic energy, which are measured by a pair of PROPERTIES (see section 2.9), namely, mass and height for PE and mass and velocity for KE, internal energy is measured by mass and temperature. The specific internal energy, i.e. internal energy per unit mass, is denoted by the symbol u.

CHEMICAL ENERGY Chemical energy is energy released due to changes in chemical composition. The most common case of release of chemical energy is that of combustion, i.e. the oxidation reaction of certain solid, liquid or gaseous fuels (see Chapter 13).

ATOMIC ENERGY Atomic energy is energy released due to changes in atomic structure, such as nuclear fission or fusion. This will not be discussed in the present book.

The units of energy are clearly those of work, i.e. energy in N-m, or specific energy in N-m/kg. The conversion constants given by equations (2.13) and (2.14) again apply when electrical energy units or heat units are to be converted to mechanical energy or work units.

EXAMPLE 2.4

Nitrous oxide (laughing gas) of density 31·684 kg/m^3 is carried in a pipe line 60·96 m above sea level and is at a temperature of 148·9°C. It flows along the pipe at the rate of 6·096 m/s and its specific heat at constant volume is 4·1868 × 0·161 = 0·6741 kJ/kg-K.

Taking sea level as datum for heights and 0°C as datum for energy involving temperature, evaluate the potential, kinetic and internal energy, and thus the total energy, of the gas.

SOLUTION

$$\text{PE} = mgz = 1 \times 9{\cdot}81 \times 60{\cdot}96 = 598{\cdot}02 \text{ J/kg} \quad \textit{Ans.}$$
$$\text{KE} = \tfrac{1}{2}mc^2 = \tfrac{1}{2}(6{\cdot}096)^2 = 18{\cdot}58 \text{ J/kg} \quad \textit{Ans.}$$
$$\text{IE} = mu = mc_v(148{\cdot}9 - 0)$$
$$u = 0{\cdot}6741 \times 148{\cdot}9 = 100{\cdot}373 \text{ kJ/kg}$$
$$= 100\,373 \text{ J/kg} \quad \textit{Ans.}$$

Therefore, neglecting the chemical energy,

$$\text{Total energy } E \text{ of the gas} = (598{\cdot}02 + 18{\cdot}58 + 100\,373) \text{ J/kg}$$
$$= 100\,989{\cdot}60 \text{ J/kg}$$
$$= 100{\cdot}99 \text{ kJ/kg} \quad \textit{Ans.}$$

Notes

(i) The kinetic energy of the gas is almost negligible and the potential energy is also very small compared with the internal energy of the gas.
(ii) In very many cases in engineering where gases flow into and out of an engine or turbine or boiler, the changes in potential and kinetic energy between entry and exit are negligible when compared with the change in internal and flow energy taking place (see Chapter 4).

2.5 The System (*Fig*. 2.10)

Any thermodynamic analysis is concerned with energy transfers to or from, and property changes within, a certain region. This region is defined as the system, and a number of other definitions follow directly from this.

(*a*) *System*

THE SYSTEM IS A SPECIFIC REGION, NOT NECESSARILY OF CONSTANT VOLUME, NOR FIXED IN SPACE, WHERE TRANSFER OF MASS AND ENERGY ARE TO BE STUDIED.

Since the system is limited in extent, a boundary has to be introduced—

(*b*) *Boundary*

THE ACTUAL OR HYPOTHETICAL ENVELOPE ENCLOSING THE SYSTEM IS THE BOUNDARY OF THE SYSTEM.

(*c*) *Surroundings*

THE REGION OUTSIDE THE SYSTEM IS THE SURROUNDINGS.

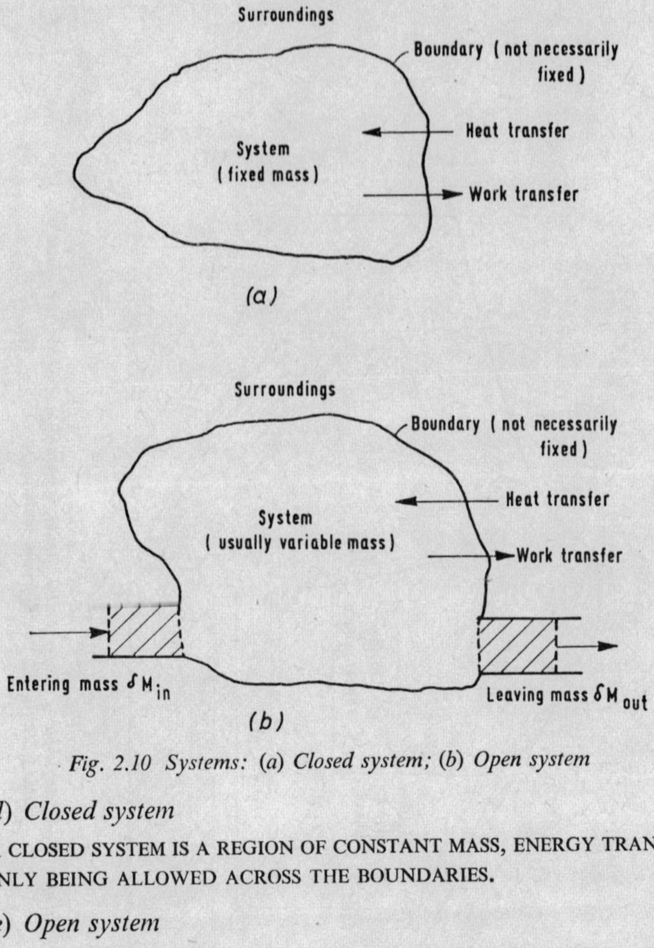

Fig. 2.10 Systems: (a) Closed system; (b) Open system

(*d*) *Closed system*

A CLOSED SYSTEM IS A REGION OF CONSTANT MASS, ENERGY TRANSFER ONLY BEING ALLOWED ACROSS THE BOUNDARIES.

(*e*) *Open system*

AN OPEN SYSTEM HAS MASS AS WELL AS ENERGY TRANSFER ACROSS ITS BOUNDARIES, AND THE MASS WITHIN THE SYSTEM IS NOT NECESSARILY CONSTANT.

(*f*) *Isolated system*

AN ISOLATED SYSTEM CANNOT TRANSFER EITHER MASS OR ENERGY TO OR FROM THE SURROUNDINGS.

2.6 Homogeneous and Heterogeneous Systems

A quantity of matter, homogeneous in chemical composition and physical structure, is called a PHASE (see also *Fig.* 2.11).

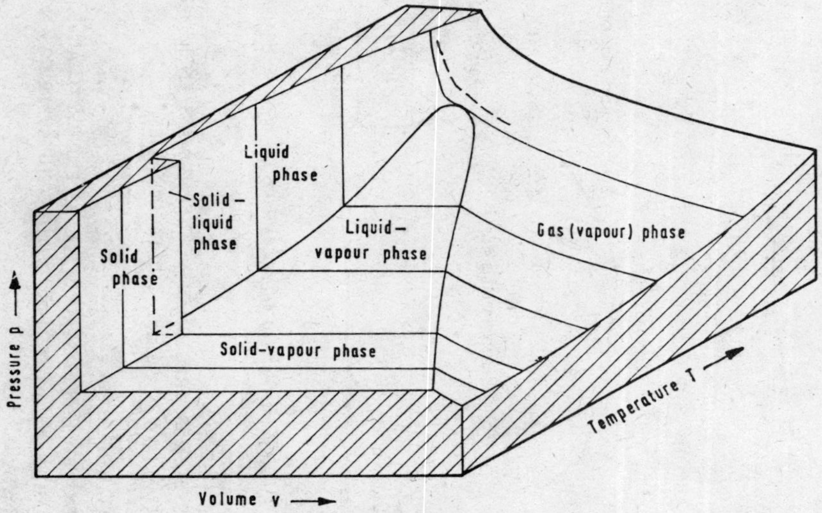

Fig. 2.11 p–v–T surface

This illustrates three single-phase states (solid, liquid, gaseous) and three two-phase states (solid/liquid, liquid/vapour, solid/vapour)

A system comprising a single phase is called a homogeneous system while one consisting of two or more phases is a heterogeneous system.

2.7 Components, Constituents and Pure Substance

The two terms, components and constituents, may be used synonymously in the absence of chemical changes.

A pure substance is a substance with one chemical structure, but may exist in one or more phases.

2.8 Equilibrium

A system in thermodynamic equilibrium is incapable of spontaneous change; it is in complete balance with the surroundings, e.g. a cylinder containing gas at uniform pressure and temperature equal to those of the surroundings.

2.9 Property, State and Process

(*a*) PROPERTIES

Any observable characteristics of a system, such as mass m, pressure p, temperature T (see section 2.12), volume V, internal energy U, may be regarded as internal, or thermodynamic, properties. These properties are appropriate to closed systems.

For open systems, or closed systems in motion, there is a second class of properties, namely external or mechanical, which are characteristics of either the motion or the position of the system in the gravitational field, namely velocity c and height z, i.e. of kinetic and potential energy (see section 2.4).

INTENSIVE PROPERTIES are independent of the mass of the system, e.g. pressure, temperature, viscosity and height.

EXTENSIVE PROPERTIES are related to the mass, e.g. volume, and energies of all kinds.

(*b*) STATE

State is a condition of the system as described or measured by the properties.

Two states of a substance are called identical if every property is the same in both cases. When the state alters, that change is described in terms of END STATES. The process involved in a change of state is described in terms of the PATH (i.e. the values of properties during the change) and the method by which the path is described.

(*c*) PROCESS

A process occurs whenever the system undergoes a change in state or an energy transfer at a steady state.

A process in which the initial and end states are identical is called a CYCLE.

2.10 Enthalpy

The thermodynamic properties mentioned so far are pressure p, specific volume v, absolute temperature T and specific internal energy u. New properties may be defined mathematically in terms of combinations of established properties, so long as dimensional homogeneity is satisfied. We shall see later that in open system calculations, the combination of specific internal energy u with the specific flow work pv (equation (2.11)) occurs very frequently. We therefore define a new property, enthalpy h, as

$$h = u + pv \tag{2.17}$$

or, applied to mass m,

$$H = U + pV \tag{2.17a}$$

2.11 Relationships between Thermodynamic Properties

It is an experimental fact that, for the pure substance, any thermodynamic property is defined by any other two properties. The latter may be regarded as independent, the former as dependent, properties or variables.

Thus, for pressure p, specific volume v and temperature T

$$f(p, v, T) = 0 \tag{2.18}$$

but the actual choice of properties is a matter of convenience. Thus, we could write equally well

$$f(u, v, T) = 0 \tag{2.18a}$$

or

$$f(u, p, T) = 0 \tag{2.18b}$$

or

$$f(u, p, v) = 0 \tag{2.18c}$$

Equations (2.18) are EQUATIONS OF STATE and may be represented by three-dimensional surfaces (*Fig.* 2.11).

Fig. 2.11 also illustrates the various single- and two-phase states (see section 2.6), namely liquid, solid, gaseous, and solid–liquid, liquid–vapour, solid–vapour.

2.12 Heat and Temperature

In section 2.1.1 work was defined as the capacity to exert a force through a distance.

In relation to the system (section 2.5) which is the framework within which all thermodynamic processes are considered, work may be defined as follows—

WORK IS ENERGY TRANSFERRED, WITHOUT TRANSFER OF MASS, ACROSS THE BOUNDARY OF A SYSTEM DUE TO THE MOVEMENT OF PART OF THE BOUNDARY UNDER THE ACTION OF A FORCE.

By analogy with this definition of work, we may now define heat as follows—

HEAT IS ENERGY TRANSFERRED WITHOUT TRANSFER OF MASS ACROSS THE BOUNDARY OF A SYSTEM DUE TO A DIFFERENCE IN TEMPERATURE BETWEEN THE SYSTEM AND ITS SURROUNDINGS.

Temperature has not been formally defined as yet, and will be treated in detail in Chapter 3. For the moment, however, the following definition is adequate—

TWO SYSTEMS HAVE EQUAL TEMPERATURE IF NO CHANGE IN ANY OBSERVABLE CHARACTERISTIC OR CHARACTERISTICS (AS MEASURED BY ANY SUITABLE PROPERTY OR PROPERTIES) TAKES PLACE WHEN THEY ARE BROUGHT INTO CONTACT WITH EACH OTHER.

Thus, when a perfectly insulated cylinder (i.e. one impervious to heat flow) containing gas at pressure p and volume V, sealed by a perfectly insulating piston, has the latter replaced (without escape or change of state of the gas) by a perfectly conducting piston, and no change in state takes place subsequently, i.e. pressure p and volume V remain unchanged, the system (the gas within the cylinder and the containing walls including the piston) and the surroundings (atmosphere) are clearly at the same temperature.

If, on the other hand, the temperature of the surroundings is lower than that of the gas in the cylinder, the temperature of the gas will fall and, assuming the volume of the cylinder to remain fixed, this would result in a drop in pressure p. Alternatively, assuming that the pressure in the cylinder is initially in equilibrium with the surroundings and will remain so subsequently, a reduction in volume

V will be observed. In either case a change in an observable characteristic, as measured by pressure *p* in the first instance or volume *V* in the second instance, has taken place (see *Fig*. 2.12).

The measurement of temperature as such is discussed in Chapter 3. It is performed with the aid of a THERMOMETER, which may itself be regarded as a thermodynamic system having an observable

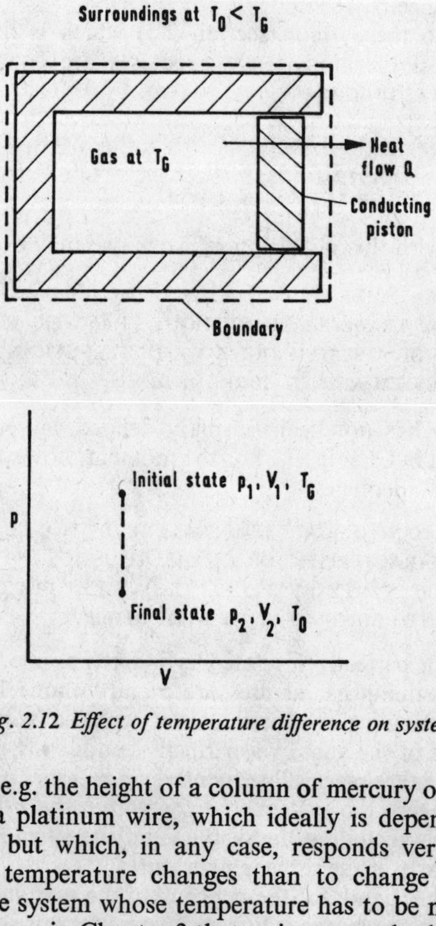

Fig. 2.12 Effect of temperature difference on system

characteristic, e.g. the height of a column of mercury or the electrical resistance of a platinum wire, which ideally is dependent on temperature only but which, in any case, responds very much more sensitively to temperature changes than to change in any other property of the system whose temperature has to be measured.

It will be shown in Chapter 3 that, using a standard thermometer, we may define a TEMPERATURE SCALE dividing the temperature interval between two uniquely defined fixed points into a stated number of smaller (and equal) intervals. On the CELSIUS scale which we shall mainly use in the present text the two fixed points are the ICE POINT (0°C) and the BOILING POINT (100°C) of water at standard atmospheric pressure, the interval thus being 100°C. Having formulated a temperature scale, it is now possible to define a unit of heat. In the Celsius system, the unit of heat is the kilocalorie (kcal) defined as follows—

ONE KILOCALORIE IS THAT QUANTITY OF HEAT (i.e. energy transferred in virtue of a difference in temperature) WHICH WILL RAISE THE TEMPERATURE OF 1 KG OF WATER FROM 14·5°C TO 15·5°C AT STANDARD ATMOSPHERIC PRESSURE.

It should be noted that not only has the magnitude of the temperature interval as such been defined, but also the particular interval on the temperature scale (14·5° to 15·5°C) and the pressure of the water.* The reason is, of course, that the thermodynamic property of water known as the SPECIFIC HEAT AT CONSTANT PRESSURE c_p (see Chapter 3, section 3.5) varies slightly.

Due to the equivalence, in a limited sense, of heat and work (see section 2.3), the same units are clearly appropriate to the measurement of both, although in practice heat is measured in heat units (kcal) and work in work units (kilojoules). The conversion constant, known as the MECHANICAL EQUIVALENT OF HEAT (see Chapter 4, section 4.1), has already been given in section 2.3 as

$$J = 4{\cdot}1868 \text{ kJ/kcal} \qquad (2.19a)$$

In many problems in thermodynamics, heat quantities (Q) and work quantities (W) occur simultaneously. In formulating the relevant equations, it will be assumed that a consistent system of units is employed. It is left to the student to attach appropriate units to the quantities denoted by symbols in the many equations of the following chapters.

2.13 Heat and Work in Relation to Thermodynamic Systems

We may show that heat and work, unlike internal, kinetic or potential energy, are not properties since they are not characteristics of the system itself, but appear only as boundary effects during a process.

For a given change of state, defined by the initial and final values of the properties of the system, the changes in the various properties

* In the British system, the unit of heat is the BRITISH THERMAL UNIT, defined as that quantity of heat required to raise the temperature of 1 lb of water from 60°F to 61°F, where

$$1\text{°F (Fahrenheit)} = \frac{1}{1{\cdot}8}\text{°C}$$

and the temperature of the ice and boiling points, corresponding to 0°C and 100°C, are, respectively, 32°F and 212°F.

are clearly independent of the actual path taken, since they depend solely on the end states, while the heat and work transfers are functions of the process path.

In mathematical terms the property equations (2.18) in section 2.11 lead to perfect differentials, e.g.

$$\text{from } f(p, v, T) = 0 \tag{2.19}$$

$$dp = \left(\frac{\partial p}{\partial v}\right)_T dv + \left(\frac{\partial p}{\partial T}\right)_v dT \tag{2.19a}$$

$$= \alpha\, dv + \beta\, dT \tag{2.19b}$$

where α and β are in turn functions of v and T.

Since, for a functional relationship like that of equation (2.19),

$$\frac{\partial^2 p}{\partial v \partial T} = \frac{\partial^2 p}{\partial T \partial v}$$

$$\left(\frac{\partial \alpha}{\partial T}\right)_v = \left(\frac{\partial \beta}{\partial v}\right)_T \tag{2.20}$$

Equations (2.19a) and (2.19b) constitute perfect or exact differentials and result from the unique relationship between the three variables p, v, T given by equation (2.18).

Equation (2.20) may be regarded as the mathematical test for the existence or otherwise of a perfect or exact differential.

It is evident that we may *not* write,

$$\text{for heat} \qquad f(Q, v, T) = 0 \tag{2.21}$$

$$\text{or for work} \qquad f(W, v, T) = 0 \tag{2.22}$$

since, in fact, heat and work can only be defined in terms of a process path joining two states, not in terms of the states themselves.

It therefore follows immediately that

$$dQ \neq \left(\frac{\partial Q}{\partial v}\right)_T dv + \left(\frac{\partial Q}{\partial T}\right)_v dT \tag{2.21a}$$

and

$$dW \neq \left(\frac{\partial W}{\partial v}\right)_T dv + \left(\frac{\partial V}{\partial T}\right)_v dT \tag{2.22a}$$

Similarly the test for the exact differential breaks down, i.e. equation (2.20) does not hold when applied to heat and work.

The full implications of the above statements will become apparent only from a close study of Chapters 4, 5, 6 and 7. The mathematical relationships between thermodynamic properties (the so-called thermodynamic relationships) are discussed fully in Appendix II.

2.14 Sign Convention for Heat and Work (*Fig.* 2.13)

(*a*) HEAT Q

When heat is transferred *from* the surroundings *to* the system (as the result of an excess of temperature of the surroundings over the system) the heat received by the system is taken to be POSITIVE,

$$\text{Heat received } by \text{ system} = +\,Q \tag{2.23a}$$

Conversely, when heat is transferred *from* the system *to* the surroundings, the heat given by the system is taken to be NEGATIVE,

$$\text{Heat given up by system} = -\,Q \tag{2.23b}$$

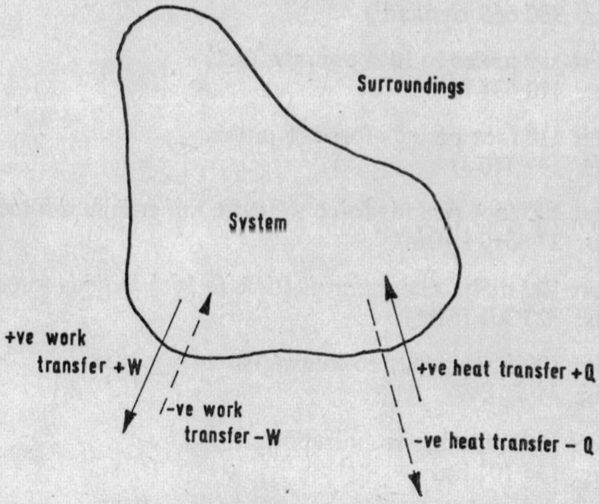

Fig. 2.13 Heat and work sign convention

(*b*) WORK W

When the system does work *on* the surroundings, i.e. energy is transferred in the form of work *from* the system to the surroundings, the work done *by* the system is taken to be positive,

$$\text{Work output of system} = +\,W \tag{2.24a}$$

Similarly, when the surroundings do work on the system, i.e. energy is transferred in the form of work *from* the surroundings to the system, the work done *on* the system is taken to be negative,

$$\text{Work input to system} = -\,W \tag{2.24b}$$

REFERENCES

2.1 Anderton, P., and Bigg, P. H., *Changing to the Metric System* (HMSO, 1965).
2.2 *The International System (SI) Units*, BS 3763: 1964, UDC 389.151 (British Standards Institution).
2.3 *The Use of SI Units*, PD 5686 (British Standards Institution, January 1969).
2.4 Mayhew, Y. R., and Rogers, G. F. C., *Thermodynamic and Transport Properties of Fluids*, SI Units, 2nd ed. (Oxford, Blackwell, 1967).

EXERCISES

2.1 Convert 42 feet of water to pounds force per square inch.
(*Ans:* 18·2082 lbf/in.^{2})

2.2 Convert 1 centimetre of water to dynes per square centimetre.
(*Ans:* 980·665 dyn/cm^2)

2.3 Convert 150 joules to foot pounds-force.
(*Ans:* 110·634 ft-lbf)

2.4 Convert 110 foot pounds-force to joules.
(*Ans:* 149·140 J)

2.5 Convert 200 foot pounds-force to metre kilogrammes-force.
(*Ans:* 27·6510 kgf-m)

2.6 Convert 100 metre kilogrammes-force to foot pounds-force.
(*Ans:* 723·301 ft-lbf)

2.7 Convert 100 horsepower to kilowatts.
(*Ans:* 74·57 kW)

2.8 Convert 100 kilowatts to horsepower.
(*Ans:* 134·102 hp)

2.9 The change in temperature was from −450°F (−268°C) to 300°F (148·9°C). Check the equivalents in degrees Celsius.

2.10 In the International System of Units (SI) the quantity of heat is expressed in kilojoules. Express the following quantities of British thermal units in kilojoules: 100, 500, 1000, 2000.
(*Ans:* 105·51 kJ; 527·55 kJ; 1055·06 kJ; 2110·11 kJ)

2.11 What are the same quantities of British thermal units expressed as kilocalories?
(*Ans:* 25·200 kcal; 125·998 kcal; 251·996 kcal; and 503·992 kcal)

2.12 Saturated steam at 20 bar has a specific volume of 0·09957 m^3/kg and internal energy 2600 kJ/kg. Calculate the enthalpy of this steam in (*a*) kJ/kg, (*b*) kcal/kg, (*c*) Btu/lb.
(*Ans:* 2799 kJ/kg, (*b*) 669 kcal/kg, (*c*) 1201 Btu/lb)

2.13 Freon-12 (CCl_2F_2) in the saturated state at 20°C has a pressure of 5·673 bar and a specific volume of 0·0308 m³/kg, the enthalpy being 195·78 kJ/kg. Calculate the internal energy in (*a*) kJ/kg, (*b*) kcal/kg, (*c*) Btu/lb.

(*Ans:* (*a*) 177·99 kJ/kg, (*b*) 42·50 kcal/kg, (*c*) 76·45 Btu/lb)

2.14 A manometer containing paraffin of specific gravity 0·81 is used to measure the pressure drop across an orifice which measures the air consumption of an internal combustion engine. Express a difference in level of 0·304 m in the manometer in bar if the density of the air is 1·235 kg/m³ and of water 10³ kg/m³.

(*Ans:* 0·00219 bar)

2.15 The vacuum gauge on the condenser of a steam turbine reads 0·66 m.Hg. What is the absolute pressure in the condenser in bar at an atmospheric pressure of 1·013 bar.

(*Ans:* 0·088 bar)

2.16 What is the work in kilojoules in blowing up a balloon till it is just taut and of capacity 0·568 m³ at an atmospheric pressure of 0·75 m.Hg? What is the equivalent value in ft-lbf. Assume the balloon to be made of light fabric which is unstressed and inelastic.

(*Ans:* 56·8 kJ = 41 800 ft-lbf)

2.17 A swinging field dynamometer is used to determine the brake horse-power of a gas turbine. A 227 kg load at a radius of 0·61 m rests on a weighbridge which is balanced at 5·67 kgf recorded on the balance arm. The dynamometer shaft speed is 4000 rev/min. Calculate the output of the turbine (*a*) in hp, (*b*) in kW.

(*Ans:* (*a*) 37·10 hp, (*b*) 27·67 kW)

2.18 A horizontal cylinder of 0·406 m diameter is fitted with a frictionless piston and the fluid in the cylinder is stirred mechanically over a period of 12 min. During this time the piston is observed to move slowly through a distance of 0·559 m against the atmospheric pressure of 1·013 bar. The net work done by the fluid during this time is 2·44 kJ. Calculate the shaft torque and the power of the electric motor driving the stirrer.

(*Ans:* (*a*) 0·649 m-N, (*b*) 0·0091 hp)

Chapter 3

Temperature and the Ideal Gas

3.1 Definition

Temperature is that property which determines the ability of the system to transfer heat.

3.2 Temperature Scales and Thermometers

A temperature scale is an arbitrary set of numbers and a method for assigning each number to a definitive level of temperature, e.g. the melting point of ice and the boiling point of water at standard atmospheric pressure.

A thermometer is a measuring device yielding a number at each temperature level, the number being functionally related to the temperature.

Actual thermometers are based on changes of certain properties with temperature, e.g.

(*a*) volumetric expansion of gases, liquids and solids
(*b*) pressure exerted by gases
(*c*) electrical resistance of solids
(*d*) vapour pressure of liquids
(*e*) thermoelectricity.

The number indicated by the thermometer depends on

(*i*) the particular substance used
(*ii*) the property or properties being measured
(*iii*) the design and construction of the instrument.

The basic difficulty is that, as a rule, the measured property varies not only with temperature, but also with other properties, e.g. with pressure, which cannot necessarily be held constant. A temperature scale subject to these limitations is an INCOMPLETE TEMPERATURE SCALE.

Thus, if x is the measured property, and it is a function of both temperature and pressure, we may write

$$x = f(T, p), \quad \text{i.e.} \quad dx = \left(\frac{\partial x}{\partial T}\right)_p dT + \left(\frac{\partial x}{\partial p}\right)_T dp$$

whence

$$dT = \frac{dx - \left(\dfrac{\partial x}{\partial p}\right)_T dp}{\left(\dfrac{\partial x}{\partial T}\right)_p} \tag{3.1}$$

Hence an unambiguous reading is obtained only when p is held constant.

The mercury-in-glass thermometer provides an example of the observed quantity, namely the height of the column of mercury, being dependent not only on temperature, but also to a significant extent on the external pressure acting on the bulb and stem.

3.3 The Gas Scale of Temperature

(See also section 3.5)

The equation of state of a perfect gas applied to unit mass may be written, in any system of coherent units (see section 2.1.1), in the form

$$pv = RT \tag{3.2}$$

where, in SI units,

p = pressure (N/m^2)
v = specific volume (m^3/kg)
R = gas constant (J/kg-K)
T = absolute temperature (K).
(K is the symbol for degrees Kelvin.)

In practice, as already stated in section 2.1.2, pressure is usually measured in bars where

$$1 \text{ bar} = 10^5 \text{ N/m}^2$$

and energy is more conveniently measured in kilojoules (kJ) where

$$1 \text{ kJ} = 10^3 \text{ J}$$

Hence equation (3.2) has to be modified by the introduction of a constant, i.e.

$$\begin{aligned} p(\text{N/m}^2) \times v(\text{m}^3/\text{kg}) &\equiv 10^5 p(\text{bar}) \times v(\text{m}^3/\text{kg}) \\ &\equiv R(\text{J/kg-K}) \times T(\text{K}) \\ &\equiv 10^3 R(\text{kJ/kg-K}) \times T(\text{K}) \end{aligned}$$

i.e.
$$10^5 p(\text{bar}) \times v(\text{m}^3/\text{kg}) = 10^3 R(\text{kJ/kg-K}) \times T(\text{K})$$

Therefore a constant of 10^2 appears, giving the following standard form of the perfect gas equation:

$$10^2 p(\text{bar}) \times v(\text{m}^3/\text{kg}) = R(\text{kJ/kg-K}) \times T(\text{K}) \qquad (3.2\text{a})$$

i.e.
$$10^2 pv = RT \qquad \text{or} \qquad pv = 10^{-2} RT$$

The perfect gas is an abstraction, but experiment shows that at low pressures the behaviour of the so-called permanent gases, e.g. O_2, N_2, H_2, approximates to that expressed by equation (3.2), i.e.

$$T = \left(\frac{pv}{R}\right)_{p(\text{real gases}) \to 0} \qquad (3.2\text{b})$$

Equation (3.2b) constitutes the ideal or perfect gas scale of temperature, and *Fig.* 3.1 illustrates the gas thermometer which uses this scale.

This thermometer measures the increase in volume with temperature of a quantity of gas trapped by a suitable liquid in a U-tube

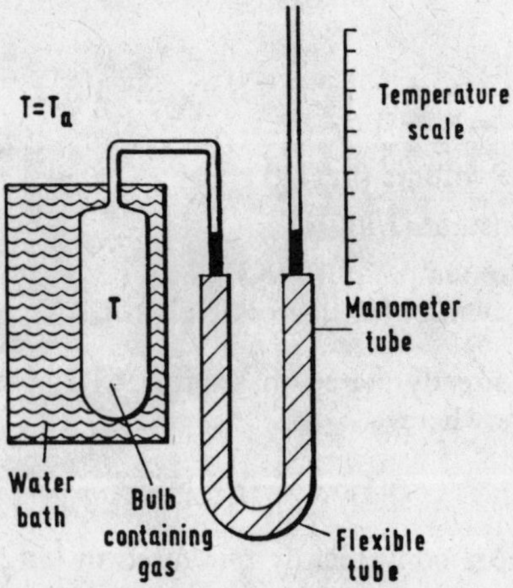

Fig. 3.1 Perfect gas thermometer

when the pressure is kept constant by ensuring that the liquid levels in the two limbs of the U-tube are equal, i.e. from equation (3.2b),

$$\Delta T = \frac{p}{R}\Delta V \qquad (3.2c)$$

i.e. the temperature increment is proportional to the observed volume increment.

Alternatively, if instead of adjusting the position of the outer limb of the U-tube, the liquid is allowed to find its own level, the pressure will change, the volume of trapped gas being kept constant—

$$\Delta T = \frac{V}{R}\Delta p \qquad (3.2c)$$

i.e. the temperature increment is proportional to the observed pressure increment.

The two forms of the gas thermometer are known respectively as the constant pressure and the constant volume thermometer. On this gas scale of temperature the triple point of water (equilibrium condition of vapour, liquid and solid) occurs at 273·16 K. (K refers to the absolute or Kelvin scale, which is later shown to be identical with the perfect gas scale.) The ice point (freezing point of water at standard atmospheric pressure) occurs at 273·15 K and the steam point (boiling point of water at standard atmospheric pressure) at 373·15 K. The interval is thus 100·00° and this is the basis of the Celsius scale, whose datum is the ice point.

Hence

$$t\,(^\circ\mathrm{C}) = T\,(\mathrm{K}) - 273{\cdot}15 \qquad (3.3)$$

On the Fahrenheit scale the interval between the ice and steam point is divided into 180°, while the datum is the freezing point of an eutectic mixture of salt and water. Thus the ice point becomes 32°F and the steam point 212°F. The absolute scale corresponding to the Fahrenheit scale is the Rankine scale, such that, at the ice point

$$32^\circ\mathrm{F} = 491{\cdot}67^\circ\mathrm{R}$$

whence

$$t(^\circ\mathrm{F}) = T(^\circ\mathrm{R}) - 459{\cdot}67 \qquad (3.4)$$

The relationship between the Fahrenheit and Celsius scales has been fixed by international agreement, in the form of the INTERNATIONAL SCALE OF TEMPERATURE, specifying numbers at different fixed points (*Fig.* 3.2).

The definition of absolute zero on the perfect gas scale is that temperature at which the volume of the gas becomes zero.

EXAMPLE 3.1

(*a*) The following data were extracted from physical tables. Express the temperatures in degrees Fahrenheit.

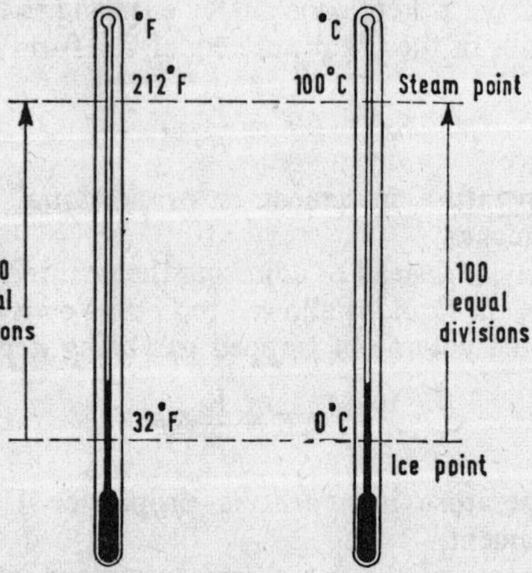

Fig. 3.2 Fahrenheit (left) *and Celsius scales* (*mercury-in-glass thermometer*)

At atmospheric pressure (1·01325 bar) the boiling point of water is 100°C and that of helium is − 268·8°C.

(*b*) The following fixed points at one atmosphere pressure (1·01325 bar) are given in degrees Fahrenheit; express them in degrees Celsius.

Oxygen point	−297·35°F
Sulphur point	832·28°F
Antimony point	1166·90°F
Silver point	1761·4°F
Gold point	1945·4°F

SOLUTION

(*a*) $$100°\text{C} = 100 \times \tfrac{9}{5} + 32 = 212°\text{F}$$

$$-268{\cdot}8°\text{C} = -268{\cdot}8 \times \tfrac{9}{5} + 32 = 451{\cdot}8°\text{F}$$

(*b*) $$-297{\cdot}35°\text{F} = (-297{\cdot}35 - 32) \times \tfrac{5}{9} = -183{\cdot}1°\text{C}$$

$$832{\cdot}28°\text{F} = (832{\cdot}28 - 32)\tfrac{5}{9} = 444{\cdot}6°\text{C}$$

$$1166{\cdot}90°\text{F} = (1\,166{\cdot}90 - 32)\tfrac{5}{9} = 630{\cdot}50°\text{C}.$$

Similarly 1761·4°F = 960·8°C

and 1945·4°F = 1 063·0°C

3.4 The Thermodynamic (or Absolute) Scale of Temperature

The second law of thermodynamics (Chapter 6) gives rise to an absolute temperature scale independent of the properties of particular substances. This is clearly a great advantage in that it provides the basis for a true absolute temperature scale. Numerically, however, this thermodynamic or Kelvin scale agrees with the perfect gas scale, as discussed above.

3.5 Note on the Perfect Gas

(For a more detailed treatment see Chapter 10)

The perfect gas is an abstraction whose behaviour is approached by real gases under certain conditions. All pure substances may exist in one of three states of aggregation, solid, liquid or vapour. These PHASES are separated by phase boundaries, and may be represented conveniently in a pressure–volume (p–v) diagram (*Fig*. 3.3).

All phase diagrams show basically the same form, and all have a critical point C where the saturated liquid and saturated vapour boundaries meet. Lines of constant temperature (isothermals) may be superimposed on the p–v diagram, the isothermal passing

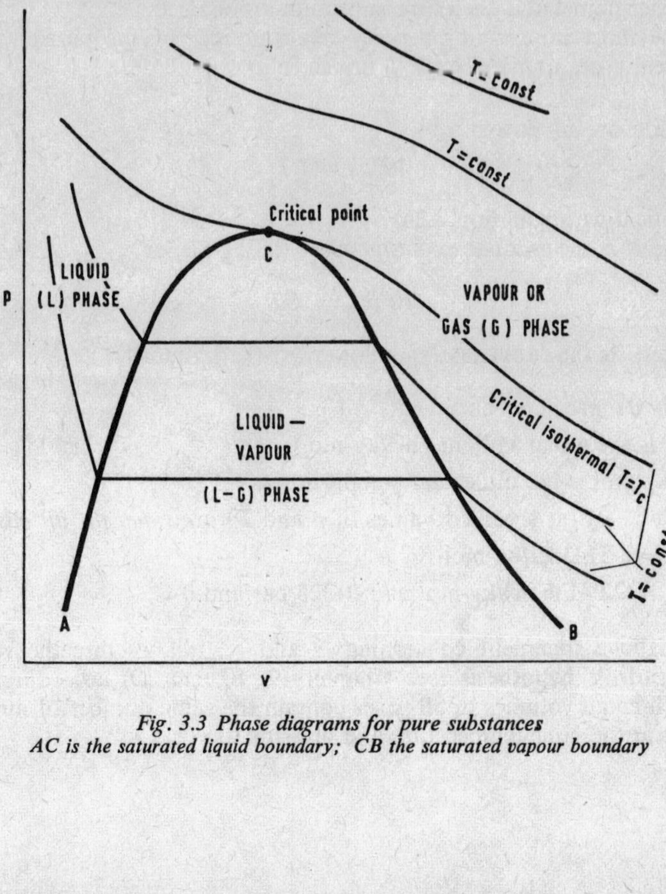

Fig. 3.3 Phase diagrams for pure substances
AC is the saturated liquid boundary; CB the saturated vapour boundary

through C being the critical isothermal T_c. Isothermals below T_c have two discontinuities, and pass horizontally, i.e. at constant pressure, through the liquid–vapour or wet vapour phase. Within that phase, pressure p_s and temperature T_s are uniquely related (saturation pressure and temperature) independently of the degree of "wetness' of the fluid. Above T_c, the isothermals lie entirely within the vapour (or superheated vapour) region, and when $T > 2T_c$, the perfect gas condition, equation (3.2), pv = constant when T = constant, is approached. For steam, $p_c = 221{\cdot}2$ bar $T_c = 374{\cdot}15°C = 647{\cdot}30$ K, so that, at normal pressures, steam approaches the perfect gas condition only at very high temperatures ($>$ 100 K). Perfect gas behaviour is also approached below T_c at very low pressures, so that vapour in the atmosphere once again satisfies the condition $pv = RT$. The so-called permanent gases (O_2, N_2, H_2, air) have critical temperatures in the region 33·3 K (H_2) to 155 K (O_2) so that at normal temperatures $T \gg 2T_c$. Hence they again satisfy the perfect gas condition approximately. The use of one of the permanent gases as the thermometric substance for the gas thermometer is therefore substantially justified.

The most important property relationships for the perfect gas (or real gases above $T \simeq 2T_c$) are as follows—

EQUATION OF STATE

$$10^2 pv = RT \tag{3.5}$$

(identical with equation 3.2a)
where R is the SPECIFIC GAS CONSTANT (kJ/kg-K), or

$$10^2 pv_m = GT \tag{3.6}$$

where G is the UNIVERSAL GAS CONSTANT (kJ/kg-mol-K),

p is the pressure (bar),
v_m is the molar volume (m^3/kg-mol),
1 kg-mol is the molecular weight in kg,
G and v_m (at specified values of p and T) *are equal for all gases,*
$G = 8{\cdot}3143$ kJ/kg-mol-K,
$v_m = 22{\cdot}4136$ m^3/kg-mol at 1·01325 bar and 0°C.

The above statement concerning G and v_m follows directly from Avogadro's hypothesis (see Chapters 9, 10 and 12) according to which equal volumes of all gases contain the same number of molecules at the same temperature and pressure.

From equations (3.5) and (3.6)

$$R = \frac{G}{M} \tag{3.7}$$

where M = molecular weight of a particular gas.

EXAMPLE 3.2

The universal gas constant G is 8·3143 kJ/mol-K; express this in Btu/lb-mol-°R, chu/lb-mol-K, ft-lbf/lb-mol-°R and ft-lbf/lb-mol-K.

SOLUTION

1 ft-lbf = 1·35582 J

8314·3 J/kg-mol-K = 8314·3/1·35582 ft-lbf/kg-mol-K
= 6132·3 ft-lbf/kg-mol-K

v_m = 22·4136 m³/kg-mol = 10·1657 m³/lb-mol

8·3143 kJ/kg-mol-K = 6132·3 × (10·1657/22·4136)
= 2780 ft-lbf/lb-mol-K *Ans.*
= $\frac{5}{9}$ × 2780 = 1545 ft-lbf/lb-mol-°R *Ans.*

1 ft-lbf = 1/778 Btu = 1/1400 chu

2780 ft lbf/lb mol-K = 2780/1400 = 1·986 chu/lb-mol-K *Ans.*

1545 ft-lbf/lb-mol-°R = 1545/778 = 1·986 Btu/lb-mol-°R *Ans.*

INTERNAL ENERGY AND SPECIFIC HEAT AT CONSTANT VOLUME

A further property of the perfect gas, known as Joule's law, is that internal energy of the perfect gas is a function of temperature only, i.e.

$$u = f(T) \quad \text{(Joule's law)} \tag{3.8}$$

We shall see in Chapter 4 that, during a constant volume heating process, the heat input equals the change of internal energy. But the heat input per degree temperature rise for unit mass is the specific heat at constant volume c_v.

Hence
$$c_v = \left(\frac{dq}{dT}\right)_{const.\ vol.} = \left(\frac{du}{dT}\right)_{const.\ vol.}$$
$$= \text{constant for any perfect gas} \tag{3.9}$$

Integrating equation (3.9) with $u = 0$, at $T = 0$

$$u = c_v T \tag{3.9a}$$

The perfect gas is defined by the equation of state (3.5) and the internal energy equation (3.9a), the latter being derived from the assumption that the specific heat at constant volume c_v is constant, equation (3.9).

The so-called semi-perfect gas, defined by the same equation of state (3.5) and equation (3.8), in which the internal energy u, whilst still a function of temperature only, is based on a specific heat c_v which is now itself a function of temperature. Although the semi-perfect gas, like the perfect gas, is an artifice, it conforms more closely to the behaviour of real gases (see Chapter 10).

Enthalpy and specific heat at constant pressure

In section 2.10 enthalpy was defined as

$$h = u + pv$$

Now from equation (3.5) $pv = f_1(T)$ and from (3.8), $u = f_2(T)$ whence

$$h = f_3(T) = F(T) \tag{3.10}$$

Thus enthalpy, like internal energy, is a function of temperature only for the perfect gas.

We shall see in Chapter 4 that during constant pressure heating the heat input equals the change of enthalpy. But heat input per unit mass and per degree temperature rise is defined as the specific heat at constant pressure c_p.

Hence $$c_p = \left(\frac{dq}{dT}\right)_{const.\ press.} = \left(\frac{dh}{dT}\right)_{const.\ press.}$$

$$= \text{constant for any perfect gas} \tag{3.11}$$

Integrating equation (3.11), with $h = 0$ at $T = 0$,

$$h = c_pT \tag{3.11a}$$

Relationship between c_p and c_v

From equations (3.9a) and (3.11a), with $h = u + pv$, and equations (3.5) and (3.7)

$$c_pT = c_vT + RT$$

whence $$c_p - c_v = R = \frac{G}{M} \tag{3.12}$$

i.e. the difference of specific heats is equal to the specific gas constant.

The ratio of specific heats c_p/c_v for the perfect gas is constant, from (3.9a) and (3.11a), and is usually written

$$\gamma = \frac{c_p}{c_v} \tag{3.13}$$

From equation (3.12),

$$(\gamma - 1)c_v = R \quad \text{or} \quad \left(\frac{\gamma - 1}{\gamma}\right) c_p = R \tag{3.12a}$$

Other perfect gas relationships will be derived later as and when required.

EXAMPLE 3.3

The following table gives the gas constant R (kJ/kg-K) for gases commonly met in engineering, as well as the specific heat at constant pressure c_p (kJ/kg-K), molecular weight M and specific heat ratio γ at 15°C. Complete the table by determining the specific heat at constant volume c_v for the gases listed.

	R (kJ/kg-K)	c_p (kJ/kg-K)	M	γ
Air	0·2870	0·992	28·97	1·402
Oxygen O_2	0·2598	1·013	32	1·40
Nitrogen N_2	0·2967	0·984	28·02	1·40
Hydrogen H_2	4·1240	14·319	2·016	1·41
Carbon monoxide CO	0·2968	1·013	28·01	1·401
Carbon dioxide CO_2	0·1890	0·842	44	1·306

SOLUTION

$$c_p - c_v = R = G/M$$

$$\gamma = c_p/c_v$$

Therefore

$$R = c_v(\gamma - 1) = c_p(\gamma - 1)/\gamma$$

Air $\quad 0{\cdot}2870 = c_v(1{\cdot}402 - 1)$

$$c_v = 0{\cdot}2870/0{\cdot}402 = 0{\cdot}714 \text{ kJ/kg-K}$$

O_2 $\quad 0{\cdot}2598 = c_v(1{\cdot}40 - 1)$

$$c_v = 0{\cdot}2598/0{\cdot}40 = 0{\cdot}6495 \text{ kJ/kg-K}$$

N_2 $\quad 0{\cdot}2967 = c_v(1{\cdot}41 - 1)$

$$c_v = 0{\cdot}2967/0{\cdot}41 = 0{\cdot}7237 \text{ kJ/kg-K}$$

H_2 $\quad 4{\cdot}1240 = c_v(1{\cdot}41 - 1)$

$$c_v = 4{\cdot}1240/0{\cdot}41 = 10{\cdot}059 \text{ kJ/kg-K}$$

CO $\quad 0{\cdot}2968 = c_v(1{\cdot}401 - 1)$

$$c_v = 0{\cdot}2968/0{\cdot}401 = 0{\cdot}740 \text{ kJ/kg-K}$$

CO_2 $\quad 0{\cdot}1890 = c_v(1{\cdot}306 - 1)$

$$c_v = 0{\cdot}1890/0{\cdot}306 = 0{\cdot}618 \text{ kJ/kg-K}$$

3.6 Note on Vapours*

(For detailed treatment see Chapter 8)

Vapours may exist as one of the following

(*a*) *Wet vapours* (see *Fig.* 3.3), i.e. mixtures of liquid and vapour components characterized by dryness fraction x, the proportion of dry vapour in mixture;
(*b*) *Saturated vapours*, i.e. existing in any state along the saturated vapour boundary *CB*.
(*c*) *Superheated vapours*, i.e. vapour which has been heated beyond the saturation temperature T_g for a particular pressure p_g.

Changes are best understood by considering a heating process for 1 kg of water from an initial temperature T_1 at constant pressure p to a final condition p_3, v_3 in the (superheated) vapour phase (see *Fig.* 3.4).

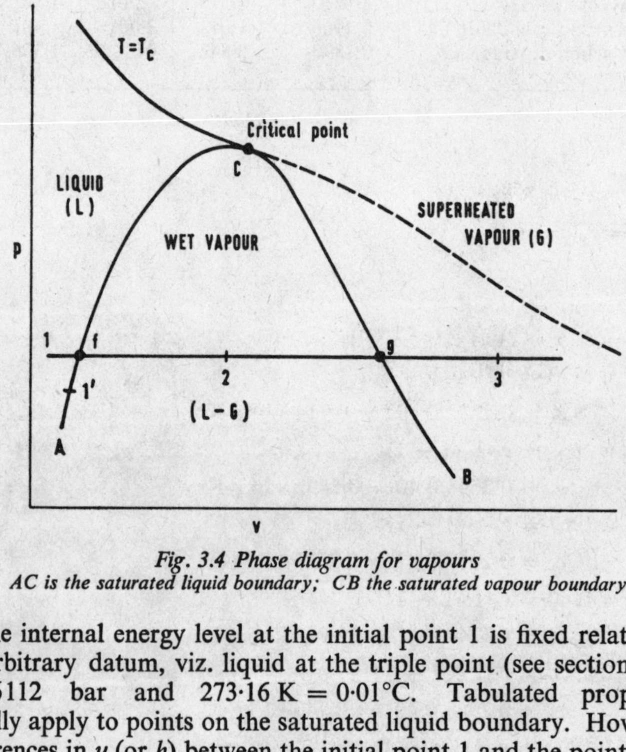

Fig. 3.4 Phase diagram for vapours
AC is the saturated liquid boundary; CB the saturated vapour boundary

The internal energy level at the initial point 1 is fixed relative to an arbitrary datum, viz. liquid at the triple point (see section 3.3), 0·006112 bar and 273·16 K = 0·01°C. Tabulated properties usually apply to points on the saturated liquid boundary. However, differences in u (or h) between the initial point 1 and the point 1′ on

* This section serves as an introduction to the use of steam tables.

the saturated liquid boundary having the same temperature, i.e. $T_1 = T_{1'}$, are negligible. Hence, knowing T_1, use tables to find the saturated liquid condition $(u_f)_1$ or $(h_f)_1$.

1–*f*: constant pressure heating of liquid to boiling point *f*. From equation (3.11)

$$\text{Heat added} = (h_f - h_1) = (h_f - h_{1'}) \qquad \text{(i)}$$

f–*g*: evaporation, still at constant pressure, i.e. transition from liquid to saturated vapour state.

$$\text{Heat added} = h_g - h_f = h_{fg}$$
$$= \text{latent heat of evaporation} \qquad \text{(ii)}$$

$$\text{Change in volume} = v_g - v_f = v_{fg} \text{ (but } v_f \ll v_g) \qquad \text{(iii)}$$

f–2: partial evaporation. State at 2 is expressed by dryness fraction

$$x = \frac{\text{mass of dry saturated vapour (at state } g)}{\text{mass of liquid (state } f) + \text{mass of vapour (state } g)}$$

i.e. Heated added $= xh_{fg}$ (iv)

Enthalpy at 2 $h_2 = h_f + xh_{fg}$ (v)

Specific volume at 2 $v_2 = v_f + xv_{fg}$ (vi)

g–3: constant pressure heating in superheated vapour phase.

$$\text{Heat added} = h_3 - h_g \qquad \text{(vii)}$$

(h_3 found in superheat tables for specific volume at $3 = v_3$ at given p_3.)

Tables also list the internal energy u_f, u_g. In the superheat region u is found from $u = h - pv$ (definition of enthalpy).

EXAMPLE 3.4

(1) Evaluate the enthalpy and specific volume of steam at 28 bar and (*a*) saturated, (*b*) 230°C, dryness fraction $x = 0{\cdot}8$, and (*c*) at 350°C.

SOLUTION

Using abridged property tables by Mayhew and Rogers, SI units, ref. 3.1:
(*a*) The answer is obtained from saturated steam tables,

$h = h_g = 2803$ kJ/kg *Ans.*
$v = v_g = 0{\cdot}07142$ m³/kg *Ans.*

(*b*) $h = h_f + xh_{fg} = h_g - (1 - x)h_{fg}$
$= 991 + 0{\cdot}8 \times 1812 = 2803 - 0{\cdot}2 \times 1812$
$= 2440{\cdot}6$ kJ/kg *Ans.*
$v = v_f + xv_{fg} = v_g - (1 - x)v_{fg}$

Approximately,

$$xv_g = 0{\cdot}8 \times 0{\cdot}07142$$
$$= 0{\cdot}05714 \text{ m}^3/\text{kg} \quad \textit{Ans.}$$

(*c*) The steam is superheated and it is necessary to interpolate the abridged tables thus for pressure:

$$h_{20} = 3\,138 \text{ kJ/kg}$$
$$h_{30} = 3\,117$$

Therefore

$$h_{28} = 3\,138 - 0{\cdot}8(3\,138 - 3\,117)$$
$$= 3\,121{\cdot}2 \text{ kJ/kg} \quad \textit{Ans.}$$
$$v = 0{\cdot}1386 - 0{\cdot}8(0{\cdot}1386 - 0{\cdot}0905)$$
$$= 0{\cdot}1001 \text{ m}^3/\text{kg} \quad \textit{Ans.}$$

Note: Had the temperature been 370°C it would have been necessary to interpolate for temperature as well as pressure, thus:
As above

$$h_{20} \text{ and } 400°\text{C} = 3248 \text{ kJ/kg}$$
$$h_{30} \text{ and } 400°\text{C} = 3231 \text{ kJ/kg}$$

Therefore

$$h_{28} \text{ and } 400°\text{C} = 3248 - 0{\cdot}8(3248 - 3231)$$
$$= 3234{\cdot}4 \text{ kJ/kg}$$

and

$$h_{28} \text{ and } 350°\text{C} = 3\,121{\cdot}2 \text{ kJ/kg}$$

Therefore

$$h_{28} \text{ and } 370°\text{C} = 3\,121{\cdot}2 = 20/50(3234{\cdot}4 - 3\,121{\cdot}2)$$
$$= 3\,166{\cdot}48 \text{ kJ/kg} \quad \textit{Ans.}$$

(2) Evaluate the internal energy of saturated steam at 36 bar knowing that the enthalpy is 2802 kJ/kg and the specific volume is 0·05544 m³/kg.

$$u = h - pv$$
$$= 2802 - 100 \times 36 \times 0{\cdot}05544$$
$$= 2602 \text{ kJ/kg} \quad \textit{Ans.}$$

Note: This calculation is more necessary when the steam is superheated as tables do not usually give the internal energy but only specific volume, enthalpy and entropy. The calculation in such cases is as above.

REFERENCE

3.1 MAYHEW, Y. R., and ROGERS, G. F. C., *Thermodynamic Properties of Fluids and Other Data*, SI units (Oxford, Blackwell, 1959) 3rd edition.

EXERCISES

3.1 Discuss what is meant by an incomplete temperature scale. How can such a scale be made complete?

3.2 Define a perfect gas. Under what circumstances can such gases as air, carbon dioxide, steam, etc., be considered perfect?

3.3 A certain thermometer, using pressure as the thermometric property, gives values of p of 0·006112 and 1·01325 at the ice point and steam point respectively. The temperatures at these two points are assigned the numbers 0 and 100 respectively. Determine the temperature corresponding to $p = 0{\cdot}25$ if the temperature t is defined in terms of p by $t = a \ln p + b$ where a and b are constants.
(*Ans:* 73·24)

3.4 The temperature t on a thermometric scale is defined in terms of a property P by the relation

$$t = a \ln P + b$$

where a and b are constants. Experimental values of P for the ice point 32°F and the steam point 212°F were found to be 1·86 and 6·81 respectively. Find the temperature corresponding to a value of $P = 3{\cdot}038$ on the thermometer. Express the answer in degrees Celsius.
(*Ans:* 37·8°C, 100°F)

3.5 At very low pressures gases behave like the ideal or perfect gases and this gives rise to two Ideal Gas Temperature scales namely

$$p = p_i\{1 + \tfrac{1}{492}(t_f - 32)\}$$
$$p = p_i(1 + \tfrac{1}{273}t_c)$$

Use these equations to determine the absolute zero of temperature in the Rankine and Celsius scales of temperature.
(*Ans:* −460°F, −273°C, when $p = 0$)

3.6 What is the mass of 0·6 m³ of oxygen when at 7 bar pressure and 280°C temperature?
(*Ans:* 2·92 kg)

3.7 0·9 m³ of gas, initially at 1 bar, are compressed to 0·3 m³, the temperature remaining constant. What is the final pressure of the gas? If the temperature increases in the ratio 1·2:1, recalculate the final pressure.
(*Ans:* 3 bar, 3·6 bar)

3.8 A 3 m³ tank contains oxygen at 4 bar and 15°C. Oxygen is pumped into the tank until the pressure has doubled and the temperature is 30°C. How much oxygen was pumped into the tank.
(*Ans:* 14·48 kg)

3.9 1 kg of hydrogen, at a pressure of 12 bar and a temperature of 4°C, is to be stored in a tank. Calculate the volume of the tank. Repeat for 1 kg of the following gases:

(*a*) oxygen
(*b*) methane
(*c*) carbon dioxide.
(*Ans:* 0·95 m^3; 0·06 m^3; 0·12 m^3; 0·04 m^3)

3.10 1 kg of oxygen at a temperature of 15°C is raised to 150°C at constant volume. Find the internal energy and enthalpy for both states if the datum of internal energy is 0°C. State also the values of the corresponding specific heats ($\gamma = 1{\cdot}4$).

(*Ans:* $u_1 = 9{\cdot}743$ kJ/kg; $u_2 = 97{\cdot}43$ kJ/kg
$h_1 = 13{\cdot}64$ kJ/kg; $h_2 = 136{\cdot}40$ kJ/kg
$c_v = 0{\cdot}6495$ kJ/kg-K; $c_p = 0{\cdot}9093$ kJ/kg-K)

3.11 Repeat Exercise 3.10 for carbon dioxide taking the ratio of specific heats as 1·33.

(*Ans:* $u_1 = 8{\cdot}591$ kJ/kg; $u_2 = 85{\cdot}91$ kJ/kg
$h_1 = 11{\cdot}426$ kJ/kg; $h_2 = 114{\cdot}26$ kJ/kg
$c_v = 0{\cdot}5727$ kJ/kg-K; $c_p = 0{\cdot}7617$ kJ/kg-K)

3.12 Evaluate the change in enthalpy and internal energy of steam between the following pairs of conditions, all at a pressure of 15 bar:

(*a*) 0·6 dry
(*b*) dry saturated
(*c*) superheated at 800°F.

Compare the changes from (*b*) to (*c*) with those to be expected if steam were a perfect gas and comment on the result.

$\gamma = 1{\cdot}3$ and $R = 0{\cdot}4615$ kJ/kg-K
(*Ans:* $\Delta h_{ab} = 779$ kJ/kg, $\Delta u_{ab} = 701$ kJ/kg
$\Delta h_{bc} = 572$ kJ/kg, $\Delta u_{bc} = 440$ kJ/kg
Perfect gas: $\Delta R_{bc} = 503$ kJ/kg, $\Delta u_{bc} = 387$ kJ/kg)

3.13 Evaluate the change in enthalpy and internal energy of freon-12 between the following pairs of conditions at 25°C:

(*a*) saturated liquid
(*b*) 0·6 dry
(*c*) saturated vapour.
(*Ans:* $\Delta h_{ab} = 113{\cdot}63$ kJ/kg; $\Delta u_{ab} = 167{\cdot}03$ kJ/kg
$\Delta h_{bc} = 59{\cdot}70$ kJ/kg; $\Delta u_{bc} = -4{\cdot}40$ kJ/kg)

Chapter 4

The First Law of Thermodynamics*

The first law of thermodynamics is a particular statement of the general principle of conservation of energy applied to heat and work, already defined in section 2.13. Together with the second law it governs those processes whereby we seek to turn heat into work, i.e. in all forms of engine and equally the converse processes carried out in so-called reversed heat engines (refrigerators or heat pumps).

4.1 Mechanical Equivalent of Heat

Joule, in the course of experiments performed between 1840 and 1849, found that heat and work can bring about equal effects and concluded that, in a limited sense, they must be equivalent.

His experiment was in two parts. In the first part, water in a calorimeter was heated by a given amount, expressed as a temperature rise ΔT, and the corresponding heat quantity Q was evaluated. In the second part, the water was stirred by a paddle wheel, the amount of work transferred to the water, $-W$, being equal to $-m \,.\, \Delta z$ where m was the mass, and $-\Delta z$ the distance fallen by the weight. Joule found that, for a certain work transfer W, the effect produced on the system, i.e. the temperature rise ΔT, was exactly the same as with the original heat transfer Q, and that this was the *only* effect. He therefore concluded that heat and work are equivalent, at least in a limited sense (since the experiment cannot be performed in reverse, i.e. cooling of the water

* Throughout this chapter, lower case letters refer to extensive properties, as well as to heat and work, applied to unit mass, while capital letters refer to the same quantities applied to mass m of working fluid

Quantity	*Unit mass*	*Mass m*
Heat	q	Q
Work	w	W
Enthalpy	h	H
Internal energy	u	U
Volume	v	V

will produce heat flow from the system to a heat sink, but *not* work transfer from the system to the surroundings by raising the mass m).

This equivalence is expressed by the relationship

$$W = \text{constant} \times Q \tag{4.1}$$

the value of the constant being known as the mechanical equivalent of heat, denoted by J,

$$J = 4{\cdot}1868 \text{ kJ/kcal} \tag{4.1a)*}$$

(see also equation (2.14).)

Note that equation (4.1) expresses an empirical fact, and cannot be deduced logically.

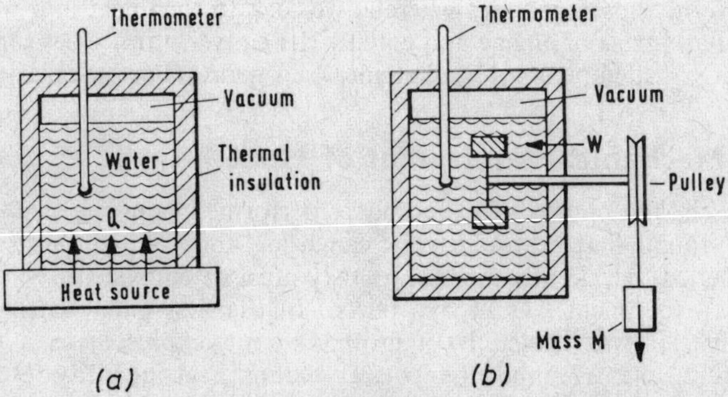

Fig. 4.1
(a) Joule's experiment: heating (b) Joule's experiment: paddling

4.2 Cyclic Processes

If the two parts of Joule's experiment are combined, with the first part (the heating process) reversed to become a cooling process, and arranged to follow the paddle wheel process, the water may be returned to its original state. The two processes together constitute a cycle and in view of equations (4.1) and (4.1a), and using the symbol $\sum$ to denote the summation round the cycle of energy transfer in the form of heat and work we may write

$$\sum \delta W = J \sum \delta Q \tag{4.2}$$

* In the SI system, units of heat and work are identical, viz. joules (J) or in practice kilojoules (kJ). Hence, $J = 1{\cdot}0$. In physics, however, heat is usually measured in calories (cal) or kilocalories (kcal) in which case equation (4.1a) applies.

The above cyclic process is merely one of an infinitely large number. However, the relationship expressed by equation (4.2) has been found to be true in all cases investigated so far involving cyclic processes. Its generalization becomes the first law of thermodynamics.

4.3 First Law of Thermodynamics

This may now be formulated as follows—

WHEN A SYSTEM EXECUTES A CYCLIC PROCESS, THE ALGEBRAIC SUM OF THE WORK TRANSFERS IS PROPORTIONAL TO THE ALGEBRAIC SUM OF THE HEAT TRANSFERS.

This purely experimental result may be expressed, in mathematical form as

$$\oint dW = J\oint dQ$$

or

$$J\oint dQ - \oint dW = 0 \tag{4.2a}$$

where $\oint dZ$ denotes the cyclic integral of the quantity Z.

Since $J = 4{\cdot}1868$ kJ/kcal it is evident that the maximum amount of work obtained from a system working in a cycle and receiving 1 cal is 4·1868 kJ. In fact, the second law of thermodynamics informs us that this can never be achieved, since inevitably part of the heat received is rejected during the same cycle (see Chapter 6).

Thus, as already stated, heat and work are equivalent only in a limited sense, the *complete* conversion of work into heat being possible as in Joule's paddle wheel experiment, but the reverse being impossible.

Since, when using SI units for both heat and work, the constant J reduces to unity, equation (4.2a) may be written more conveniently as

$$\oint(dQ - dW) = 0 \tag{4.2b}$$

4.4 The First Law Applied to Closed System Processes

Equation (4.2a), the mathematical statement of the first law, applies to cyclic processes. This presentation has the advantage of uniformity with the corresponding presentation of the second law, which is usually expressed in a form restricted to cyclic processes.

However, the first law, expressing conservation of energy, may be expressed in an alternative and often more useful form applicable to non-cyclic processes, deduced either from common sense arguments or else derived from equation (4.2a) or (4.2b).

(*a*) Empirical derivation

Referring to the Joule experiment in section 4.1, both the direct heating process and the stirring process constitute energy addition to the system, resulting in an increase in the energy level of the system, measured by the increase in temperature. It is evident that, had the two processes taken place simultaneously, the rise in energy level would have been even greater. Again, by applying the principle of conservation of energy, the sum of the energy transfers across the boundaries of the system must be equal to the increase in stored energy. The only significant form of stored energy in the Joule experiment (and the vast majority of experiments performed) is internal energy U, but if the system as a whole were in motion, there would also be changes in kinetic energy KE and potential energy PE (or in magnetic and electrical energy).

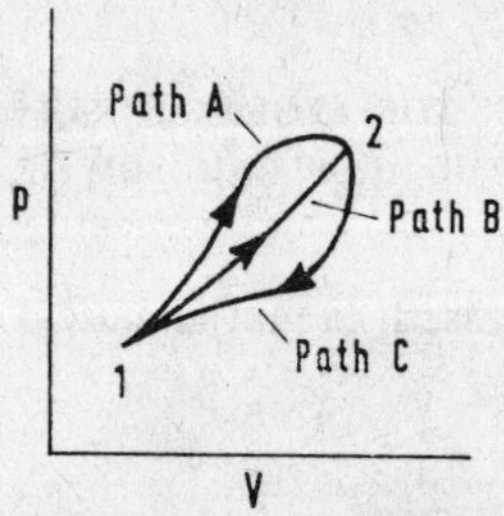

Fig. 4.2 Change of state

Hence we may write, for any process 1–2,

$$Q_{12} - W_{12} = \Delta U_{12} + \Delta(\text{KE})_{12} + \Delta(\text{PE})_{12} = \Delta E_{12} \qquad (4.3)$$

where Q_{12} = heat transferred to system

$-W_{12}$ = work transferred to system (*N.B.* negative sign)

$\Delta U_{12} = m(u_2 - u_1)$ = change of internal energy

$\Delta(\text{KE})_{12} = \frac{1}{2}m(c_2^2 - c_1^2)$ = change of kinetic energy

$\Delta(\text{PE})_{12} = mg(Z_2 - Z_1)$ = change of potential energy

$$\Delta E_{12} = \Delta U_{12} + \Delta(\text{KE})_{12} + \Delta(\text{PE})_{12} = \text{change of total energy} \qquad (4.3a)$$

We may express this in words as follows—

Corollary 1: THERE EXISTS A PROPERTY OF A SYSTEM, CALLED ENERGY, SUCH THAT A CHANGE IN ITS VALUE IS EQUAL TO THE DIFFERENCE BETWEEN THE HEAT SUPPLIED AND THE WORK DONE DURING ANY CHANGE IN STATE.

(*b*) Formal derivation (*Fig.* 4.2)

Let a system change its state first along path A, then along path B, from 1 to 2 and in each case let the system return from 2 to 1 in path C. Thus, from equation (4.2b), i.e. the first law applied to the cyclic process $1A2C1$ we have

$$(\sum\delta Q - \sum\delta W)_{AC} = 0$$

i.e.
$$\sum_1^2(\delta Q - \delta W)_A + \sum_2^1(\delta Q - \delta W)_C = 0 \qquad \text{(i)}$$

while, applying equations (4.2b) to the cyclic process $1B2C1$ we have

$$(\sum\delta Q - \sum W)_{BC} = 0$$

i.e.
$$\sum_1^2(\delta Q - \delta W)_B + \sum_2^1(\delta Q - \delta W)_C = 0 \qquad \text{(ii)}$$

Comparing (i) with (ii) it is obvious that

$$\sum_1^2(\delta Q - \delta W)_A = \sum_1^2(\delta Q - \delta W)_B \qquad \text{(iii)}$$

The argument could be applied to any other path, from which it must be concluded that the quantity

$$\sum_1^2(\delta Q - \delta W)$$

is independent of the path.

But this behaviour is characteristic of a property (see section 2.13). Hence we may deduce that

$$\sum_1^2(\delta Q - \delta W) = E_2 - E_1 \qquad (4.4)$$

i.e.
$$Q_{12} - W_{12} = \Delta E_{12} \qquad (4.4a)$$

where E is this new property, i.e. the energy of the system (composed of internal, kinetic and potential energy).

Since, in a closed system, kinetic and potential energy effects generally play a negligible part, equation (4.4a) may be simplified still further to the form known as the

NON-FLOW ENERGY EQUATION (N.F.E.E.)

$$Q_{12} - W_{12} = \Delta U_{12} \qquad (4.4b)$$

If energy transfers across the system boundary are ruled out, equation (4.4) indicates that E cannot change. Hence we may conclude—

Corollary 2: THE ENERGY OF THE ISOLATED SYSTEM IS CONSTANT.

EXAMPLE 4.1

A stationary mass of oxygen is compressed, without friction, from an initial state of 0·14 m³, 1·5 bar and 15°C to a final state of 0·06 m³ and 1·5 bar. During this process 0·6 kJ of heat are lost to the surroundings. What is the change in internal energy and the final temperature of the oxygen? (Specific heat at constant volume $c_v = 0{\cdot}649$ kJ/kg-K).

SOLUTION

Initially $\quad pV = mRT = mGT/M$

$$100 \times 1{\cdot}5 \times 0{\cdot}14 = m \times (8{\cdot}314/32) \times 288$$

$$m = 0{\cdot}2087 \text{ kg}$$

$$U_1 = mc_v(t - 0)$$

$$= 0{\cdot}2807 \times 0{\cdot}649 \times 15 = 2{\cdot}7326 \text{ kJ} \quad \textit{Ans.}$$

During the change at constant pressure:

$$\text{Work done in compression} = p(V_2 - V_1)$$

$$= 100 \times 1{\cdot}5 \times (0{\cdot}06 - 0{\cdot}14)$$

$$= -12{\cdot}00 \text{ kJ}$$

$$Q_{12} - W_{12} = \Delta U_{12}$$

$$-0{\cdot}6 + 12 = U_2 - U_2 = mc_v(t_2) - 2{\cdot}7326 \text{ kJ}$$

$$t_2 = 14{\cdot}133/0{\cdot}2807 \times 0{\cdot}649 = 77{\cdot}8°\text{C} \quad \textit{Ans.}$$

Note: In very many cases in engineering the kinetic energy and potential energy terms are negligible. For example, if this oxygen is raised through 7 m and at that level has an 86 km/h velocity imposed on it, the additional energy due to these new conditions is

$$\tfrac{1}{2}mc^2 + (mg \times \text{height})$$

Now $\quad 86 \text{ km/h} = 86000/3600 = 23{\cdot}89 \text{ m/s}$

$$\tfrac{1}{2}mc^2 = \tfrac{1}{2}(0{\cdot}2807 \times 23{\cdot}89^2) \times 10^{-3} = 0{\cdot}0799 \text{ kJ}$$

$$m \times \text{height} = 0{\cdot}2807 \times 9{\cdot}81 \times 10^{-3} = 0{\cdot}0193 \text{ kJ}$$

Therefore

$$\tfrac{1}{2}mc^2 + mgz = 0{\cdot}0799 + 0{\cdot}0193$$

$$= 0{\cdot}0992 \text{ kJ}$$

4.5 Particular Closed System Processes

Closed systems may undergo many different kinds of processes, some of which will be listed here. It will be assumed that all these processes are non-dissipative, i.e. reversible (the term reversible is discussed in Chapter 5), that work always takes the form of displacement work (section 2.2*a*) not shear work or shaft work (see section 2.2*b*, *c*), and that kinetic and potential energy effects are negligible. All processes listed below apply to mass m of working fluid contained within the system boundaries.

(*a*) **Constant volume process** (*Fig.* 4.3)

Work done, from equation (2.8a), with $\Delta V_{12} = 0$,

$$W_{12} = 0$$

Heat transferred: applying equation (4.4),

$$Q_{12} - W_{12} = \Delta E_{12} = \Delta U_{12}$$

whence $$Q_{12} = \Delta U_{12} \tag{4.5}$$

i.e.

HEAT ADDED = CHANGE OF INTERNAL ENERGY

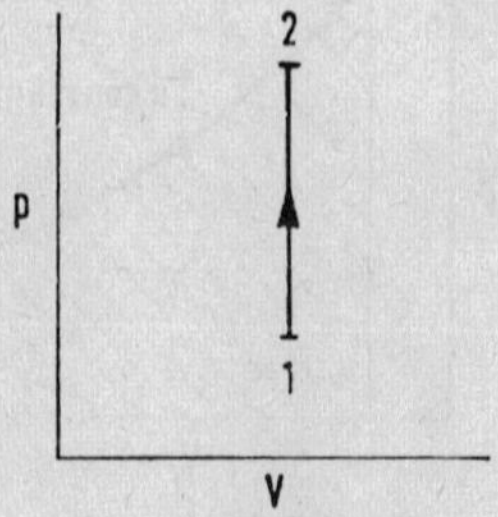

Fig. 4.3 Constant volume process

∴ for the perfect gas only (see equation (3.9), section 3.5)

$$Q_{12} = mc_v(T_2 - T_1) \tag{4.5a}$$

(*b*) **Constant pressure process** (*Fig.* 4.4)

Work done, from equation (2.8a),

$$W_{12} = p\Delta V_{12}$$

Heat transferred: applying equation (4.4),

$$Q_{12} - W_{12} = \Delta U_{12}$$

i.e. $$Q_{12} = \Delta U_{12} + p\Delta V_{12}$$

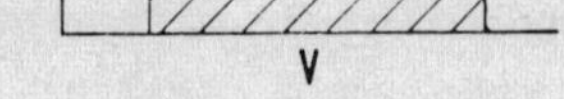

Fig. 4.4 Constant pressure process

But from the definition of enthalpy h

$$h = u + pv$$

or $$H = U + pV$$ (equation (2.17), section 2.10)

$$Q_{12} = \Delta U_{12} + p\Delta V_{12} + V\underset{\uparrow}{\Delta p_{12}} = \Delta H_{12} \tag{4.6}$$

(= 0 in a constant pressure process)

i.e. HEAT ADDED = CHANGE OF ENTHALPY

For the perfect gas only (see equation (3.11), section 3.5)

$$Q_{12} = mc_p(T_2 - T_1) \tag{4.6a}$$

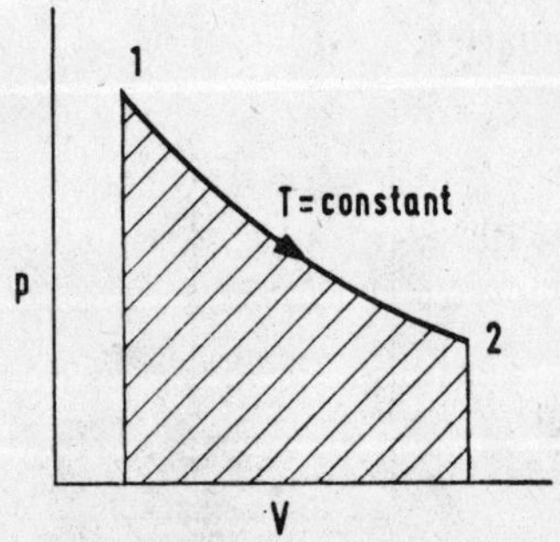

Fig. 4.5 Isothermal process

(*c*) **Isothermal process** (constant temperature) (*Fig.* 4.5)

Work done, from equation (2.8a),

$$W_{12} = \int_1^2 p\,dV$$

Heat transferred: applying equation (4.4)

$$Q_{12} - W_{12} = \Delta U_{12} \tag{4.7}$$

i.e. $$Q_{12} - \int_1^2 p\,dV = \Delta U_{12}$$

For the perfect gas only, the internal energy U is a function of temperature only (see equation (3.8), section 3.5) whence

$$\Delta U_{12} = 0$$

$$\therefore Q_{12} = W_{12} \tag{4.7a}$$

Now $$W_{12} = \int_1^2 p\,dV$$

where, from the perfect gas equation $p = mRT/V$ (equation (3.5), section 3.5)

$$W_{12} = \int_1^2 \frac{mRT\,dV}{V} = mRT \ln\left(\frac{V_2}{V_1}\right)$$

Substituting in equation (4.7a)

$$Q_{12} = W_{12} = mRT \ln\left(\frac{V_2}{V_1}\right) \tag{4.7b}$$

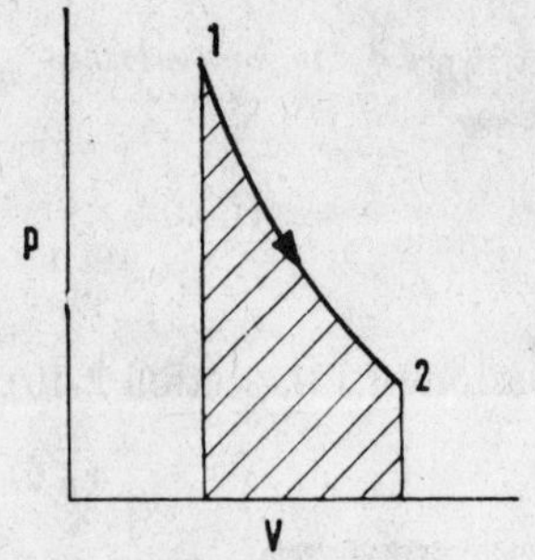

Fig. 4.6 Adiabatic process

(*d*) **Adiabatic process** (*Fig.* 4.6)

An adiabatic expansion is one in which only work transfer, but *not* heat transfer, occurs across the system boundary. Hence applying equation (4.4)

$$Q_{12} - W_{12} = \Delta U_{12}$$

i.e.

$$W_{12} = -\Delta U_{12} \tag{4.8}$$

i.e. WORK DONE BY SYSTEM = CHANGE OF INTERNAL ENERGY

For the perfect gas only, $u = c_v T$ (equation (3.9a), section 3.5) or

$$U = mc_v T$$

$$\therefore \quad \int_1^2 p\,dV = -\,mc_v(T_2 - T_1)$$

or

$$p\,dV = -\,mc_v\,dT \tag{i}$$

where from the perfect gas equation $p = mRT/V$ (equation (3.5))

$$mc_v\,dT = \frac{mRT\,dV}{V} = 0 \quad \text{or} \quad \frac{dT}{T} = -\frac{R}{c_v}\frac{dV}{V}$$

But from equation (3.12), section 3.5,

$$\frac{R}{c_v} = \frac{c_p - c_v}{c_v} = (\gamma - 1) \tag{ii}$$

i.e.

$$\frac{dT}{T} = -(\gamma - 1)\frac{dV}{V} \quad \text{or} \quad \frac{T_2}{T_1} = \left(\frac{V_2}{V_1}\right)^{-(\gamma-1)} \tag{iii}$$

Using the perfect gas equation to replace V_2/V_1 by T_2p_1/T_1p_2 in equation (iii) we have

$$\frac{T_2}{T_1} = \left(\frac{p_2}{p_1}\right)^{(\gamma-1)/\gamma} \tag{iv}$$

and combining equation (iii) and (iv) we have

$$\frac{p_2}{p_1}\left(\frac{V_2}{V_1}\right)^{\gamma} = 1 \quad \text{or} \quad pV^{\gamma} = \text{constant} \tag{4.9}$$

Alternative forms of equation (4.9) are (from (iii) and (iv) above)

$$TV^{(\gamma-1)} = \text{constant} \tag{4.9a}$$

$$Tp^{(-\gamma+1)/\gamma} = \text{constant} \tag{4.9b}$$

For reasons connected with the second law, and the fact that during an adiabatic reversible change of state the property *entropy* remains constant, such processes are known as *isentropic* processes (see Chapter 7).

(*e*) **Polytropic process** (*Fig.* 4.7)

Processes (*a*), (*b*), (*c*) and (*d*) above are represented respectively by pV^∞ = constant, pV^0 = constant, pV^1 = constant and pV^γ = constant (the latter two expressions being confined to the perfect gas). The general process may therefore be represented by

$$pV^n = \text{constant} \quad (4.10)$$

where n is the POLYTROPIC EXPONENT.

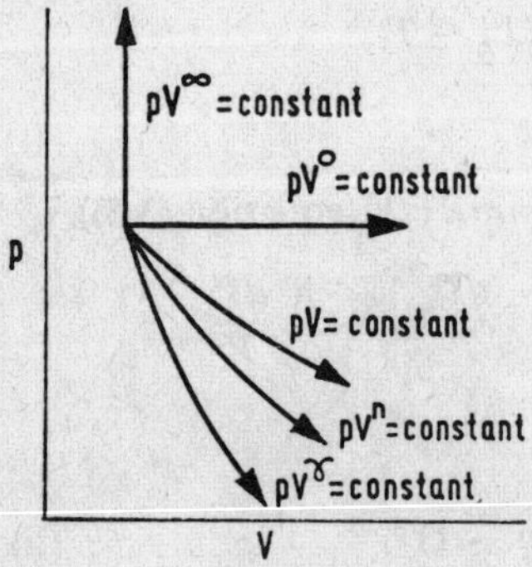

Fig. 4.7 Polytropic process

Substituting in equation (4.4)

$$Q_{12} - W_{12} = \Delta U_{12}$$

i.e. $$Q_{12} - \int_1^2 p\,dV = \Delta U_{12}$$

where pV^n = constant.

$$\therefore \quad Q_{12} - p_1 V_1^{\,n} \int_1^2 \frac{dV}{V^n} = \Delta U_{12}$$

$$Q_{12} - \frac{p_2V_2 - p_1V_1}{1-n} = U_2 - U_1 \quad (4.11)$$

For the *perfect gas* only, equation (4.11) becomes

$$Q_{12} = mc_v(T_2 - T_1) + mR\frac{(T_2 - T_1)}{1-n} \quad \text{(i)}$$

and using equation (3.12), section 3.5, i.e. $c_v(\gamma - 1) = R$, and substituting in (i)

$$Q_{12} = \frac{\gamma - n}{-1 + n} \cdot \frac{mR(T_1 - T_2)}{\gamma - 1} \quad (4.11a)$$

or $$Q_{12} = \frac{\gamma - n}{\gamma - 1} W_{12} \quad (4.11b)$$

Equations (4.11a) and (4.11b) show that if $n = \gamma$, then $Q_{12} = 0$, the condition for the adiabatic reversible process described by equations (4.9) and (4.9b).

EXAMPLE 4.2

(1) *Constant pressure process*

In an oil engine, the pressure at the end of compression is 34 bar and the volume is 0·0025 m³. The combustion process occurs at constant pressure until the volume of the working fluid is doubled, by which time 24 kJ of heat have been supplied. Calculate the change in internal energy due to combustion.

SOLUTION (*Fig.* 4.8a)

$$W = p(V_2 - V_1)$$
$$= 100 \times 34 \times 0{\cdot}0025$$
$$= 8{\cdot}492 \text{ kJ}$$
$$Q - W = U_2 - U_1$$
$$24 - 8{\cdot}492 = U_2 - U_1 = 15{\cdot}508 \text{ kJ} \quad \textit{Ans.}$$

Change is a gain of 15·508 kJ of internal energy.

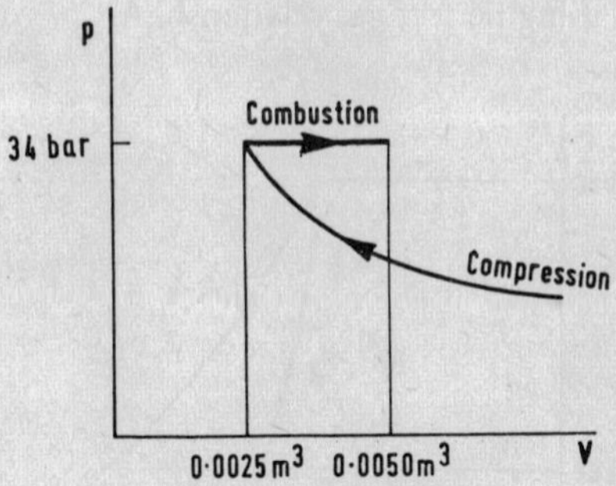

Fig. 4.8a

(2) *Constant volume process*

0·17 m³ of nitrogen at 17 bar and 150°C is cooled, at constant volume, until the temperature becomes 15°C. Determine the amount of heat transferred, the final pressure and the change in enthalpy of the gas, using the perfect gas relationships. ($c_v = 0{\cdot}724$ kJ/kg-K, $c_p = 0{\cdot}984$ kJ/kg-K)

SOLUTION

$Q - W = \Delta U$ and $W = 0$, since the volume is constant.

Therefore

$$Q_{12} = \Delta U_{12} = mc_v(T_2 - T_1)$$
$$m = pv/RT = (17 \times 0{\cdot}17 \times 100)/0{\cdot}2967 \times 423$$
$$= 2{\cdot}301 \text{ kg}$$
$$Q = 2{\cdot}301 \times 0{\cdot}724 \times (15 - 150) = -224{\cdot}5 \text{ kJ} \quad \textit{Ans.}$$

The negative sign means that heat is lost to the surroundings. Now, $p_2 = p_1T_2/T_1$, and therefore

$$p_2 = 17 \times 423/288 = 25 \text{ bar} \quad \textit{Ans.}$$

$$H_1 = mc_pT_1 \qquad H_2 = mc_pT_2$$

$$\Delta H = H_2 - H_2$$

$$= 2{\cdot}301 \times 0{\cdot}984\,(288 - 423)$$

$$= -306 \text{ kJ} \quad \textit{Ans.}$$

(3) *Isothermal and adiabatic process*

Atmospheric air at 1·06 bar and 15°C is compressed isothermally to 14 bar and is thereafter expanded adiabatically and without friction back to atmospheric pressure. What is the final temperature and specific volume of the air? Calculate also the net work done and the heat exchange with the surroundings, using perfect gas relationships.

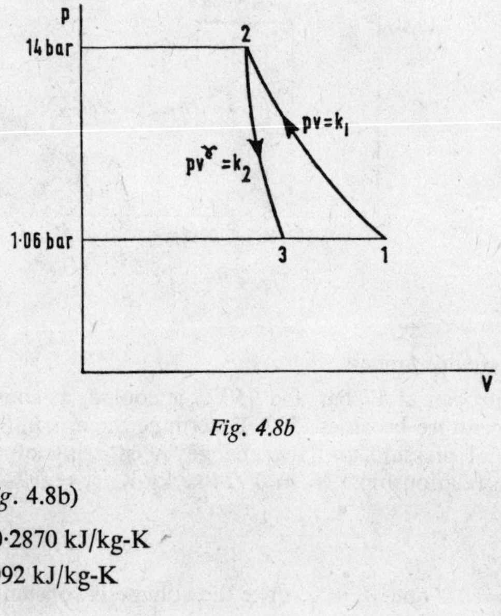

Fig. 4.8b

SOLUTION (*Fig.* 4.8b)

$$R_{air} = 0{\cdot}2870 \text{ kJ/kg-K}$$

$$c_p = 0{\cdot}992 \text{ kJ/kg-K}$$

Therefore

$$c_v = c_p - R = 0{\cdot}992 - 0{\cdot}2870 = 0{\cdot}705 \text{ kJ/kg-K}$$

$$\gamma = c_p/c_v = 0{\cdot}992/0{\cdot}705 = 1{\cdot}41$$

$$T_2 = T_1 \qquad T_3/T_2 = (p_3/p_2)^{(\gamma-1)/\gamma}$$

$$T_3 = 288(1{\cdot}06/14)^{0{\cdot}41/1{\cdot}41} = 288/2{\cdot}09 = 137{\cdot}8 \text{ K} = -135{\cdot}2^\circ\text{C} \quad \textit{Ans.}$$

$$p_2v_2/T_2 = p_3v_3/T_3 \text{ and } v_2 = RT_2/p_2$$

Therefore

$$v_2 = (0{\cdot}2780 \times 288)/14 \times 10^2 = 5{\cdot}72/10^2 = 0{\cdot}0572 \text{ m}^3/\text{kg}$$

$$v_3 = (14 \times 0{\cdot}0572 \times 137{\cdot}8)/288 \times 1{\cdot}06 = 0{\cdot}361 \text{ m}^3/\text{kg} \quad \textit{Ans.}$$

Work done in isothermal compression (equation (4.7b),)

$$= RT \ln \left(\frac{v_2}{v_1}\right) = RT \ln \left(\frac{p_1}{p_2}\right)$$

$$= 0{\cdot}2870 \times 288 \ln (14/1{\cdot}06)$$

$$213{\cdot}3 \text{ kJ/kg}$$

Work done in adiabatic frictionless expansion

$$= \frac{p_2v_2 - p_3v_3}{\gamma - 1} = \frac{R(T_2 - T_3)}{\gamma - 1}$$

$$= 0{\cdot}2870(288 - 137{\cdot}8)/0{\cdot}41 = 105{\cdot}1 \text{ kJ/kg}$$

Net work done 1–2–3

$$= 105{\cdot}1 - 213{\cdot}3 = -108{\cdot}2 \text{ kJ/kg} \quad \textit{Ans.}$$

$$Q_{123} - W_{123} = \Delta U_{13} = c_v(T_3 - T_1)$$

$$= 0{\cdot}705(137{\cdot}8 - 288) = -105{\cdot}9 \text{ kJ/kg}$$

Therefore

$$Q_{123} = -108{\cdot}2 - 105{\cdot}9 = -214{\cdot}1 \text{ kJ/kg} \quad \textit{Ans.}$$

The −ve sign indicates that heat is given out to the surroundings in the process. Since no heat transfer occurs during process 2–3, Q_{123} should equal Q_{12}. The slight discrepancy is due to numerical inaccuracies.

Note: This question has assumed that the specific heats remained constant over the range of temperature of the question.

4.6 The First Law Applied to Open or Flow Processes

The open system, unlike the closed system, experiences transfers of mass as well as energy across its boundaries. In drawing up an energy balance equivalent to equation (4.4) for the closed system it is essential to take into account for any given process—

(*a*) the change of energy associated with matter contained within the boundary of the system,
(*b*) the change of energy associated with matter crossing the boundary of the system,
(*c*) heat and work transfer across the system boundary.

Two kinds of processes occur—

(1) STEADY FLOW PROCESSES in which the mass inflow into the system, during any one period, equals the mass outflow

$$m_{in} = m_{out}$$

and in which there is no change of energy stored within the system, so that energy changes under (*a*) above vanish.

(2) UNSTEADY FLOW PROCESSES in which mass inflow and outflow are not balanced

$$m_{in} \neq m_{out}$$

and in which there is a change of energy stored within the system.

The latter is the more general case and will be discussed first.

The non-flow energy equation (4.4b) no longer applies and must be extended to include energy transfers associated with matter crossing the boundary of the system, i.e.

$$Q_{12} - W_{12} = \underbrace{\Delta E_{12}}_{\substack{storage\\energy}} + \underbrace{\Delta E_{(outflow)} - \Delta E_{(inflow)}}_{flow\ energy} \qquad (4.12)$$

The storage energy term is given by equation (4.3a),

$$\Delta E_{12} = m_2[u_2 + (\mathrm{KE})_2 + (\mathrm{PE})_2] - m_1[u_1 + (\mathrm{KE})_1 + (\mathrm{PE})_1] \qquad \text{(i)}$$

where suffixes 1 and 2 refer to initial and final conditions for the time interval Δt under consideration. The flow energy term is assessed as follows (see *Fig.* 4.9). Let m_{in} be the mass of fluid crossing the boundary of the system during the time interval Δt. In *Fig.* 4.9 this mass element is shown in the position it occupies at the beginning

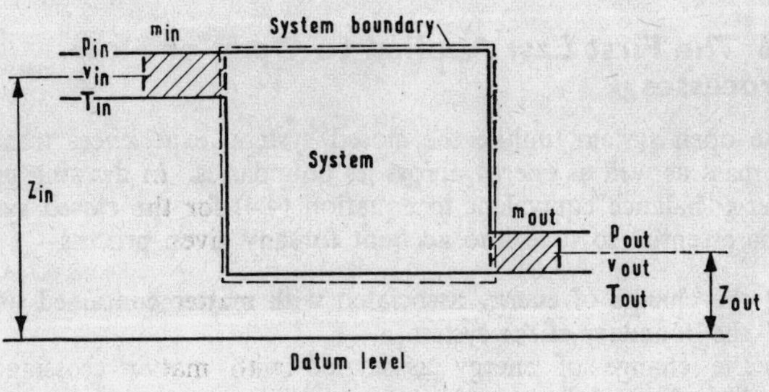

Fig. 4.9 Open system

of the interval. The energy content of this mass element is given, in accordance with equation (i) above, by

$$\Delta E_{in} = m_{in}[u_{in} + (\mathrm{KE})_{in} + (\mathrm{PE})_{in}] \tag{ii}$$

In accordance with equation (2.11), section 2.2, the mass element entering will perform flow work on the system, given by

$$(W_{(flow)})_{in} = m_{in}(p_{in}v_{in}) \tag{iii}$$

so that the total energy associated with m_{in} may be obtained by addition of equations (ii) and (iii),

$$\Delta E_{(inflow)} = m_{in}[u_{in} + (\mathrm{KE})_{in} + (\mathrm{PE})_{in} + p_{in}v_{in}] \tag{iv}$$

But in section 2.10 the combination of properties $(u + pv)$ has been defined as the enthalpy h and its importance is now evident. Equation (iv) may therefore be written

$$\Delta E_{(inflow)} = m_{in}[h_{in} + (\mathrm{KE})_{in} + (\mathrm{PE})_{in}] \tag{v}$$

Similarly for outflow,

$$\Delta E_{(inflow)} = m_{out}[h_{out} + (\mathrm{KE})_{out} + (\mathrm{PE})_{out}] \tag{vi}$$

Substituting for equations (i), (v) and (vi) in equation (4.12) we obtain

$$\begin{aligned} &Q_{12} - W_{12} \\ &= \{m_2[u_2 + (\mathrm{KE})_2 + (\mathrm{PE})_2] - m_1[u_1 + (\mathrm{KE})_1 + (\mathrm{PE})_1]\}_{\Delta E(storage)} \\ &+ \{m_{out}[h_{out} + (\mathrm{KE})_{out} + (\mathrm{PE})_{out}] - m_{in}[h_{in} + (\mathrm{KE})_{in} + (\mathrm{PE})_{in}]\}_{\Delta E(flow)} \end{aligned} \tag{4.12a}$$

or

$$\begin{aligned} &Q_{12} - W_{12} \\ &= \{m_2(u_2 + \tfrac{1}{2}c_2^2 + gZ_2) - m_1(u_1 + \tfrac{1}{2}c_1^2 + gZ_1)\}_{\Delta E(storage)} \\ &+ \{m_{out}(h_{out} + \tfrac{1}{2}c_{out}^2 + gZ_{out}) - m_{in}(h_{in} + \tfrac{1}{2}c_{in}^2 + gZ_{in})\}_{\Delta E(flow)} \end{aligned} \tag{4.12b}$$

Equation (4.12a) or (4.12b) is a general expression of the energy equation applied to open system processes allowing for both mass and energy changes within the system. By the above definition, this constitutes an unsteady flow process. Hence equation (4.12a) or (4.12b) is referred to as the

UNSTEADY FLOW ENERGY EQUATION (U.S.F.E.E.)

In a steady flow process, the storage energy term in equation (4.12a) vanishes since, by definition, neither mass nor energy changes occur within the system. Furthermore, the mass inflow m_{in} is exactly balanced by the mass outflow m_{out}, so that, substituting

$$m_{in} = m_{out} = m$$

in equations (4.12a) and (4.12b), and replacing the now redundant suffixes *in* and *out* by the more convenient suffixes 1 and 2 we obtain

$$Q_{12} - W_{12} = m[h_2 + (\mathrm{KE})_2 + (\mathrm{PE})_2] - m[h_1 + (\mathrm{KE})_1 + (\mathrm{PE})_1] \qquad (4.13a)$$

or $$Q_{12} - W_{12} = m[(h_2 + \tfrac{1}{2}c_2^2 + gZ_2) - (h_1 + \tfrac{1}{2}c_1^2 + gZ_1)] \qquad (4.13b)$$

Equations (4.13a) and (4.13b) are referred to as the

STEADY FLOW ENERGY EQUATION (S.F.E.E.)

EXAMPLE 4.3

In a steady flow system, the mass crossing the boundary is the same at entry and exit.

Neglecting the small mass of fuel injected into the cylinder of an engine, the mass induced and exhausted is the same, and the contents of the cylinder are almost entirely air so that little error will be introduced in assuming pure air as the working fluid.

If the pressure and temperature of the air is measured at entry and exit from the cylinder as 0·98 bar and 15°C and 4·2 bar and 54°C respectively, what is the net flow work of the system?

SOLUTION

From the definition of flow work

$$(W_{flow})_{net} = (W_{flow})_{out} - (W_{flow})_{in} = m_{out}p_{out}v_{out} - m_{in}p_{in}v_{in}$$

$$= m(p_{out}v_{out} - p_{in}v_{in}) \qquad \text{(i)}$$

$$= mR(T_{out} - T_{in}) \qquad \text{(ii)}$$

Substituting in (ii),

$$(W_{flow})_{net} = 0{\cdot}2870(54 - 15) = 11{\cdot}19 \text{ kJ/kg} \quad \textit{Ans.}$$

Alternatively

$$v_{out} = \frac{RT_{out}}{p_{out}} = \frac{0{\cdot}2870 \times 327}{10^2 \times 4{\cdot}2} = 0{\cdot}224 \text{ m}^3\text{/kg}$$

$$v_{in} = \frac{RT_{in}}{p_{in}} = \frac{0{\cdot}2870 \times 288}{0{\cdot}98 \times 10^2} = 0{\cdot}845 \text{ m}^3\text{/kg}$$

Substituting in (i),

$$(W_{flow})_{net} = 10^2(4{\cdot}2 \times 0{\cdot}224 - 0{\cdot}98 \times 0{\cdot}845) = 11{\cdot}26 \text{ kJ/kg} \quad \textit{Ans.}$$

NOTE ON CONTROL VOLUME AND CONTROL SURFACE

Some writers, in considering the open or flow system, introduce in addition to the concept of system and system boundary the concepts of control volume and control surface. The system then becomes a region bounded by the system boundary, containing a fixed amount of matter and experiencing no mass transfer but free to move in space and to undergo changes in the form of its boundaries, while the control surface enclosing the volume is of constant form and is fixed in space, though it experiences mass transfer, and possibly changes in the amount of matter, contained within it. The distinction thus made between a region fixed in space but experiencing transfer of mass and one moving in space with the group of particles of matter under consideration and experiencing no mass transfer clarifies the formulation of the equations of continuity, momentum and energy in complicated three-dimensional flow problems. However, it is not considered necessary or advantageous to introduce the concept of control volume and control surface into the present much simpler treatment which is not related to any particular system of space co-ordinates.

Hence, in the present treatment, the open or flow system, unlike the closed system, experiences transfer of mass across its boundaries. The latter are neither of fixed shape nor fixed in space, except in special cases, as also occurred with the closed system.

4.7 Application of the Steady Flow Energy Equation to Special Cases

In engineering thermodynamics, the steady flow energy equation (4.13a) (S.F.E.E.) is of much greater importance than the more complicated unsteady flow energy equation (U.S.F.E.E.). Thus the behaviour of the components, and of the plant as a whole, in heat engines operating under conditions of steady flow, e.g. steam plant or gas turbines, is governed by the S.F.E.E. A number of examples are now given, with heat and work expressions both for mass m and for unit mass of working fluid.

(*a*) BOILER (*Fig.* 4.10)

Neglecting KE and PE effects and putting $W_{12} = 0$

$$Q_{12} = m(h_2 - h_1) \tag{4.14}$$

$$q_{12} = (h_2 - h_1) \tag{4.14a}$$

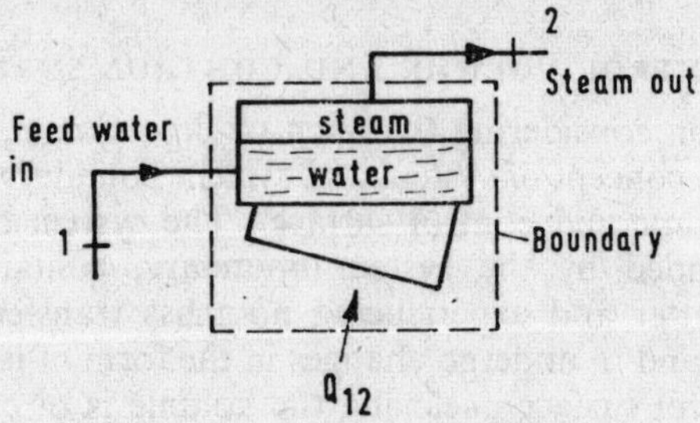

Fig. 4.10 Boiler

(*b*) COMPRESSORS (any type, i.e. reciprocating, aerodynamic or rotary positive displacement) (*Fig.* 4.11)

Work is done on the system and heat is usually lost by the system (for cooling water and radiation) so that, neglecting KE and PE effects,

$$Q_{12} - W_{12} = m(h_2 - h_1) \tag{4.15}$$

or

$$q_{12} - w_{12} = h_2 - h_1 \tag{4.15a}$$

where W_{12} is negative and Q_{12} is usually negative.

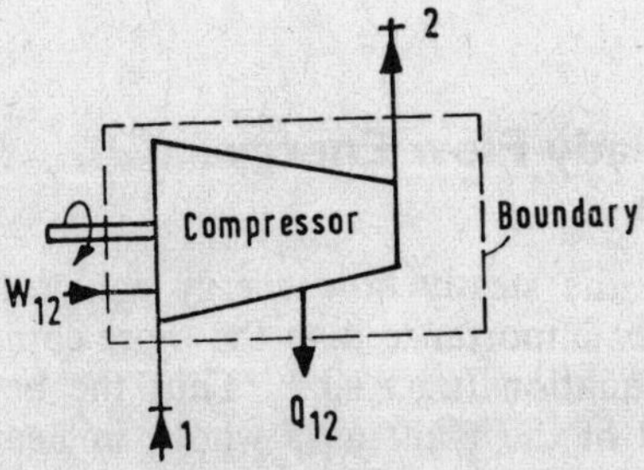

Fig. 4.11 Compressor

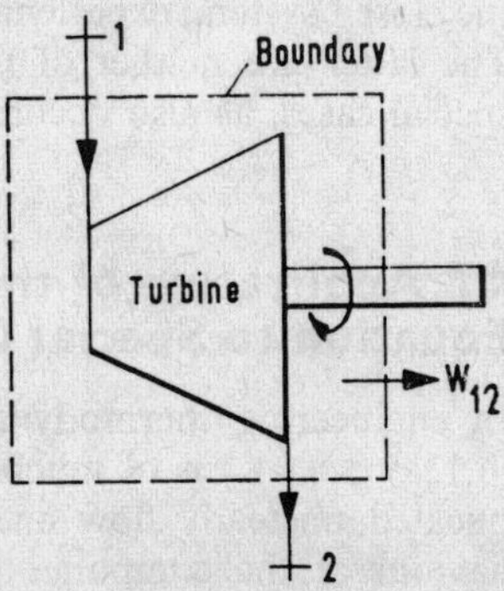

Fig. 4.12 Turbine

(*c*) TURBINE (*Fig.* 4.12)

Work is done by the system, and a small amount of heat may be lost from the system. The KE effect may generally be neglected. Hence, applying equation (4.13b)

$$Q_{12} - W_{12} = m(h_2 - h_1) \tag{4.16}$$

or

$$w_{12} - q_{12} = h_1 - h_2 \tag{4.16a}$$

Neglecting heat losses

$$w_{12} = h_1 - h_2 \tag{4.16b}$$

(*d*) NOZZLE (*Fig.* 4.13)

There is neither heat nor work transfer across the boundary of the system

$$Q_{12} = W_{12} = 0$$

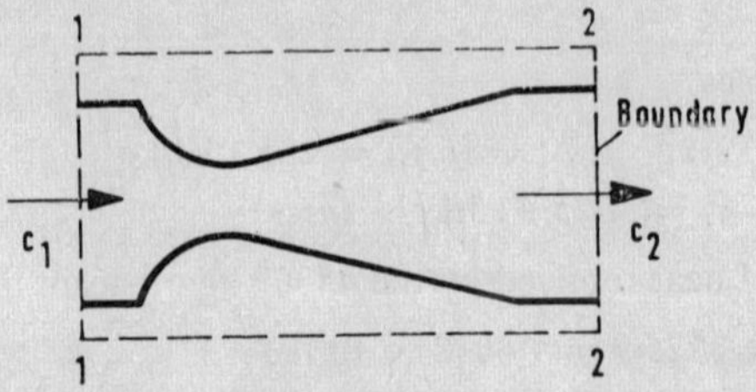

Fig. 4.13 Nozzle

However, in a nozzle, KE effects cannot be neglected. Hence applying equation (4.13b)

$$Q_{12} - W_{12} = 0 = m[(h_2 - h_1) + \tfrac{1}{2}(c_2^2 - c_1^2)]$$

or

$$c_2^2 = c_1^2 + 2(h_1 - h_2) \tag{4.17}$$

(*e*) THROTTLE VALVE (*Fig.* 4.14)

Again there is neither heat nor work transfer, i.e. $Q_{12} = W_{12} = 0$. Furthermore, both the initial and final velocities are usually low, so that KE effects may be neglected. Hence, applying equation (4.13b)

$$Q_{12} - W_{12} = 0 = m(h_2 - h_1 + 0)$$

or

$$h_2 = h_1 \tag{4.18}$$

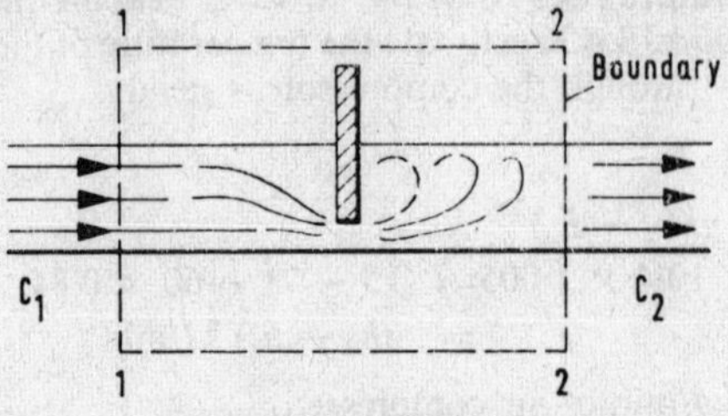

Fig. 4.14 Throttle valve

Thus, in spite of the pressure drop ($p_1 - p_2$) the enthalpy remains constant.

EXAMPLE 4.4

(1) *Constant pressure heating*

A boiler receives feed water at 49°C and delivers steam at 25 bar and 500°C. Estimate the heat supplied in the boiler per kg of steam. If the furnace is oil fired, the calorific value of the oil being 10250 kcal/kg and 4132 kg oil are burned while 45359 kg are produced, estimate the efficiency of the boiler.

SOLUTION

From steam tables,

$$h_2 = 3461 \text{ kJ/kg} \qquad h_1 = (h_f)_{49} = 205{\cdot}1 \text{ kJ/kg}$$

$$q_{12} = h_2 - h_1 = 3225{\cdot}9 \text{ kJ/kg} \quad \textit{Ans.}$$

Total amount of heat received by the 45359 kg of steam in the boiler

$$= 45359 \times 3225{\cdot}9 = 1{\cdot}4632 \times 10^8 \text{ kJ}$$

The amount of heat from 4132 kg of oil

$$= 4132 \times 10250 = 4{\cdot}2353 \times 10^8 \text{ kJ}$$

$$= 4{\cdot}1868 \times 4{\cdot}2353 \times 10^7 = 1{\cdot}7775 \times 10^8 \text{ kJ}$$

Therefore

$$\text{Efficiency of boiler} = \frac{1{\cdot}4632 \times 10^8}{1{\cdot}7775 \times 10^8} \times 10^2$$

$$= 82{\cdot}3\% \quad \textit{Ans.}$$

(2) *Polytropic compression*

A reciprocating air compressor receives air at the air filter at 34°C and delivers it to a receiver at 12·7 bar. It is driven by an electric motor which absorbs 18 kW and the efficiency of the drive is 84%. The receiver into which the air is pumped has a capacity of 1·2 m³ and the pressure rises from 3·3 bar to 12·7 bar, while the temperature of the air changes from 15°C to 49°C the change occurring in an interval of 5 min.

Water jacket cooling is used and 10·4 kg/min pass through the jacket while the temperature rises from 7°C to 15°C. Taking the specific heat of the air to be 1·005 kJ/kg-K estimate the temperature of the air as delivered, treating the flow through the compressor as steady.

SOLUTION

$$Q - W = m(h_2 - h_2)$$

$$-4{\cdot}1868 \times 10{\cdot}4 \times 1{\cdot}005 \times (15 - 7) + 60 \times 0{\cdot}84 \times 18$$

$$= m(h_2 - h_1) \text{ kJ/min} \qquad (1)$$

where m is the kg/min of air compressed.

At the receiver,

$$m_1 = p_1v_1/RT_1 \qquad m_2 = p_2v_2/RT_2$$

$$m = \tfrac{1}{5}(m_2 - m_1) = \frac{V}{R}\left(\frac{p_2}{T_2}\right) - \left(\frac{p_1}{T_1}\right)\frac{1}{5}$$

$$= \frac{1{\cdot}2}{0{\cdot}2870}\left(\frac{12{\cdot}7}{322} - \frac{3{\cdot}3}{288}\right)\frac{100}{5}\ \text{kg/min}$$

$$= 2{\cdot}740\ \text{kg/min} \quad \textit{Ans.}$$

$$m(h_2 - h_1) = 2{\cdot}740 \times 1{\cdot}005(t_2 - 34) \qquad (2)$$

Equating (1) and (2) we find

$$(t_2 - 34) \times 2{\cdot}740 \times 1{\cdot}005 = -348{\cdot}3 + 907{\cdot}2$$

$$t_2 - 34 = 203{\cdot}6$$

$$t_2 = 237{\cdot}0°\text{C} \quad \textit{Ans.}$$

(3) *Adiabatic expansion*

(*a*) A steam turbine receives steam at 20 bar and 250°C and exhausts at 1 bar, 0·85 dry. Neglecting heat losses, estimate the work done by the turbine in kJ/kg steam.

(*b*) If, when allowance is made for friction, radiation, and leakage losses, the actual work obtained is 80% of that estimated above, calculate the horsepower output of the turbine when consuming 2268 kg steam per min.

SOLUTION

(*a*)

$$q - w = h_2 - h_1$$

$$0 - w = h_2 - h_1 \qquad w = h_1 - h_2$$

From steam tables,

$$h_1 = 2904\ \text{kJ/kg} \qquad h_2 = 417 + 0{\cdot}85 \times 2258$$

$$= 2336\ \text{kJ/kg}$$

$$h_1 - h_2 = 568\ \text{kJ/kg} \quad \textit{Ans.}$$

(*b*) *Note.* This is the greatest amount of work to be obtained per kilogramme of steam in the turbine. When allowance is made for friction and other losses these are at the expense of this work and give rise to the term EFFICIENCY RATIO which is the ratio of actual work to ideal work.

$$\text{Actual work} = 0{\cdot}8 \times 568 = 454\ \text{kJ/kg}$$

$$= 2268 \times 454\ \text{kJ/min}$$

$$= (2268 \times 454)/60 = 17161\ \text{kJ/s or kW}$$

$$= 0{\cdot}746 \times 17161 = 12802\ \text{hp} \quad \textit{Ans.}$$

(4) *Throttling process*

Steam at 14 bar, 0·95 dry is throttled to 0·050. What is the quality of the steam after throttling?

SOLUTION

From steam tables

$$h_1 = 2790 - 0{\cdot}05 \times 1960 = 2702 \text{ kJ/kg}$$

$$h_2 = 2702 \text{ kJ/kg at } 0{\cdot}05 \text{ bar which exceeds}$$

$$h_g = 2561 \text{ kJ/kg at that pressure}$$

Therefore, the steam is superheated and the temperature lies between 100°C and 150°C and will be at

$$t = 100 + (14/96) \times 50 = 127{\cdot}1\text{°C}$$

Therefore, the steam is superheated 127·1 − 32·9 = 94·2 degrees *Ans.*

4.8 The Steady Flow Energy Equation and Cyclic Processes

Consider the simple steam plant illustrated in *Fig.* 4.15. Assuming that the states of the entering feed water and leaving condensate are identical, and applying equation (4.13b) to the various components of the complete plant arranged in series, we have

$$\Sigma\delta Q - \Sigma\delta W = m[\Delta h + \Delta(\text{KE}) + \Delta(\text{PE})] = 0 \qquad (4.19)$$

i.e.

$$\Sigma\delta Q = \Sigma\delta W \qquad (4.19\text{a})$$

By allowing the condensate at 2 to return to the system at 1, the flow system is converted into a closed system and equation (4.19a) becomes

$$\Sigma\delta Q - \Sigma\delta W = 0 \qquad (4.19\text{b})$$

or

$$\oint dQ - \oint dW = 0 \qquad (4.19\text{c})$$

This is identical with the original first law equation (4.2b) for a cyclic process.

The example of the complete steam plant is important in that it not only illustrates the concept of the working cycle as applied to a system composed of steady flow elements, in this case

(*i*) feed pump (*ii*) boiler (*iii*) turbine
(*iv*) condenser (*v*) condensate pump

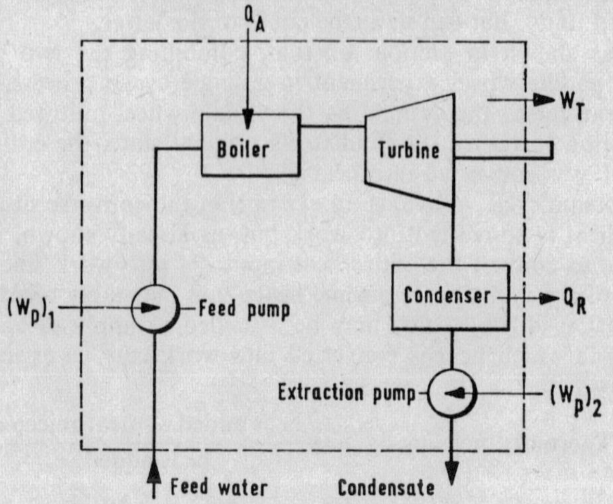

Fig. 4.15 Steam plant illustrating cyclic processes

but it introduces at the same time the most important concept of thermal efficiency. This may be defined as follows—

$$\text{Thermal efficiency } (\eta_{th}) = \frac{\text{useful work done/kg of fluid circulating}}{\text{GROSS heat added/kg of fluid circulating}}$$

Taking the simple steam cycle illustrated in *Fig.* 4.15,

Useful work done $W_{net} = W_T - (W_p)_1 - (W_p)_2$

Gross heat added $Q_{gross} = Q_A$

$$\therefore \quad \eta_{th} = \frac{W_T - (W_p)_1 - (W_p)_2}{Q_A} \tag{4.20}$$

From equation (4.19b),

$$W_{net} = \sum W = \sum Q = Q_A - Q_R$$

whence, substituting in equation (4.20)

$$\eta_{th} = \frac{W_T - (W_p)_1 - (W_p)_2}{Q_A} = \frac{Q_A - Q_R}{Q_A} = 1 - \frac{Q_R}{Q_A} \tag{4.20a}$$

It has been established experimentally that in all cyclic processes a fraction of the heat added is inevitably rejected, i.e. regarding equation (4.20a) as a general statement for all cycles—

$$Q_R \neq 0 \qquad Q_R < Q_A \tag{4.20b}$$

or

$$0 < \frac{Q_R}{Q_A} < 1 \tag{4.20c}$$

Hence the efficiency of all cyclic processes lies somewhere between zero and unity, but can never be equal to the latter.

It was shown in section 4.2 that, combining the two parts of Joule's paddle wheel experiment in a single cyclic process, namely work transfer to the system via the paddle wheel followed by heat abstraction to return the fluid to its original state, the entire work input W was converted into heat Q.

The steam cycle, above, is an example of the converse situation in which heat is converted into work but, as already shown, it is not possible to convert the entire heat input Q_A into work since in the course of the cyclic process some heat Q_R is inevitably rejected.

Equation (4.20a) which may be extended to apply to any cyclic process in which heat is converted into work may be expressed as follows—

$$\text{Thermal efficiency } (\eta_{th}) = \frac{\text{heat added} - \text{heat rejected}}{\text{heat added}}$$

$$= \frac{\Sigma\delta Q_A - \Sigma\delta Q_R}{\Sigma\delta Q_A} = \frac{\sum\delta Q}{\sum\delta Q_A} = \frac{\sum\delta W}{\sum\delta Q_A} \qquad (4.21)$$

The first law of thermodynamics provides no indication of the value of the thermal efficiency, except in so far as it cannot exceed unity.

It is known from experience that the efficiency, for any given cycle, is a maximum when the individual processes constituting the cycle are free from imperfections or, using the generally accepted term, from IRREVERSIBILITIES. These will be discussed in Chapter 5.

However, even if the cycle is completely free from all irreversibilities, the efficiency is still less than unity. The conditions which have to be satisfied in order to convert heat into work with maximum efficiency are established by applying the second law of thermodynamics, which will be discussed in Chapters 6 and 7.

4.9 Comparison of the Steady Flow Energy Equation and the Euler and Bernoulli Equations in Fluid Mechanics (see also Chapter 11)

It is clear that flow processes such as those described in sections 4.7(*d*) and (*e*) are subject, in addition to the laws of thermodynamics, also to those of fluid mechanics. The latter, as the name suggests, are based on purely mechanical consideration, in particular Newton's second law, expressed as the so-called momentum equation, and the law of conservation of mass, expressed as the so-called continuity equation. Because of the similarity in form of the resultant equations, known as the Euler and Bernoulli equations, to the steady flow energy equation, a brief discussion of the relationship between these equations appears appropriate.

THE EULER EQUATION

Consider a steady flow open system whose boundaries are formed by the duct walls (solid lines) and the imaginary sections 1 and 2 (dotted lines), the boundaries being fixed in space (*Fig*. 4.16).

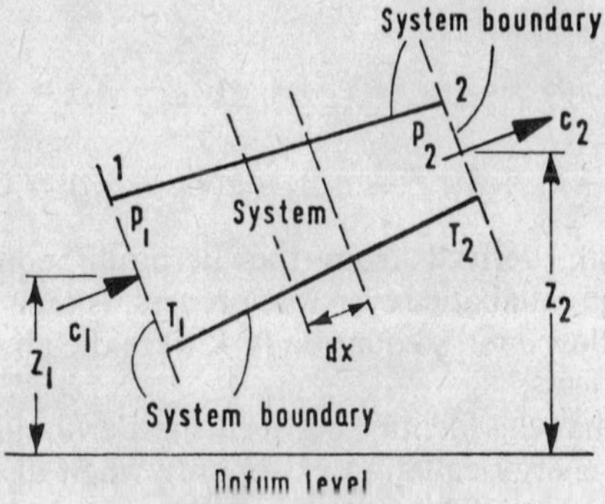

Fig. 4.16 Flow element

Applying Newton's second law to a small flow element of area A and length dx, we obtain the equation

$$v\,dp + c\,dc + g\,dZ = 0 \tag{4.22}$$

Equation (4.22) is known as the Euler equation and applies only when the following conditions are satisfied—

(*i*) the flow must be steady,
(*ii*) dissipative effects arising from shear forces must be absent, i.e. the flow must be frictionless,
(*iii*) electrical or magnetic field effects must be absent.

THE BERNOULLI EQUATION

The integrated form of the Euler equation for a finite system, e.g. the system bounded by the sections 1 and 2 in *Fig*. 4.16, is known as the Bernoulli equation.

Thus, from equation (4.22)

$$\int_1^2 v\,dp + \tfrac{1}{2}(c_2^{\,2} - c_1^{\,2}) + g(Z_2 - Z_1) = 0 \tag{4.23}$$

It will be shown in Chapter 7 (section 7.4, equation (7.8b)) that for an ADIABATIC REVERSIBLE process only (for which the new thermodynamic property, entropy s, arising from the second law of thermodynamics, has a constant value) the change of enthalpy dh for an infinitesimal pressure increment dp is given by

$$dh = v\,dp$$

Hence, substituting in equation (4.23)

$$\int_1^2 dh + \tfrac{1}{2}(c_2^2 - c_1^2) + g(Z_2 - Z_1) = 0$$

or

$$h_2 - h_1 + \tfrac{1}{2}(c_2^2 - c_1^2) + g(Z_2 - Z_1) = 0 \qquad (4.23a)$$

Equation (4.23a), derived from the Bernoulli equation for the special case of an adiabatic reversible process, is seen to be identical with the steady flow energy equation (4.13b) in the absence of heat or work transfer effects.

It is evident that the identity between the Bernoulli equation and the steady flow energy equation exists only when dissipative effects are completely absent. This restriction does *not* apply to the steady flow energy equation which, in the form

$$h_2 - h_1 + \tfrac{1}{2}(c_2^2 - c_1^2) + g(Z_2 - Z_1) = 0 \qquad (4.24)$$

applies quite generally to any steady flow process, whether dissipative or otherwise, provided that heat and work transfer effects are absent.

THUS THE STEADY FLOW ENERGY EQUATION IS GENERAL; THE EULER AND BERNOULLI EQUATIONS ARE NOT.

EXERCISES

First law and closed systems

For each example draw a diagrammatic sketch of the system. Label: system, surroundings, boundary; show direction of heat and work by arrows.

4.1 The energy of a system increases by 120 kJ while 150 kJ of work is transferred to the surroundings. Is heat added or taken away from the system?
(*Ans:* 270 kJ to system)

4.2 (*a*) The energy of a system increases by 25 kJ while the system is receiving 20 kJ of work. How much heat is transferred and in what direction?

(*b*) The energy of a system decreases by 50 kJ while the system is receiving 60 kJ of work. How much heat is transferred?

(*c*) During an expansion process the work done, and the heat received, by a system are respectively 20 kJ and 50000 J. Calculate the change in energy of the system.

(*Ans:* (*a*) 5 kJ to system, (*b*) 110 kJ from system, (*c*) +30 kJ)

4.3 A closed system consists of a cylinder of water stirred by a paddle wheel. For the process, the work was 34 kJ, the initial internal energy was 120 kJ, and the final internal energy after 1 h of stirring was 144 kJ. Find the heat transferred in kJ/h. Is the temperature of the system rising or falling?

(*Ans:* 10 kJ from system, rising)

4.4 During a constant pressure heating process undergone by the gas in a reciprocating engine cylinder, the following energy transfers occurred:

Work done by gas	3·20 kJ
Heat added to gas	6·20 kJ
Friction between piston and cylinder wall	0·50 kJ

Neglecting external pressure on the piston and assuming friction to be converted into heat of which half is ultimately conducted to the gas, whilst the other half appears as increased metal temperature, find:

(*a*) the external useful work
(*b*) the change of internal energy of the gas
(*c*) the change of internal energy of the cylinder and piston.
(*Ans:* (*a*) 2·70 kJ, (*b*) +3·25 kJ, (*c*) 0·25 kJ.

4.5 During an adiabatic compression process 20 kJ of work are transferred.

(*a*) What are the value and direction of the change in energy of the system?

(*b*) If the surrounding atmosphere has a constant pressure of 1·05 bar and the volume change of the system is 0·14 m^3 show the division of work transfer and explain.

(*c*) Repeat the above for an equivalent expansion process.

(*Ans:* (*a*) 20 kJ gain
(*b*) 14·9 kJ from atmosphere, 5·1 kJ external work
(*c*) *a* and *b* with change of sign)

4.6 Gas in a cylinder sealed by a freely sliding piston receives 5 kJ of heat and expands by 0·14 m^3 against an atmosphere of 1·4 bar. Through a lever system actuated by the piston a mass of 45 kg is also raised through 1·5 m.

(*a*) Find the change of internal energy of the gas.

(*b*) The piston returns to its original position by an adiabatic process requiring 14 kJ of external work input. Find the change of internal energy of the gas.

(c) Determine the total change of energy both for the gas and the system as a whole for the combined processes (a) and (b).

(*Ans:* (a) $-15{\cdot}26$ kJ, (b) $+33{\cdot}60$ kJ, (c) $+18{\cdot}34$ kJ for gas and system)

4.7 Calculate the enthalpy of 1 kg of fluid which occupies a volume of 0·34 m^3 if the internal energy is 815 kJ/kg and the pressure is 2·1 bar.

(*Ans:* 886·4 kJ/kg)

4.8 During the expansion of 1 kg of a fluid the internal energy changes from an initial value of 1162 kJ/kg to a final value of 1025 kJ/kg. If 85 kJ of shaft work are obtained, calculate the quantity and direction of the heat exchange between the fluid and the surroundings. Assume expansion is frictionless.

(*Ans:* 52 kJ outwards)

4.9 In a certain non-flow process with air as the working fluid, 1000 kJ are removed from the system and the internal energy decreases by 200 kJ. Is the process an expansion or compression? What work is performed?

(*Ans:* 800 kJ compression work)

4.10 An engineering workshop loses 76000 kcal/h in cold weather. The power supplied to the machinery is 75 kW; fifty 100-W lamps are burned continuously. What additional heat is required to maintain the temperature of the workshop constant?

(*Ans:* 30200 kJ/h = 7220 kcal/h)

First law and steady flow open system

4.11 An open system has a mass flow rate of 1 kg/s. The enthalpy, velocity and height at entry are 232 kJ/kg, 91 m/s, and 98 m respectively. At exit these quantities are 230 kJ/kg, 15 m/s and 3·35 m. If the heat received by the system is 10 kJ/s, what is the work done by this system per minute?

(*Ans:* 1020 kJ/min)

4.12 A steady flow of steam enters a condenser with an enthalpy of 2400 kJ/kg and a velocity of 366 m/s. The condensate leaves the condenser with an enthalpy of 162 kJ/kg and a velocity of 6 m/s. What is the heat transferred to the cooling water per kg steam condensed?

(*Ans:* 2305 kJ/kg)

4.13 Steam enters a radiator at 1 bar with a specific volume of 1·694 m^3/kg and an internal energy of 2506 kJ/kg. After condensing in the radiator, the water leaves at a pressure of 1 bar. The specific volume and enthalpy are $1{\cdot}044 \times 10^{-3}$ m^3/kg and 417 kJ/kg respectively. If the entry and exit velocities are equal, what is the heat output of the radiator per kg of steam?

(*Ans:* 2258 kJ/kg)

4.14 An air compressor delivers 4·5 kg of air per minute at a pressure of 7 bar and a specific volume of 0·17 m^3/kg. Ambient conditions are: pressure 1 bar and specific volume 0·86 m^3/kg. The initial and final internal energy values for the air are 28 kJ/kg and 110 kJ/kg

respectively. Heat rejected to the cooling jacket is 76 kJ/kg of air pumped. Neglecting changes in kinetic and potential energies, what is the shaft power required to drive the compressor?
(*Ans:* 14·3 kW)

4.15 Steam at a pressure of 0·14 bar and dryness fraction 0·9 enters a jet condenser where it is condensed by direct contact with cold water. The steam enters the condenser at the rate of 270 kg/h. The cold water is supplied at a temperature of 14°C and the mixture of cooling water and condensed steam leaves at a temperature of 13°C below the saturation temperature corresponding to the pressure of the entering steam. If the rate of heat transfer from the condenser to the surroundings is 1700 kcal/h, determine the quantity of cooling water required per hour.
(*Ans:* 4125 kg/h)

4.16 A valve in a pipeline containing ammonia (NH_3) at 14·70 bar, 0·9 dry, is only partly closed to allow the ammonia to leak past the valve to a region of pressure 1·32 bar. What is the condition of the ammonia downstream of the valve and what are the temperatures both upstream and downstream? Treat as a throttling process.
(*Ans:* $x = 0{\cdot}965$, upstream 38°C, downstream −28°C)

First law and unsteady flow open system

4.17 State the energy equation for the open system under conditions of unsteady flow and briefly explain the significance and relative importance of the various terms.

A storage bottle of volume 0·14 m³ contains nitrogen at 98 bar and 18°C. It is then connected to a supply line delivering nitrogen at 141 bar and 27°C until the pressure in the receiver equals that of the mains. Assuming the process to be adiabatic, find the final temperature in the receiver and the mass of nitrogen admitted from the mains, using the perfect gas equation.

For nitrogen; gas constant $R = 0{\cdot}2951$ kJ/kg-K
ratio of specific heats $\gamma = 1{\cdot}4$

(*Ans:* 48°C, 4·77 kg)

4.18 A receiver of volume 1·4 m³ contains dry air at 7 bar, and 27°C. Dry air from a high pressure mains at 21·3 bar and 49°C is then admitted to the receiver until the pressure in the latter equals that in the mains. Assuming the process to be adiabatic, find the final temperature in the receiver and the mass of air admitted from the mains.
(*Ans:* 113·3°C, 14·56 kg)

4.19 Steam from a supply at 15 bar and 250°C enters a drum having a volume of 1·4 m³ containing steam at 3·5 bar, having a dryness fraction of 0·9. If the final pressure in the drum is 10 bar, show that, assuming adiabatic conditions throughout, the final temperature of the trapped steam is approximately 232°C. Hence find the change of mass content of the drum.
(*Ans:* 2·8 kg)

Chapter 5

Reversible and Irreversible Processes

5.1 General Remarks

It has already been seen (section 4.1) that work can be completely converted into heat, the converse being far from true. Thus in thermodynamic cycles only a fraction of the heat supplied to the cycle is converted into work.

Work therefore is a more "precious" form of energy transfer than heat. The ability of a system to do work, either in a process or a cycle, is exploited to maximum advantage only when the process or cycle are REVERSIBLE.

When a process occurs between an initial state such that external work can be done on the surroundings, to a dead state identical with that of the surroundings, the maximum amount of work which can be done when that process is ideally performed is known as the available energy or AVAILABILITY (Chapter 14). In any non-ideal or irreversible process this amount of work will be less.

We may express this in the form—

A PROCESS IS IDEALLY PERFORMED WHEN NEITHER WORK NOR AVAILABLE ENERGY IS USED TO CAUSE A CHANGE THAT COULD HAVE BEEN ACCOMPLISHED IN WHOLE OR IN PART BY TRANSFER OF HEAT: SUCH A PROCESS IS KNOWN AS A REVERSIBLE PROCESS.

The implications of the term "reversible" are not immediately obvious. They arise from the impossibility of completely converting heat into work, while the converse is possible. Any imperfections in the process imply the dissipation of some work in the form of heat, dissipative processes being of necessity irreversible. The Joule paddle wheel experiment is an extreme example of this.

Certain consequences follow—

(1) A REVERSIBLE PROCESS MUST PASS THROUGH THE SAME STATES ON THE REVERSED PATH AS WERE INITIALLY VISITED ON THE FORWARD PATH.

(2) A REVERSIBLE PROCESS, WHEN UNDONE, WILL LEAVE NO EVIDENCE OF THE EVENTS IN THE SURROUNDINGS OR THE SYSTEM.

(3) A REVERSIBLE PROCESS MUST PASS THROUGH A CONTINUOUS SERIES OF EQUILIBRIUM STATES.

Irreversibilities or imperfections can be of two kinds, and may be considered under two headings—

(*a*) External irreversibilities associated with dissipative effects outside the working fluid,
(*b*) Internal irreversibilities associated with dissipative effects within the working fluid.

The following cases each illustrate a particular kind of irreversibility.

5.2 Mechanical Friction (External Irreversibility) (*Fig.* 5.1)

Consider a gas undergoing an adiabatic process and contained by a piston in a cyclinder. During the outward journey the frictional

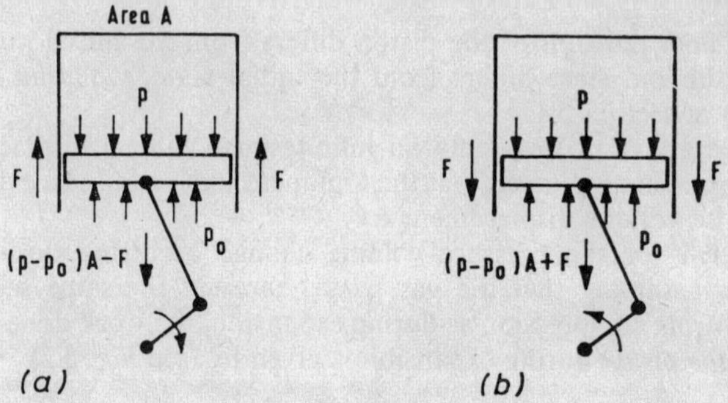

Fig. 5.1 Mechanical friction

force F and atmospheric pressure force p_0A opposes the pressure force acting on the piston, pA, the effective force therefore being

$$[(p - p_0)A - F]$$

Hence the work done *by the system*, bounded by the cylinder walls and the underside of the piston, on the surroundings, during a downward displacement dx is given by

$$(dW)_{outward} = [(p - p_0)A - F]\,dx \qquad \text{(i)}$$

During the return journey the friction force F is reversed, and the force which has to be exerted on the piston in order to move it up is $[(p - p_0)A + F]$. Hence to return the piston to its original position the amount of work done *on the system* is given by

$$(dW)_{inward} = [(p - p_0)A + F]\,dx \qquad \text{(ii)}$$

where, from (i) and (ii)

$$(dW)_{inward} > (dW)_{outward}$$

or $\qquad$ work done on system > work done by system $\qquad$ (5.1)

Hence evidence of the process will be left in the surroundings, from which work will have been abstracted, i.e. condition (2) is violated.

Alternatively, if

$$(dW)_{inward} = (dW)_{outward}$$

$$[(p - p_0)A - F]\,dx = [(p - p_0)A + F]\,dx'$$

or

$$dx' = \frac{(p - p_0)A - F}{(p - p_0)A + F}\,dx \qquad (5.2)$$

i.e. the final position of the piston differs from the initial position. Hence the end state differs from the initial state, violating condition (1) of section 5.1.

In the above process only an infinitesimal volume displacement $A\,dx$ has been considered, but the argument may readily be extended to a finite volume displacement ΔV.

In terms of the pressure–volume change occurring during the process, assuming that the gas passes through the same series of states during compression as during expansion, the work done by the gas *on the piston* during expansion is given by (see *Fig.* 5.2)

$$W_{piston} = \int_1^2 p\,dV$$

but the useful work done by the gas on the surroundings is less than this by the amount $F(x_2 - x_1) + Ap_0(x_2 - x_1)$. Similarly, the work done *on* the gas *by* the piston during compression is given by

$$W_{piston} = +\int_2^1 p\,dV = -\int_1^2 p\,dV$$

but the effective work done *by* the surroundings *on* the system exceeds W_{piston} by $(-Ap_0 + F)(x_2 - x_1)$. The work done in overcoming friction during the two processes, i.e. $2F(x_2 - x_1)$, is, of course, turned into heat. The work done against the atmosphere $Ap_0(x_2 - x_1)$ during 1–2 is reversed during the return stroke 2–1 and is therefore returned to the system. Since the gas itself is stipulated to undergo an adiabatic process, and to retrace its path during compression, the heat must be absorbed either by the piston and cylinder (without reaching the gas) or by the surroundings. In either case the process has left evidence of its occurrence in the surroundings, thus violating condition (2) above for a reversible

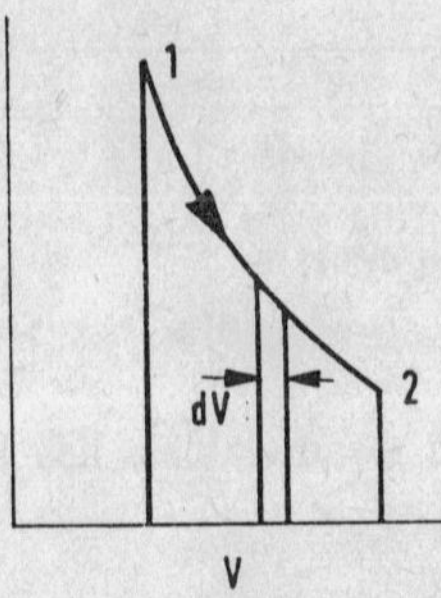

Fig. 5.2 *Displacement work on piston*

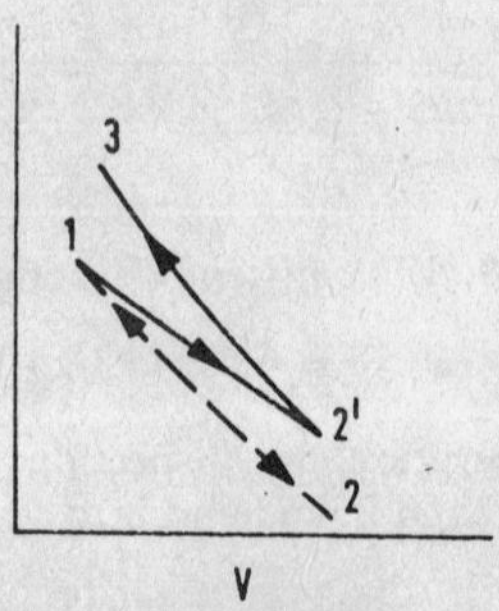

Fig. 5.3 *Expansion and compression with friction*

process. If, on the other hand, the heat is absorbed by the gas itself, the expansion path will differ from the compression path, in which case evidence of the process occurs in the system itself, thus again violating condition (2) above (see *Fig.* 5.3).

EXAMPLE 5.1

Steam is compressed adiabatically from 1 bar, 0·9 dry, to 10 bar (*a*) without friction or any other losses; (*b*) with friction accounting for 15% of the isentropic enthalpy change. Determine the end condition in the two cases, the work done per kg steam and the increase in entropy in case (*b*).

SOLUTION (*Fig.* 5.4)

Note: the solution of this problem involves the introduction of a new property known as ENTROPY which provides quantitative measure of irreversibility and is discussed in Chapter 7. Its application to steam is further discussed in Chapter 8.

From steam tables,

$$h_1 = 2675 - 0{\cdot}1 \times 2258 = 2249{\cdot}2 \text{ kJ/kg}$$

$$s_1 = 7{\cdot}359 - 0{\cdot}1 \times 6{\cdot}056 = 6{\cdot}7534 \text{ kJ/kg-K}$$

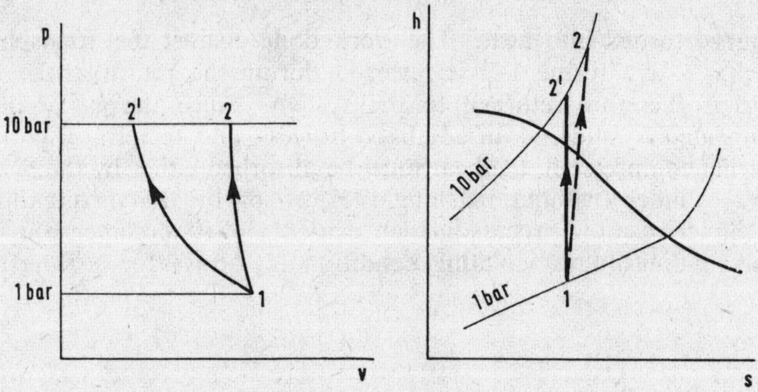

Fig. 5.4

(*a*) *No friction* (isentropic change, see section 4.5d)

$$s_{2'} = s_1 = 6{\cdot}7534 \text{ kJ/kg-K}$$

Temperature, by inspection of superheated steam tables, lies between 200°C and 250°C and is at

$$200 + \frac{6{\cdot}7534 - 6{\cdot}695}{6{\cdot}926 - 6{\cdot}695} \times 50 = 212{\cdot}6°\text{C} \quad \textit{Ans.}$$

Enthalpy after compression will be

$$h_{2'} = 2829 + \frac{12{\cdot}6}{50}(2944 - 2829)$$
$$= 2857{\cdot}98 \text{ kJ/kg}$$

Therefore,

$$\text{Work done in case } (a) = 2857{\cdot}98 - 2249{\cdot}2$$
$$= 608{\cdot}78 \text{ kJ/kg}$$

(*b*) *With friction*

The work converted to friction, which the compressor will require to supply in addition to that in case (*a*), is

$$0{\cdot}15 \times 608{\cdot}78 = 91{\cdot}32 \text{ kJ/kg}$$

Actual work of compression

$$= 608{\cdot}78 + 91{\cdot}32 = 700{\cdot}10 \text{ kJ/kg}$$

Enthalpy at point 2 after compression

$$= 2249{\cdot}2 + 700{\cdot}10 = 2949{\cdot}3 \text{ kJ/kg}$$

Temperature at point 2

$$= 250 + \frac{2949{\cdot}3 - 2944}{3052 - 2944} \times 50$$

$$= 250 + \frac{5{\cdot}3}{108} \times 50 = 252{\cdot}45°\text{C} \quad \textit{Ans.}$$

Entropy at point 2

$$= 6{\cdot}926 + \frac{5{\cdot}3}{108}(7{\cdot}124 - 6{\cdot}926) = 6{\cdot}9367$$

Change of entropy due to compression

$$= 6{\cdot}9367 - 6{\cdot}7534 = 0{\cdot}1833 \text{ kJ/kg-K} \quad \textit{Ans.}$$

5.3 Heat Transfer through Finite Temperature Gradient (External Irreversibility) (*Fig.* 5.5)

Consider again a gas in a cylinder sealed by a piston which, however, is assumed to be frictionless, thus removing external irreversibility due to mechanical friction. Let the gas undergo an isothermal expansion at temperature T_g equal to the temperature of the surroundings T_0. Such a process requires continuous heat addition in accordance with the first law statement

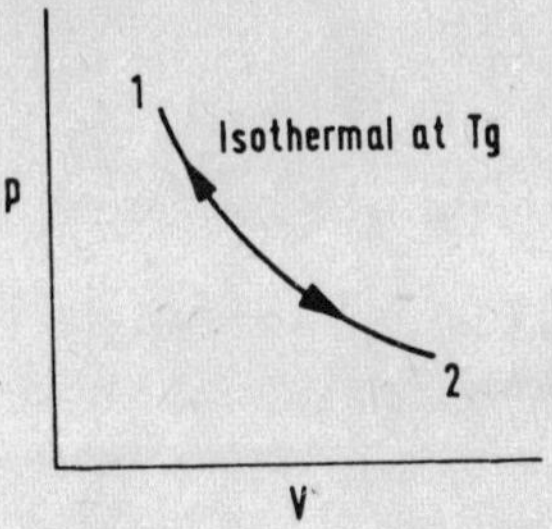

Fig. 5.5 Heat transfer with finite temperature difference

$$dQ = dW + dU$$

where $dU = 0$ for a perfect gas when T is constant.

This heat is supplied by the surroundings and, *if the process is to proceed at a finite rate*, requires a finite positive temperature gradient, i.e. the surroundings have to be at a temperature $T_g + \Delta T$. Hence, the surroundings have experienced a change in order to make it possible for the gas to undergo a reversible process. Thus, an irreversibility of the external kind is present.

Alternatively, if the temperature of the surroundings is fixed to T_0, and isothermal expansion of the gas proceeds at a rate requiring a constant heat addition rate dQ/dt, the gas, neglecting transient effects, will assume a temperature $(T_0 - T)$. Similarly, during compression, requiring a constant heat abstraction rate $-dQ/dt$, the gas will assume a temperature $(T_0 + T)$. Hence the gas does not pass through the same states on the reversed path as those visited during the outward path, thus violating condition (1) for the reversible process.

5.4 Viscosity and Inertia (Internal Irreversibility)

(*Fig.* 5.6)

Consider again a gas contained by a piston in a cylinder and undergoing an adiabatic process.

During the expansion process (outward journey) viscosity and inertia effects both lead to a negative pressure gradient from the

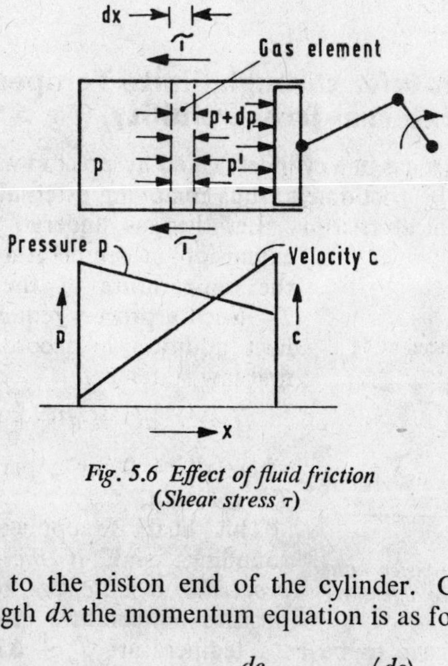

Fig. 5.6 Effect of fluid friction
(*Shear stress* τ)

cylinder head to the piston end of the cylinder. Considering an element of length dx the momentum equation is as follows—

$$\underset{\text{net pressure force}}{[p - (p + dp)]A} - \underset{\text{inertia force}}{A\rho\, dx \frac{dc}{dt}} = \underset{\text{viscous shearing force}}{\pi D\, \mu \left(\frac{dc}{dy}\right)_w dx} = 0 \qquad \text{(i)}$$

where $\mu\left(\frac{dc}{dy}\right)_w$ = shear stress = τ, A = area, μ = coefficient of viscosity, ρ = density, c = local gas velocity, D = diameter, $(dc/dy)_w$ = transverse velocity gradient at cylinder wall.

Clearly both the inertia and viscous shearing forces oppose the motion of the gas towards the piston end of the cylinder, i.e.

$$\frac{dp}{dx} = -\left[\rho\frac{dc}{dt} + \frac{\pi D}{A}\mu\left(\frac{dc}{dy}\right)_w\right] \tag{ii}$$

This negative pressure gradient results in the pressure on the piston, p', being less than the mean gas pressure p. Hence the work done by the gas on the piston for a small volume displacement dV is given by

$$(dW)_{piston} = p'\,dV < p\,dV \tag{iii}$$

where p is the mean gas pressure. It is evident that reversal of the direction of motion, i.e. changing from expansion to compression, results in reversal of the pressure gradient due to reversal of the inertia and viscous shearing forces. Hence the pressure on the piston, p'', is now greater than the mean pressure p

$$(dW)_{piston} = p''\,dV > p\,dV \tag{iv}$$

It is evident, therefore, that, as with mechanical friction, the work required for the compression process exceeds that for the expansion process, i.e.

$$(dW)_{inward} > (dW)_{outward} \tag{5.3}$$

Alternatively, if

$$(dW)_{inward} = (dW)_{outward}$$

then

$$p'\,dx = p''\,dx' \quad \text{or} \quad dx' = \frac{p'}{p''}\,dx \tag{5.4}$$

where the difference between p' and p'' is related to the pressure gradient dp/dx given by equation (ii).

Both equations (5.3) and (5.4) suggest that the process is irreversible, in that either more work is absorbed from, than is given to, the surroundings or, alternatively, in that the piston does not return to its original position. The process is described as internally irreversible, since the cause of the effect lies in the properties of the working fluid, i.e. density and viscosity, rather than in external effects such as mechanical friction or a temperature gradient.

5.5 Unresisted Expansion (Internal Irreversibility)
(*Fig.* 5.7)

Consider a gas stored under pressure in a cylinder sealed by a series of partitions 1, 2, 3, 4, etc. The initial state of the gas is p_1, V_1. Partition 1 is then suddenly withdrawn, the gas expanding to V_2 and ultimately reaching the state p_2, V_2. Withdrawal of partitions 2, 3, 4, etc., in turn leads to the states p_2, V_2; p_3, V_3; p_4, V_4; etc. The process path may be represented by the curve 1–2–3–4–etc., in the pressure–volume diagram. It is clear, however, that the area under this curve, i.e. $\int p\,dV$, cannot represent work done, since in fact no work whatever is done as a result of withdrawing successive partitions.

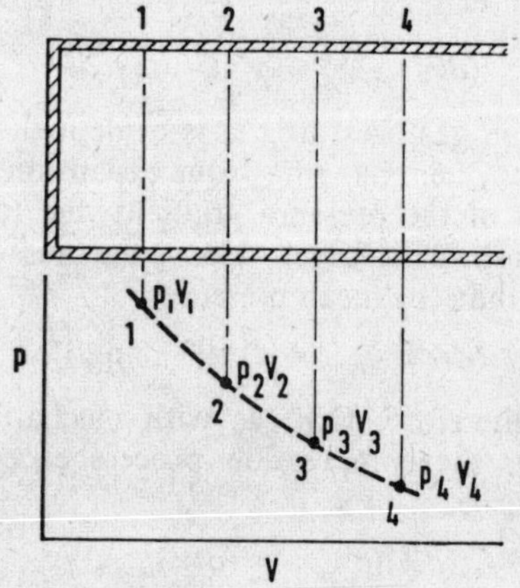

Fig. 5.7 Unresisted expansion

On the other hand, if the initial state p_1, V_1 were to be restored, work transfer and possibly heat transfer across the system boundaries would have to take place. The process is thus irreversible.

EXAMPLE 5.2

Consider a piston and cylinder of 1 m^3 capacity, when the piston is fully withdrawn, connected a valve to a similar volume of 1 m^3 containing oxygen at 8 bar abs and 175°C. The cylinder is completely evacuated on the withdrawal of the piston and, at that instant, the valve opens, allowing the oxygen to expand freely into the cylinder.

(*a*) Assuming adiabatic conditions, determine the change in entropy of the gas.

(*b*) If now the piston recompresses the gas into the original volume of 1 m^3 and it is assumed that this is done adiabatically and without friction, determine the new pressure and temperature of the gas and its final entropy change from the original condition.

SOLUTIONS

Note: the solution of this problem again involves the use of entropy, in this case applied to the perfect gas.

(*a*) During the free expansion into the larger volume of 2 m³ the piston is stationary and no work is done by or on the gas. Also the expansion is adiabatic. Therefore from the first law of thermodynamics the internal energy is constant.

$$mc_v(T_2 - T_1) = 0, \quad T_1 = T_2$$

but since the mass is constant pV/T = constant and $V_2 = 2V_1 = 2\ \text{m}^3$; therefore $p_2 = \frac{1}{2}p_1 = 4$ bar.

The entropy initially (see equation (7.9c))

$$S_1 = m_1(c_p)_1 \ln\left(\frac{T_1}{T_0}\right) - m_1R \ln\left(\frac{p_1}{p_0}\right)$$

where T_0 and p_0 refer to a datum state.

After expansion the entropy

$$S_2 = m_1(c_p)_1 \ln\left(\frac{T_2}{T_0}\right) - m_1R \ln\left(\frac{p_2}{p_0}\right)$$

on the assumption that the specific heats and gas constants do not change. Change of entropy due to the expansion

$$\Delta S = S_2 - S_1 = mc_p \ln\left(\frac{T_2}{T_1}\right) - mR \ln\left(\frac{p_2}{p_1}\right)$$

$$= 0 - mR \ln\left(\frac{p_2}{p_1}\right)$$

$$\Delta S = S_2 - S_1 = -\ mR \ln \tfrac{1}{2}$$

Now $$m = \frac{p_1V_1}{RT} = \frac{10^2 \times 8 \times 1}{0{\cdot}2598 \times 448} = 6{\cdot}89\ \text{kg}$$

$$\Delta S = -6{\cdot}89 \times 0{\cdot}2598\ (-0{\cdot}69316)$$

$$= +1{\cdot}24\ \text{kJ/K}$$

showing there has been a gain of entropy and so the expansion is irreversible.

(*b*) Since the compression is both adiabatic and frictionless it is isentropic, therefore we would expect that on compression there would be no change in entropy. We may check this.

The law of compression will be pv^γ = constant where $\gamma = c_p/c_v = 1{\cdot}40$ for oxygen.

Also since the mass is constant and the gas constant is constant

$pv/T = \text{constant}.$

It follows that

$$\frac{T_3}{T_2} = \left(\frac{v_2}{v_3}\right)^{\gamma-1} = \left(\frac{p_3}{p_2}\right)^{(\gamma-1)/\gamma}$$

where subscript 3 refers to the third and final condition. Hence

$$T_3 = T_2 \times 2^{0\cdot4} = 1\cdot32 \times T_2$$

and

$$p_3 = p_2 \times 2^{1\cdot4} = 2\cdot64\, p_3$$

The change in entropy due to the compression

$$\Delta S = S_3 - S_2 = mc_p \ln\left(\frac{T_3}{T_2}\right) - mR\ln\left(\frac{p_3}{p_2}\right)$$

$$\Delta S = S_3 - S_2 = m\frac{\gamma}{\gamma-1}R\ln\left(\frac{T_3}{T_2}\right) - mR\ln\left(\frac{p_3}{p_2}\right)$$

$$= mR\left(\frac{\gamma}{\gamma-1}\ln 1\cdot32 - \ln 2\cdot64\right)$$

$$= mR(0\cdot9717 - 0\cdot9717)$$

$$= 0$$

So that the oxygen in its final condition has a net gain of entropy entirely due to the free expansion which took place.

5.6 The Irreversible Process and the Meaning of the Integral $\int p\, dV$

Consider, for example, two adiabatic processes, 1–2′ being reversible and 1–2 irreversible (*Fig.* 5.8).

1–2′ represents a frictionless, i.e. reversible, adiabatic expansion from V_1 to V_2.

1–2 represents an adiabatic expansion with friction (internal or external) from V_1 to V_2.

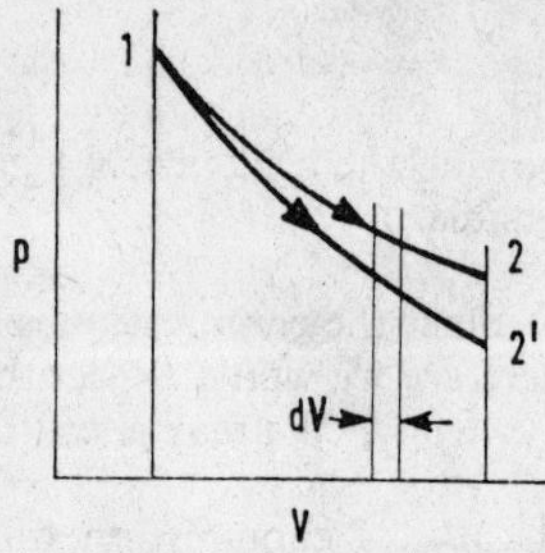

Fig. 5.8 Meaning of $\int p\, dV$

Since the irreversibility in 1–2 is equivalent to an effect which could have been produced by the direct addition of heat, the mean expansion temperature during 1–2 is higher than during 1–2′ and hence path 1–2 lies above 1–2′. Consequently

$$\int_1^2 (p\,dV)_{irrev} > \int_1^{2'} (p\,dV)_{rev} \qquad \text{(i)}$$

Now for a reversible process $\int_1^{2'} p\,dV$ = external work done.

It can be shown that for the irreversible process $\int_1^2 p\,dV$ does not represent the external work done.

$$\text{For 1–2}' \quad \int_1^{2'} \underset{\uparrow = 0}{dQ} - \int_1^{2'} dW = \int_1^{2'} dU = U_{2'} - U_1$$

$$\text{i.e.} \qquad (W_{rev})_{1-2'} = -U_{2'} + U_1 \qquad \text{(ii)}$$

$$\text{For 1–2} \quad \int_1^{2} \underset{\uparrow = 0}{dQ} - \int_1^{2} dW = \int_1^{2} dU = U_2 - U_1$$

$$\text{i.e.} \qquad (W_{irrev})_{1-2} = -U_2 + U_1 \qquad \text{(iii)}$$

the N.F.E.E. (4.4) being applicable equally to reversible and irreversible processes (see also section 5.7).

But since from above $T_2 > T_{2'}$, $U_2 > U_{2'}$, i.e. $(U_1 - U_2) < (U_1 - U_{2'})$

$$\text{or} \qquad (W_{rev})_{1-2'} > (W_{irrev})_{1-2} \qquad \text{(iv)}$$

Yet as already shown

$$\int_1^2 (p\,dV)_{irrev} > \int_1^{2'} (p\,dV)_{rev}$$

so that clearly the area under the p–V curve does *not* represent work done, i.e.

$$\int_1^2 p\,dV_{irrev} \neq (W_{12})_{irrev} \qquad (5.5)$$

However, had the process 1–2 been described reversibly by continuous heat addition we should have had

$$\int_1^2 (dQ)_{rev} - \int_1^2 (dW)_{rev} = U_2 - U_1 \qquad \text{(v)}$$

whence

$$(Q_{rev})_{1-2} - (W_{rev})_{1-2} = U_2 - U_1 = -(W_{irrev})_{1-2} \text{ (from (iii))} \qquad (5.6)$$

or
$$\int_1^2 p\,dV = (W_{rev})_{1-2} = (W_{irrev})_{1-2} + (Q_{rev})_{1-2} \qquad (5.6a)$$

Hence the heat added in the reversible expansion 1–2 may be regarded as the difference between the maximum work available for that particular process 1–2, and the work actually obtained, i.e. it represents an energy dissipation effect.

In most practical cases, comparison is made between an irreversible and a reversible process carried out in an otherwise like manner, i.e. we are comparing processes 1–2 and 1–2′. The latter is the ideal path and gives the maximum amount of work available when the gas expands adiabatically (and reversibly) from $p_1\,V_1$ to $V_{2'}$ while the path 1–2 is an irreversible adiabatic path along which less work is done.

Thus

$$\int_1^{2'} p\,dV = (W_{rev})_{1-2'} = (W_{irrev})_{1-2} + \text{loss of available energy} \qquad (5.7)$$

while from equations (5.6a) and (5.7)

$$\int_1^2 p\,dV - \int_1^{2'} p\,dV = (W_{irrev})_{1-2} - (W_{rev})_{1-2'} + (Q_{rev})_{1-2} \qquad (5.8)$$

i.e. area 122′ in *Fig.* 5.8 represents the difference between the actual and ideal work plus the heat transferred during a hypothetical reversible process following the actual process path 1–2.

EXAMPLE 5.3

Air is expanded in a piston and cylinder arrangement, from 7 bar and 93°C to 1·05 bar.

(*a*) polytropically and reversibly, the index of the expansion law being 1·3.
(*b*) irreversibly, during which 10 per cent of the work done by the air in (*a*) is dissipated by friction but external heat addition is unchanged.

Compare the work done by the air in the two cases and illustrate the fallacy of using a pressure–volume diagram in the second case.

SOLUTION

(*a*) Since the expansion is reversible the work done is

$$\int_1^{2'} p\,dv$$

and is represented by the figure 12′34 (*Fig.* 5.9*a*).

$$T_{2'} = T_1 \times \left(\frac{p_2}{p_1}\right)^{(n-1)/n} = 366 \times \left(\frac{1{\cdot}05}{7}\right)^{0{\cdot}3/1{\cdot}3} = \frac{366}{1{\cdot}548} = 236{\cdot}43\ \text{K}$$

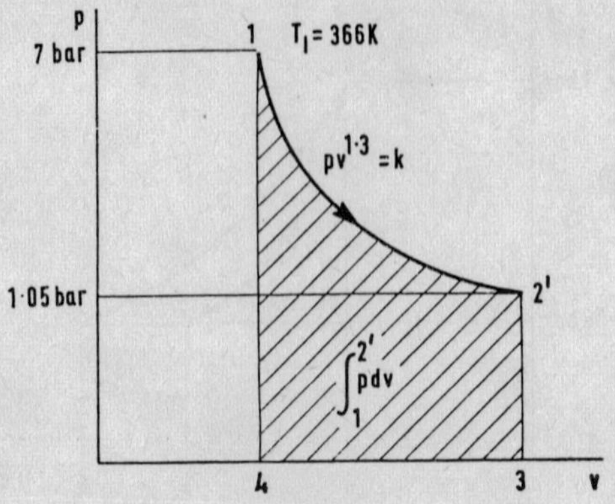

Fig. 5.9a

The work done $W_{12'} = \frac{p_2v_{2'} - p_1v_1}{1 - n} = \frac{R(T_1 - T_{2'})}{n - 1}$

$= 0{\cdot}2870(366 - 236{\cdot}43)/0{\cdot}3$

$= 124{\cdot}0$ kJ/kg *Ans.*

(*b*) Friction appears as heat amounting to

$0{\cdot}10 \times 124 = 12{\cdot}4$ kJ

$\therefore (T_2 - T_{2'})c_p = 12{\cdot}4$

$T_2 = \frac{12{\cdot}4}{0{\cdot}992} + 236{\cdot}43 = 248{\cdot}93$ K

So on the p–v diagram T_2 lies to the right of $T_{2'}$ as shown (*Fig.* 5.9*b*).

Now apply the first law to this case and denote the two expansions by subscripts 12′ and 12.

$$Q_{12'} = W_{12'} + \Delta U_{12'} \quad \text{(i)}$$

$$Q_{12} = W_{12} + \Delta U_{12} \quad \text{(ii)}$$

where $Q_{12} = Q_{12'}$ (assumption) and

$$\Delta U_{12} = \Delta U_{12'} + c_v(T_2 - T_{2'})$$
$$= \Delta U_{12'} + 8{\cdot}89$$

Substituting in (ii) and subtracting from (i)

$$0 = (W_{12'} - W_{12}) - 8{\cdot}89$$

$\therefore \quad W_{12} = 124{\cdot}0 - 8{\cdot}89 = 115{\cdot}11$ kJ/kg

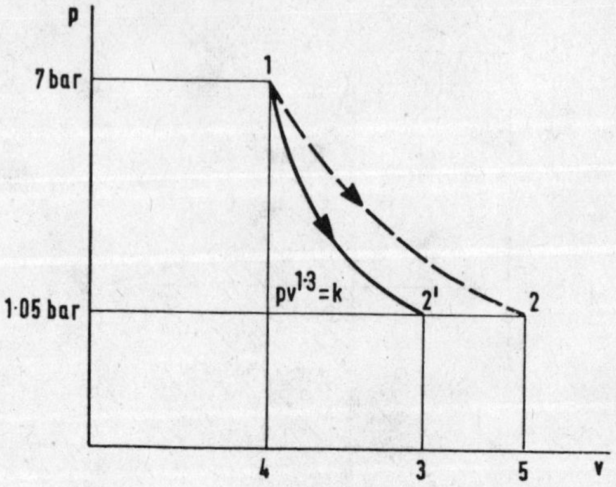

Fig. 5.9b

As is to be expected, the work done in the second expansion is less than that in the first by the amount of heat generated in overcoming friction

$$W_{12} < W_{12'}$$

If we consider the p–v diagram we see that $\int_1^{2'} p\,dv$ is given by area 12′34 as before. The $\int_1^{2} p\,dv$ is shown to be area 1254, which is greater than 12′34. It therefore cannot represent the work done W_{12}, since, if it did, the area would be less than 12′34.

5.7 The Non-flow Energy Equation Applied to Irreversible Processes

The non-flow energy equation (4.4a) was derived in section 4.4 from considerations of reversible cyclic processes.

The resultant equation

$$Q_{12} - W_{12} = \Delta E_{12} \tag{5.9}$$

is nevertheless applicable to irreversible as well as to reversible processes. This follows from the fact that the quantity E, the energy of the system, was shown to be a property, so that any change in its value is uniquely defined by the end states 1 and 2, no matter whether the path joining 1 and 2 be reversible or irreversible.

Hence the N.F.E.E. is more general than equation (5.5), since the former applies to any kind of closed system process, while the latter yields an equality only when the process is fully reversible.

5.8 The Steady Flow Energy Equation Applied to Irreversible Processes

Since the steady flow energy equation (4.13), like the non-flow energy equation, equates the difference between heat and work transfer to a property term, i.e.

$$Q_{12} - W_{12} = m[(h_2 + \tfrac{1}{2}c_2{}^2 + gZ_2) - (h_1 + \tfrac{1}{2}c_1{}^2 + gZ_1)] \quad (5.10)$$

it is again generally applicable to both reversible and irreversible processes.

It was shown in section 4.9 that this is not true of the Bernoulli equation (4.23) which, in fact, is the integrated momentum, or Euler, equation (4.22)

$$\int_1^2 v\,dp + \int_1^2 c\,dc + g\int_1^2 dZ = 0 \quad (5.11)$$

In the absence of heat transfer, and only for a fully reversible process, equation (5.11) reduces to the S.F.E.E. (5.10) with Q_{12} equal to zero. This follows from the fact that, for reversible adiabatic processes only,

$$\int_1^2 v\,dp = (h_2 - h_1) \quad (5.12)$$

(see equation (7.8b), section 7.4).

Substituting for $(h_2 - h_1)$ from equation (5.12) in equation (5.10), the following result is obtained for *reversible adiabatic processes only*,

$$-m\int_1^2 v\,dp = -\int_1^2 V dp = W_{12} + m[\tfrac{1}{2}(c_2{}^2 - c_1{}^2) + g(Z_2 - Z_1)] \quad (5.13)$$

Equation (5.13), representing the sum of work transfer and kinetic and potential energy changes for an adiabatic reversible flow process in terms of the area under the p–V curve (see *Fig.* 5.10), thus corresponds

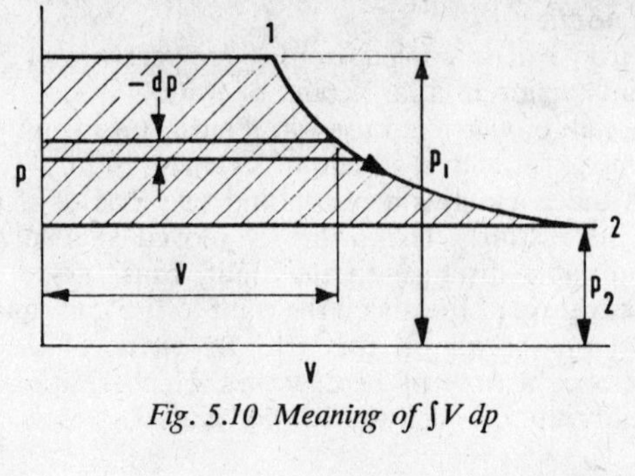

Fig. 5.10 Meaning of ∫V dp

to equation (5.6a) for the reversible non-flow process, in which the area under the p–V curve, integrated between the volume limits V_1 and V_2 instead of the pressure limits p_2 and p_1, again represents the work done. Neither statement is true for irreversible processes.

EXERCISES

5.1 List and explain all irreversibilities that may be in a system, differentiating between internal and external irreversibilities.

5.2 State whether the following processes are reversible or irreversible. In each reversible one justify your conclusion by briefly explaining how the effects of the process can be completely undone. If irreversible, state why it is so.

(Neglect heat transfer unless specifically told otherwise.)

(*a*) Gas is compressed slowly behind a piston in a cylinder with non-conducting walls.

(*b*) The temperature of a mass of gas confined in a rigid container is raised by rotating a fan impeller inside the mass of gas.

(*c*) Some water confined under a piston of constant weight is stirred until half of it evaporates.

(*d*) Some water confined as in (*c*) is evaporated by a slow addition of heat.

(*e*) Gas is expanded slowly behind a piston in a cylinder with conducting walls which are surrounded by a constant temperature bath.

(*f*) Gas expands slowly through a small orifice from one chamber into an adjacent chamber where the pressure is markedly lower.

(*g*) A heavy piston confining gas in a vertical cylinder is held in position by stops. When the stops are removed the piston falls because it has much more weight than the initial pressure of the gas can support.

(*h*) Steam on its way to a steam turbine expands adiabatically through a reducing valve.

(*i*) Steam expands adiabatically in steady flow through a frictionless nozzle.

(*j*) An iron ball is cooled from a temperature of 50°C to 30°C by immersion in a large tank of water.

(*k*) A stream of water is raised in temperature from 15°C to 95°C by mixing it with a stream of steam at 100°C.

(*l*) A cylinder contains a piston and two masses of air separated by the piston. The piston is moved suddenly through an appreciable distance to the right, causing the pressure and temperature of the air on the right to rise and causing a reverse effect on the air on the left. By virtue of the temperature difference a flow of heat occurs which restores temperature equilibrium between the two air masses.

5.3 For a reversible process in a closed system, the increase in total energy is 12 kJ, while heat of amount 20 kJ is abstracted. Compute the work and explain the algebraic sign.
(*Ans:* 32 kJ *on* system)

5.4 Calculate, by means of the energy equation, the heat transferred in the following processes:
(*a*) reversible constant pressure—closed system
(*b*) reversible steady flow
(*c*) irreversible steady flow with ΔW, ΔKE and ΔPE $= 0$.
(*Ans:* (*a*) $+\Delta H$, (*b*) $\Delta W + \Delta H + \Delta\text{KE} + \Delta\text{PE}$, (*c*) ΔH)

5.5 A reciprocating air compressor has a 50°C temperature difference between the air in the cylinder and the cooling water. If the system is the air enclosed in the cylinder, can the compression process be considered to approach reversibility? Discuss.

5.6 A reversible process of a closed system has pressure and specific volume related by $pv^n = c$. Derive a formula for work done.
An adiabatic irreversible process having the same law $pv^n = c$ is now performed. Find an expression for the work done under these conditions and compare with reversible work.

(*Ans.* $\dfrac{p_1v_1 - p_2v_2}{n-1}$; $\dfrac{p_1v_1 - p_2v_2}{n-1} - (q_{12})_{rev}$)

5.7 The work done in a reversible compression is 16 kJ while 10 kJ of heat are rejected to the surroundings. In an irreversible process which passes through the same series of states as the reversible process, 12 kJ of heat are rejected to the surroundings. What is the work done?
(*Ans:* -18 kJ)

5.8 Determine the indicated work per cycle of a piston engine which has a bore of 76 mm. and a stroke of 100 mm if the area of the indicator diagram is 12 cm². The reducing motion that drives the drum of the indicator has a ratio of 2 to 1 and the spring has a "stiffness" of 2·76 bar per cm deflection. Discuss the assumptions made.
(*Ans:* 294·7 J)

5.9 Air is compressed in a reciprocating compressor from 1·05 bar and 21°C to 10 bar. Calculate the work required to compress the air
(*a*) polytropically and reversibly, the index of compression being 1·28,
(*b*) irreversibly, during which 15 per cent of the work of compression in (*a*) is lost in friction.

Compare the works of compression in the two cases.
$c_p = 0{\cdot}992$ kJ/kg-K
$c_v = 0{\cdot}716$ kJ/kg-K
(*Ans:* (*a*) 195·9 kJ/kg, (*b*) 216·9 kJ/kg)

5.10 Illustrate, with suitable sketches, the fundamental nature of the irreversibility in the case of:

(*a*) friction between a cylinder wall and a piston, the cylinder containing a perfect gas;
(*b*) a frictionless piston sliding in a cylinder containing a gas having both viscosity and inertia.

A cylinder sealed by a piston contains 0·057 m^3 of air, having a pressure and temperature of 3·45 bar and 289 K respectively. The air expands to a volume of 0·072 m^3, the work done against friction between the piston and cylinder being 24 kJ per m^3 change in volume.

Assuming that the work done by the piston against the external atmospheric pressure is negligible, that the energy dissipated during the return motion equals that dissipated during the outward motion, and that the gas itself undergoes an adiabatic reversible process, find the final state of the gas in the cylinder if, subsequent to the original expansion, the air is re-compressed. During this re-compression the system receives from the surroundings an amount of work equivalent to that previously received by the surroundings.
(*Ans:* 0·585 m^3, 286 K, 3·32 bar)

5.11 What is meant by the terms "internal irreversibility" and "external irreversibility." Give at least two examples of each.

The expansion of 0·9 kg of dry air in a closed cylinder sealed by a piston proceeds irreversibly according to the law $pv^{1\cdot4}$ = constant, the initial pressure and temperature being 14 bar and 38°C and the final pressure being 7 bar. 10 kJ of heat are abstracted during the process. Determine the change of internal energy and the work done, and compare the latter with the corresponding quantity for an isentropic process between the same limits.
(*Ans:* $\Delta U = -36\cdot08$ kJ, $W = 26\cdot09$ kJ, c.f. 36·08 kJ)

5.12 The irreversible compression of air in an air compressor cylinder proceeds according to the law $pv^{1\cdot35}$ = constant, from 1 bar, 15°C, to 6 bar, and is accompanied by the abstraction of 10 kJ per kg of air. Treating the compression process as a flow process, compare the work done with that for

(*a*) an adiabatic process obeying the same law,
(*b*) an isothermal reversible process,

the pressure ratio and initial conditions being identical with those in the original process.
(*Ans:* 180·74 kJ/kg, (*a*) 170·74 kJ/kg, (*b*) 148·10 kJ/kg)

5.13 Discuss briefly the concept of thermodynamic irreversibility, giving two examples of both internal and external irreversibility.

Show that for an irreversible expansion process in a cylinder, whose end states are connected by the relation pv^n = constant, the work done is not equal to the integral $\int p\,dv$. Under what conditions does this integral accurately represent work done?

Hence find the work of compression when air is compressed from 1 bar, 38°C through a volumetric compression ratio of 5, according to the law $pv^{1\cdot 6}$ = constant,

(*a*) for an adiabatic irreversible process;
(*b*) for a polytropic reversible process.

(*Ans:* (*a*) 359·6 kJ/kg, (*b*) 240 kJ/kg)

Chapter 6

The Second Law of Thermodynamics

6.1 Relationship between First and Second Law

The first law, or law of conservation of energy, applies to cyclic processes, and points to the equivalence of work and heat (see equation (4.2a))—

$$J \oint dQ = \oint dW \qquad \text{(i)}$$

In words, the summated heat transfer across the boundaries of the system undergoing a cyclic process is proportional to the summated work transfer.

Regarding Joule's classical experiment for the determination of J as a special case of a cyclic process (see section 4.1), that is,

Stage (1): The falling weight does an amount of work W on the paddle wheel, raising the temperature of the water by ΔT, in the absence of external heat transfer;

Stage (2): The water is cooled to its original temperature, i.e. by $-\Delta T$, rejecting Q heat units to the surroundings,

it is clear that equation (i) is again satisfied although in fact during stage 1 there is no heat transfer, whilst during stage 2 there is no work transfer.

However, it is quite obvious from common sense considerations that this particular cyclic process could not be reversed, inasmuch as stage (1) cannot be performed in the sense of work W being done on the weight as a result of a temperature drop ΔT of the water. Equation (i) gives no indication of this limitation.

The Joule "cycle" is a limiting case, in which work abstracted from the surroundings is turned completely into heat rejected to the surroundings, the "cycle" being incapable of reversal.

Again, we know from experience that the various forms of heat engine with which we are familiar are all less than 100 per cent efficient, i.e. they abstract heat from a high temperature source, converting part into work, the balance being rejected as heat to the surroundings. Such a process again satisfies the first law statement

$$J \oint dQ = \oint dW$$

but there is no indication in that equation that $\oint dQ$ includes not only heat received but also heat rejected, just as in the case of the Joule "cycle" there was no indication that the "cycle" could only be performed in one direction.

The second law is concerned with the restrictions placed on the cyclic processes described by the simple statement $J \oint dQ = \oint dW$ and, more particularly, with the fact that, while the conversion of work into heat is possible in the sense that work received during one portion of the cycle is rejected as heat in another portion of the cycle, the converse is not true. Like the first law, it is a law of experience.

6.2 Heat Engines

As engineers, we are particularly concerned with the application of the second law to heat engines. A heat engine may be defined simply as a system operating in a cycle and producing useful work by abstracting heat from a suitable heat source.

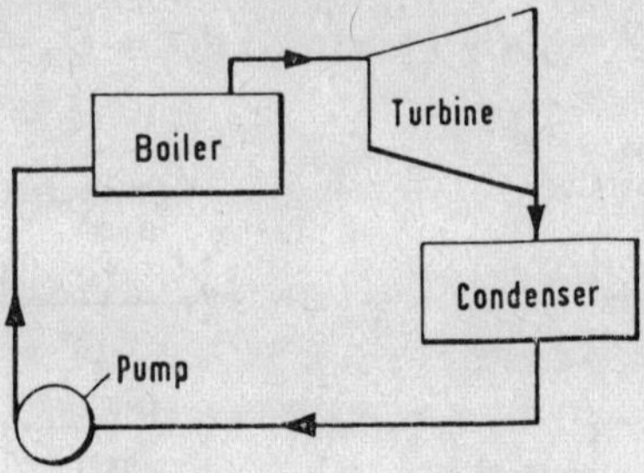

Fig. 6.1 Simple steam cycle

The second law thus relates to systems operating cyclically, and experiencing only heat and work transfer (*not* mass transfer) across the boundary. Such a system is termed, by definition, a *heat engine.* The complete steam plant (*Fig.* 6.1) comprising a boiler, turbine or engine, condenser and pump, is a good example.

(*Note* that the best known "heat engines," namely petrol and diesel engines, do not satisfy the definition, since they experience mass flow in the form of air, fuel and exhaust gas flow across the system boundary, see Chapter 13, *Figs.* 13.1 and 13.2.)

Since the object of all such heat engines is to produce work from heat received over the high temperature portion of the cycle, the most important questions, both of which are answered by the second law, are

(1) How may heat taken from the high temperature source be utilized to best advantage, i.e. how may the greatest amount of work be done for each unit of heat energy received from the hot source?
(2) What fraction of the heat energy taken from the hot source is it theoretically possible to convert into work? i.e. what is the limiting efficiency of conversion?

These questions were posed and answered by Carnot in his famous essay *Reflections on the Motive Power of Heat* written in 1824, and occasioned largely by his interest in Watt's work on the steam engine.

In discussing the second law of thermodynamics we shall also use the concept of the reversed heat engine which, together with the engine converting heat into work, is shown in the line diagram (*Fig.* 6.2).

Applying the first law to any heat engine cycle and using consistent units, we have

$$\oint dQ = \oint dW \quad \text{where} \quad \oint dQ = Q_A - Q_R \qquad \text{(ii)}$$

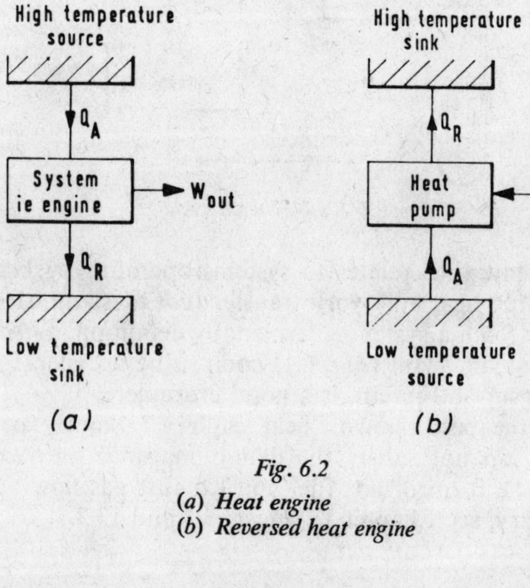

Fig. 6.2

(a) Heat engine
(b) Reversed heat engine

By definition, the efficiency of the cycle is

$$\eta = \frac{\text{net work output}}{\text{gross heat input}} = \frac{\oint dW}{Q_A} = \frac{Q_A - Q_R}{Q_A} = 1 - \frac{Q_R}{Q_A} \qquad (6.1)$$

The corresponding "efficiency" for the reversed heat engine or heat pump (*Fig.* 6.2*b*), known as the COEFFICIENT OF PERFORMANCE (C.O.P.), becomes

$$(\text{C.O.P.})_{heat\ pump} = \frac{\text{heat output}}{\text{work input}} = \frac{Q_R}{Q_R - Q_A} \qquad (6.2)$$

i.e. the C.O.P. is the reciprocal of the efficiency. The "efficiency" for a refrigerator is defined as

$$(\text{C.O.P.})_{refrig.} = \frac{\text{heat absorbed from low temp. sink}}{\text{work input}}$$

$$= \frac{Q_A}{Q_R - Q_A} \qquad (6.3)$$

From equations (6.2) and (6.3)

$$(\text{C.O.P.})_{heat\ pump} = (\text{C.O.P.})_{refrig.} + 1 \qquad (6.4)$$

EXAMPLE 6.1

A vessel of 0·06 m^3 capacity contains oxygen at 2·76 bar and 15°C. This mass of oxygen is cooled at constant volume from 1000°C to 15°C. If the amount of heat removed under these conditions were supplied to a series of perfect heat engines, what would be the maximum amount of work obtained, assuming that heat is rejected at 15°C?

Take the specific heat of oxygen, at constant volume, as

$$c_v = 0{\cdot}649 + 10^{-9} \times 40{\cdot}6T^2 \text{ kJ/kg-K}$$

SOLUTION (*Fig*. 6.3)

$$RO_2 = 8{\cdot}314/32 = 0{\cdot}2598 \text{ kJ/kg-K}$$

$$m = 100 \times 2{\cdot}76 \times 0{\cdot}06/0{\cdot}2598 \times 288 = 0{\cdot}221 \text{ kg}$$

Consider the element shaded.

The heat available for doing work = $mT\,ds$. (See equation (7.7).)

Since the cooling is at constant volume this equals $mc_v\,dT$.

The efficiency of the elemental cycle is that of the Carnot cycle working between temperature limits T and T_0, namely $(T - T_0)/T$. Therefore the maximum available work from the elemental cycle = $mc_v\,dT \times (T - T_0)/T$

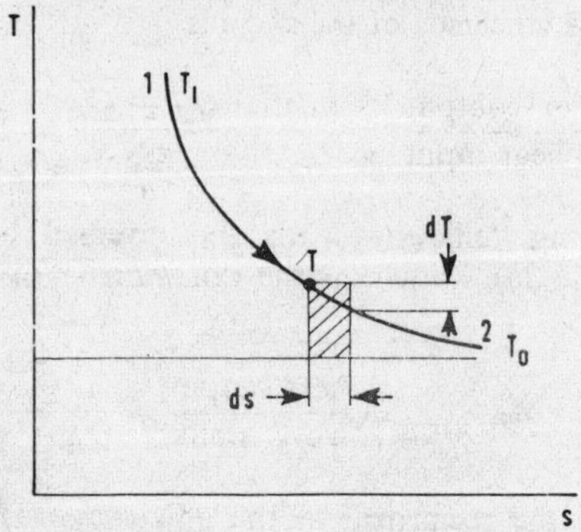

Fig. 6.3 (For a discussion of the Carnot cycle, see section 6.4)

For the whole range of temperature in the question, the maximum available work

$$m \int_{T_1}^{T_0} \frac{c_v(T - T_0)}{T} dT$$

$$= 0{\cdot}221 \int_{288}^{1273} 0{\cdot}649 \left(1 - \frac{T_0}{T}\right) dT$$

$$+ 0{\cdot}221 \int_{288}^{1273} 40{\cdot}6 \times 10^{-9} T^2 \left(1 - \frac{T_0}{T}\right) dT$$

$$= 0{\cdot}221 \times 0{\cdot}649(1273 - 288) - 0{\cdot}221 \times 0{\cdot}649(288 \ln 1273/288)$$

$$+ \tfrac{1}{3} \times 0{\cdot}221 \times 40{\cdot}6 \times 10^{-9}(1273^3 - 288^3)$$

$$- \tfrac{1}{2} \times 0{\cdot}221 \times 288(1273^2 - 288^2) \times 40{\cdot}6 \times 10^{-9}$$

$= 85{\cdot}74$ kJ *Ans.*

6.3 The Second Law of Thermodynamics

Having formulated the concept of a heat engine, and seen that such engines invariably take in heat from a high temperature source and reject part of this to a low temperature sink, we may express the second law as follows—

IT IS IMPOSSIBLE TO CONSTRUCT AN ENGINE WHICH, OPERATING IN A CYCLE, WILL PRODUCE NO OTHER EFFECT THAN THE EXTRACTION OF HEAT FROM A SINGLE HEAT RESERVOIR AND THE PERFORMANCE OF AN EQUIVALENT AMOUNT OF WORK.

This is one of several classical statements and is known as the KELVIN–PLANCK STATEMENT.

If such an engine were possible, it could extract heat from an infinite and readily available heat reservoir, such as the sea or a large river, and deliver an equivalent amount of work continuously to the surroundings.

Such an engine would be known as a PERPETUAL MOTION MACHINE OF THE SECOND KIND. Assuming that, instead of receiving its heat directly from a low temperature reservoir, such as the sea, the perpetual motion machine requires a high temperature reservoir, it would still be possible to utilize the infinite heat capacity of the low temperature reservoir without having to supply heat from any other source, by using a heat pump which, although absorbing some of the work output of the engine, would still leave a balance of useful work.

The above statement of the second law completely rules out the possibility of a perpetual motion machine of the second kind being realized.

It should be noted that a PERPETUAL MOTION MACHINE OF THE FIRST KIND, i.e. an engine working in a cycle without receiving heat from any source, but delivering work to the surroundings, violates not only the second law, as does the perpetual motion machine of the second kind, but also the first law (see equation (4.2b), section 4.3), i.e. since $\oint dQ = \oint dW$ it is impossible to obtain useful work, $\oint dW$, if the cycle heat input, $\oint dQ$, is zero.

A number of very important corollaries follow. The first, known as the Clausius statement, is applicable to heat pumps and refrigerators—

Corollary 1: HEAT CANNOT PASS SPONTANEOUSLY, I.E. WITHOUT THE ASSISTANCE OF AN EXTERNAL AGENCY, FROM A LOWER TEMPERATURE BODY TO A HIGHER TEMPERATURE BODY.

The Clausius statement may be proved in terms of the Kelvin–Planck statement, as follows.

Consider a high temperature reservoir T_1 and low temperature reservoir T_2 (see *Fig.* 6.4).

In situation (*a*) Q_1 heat units are transferred spontaneously from the low to the high temperature reservoir, in violation of the Clausius statement. In situation (*b*) Q_2 heat units are transferred from the high temperature reservoir to a heat engine, Q_1 heat units being rejected to the low temperature sink. This results in a work output

$$W_{net} = Q_2 - Q_1 \qquad \text{(i)}$$

(assuming $Q_2 > Q_1$).

If situations (*a*) and (*b*) are now combined, the net heat absorbed from the low temperature reservoir is zero, while that absorbed from

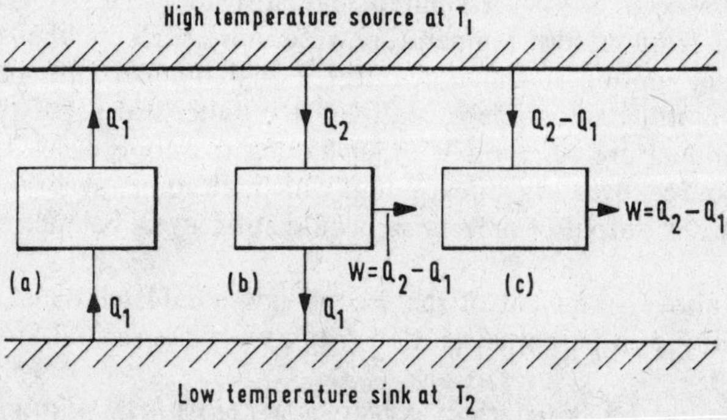

Fig. 6.4 Proof of the Clausius statement

the high temperature reservoir is $(Q_2 - Q_1)$, resulting in a useful work output

$$W_{net} = Q_2 - Q_1 \qquad \text{(ii)}$$

Hence the resultant situation, i.e. (*c*), is one in which an engine operating in a cycle produces no other effect than the extraction of heat from a single heat reservoir and the performance of an equivalent amount of work.

Thus situation (*c*) directly contravenes the Kelvin–Planck statement of the second law, so that the Clausius statement, violated in situation (*a*), is clearly true.

6.4 The Carnot Cycle (*Fig.* 6.5)

Carnot recognized that the conditions* for converting heat into work with maximum efficiency were as follows—

(*a*) The cycle must consist of a series of processes each of which is fully reversible with regard to both internal and external actions.

* It is immaterial whether the cycle consists of non-flow or steady flow processes, so long as the condition of reversibility is satisfied.

(*b*) All heat reception occurs at the maximum cycle temperature and all heat rejection occurs at the MINIMUM CYCLE TEMPERATURE.

The Carnot cycle consists of four reversible processes, two isothermal and two adiabatic—

(*i*) Isothermal expansion of the working fluid at the temperature of the hot source, T_{max}, accompanied by heat absorption from the source (1–2).
(*ii*) Adiabatic expansion with temperature drop of working fluid from T_{max} to T_{min}, the temperature of the cold sink (2–3).
(*iii*) Isothermal compression at the temperature of the cold sink, T_{min}, accompanied by heat rejection to the sink (3–4).
(*iv*) Adiabatic compression with temperature rise of working fluid from T_{min} to T_{max} (4–1).

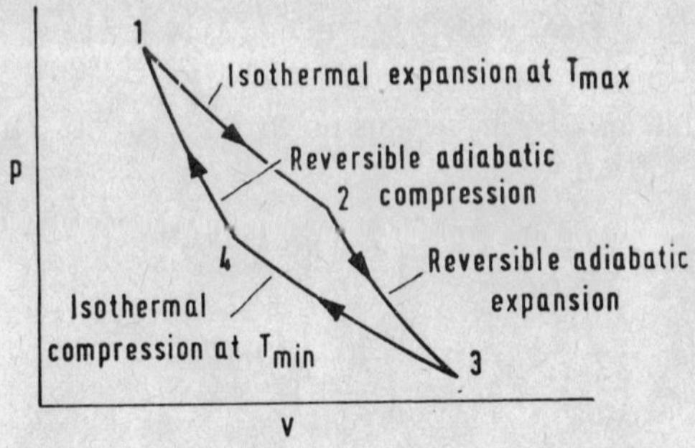

Fig. 6.5 Carnot cycle

Applied to a non-flow cycle, i.e. one carried out in a closed system, such as a cylinder sealed by a frictionless piston, the essential requirements for the cycle are

(*a*) A working fluid (nature immaterial, e.g. perfect gas),
(*b*) A frictionless piston sliding in a perfectly sealed cylinder, both piston and cylinder being perfect insulators,
(*c*) A perfectly conducting cylinder head,
(*d*) A hot source, i.e. heat reservoir of infinite capacity at the maximum temperature T_{max},
(*e*) A cold sink, i.e. heat reservoir of infinite capacity at the minimum temperature T_{min},
(*f*) Since all processes have to be fully reversible, it follows that they must be performed infinitely slowly.

It may readily be proved that, given the available maximum and minimum temperatures T_{max} and T_{min}, the Carnot engine has the highest possible efficiency, and therefore represents a standard of perfection with which other engines may be compared. Furthermore, it may be shown that the efficiency is independent of the working fluid used, which may even be a theoretical abstraction like the perfect gas. The fact that an engine working on this cycle would be wholly impracticable (due to the very large volume change involved and the need for infinitely slow processes) does not detract from its usefulness as an ideal standard.

The Carnot efficiency may most readily be deduced for the case of an engine using the perfect gas as working fluid.

The four processes applied to unit mass of working fluid then become

1–2: isothermal heat addition at T_{max} (equation (4.7b), section 4.5).

$$\text{Heat added } Q_A = RT_{max} \ln \left(\frac{v_2}{v_1}\right) \qquad \text{(i)}$$

2–3: reversible adiabatic expansion from T_{max} to T_{min} (equation (4.9b), section 4.5).

$$\text{i.e.} \quad \frac{p_2}{p_3} = \left(\frac{T_2}{T_3}\right)^{\gamma/(\gamma-1)} = \left(\frac{T_{max}}{T_{min}}\right)^{\gamma/(\gamma-1)} \qquad \text{(ii)}$$

$$\text{or} \quad \frac{v_2}{v_3} = \left(\frac{T_2}{T_3}\right)^{-1/(\gamma-1)} = \left(\frac{T_{max}}{T_{min}}\right)^{-1/(\gamma-1)} \qquad \text{(iii)}$$

3–4: isothermal heat injection at T_{min} (equation (4.7b)).

$$\text{Heat rejected } Q_R = RT_{min} \ln \left(\frac{v_3}{v_4}\right) \qquad \text{(iv)}$$

4–1: reversible adiabatic compression from T_{min} to T_{max} (equation (4.9b)).

$$\text{i.e.} \quad \frac{p_1}{p_4} = \left(\frac{T_1}{T_4}\right)^{\gamma/(\gamma-1)} = \left(\frac{T_{max}}{T_{min}}\right)^{\gamma/(\gamma-1)} \qquad \text{(v)}$$

$$\text{or} \quad \frac{v_1}{v_4} = \left(\frac{T_1}{T_4}\right)^{-1/(\gamma-1)} = \left(\frac{T_{min}}{T_{max}}\right)^{1/(\gamma-1)} \qquad \text{(vi)}$$

From (iii) and (vi)

$$\frac{v_1}{v_4} = \frac{v_2}{v_3}, \quad \text{i.e.} \quad \frac{v_2}{v_1} = \frac{v_3}{v_4} \qquad \text{(vii)}$$

whence, substituting from (vii) in (iv)

$$Q_R = RT_{min} \ln\left(\frac{v_2}{v_1}\right)$$

and, from (i),
$$Q_A = RT_{max} \ln\left(\frac{v_2}{v_1}\right)$$

But
$$\eta_{carnot} = \frac{\oint dW}{Q_A} = \frac{Q_A - Q_R}{Q_A} = 1 - \frac{Q_R}{Q_A}$$

$$= 1 - \frac{RT_{min} \ln (v_2/v_1)}{RT_{max} \ln (v_2/v_1)}$$

i.e.
$$\eta_{carnot} = 1 - \frac{T_{min}}{T_{max}} \tag{6.5}$$

Extending the above reasoning to both the Carnot heat pump and the Carnot refrigerator, we have

From equation (6.2) for the heat pump:

$$(\text{C.O.P.})_{heat\ pump} = \frac{Q_R}{Q_R - Q_A} = \frac{T_{max}}{T_{max} - T_{min}} \tag{6.6}$$

From equation (6.3) for the refrigerator:

$$(\text{C.O.P.})_{refrig.} = \frac{Q_A}{(Q_R - Q_A)} = \frac{T_{min}}{T_{max} - T_{min}} \tag{6.7}$$

Three further corollaries of the second law may now be presented—

Corollary 2. NO ENGINE, OPERATING BETWEEN TWO HEAT RESERVOIRS EACH HAVING A FIXED TEMPERATURE, CAN BE MORE EFFICIENT THAN A REVERSIBLE ENGINE OPERATING BETWEEN THE SAME TEMPERATURES.

Corollary 3. ALL REVERSIBLE ENGINES OPERATING BETWEEN TWO HEAT RESERVOIRS, EACH HAVING ITS OWN FIXED TEMPERATURE, HAVE THE SAME EFFICIENCY.

Corollary 4. THE EFFICIENCY OF ANY REVERSIBLE ENGINE OPERATING BETWEEN TWO RESERVOIRS IS INDEPENDENT OF THE NATURE OF THE WORKING FLUID AND DEPENDS SOLELY ON THE TEMPERATURE OF THE RESERVOIRS.

Each of the above statements may be proved by showing that, if not true, the second law would be violated. The argument is applied to Corollary 2 as follows.

Consider a reversible engine operating between T_{max} and T_{min} driving a reversible heat pump operating between the same temperature limits (*Fig.* 6.6).

The work output of the engine is given by equation (6.5)

$$W_{engine} = Q_A \eta = Q_A \left(1 - \frac{T_{min}}{T_{max}}\right)$$

This work output is supplied to the heat pump working on the reversed Carnot cycle between the same temperature limits. From

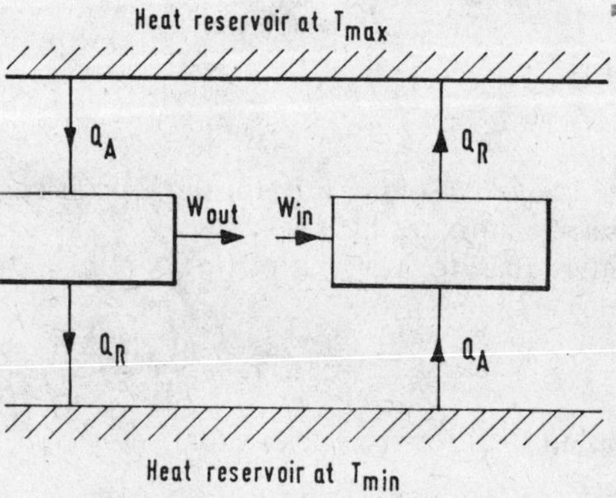

Fig. 6.6 Carnot efficiency as maximum possible efficiency

equation (6.6), the heat rejected by the heat pump to the high temperature reservoir becomes

$$Q_R = W_{in}(\text{C.O.P.})_{heat\ pump} = Q_A \left(1 - \frac{T_{min}}{T_{max}}\right)\left(\frac{T_{max}}{T_{max} - T_{min}}\right) = Q_A$$

The net effect, therefore, is that exactly as much heat is returned to the source by the heat pump as is drawn from it by the engine, the same being true of the heat sink which receives as much heat from the engine as is drawn from it by the heat pump. There is no net work input or output, i.e. the system consisting of the two reservoirs, the engine and the heat pump, suffers no net change.

Now assume that an irreversible engine has a higher efficiency than a reversible engine, and let this irreversible engine drive the reversible heat pump, as before. For the given heat quantity $(Q_R - Q_A)$ pumped by the heat pump, the work input remains unchanged. The irreversible engine, having a higher efficiency than the reversible engine, produces an amount of work $(Q_A - Q_R')$ where $Q_R' < Q_R$, for Q_A heat units drawn from the hot reservoir.

Hence the following effects occur—

(*a*) The hot reservoir gives up Q_A heat units to the irreversible engine and receives Q_A heat units from the reversible heat pump, i.e. the hot reservoir suffers no net change.

(*b*) The irreversible engine produces an amount of work $(Q_A - Q_R')$ of which only $(Q_A - Q_R)$ is required by the heat pump, leaving a surplus of $(Q_A - Q_R') - (Q_A - Q_R) = (Q_R - Q_R') > 0$ since $Q_R' < Q_R$ (see above).

(*c*) The low temperature reservoir receives Q_R' heat units from the irreversible engine and gives up Q_R heat units to the reversible heat pump, i.e. gives a net amount of heat $(Q_R - Q_R') > 0$.

Hence, taking the combination as a whole, an amount of work $(Q_R - Q_R')$ is produced by abstracting an amount of heat $(Q_R - Q_R')$ from the cold reservoir, the whole constituting an engine which, operating in a cycle, produces no other effect than the extraction of heat from a single reservoir and the performance of an equivalent amount of work. This flatly contradicts the second law and is therefore evidently untrue.

Similar arguments may be used to prove Corollaries 3 and 4.

EXAMPLE 6.2

What is the power input to a perfect reversed heat engine working between 18°C and −4°C that will produce 1200 kg of ice per hour at −4°C from water at 15°C. The specific heat of ice may be taken as 2·0934 kJ/kg-K and the latent heat 334·5 kJ/kg. The specific heat of water is 4·2 kJ/kg-K.

SOLUTION (*Fig.* 6.7)

$$\text{The coefficient of performance C.O.P.} = \frac{\text{heat absorbed}}{\text{work of compression}}$$

$$\frac{T_{min}}{T_{max} - T_{min}} = \frac{269}{291 - 269} = \frac{269}{22} = 12{\cdot}23$$

$$\text{But heat absorbed} = \frac{1200}{3600}[(15 - 0)4{\cdot}2 + 334{\cdot}5 + 2{\cdot}0934\,(0-4)$$

$$= 119{\cdot}5 \text{ kJ/s} = 119{\cdot}5 \text{ kW}$$

$$\text{Work input} = 119{\cdot}5/12{\cdot}23 = 9{\cdot}76 \text{ kW}$$

$$= 13{\cdot}1 \text{ hp} \quad \textit{Ans.}$$

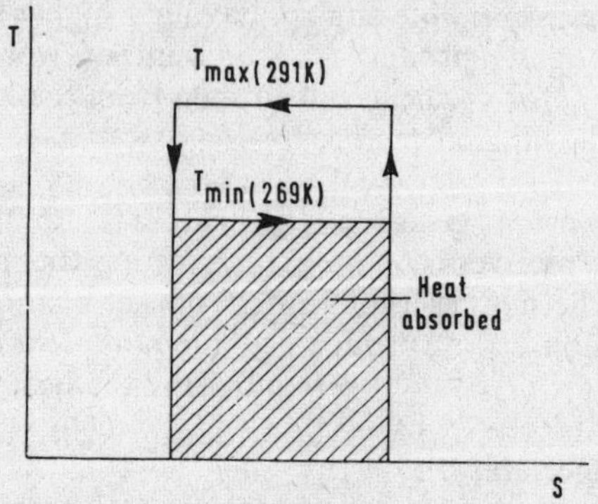

Fig. 6.7

6.5 The Thermodynamic Scale of Temperature

Any property which depends solely on temperature can be used to define a temperature scale. Various properties depending mainly on temperature, but to a lesser extent also on other effects, have already been listed in section 3.2. The perfect gas scale of temperature, based on the relationship $T = pv/R$ constitutes a complete scale, in that the product pv, i.e. the quantity measured in the perfect gas thermometer, is solely dependent on T. Nevertheless, the perfect gas itself is an abstraction, real gas behaving in a more complex manner, as indicated for instance by van der Waals' equation (see Chapter 9)

$$\left(p + \frac{a}{v^2}\right)(v - b) = RT$$

where a and b are constants.

It is clear, therefore, that any temperature scale based on a property or properties of a real substance is substantially an incomplete scale (see section 3.2). Only a temperature scale based on phenomena divorced from real substances is likely to be a complete (and hence unambiguous) scale.

In 1848 Lord Kelvin showed that, as a result of Carnot's cycle, it was possible to define an absolute temperature scale, i.e. one not dependent on the properties of any particular substance.

Let θ denote temperature measured on any scale and let the source and sink temperatures between which a reversible engine operates be θ_1 and θ_2 (*Fig.* 6.8).

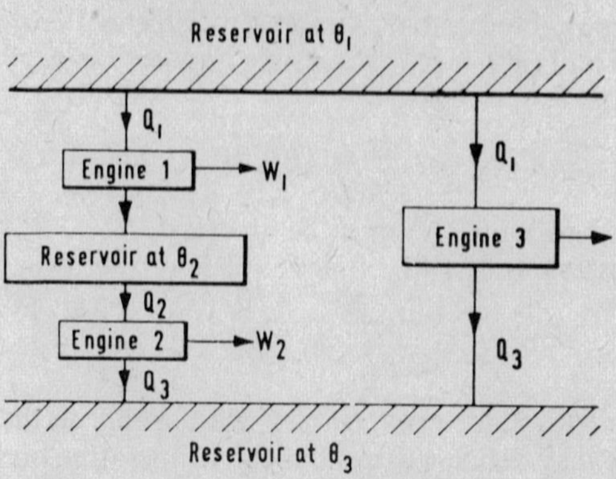

Fig. 6.8 Thermodynamic temperature scale

The efficiency of this reversible engine, which is independent of the working fluid according to Corollary 4, is a function of temperature only, according to Corollary 3, i.e. if Q_1 and Q_2 are respectively the heat quantities received from the source at θ_1, and rejected to the sink at θ_2,

$$\eta = \frac{Q_1 - Q_2}{Q_1} = f(\theta_1, \theta_2)$$

or

$$\left(1 - \frac{Q_2}{Q_1}\right) = f(\theta_1, \theta_2)$$

where f is an unspecified function.

Thus Q_2/Q_1 (or Q_1/Q_2) must be a function of θ_1 and θ_2 and we may write

$$\frac{Q_1}{Q_2} = F(\theta_1, \theta_2) \qquad \text{(i)}$$

If the heat rejected Q_2 at θ_2 is supplied to a second reversible engine, working between θ_2 and θ_3 (where $\theta_2 > \theta_3$), then by similar reasoning

$$\frac{Q_2}{Q_3} = F(\theta_2, \theta_3) \qquad \text{(ii)}$$

and multiplying (i) and (ii)

$$\frac{Q_1}{Q_3} = F(\theta_1, \theta_2) \times F(\theta_2, \theta_3) \qquad \text{(iii)}$$

But Q_1/Q_3 must be capable of representation in the same form as equations (i) and (ii), since, considering a reversible engine operating between θ_1 and θ_3

$$\frac{Q_1}{Q_3} = F(\theta_1, \theta_3) \tag{iv}$$

i.e. statements (iii) and (iv) must be identical.

This is possible only if the function F is of the form

$$F(\theta_1, \theta_2) = \frac{\phi(\theta_1)}{\phi(\theta_2)} \quad \text{or} \quad F(\theta_2, \theta_3) = \frac{\phi(\theta_2)}{\phi(\theta_3)}$$

The function ϕ is undetermined and will depend on the scale used. However, we may define $\phi(\theta)$ itself as the temperature on the absolute scale, i.e.

$$T = \phi(\theta) \tag{v}$$

Hence, reverting to equation (i), the absolute or thermodynamic temperature is defined by

$$\frac{Q_1}{Q_2} = \frac{T_1}{T_2} \tag{6.8}$$

If T_0 is a positive number assigned to a reservoir at an easily reproducible reference state (e.g. a pure substance melting at atmospheric pressure), the temperature T of any other reservoir is uniquely determined by equation (6.8)

$$T = T_0 \frac{Q}{Q_0} \tag{6.8a}$$

If $T > T_0$, Q is the heat quantity flowing from the reservoir at T to a reversible engine interposed between this reservoir and a second one at T_0, while Q_0 is the heat quantity rejected to the latter.

If $T < T_0$, Q_0 is the heat quantity received from the reservoir at the reference temperature, and Q the heat rejected by the reversible engine to the second reservoir at temperature T.

This temperature scale is completely independent of any particular substance, since from equation (6.5) the ratio of the heat quantities Q and Q_0 is a function only of the temperature ratio [in deriving equation (6.5), the perfect gas scale of temperature was actually used] i.e. taking the case $T > T_0$, and using T' to denote temperature in the perfect gas scale,

$$\eta_{carnot} = 1 - \frac{T_0'}{T'} = 1 - \frac{Q_0}{Q} \tag{6.5a}$$

$$\frac{T'}{T_0'} = \frac{Q}{Q_0} \tag{6.5b}$$

while, according to Corollary 4, the Carnot efficiency, in turn, is independent of the working fluid.

Since equation (6.5b) was derived from the perfect gas scale of temperature, comparison of equations (6.5b) and (6.8a) leads to the result

$$\frac{Q}{Q_0} = \frac{T}{T_0} = \frac{T'}{T_0'} \tag{6.9}$$

i.e. the perfect gas and thermodynamic scales of temperature are identical.

Meaning of absolute zero on the thermodynamic scale

Consider the case $T_2 < T_1$, where $T_2 = 0$. Substituting in equation (6.5a) but using the thermodynamic rather than the perfect gas scale, we have

$$\eta_{carnot} = 1 - \frac{T_2}{T_1} = 1 - \frac{0}{T_1} = 1$$

i.e. the efficiency of a reversible heat engine operating between an upper reservoir temperature T_1 and a lower temperature $T_2 = 0$ is unity. This implies that all the heat received from the high temperature reservoir is turned into work, which is contrary to the second law and therefore represents a state that cannot be attained. However, absolute zero may be approached to within a minute temperature interval, without violating the second law.

Negative temperatures on the thermodynamic (or perfect gas) scale imply Carnot efficiencies in excess of unity and clearly cannot exist.

Meaning of temperature interval on the thermodynamic scale

Consider a series of Carnot engines, operating in cascade each over a temperature interval of 1 deg, such that the first engine in the series operates with an upper reservoir temperature T.

Hence for the first engine, from equation (6.9), if $(Q_A)_1$ is the heat received from the high temperature reservoir, and $(Q_R)_1$ the heat rejected by the engine—

$$\frac{(Q_A)_1}{(Q_R)_1} = \frac{T}{T-1} \quad \text{i.e.} \quad (Q_R)_1 = (Q_A)_1 \frac{T-1}{T} = \left(1 - \frac{1}{T}\right)(Q_A)_1$$

or work output of first engine

$$W_1 = (Q_A)_1 - (Q_R)_1 = (Q_A)_1\left(1 - 1 + \frac{1}{T}\right) = \frac{(Q_A)_1}{T}$$

For the second engine, the heat received from the reservoir at $(T - 1)$ is that rejected to it by the first engine, i.e.

$$(Q_A)_2 = (Q_A)_1\left(1 - \frac{1}{T}\right)$$

and

$$\frac{(Q_A)_2}{(Q_R)_2} = \frac{T-1}{T-2}$$

$$\text{or}\quad (Q_R)_2 = (Q_A)_2\frac{(T-2)}{(T-1)} = (Q_A)_1\frac{(T-1)(T-2)}{T(T-1)}$$

or the work output of the second engine,

$$W_2 = (Q_A)_2 - (Q_R)_2 = (Q_A)_1\left(\frac{T-1}{T} - \frac{T-2}{T}\right) = (Q_A)_1\left(\frac{1}{T}\right)$$

Hence the output of the second engine is equal to that of the first engine, and likewise for all subsequent engines in the cascade.

Unit temperature interval therefore corresponds to a fixed amount of work from a reversible engine operating between reservoirs whose temperatures differ by a unit interval, no matter at what absolute level. The size of this interval is determined by fixing W and Q, but in fact the ice and steam points are chosen as temperature references as in the case of the perfect gas scale, the interval being divided into 100 equal steps.

Operating a reversible heat engine between the ice and steam point, it would be found by measurement that

$$\frac{Q_s}{Q_i} = 1{\cdot}366 \quad \text{i.e.} \quad \frac{T_s}{T_i} = 1{\cdot}366$$

Since $T_s - T_i = 100$, $T_s = 373{\cdot}15°$, $T_i = 273{\cdot}15°$. This defines the Kelvin or absolute Celsius scale (see section 3.3). If the interval $(T_s - T_i)$ is divided into 180 equal steps (Fahrenheit scale) the result becomes $T_s = 671{\cdot}67°$, $T_i = 491{\cdot}67°$ and this then defines the Rankine or absolute Fahrenheit scale.

EXAMPLE 6.3

If 20 kJ are added to a Carnot cycle at temperature 100°C and 14·6 kJ are rejected at 0°C, determine the location of absolute zero on the Centigrade scale.

SOLUTION

$$\eta = 1 - \frac{Q_R}{Q_A} = 1 - \frac{14{\cdot}6}{20} = 0{\cdot}268$$

But $$\frac{Q_R}{Q_A} = \frac{t + 0}{t + 100}$$

and $$\eta = \frac{T_1 - T_2}{T_1} = \frac{(t + 100) - (t + 0)}{(t + 100)}$$

Therefore

$$0{\cdot}268 = \frac{100}{t + 100}$$

$$t = \frac{100 - 0{\cdot}268 \times 100}{0{\cdot}268} = 273{\cdot}13$$

Therefore, absolute zero = $-273{\cdot}13$°C *Ans.*

Alternatively,

$$\frac{Q}{T} = \text{constant} \qquad \therefore \frac{20}{t + 100} = \frac{14{\cdot}6}{t}$$

from which we find that $t = 273{\cdot}13$, i.e. absolute zero at $-273{\cdot}13$°C *Ans.*

EXERCISES

6.1 Define the terms: source, sink and heat reservoir.

6.2 Prove that for constant temperature of source and sink:

(*a*) No cycle can be more efficient than a reversible cycle.
(*b*) All reversible cycles have the same efficiency.
(*c*) Efficiency is independent of the working fluid.

6.3 A new cycle is alleged to develop 10 hp for a heat addition rate of 600 kJ/min. Cycle max. and min. temperature are 1 100 K and 530 K respectively. Is this possible?
(*Ans:* No)

6.4 A reversible power cycle is used to drive a reversible refrigerator cycle. Power cycle takes in ΔQ_1 heat units at T_1 and rejects ΔQ_2 at T_2. Heat pump abstracts ΔQ_4 from sink at T_4 and discharges ΔQ_3 at T_3. Devise a general expression for the ratio $\dfrac{\Delta Q_4}{\Delta Q_1}$ in terms of temperature alone.

(*Ans:* $\Delta Q_4/\Delta Q_1 = T_4(T_1 - T_2)/T_1(T_3 - T_4)$)

6.5 A gas cycle consists of three reversible processes:

1–2 isothermal compression through a volume ratio

$V_1/V_2 = 2$, at $T_1 = 560$ K

2–3 constant pressure expansion

3–1 reversible adiabatic expansion

Find the gross heat and overall work transfer for the cycle and compare the efficiency with that of a Carnot cycle operating over the same temperature limits.

$R = 0{\cdot}2871$ kJ/kg-K, $\gamma = 1{\cdot}4$, $c_p = 1{\cdot}005$ kJ/kg-K

(*Ans:* 118·59 kJ/kg, 8·69 kJ/kg, 7·3%, cf. 17·7%)

6.6 An oil engine has a cylinder of 0·3 m bore by 0·45 m stroke. At the beginning of compression the pressure is 1 bar and the temperature is 32°C. At the end of compression the volume is 0·003 m^3 and the pressure is 31 bar. Determine the temperature at the end of compression. $R = 0{\cdot}2870$ kJ/kg-K.

(*Ans:* 894 K)

6.7 Find the change in internal energy of 0·135 kg of air when rising in temperature from −6°C to 220°C. Take γ as 1·4 and R as 0·2871 kJ/kg-K. If 30 kJ are supplied to the gas during the process, how much external work is performed?

(*Ans:* 21·88 kJ, 8·12 kJ)

6.8 A closed vessel having a volume of 0·0005 m^3 contains oxygen at a temperature of 15°C and an absolute pressure of 25 bar. Calculate the maximum pressure and temperature if 22 kJ are given to the gas. Take *cv* as 0·7076 kJ/kg-K.

(*Ans:* 186·5 bar, 1877°C)

6.9 A cylinder, closed by a piston, has a volume of 0·03 m^3 and contains air at 1 bar and 15°C. The air is compressed isentropically to one-quarter of its initial volume and then cools reversibly at constant pressure until its temperature is again 15°C. Finally, it expands isentropically to 1 bar. Determine the final temperature at the end of expansion and the net work done for the complete process. Take the mean value of γ as 1·40.

(*Ans:* −117°C, −4·78 kJ)

6.10 An ideal heat pump works according to the reversed Carnot cycle, the upper and lower temperature limits of the working fluid being 2·2°C and 87·8°C respectively. Low-grade heat is obtained from nearby river water which is drawn off at the rate of 5587 kg/h the supply and return temperatures of this water being 7·2°C and 2·2°C respectively. If the machine is to be used for heating a building, estimate how many cubic metres of air at 1 bar and 10°C can be heated at constant pressure from 10°C to 20°C. Determine also the horsepower required ideally to drive the pump. Take s.h. water as 4·1868 kJ/kg-K and R_{air} as 0·2951 kJ/kg-K.

(*Ans:* 3·398 m^3/s, 10·09 kW)

6.11 State the Carnot Principle relating to the efficiency of a reversible heat engine.

Two reversible heat engines A and B are arranged in series, A rejecting heat directly to B. Engine A receives 200 kJ at a temperature of 421°C from a hot source, whilst engine B is in communication with a cold sink at a temperature of 4·4°C. If the work output of A is twice that of B, find:

(*a*) The intermediate temperature between A and B,
(*b*) the efficiency of each engine,
(*c*) the heat rejected to the cold sink.

(*Ans:* 143·4°C, 40% and 33·5%, 80 kJ)

6.12 A heat engine operates between maximum and minimum temperatures of 671°C and 60°C respectively, with an efficiency of 50% of the appropriate Carnot efficiency. It drives a heat pump which uses river water at 4·4°C to heat a block of flats in which the temperature is to be maintained at 21·1°C. Assuming that a temperature difference of 11·1°C exists between the working fluid and the river water, on the one hand, and the required room temperature on the other, and assuming the heat pump to operate on the reversed Carnot cycle, but with a coefficient of performance of 50% of the ideal value for the same temperature limits of the working fluid, find the heat input to the engine per unit heat output from the heat pump.

Why is direct heating thermodynamically more wasteful?

(*Ans:* 0·82 kJ/kJ heat input)

6.13 What is meant by a temperature scale and what are the characteristics ideally required from such a scale?

In terms of the various properties of the perfect gas and introducing the phase diagram, explain why the perfect gas scale satisfies these conditions.

How does the thermodynamic scale of temperature differ from the perfect gas scale?

In a Carnot cycle 42·20 kJ are added at 149°C and 28·87 kJ are rejected at 15°C. Find the relationship between the Fahrenheit and the thermodynamic scales of temperature.

(*Ans:* ° thermodynamic scale = °C + 273)

6.14 What is meant by a complete temperature scale?

In terms of the various properties of the perfect gas, and using a phase diagram, explain why the perfect gas scale may be regarded as complete, and discuss the relationship between it and the thermodynamic scale of temperature.

In a Carnot cycle 52·75 kJ are added at 176·7°C and 34·51 kJ are rejected at 21·1°C. Hence find the relationship between the Fahrenheit and the thermodynamic scales of temperature.

(*Ans:* ° thermodynamic scale = °C + 273·2)

Chapter 7

Entropy

7.1 General

The first law when extended to non-cyclic processes led to the definition of energy as a property of the system. The second law by a similar extension to non-cyclic processes leads to the definition of a less obvious, but equally important property, entropy.

Just as energy, in particular internal energy, while derived logically from a macroscopic treatment of thermodynamics is nevertheless associated with the microscopic behaviour of matter, namely the kinetic energy of the individual atoms or molecules in motion (Chapter 9), so entropy is again derived from the present macroscopic analysis, but is in fact associated with the microscopic behaviour of matter. In this instance this is the total number of microstates (or thermodynamic probability) for a given macrostate, a state of undifferentiated sameness or uniformity having the highest probability or entropy (see Appendix I).

We shall see later that the use of entropy simplifies the solution of many problems in thermodynamics, and gives us a better insight particularly into irreversible processes.

We arrive at a definition of entropy by reverting to the Carnot cycle. From equation (6.5), for the Carnot cycle operating between temperatures T_{max} and T_{min}, we have

$$\frac{Q_A}{T_{max}} = \frac{Q_R}{T_{min}} \tag{i}$$

since

$$\eta = 1 - \frac{T_{min}}{T_{max}} = 1 - \frac{Q_R}{Q_A} \tag{ii}$$

Adopting the alternative sign convention

positive Q = heat transferred to system
negative Q = heat rejected by system

and integrating (i) round the cycle

$$\frac{Q_A}{T_{max}} + \frac{Q_R}{T_{min}} = 0 \tag{7.1}$$

Since heat transfer occurs only during the two isothermal processes, the above result may be written as a cyclic integral

$$\oint \frac{dQ}{T} = 0 \tag{7.1a}$$

This result may be extended to any reversible cyclic process, as follows (refer to *Fig.* 7.1).

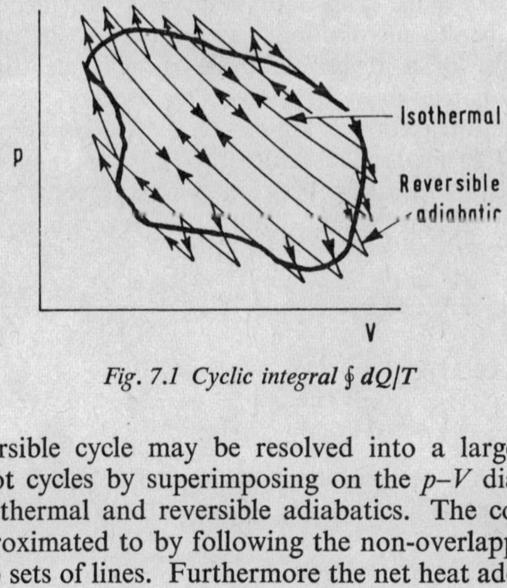

Fig. 7.1 Cyclic integral $\oint dQ/T$

Any reversible cycle may be resolved into a large number of small Carnot cycles by superimposing on the p–V diagram a network of isothermal and reversible adiabatics. The complete cycle is then approximated to by following the non-overlapping portions of these two sets of lines. Furthermore the net heat addition for the cycle is given by summating the heat added and rejected along the isothermals of each individual Carnot cycle. The overlapping portions are of no importance, since the heat rejected along the overlapping portion by the small Carnot cycle above exactly cancels the heat received by the small Carnot cycle below. The adiabatics, of course, make no contribution to the heat added. Similarly the quantity dQ/T may be integrated cyclically by again summating the quantity dQ/T for each individual Carnot cycle. But from equation (7.1a)

$$\oint \frac{dQ}{T} = 0$$

It follows, therefore, that for the reversible cycle as a whole the cyclic integral is zero, i.e.

$$\oint \left(\frac{dQ}{T}\right)_{\textit{reversible cycle}} = 0 \qquad (7.2)$$

Thus, when the working fluid has passed through any series of reversible changes which cause it to return to its initial state, the quantities of heat received and rejected are so related to the absolute temperature of the fluid at each stage as to make the integral $\oint(dQ/T)$ vanish for the cycle as a whole. In this respect the quantity dQ/T behaves like any of the established properties, p, v, T, u or h in that for each of these we have, for a cyclic process $\oint dp = 0$, $\oint dv = 0$, $\oint dT = 0$, $\oint du = 0$, $\oint dh = 0$ provided the system (*not* necessarily the surroundings) is returned to its original state. The restriction of a reversible cycle is new, i.e. the statement $\oint(dQ/T) = 0$ is true only for a *reversible* cycle.

If the cyclic integration is applied to a cycle containing irreversibilities, $\oint(dQ/T)$ no longer vanishes. Thus, for any of the small "Carnot" cycles, operating between temperatures T_1 and T_2 with heat exchange Q_1 and Q_2, and with some irreversibility,

$$\eta = \frac{Q_1 - Q_2}{Q_1} < \frac{T_1 - T_2}{T_1} \quad \text{i.e.} \quad \frac{-Q_2}{Q_1} < \frac{-T_2}{T_1}$$

whence

$$\frac{Q_2}{Q_1} > \frac{T_2}{T_1} \quad \text{i.e.} \quad \left(\frac{Q_1}{T_1} - \frac{Q_2}{T_2}\right) < 0$$

Hence, for such a small cycle $\oint(dQ/T) < 0$, and this result applies equally to the cycle as a whole.

Hence we write

$$\oint \left(\frac{dQ}{T}\right)_{\textit{irreversible cycle}} < 0 \qquad (7.3)$$

This is the Clausius inequality, usually expressed as a corollary of the second law—

Corollary 5. WHENEVER A SYSTEM UNDERGOES A CYCLE, THE INTEGRAL AROUND THE CYCLE OF dQ/T IS LESS THAN ZERO FOR AN IRREVERSIBLE CYCLE, AND EQUAL TO ZERO FOR A REVERSIBLE CYCLE.

In spite of the restriction of the statement $\oint(dQ/T) = 0$ to reversible cycles, we may tentatively define a new property by analogy with the relations $\oint dp = 0$, $\oint dT = 0$, etc. This is again expressed as a corollary, to emphasize the connexion between entropy and the second law—

Corollary 6. THERE EXISTS A PROPERTY OF A SYSTEM (denoted by the symbol S) SUCH THAT A CHANGE IN ITS VALUE IS EQUAL TO

$$S_2 - S_1 = \int_1^2 \frac{dQ}{T}$$

FOR ANY REVERSIBLE PROCESS UNDERGONE BY THE SYSTEM BETWEEN STATES 1 AND 2. THIS PROPERTY IS CALLED ENTROPY.

Consider the two reversible paths a and b joining the states 1 and 2 (*Fig.* 7.2). If path b is reversed, i.e. traversed from 2 to 1, then paths a and b together constitute a reversible cycle, i.e.

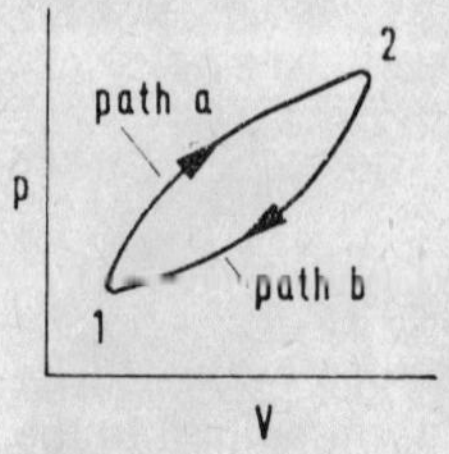

Fig. 7.2 Entropy as a property

$$\oint \frac{dQ}{T} = \int_1^2 \left(\frac{dQ}{T}\right)_{path\ a} + \int_2^1 \left(\frac{dQ}{T}\right)_{path\ b} = 0$$

$$\therefore \quad \int_1^2 \left(\frac{dQ}{T}\right)_{path\ a} = \int_1^2 \left(\frac{dQ}{T}\right)_{path\ b}$$

The same applies to any other *reversible* path joining states 1 and 2

$$\int_1^2 \left(\frac{dQ}{T}\right)_{rev} = S_2 - S_1 \tag{7.4}$$

and this is independent of the path.

Statement (7.4) is, of course, analogous to statements for other properties, e.g.

$$\int_1^2 dp = p_2 - p_1, \quad \int_1^2 dT = T_2 - T_1, \quad \text{etc.} \tag{7.5}$$

irrespective of the path joining 1 and 2, except that statements (7.5) apply to reversible and irreversible paths alike.

7.2 Evaluation of Entropy

Equation (7.4) provides the means for evaluating entropy in terms of a known initial value S. Any reversible path may be chosen, the criterion being convenience. If the entropy change along an irreversible path is required, this has to be resolved first into a series of reversible paths terminating at the same point (*Fig.* 7.3).

Thus the adiabatic irreversible path 1–2 may be replaced by the adiabatic reversible path 1–3 followed by constant volume heating 3–2. (Alternatively isothermal compression and constant volume heating or any other convenient combination.)

As with the other energy properties u and h, no absolute value can be assigned to S. Instead a datum is chosen arbitrarily, usually

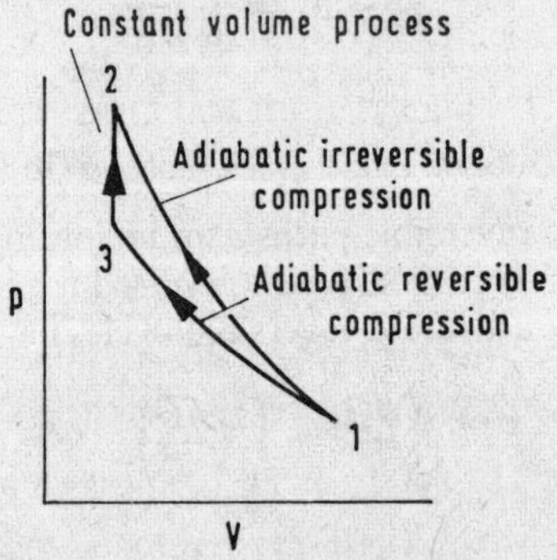

Fig. 7.3 Evaluation of entropy

0·006112 bar and 0°C. The value of entropy is then evaluated by following any suitable reversible path from the datum point 0 to the required point 1,

$$S_1 = \int_0^1 \left(\frac{dQ}{T}\right)_{reversible} \tag{7.4a}$$

EXAMPLE 7.1

Evaluate the entropy of 1 kg of air at 2 bar, 333 K given $R = 0{\cdot}287$ kJ/kg-K, $c_v = 0{\cdot}718$ kJ/kg-K, $c_p = 1{\cdot}005$ kJ/kg-K.

SOLUTION (*Fig*. 7.4)

Assume state 1 is reached by a combination of constant pressure cooling, followed by constant volume heating. Temperature T_2 is given by applying the perfect gas equation (3.2) between states 1 and 2,

$$\frac{T_2}{333} = \frac{1}{2} \quad \therefore T_2 = 166{\cdot}5 \text{ K}$$

$$s_1 = \int_0^2 \frac{c_p\,dT}{T} + \int_2^1 \frac{c_v\,dT}{T} = \left(1{\cdot}005 \ln \frac{166{\cdot}5}{273} + 0{\cdot}718 \ln \frac{333}{166{\cdot}5}\right)$$

$$= 0 \text{ kJ/kg-K}$$

State 1 therefore has the same entropy as the reference state 0.

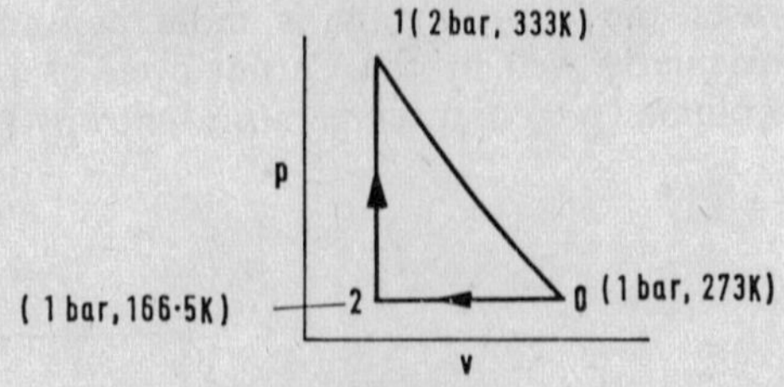

Fig. 7.4

7.3 Temperature–Entropy (T–s) Diagram

Up to now, the p–v diagram has been used to illustrate various processes. The justification for this procedure is that

(*a*) Any two properties uniquely define the state of a pure substance (existing in a single phase) (section 2.11).
(*b*) Pressure and volume are particularly suitable, because they are easily measured and at the same time provide an indication of work done, i.e.

$$dw_{rev} = p\,dv \quad \text{(for closed systems)} \tag{7.6}$$

However, any other pair of properties could be used. Thus from equation (7.4), in its most general form

$$ds = \left(\frac{dq}{T}\right)_{rev}$$

or

$$dq_{rev} = T\,ds \tag{7.7}$$

Hence, there is a close analogy between equations (7.6) and (7.7)—

Areas under the p–v curve represent work done by the system.
Areas under the T–s curve represent heat added to the system.
(Both reversible.)

Furthermore, for a cyclic process, from the first law of thermodynamics

$$\oint dq = \oint dw$$

so that the area enclosed by the loop representing the cycle gives either the net heat added to the system or the net work done by the system in a reversible cycle, both in the p–v and in the T–s diagram.

In many cases the T–s diagram is more convenient. This is illustrated particularly well by the Carnot cycle as represented in the pressure–volume (p–v) and temperature–entropy (T–s) diagram (*Fig.* 7.5)

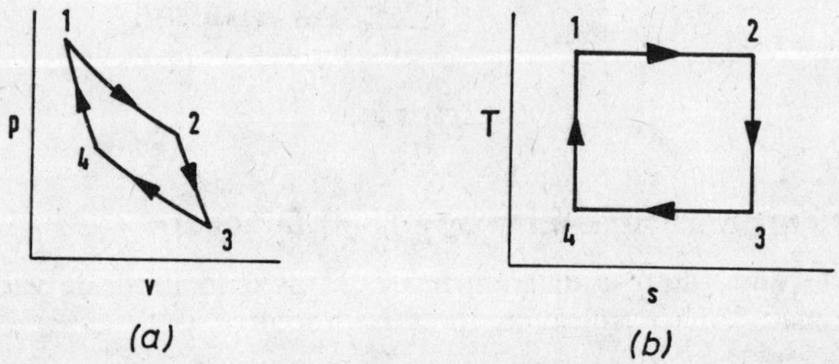

Fig. 7.5 Carnot cycle in p–v and T–s diagrams

Consider a Carnot cycle for the perfect gas in the p–v and the T–s diagrams.

1–2: *Isothermal expansion*

p–v diagram: expansion according to $p_1v_1 = p_2v_2$ (eqn. (3.2))

T–s diagram: horizontal line, $T = T_{max}$, $\Delta s_{12} = \dfrac{q_{12}}{T_{max}}$

where $q_{12} = RT_{max} \ln \left(\dfrac{v_2}{v_1}\right)$

2–3: *Adiabatic reversible (isentropic) expansion*

p–v diagram: expansion according to $p_2v_2^{\gamma} = p_3v_3^{\gamma}$ (eqn. (4.9))
T–s diagram: vertical line since

$$ds = \int_2^3 \frac{dq}{T} = 0$$

3–4: *Isothermal compression*

p–v diagram: compression according to $p_3v_3 = p_4v_4$

T–s diagram: horizontal line $T = T_{min}$, $\Delta s_{34} = \dfrac{q_{34}}{T_{min}}$

$$\Delta s_{34} = \Delta s_{12} \quad \text{since} \quad \frac{q_{12}}{T_{max}} + \frac{q_{34}}{T_{min}} = 0$$

4–1: *Isentropic compression*

p–v diagram: compression according to $p_4 v_4^{\gamma} = p_1 v_1^{\gamma}$
T–s diagram: vertical line since

$$\Delta s_{41} = \int_4^1 \frac{dq}{T} = 0$$

It is evident that the Carnot cycle is represented much more simply in the T–s than in the p–v diagram.

The isothermal and isentropic processes of the Carnot cycle are only two of a very large number of possible processes capable of representation in both p–v and T–s diagrams.

EXAMPLE 7.2

The pressure, volume and temperature at the beginning of the isothermal expansion in a Carnot cycle are 51 bar, 0·032 m³ and 1000°C. The volume at the end of the isothermal expansion is 0·08 m³. If the temperature at the end of the adiabatic expansion is 268°C, find

(*a*) the heat rejected per cycle,
(*b*) the heat added per cycle and
(*c*) the thermal efficiency of the cycle.

Assume air as the working fluid. Illustrate the cycle on both the pressure–volume and the temperature–entropy fields. It is instructive to construct the p–v cycle to scale. It is then realized that the cycle is impracticable due to its low mean effective pressure.

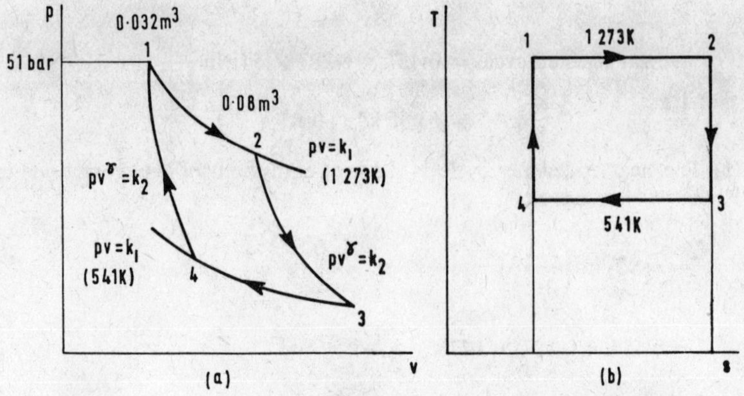

Fig. 7.6

SOLUTION (*Fig*. 7.6)

(The notation used is that of *Fig*. 7.6.)

$$\text{For air,}\quad R = \frac{8{\cdot}314}{28{\cdot}97} = 0{\cdot}287 \text{ kJ/kg-K}$$

$$pv = mRT,\quad m = \frac{100 \times 51 \times 0{\cdot}032}{0{\cdot}287 \times 1273} = 0{\cdot}457 \text{ kg}$$

$$p_2v_2 = p_1v_1,\quad p_2 = \frac{51 \times 0{\cdot}032}{0{\cdot}08} = 20{\cdot}4 \text{ bar}$$

$$\frac{T_3}{T_2} = \left(\frac{p_3}{p_2}\right)^{(\gamma-1)/\gamma} = \left(\frac{v_2}{v_3}\right)^{\gamma-1} \quad \text{and} \quad T_2 = 1273 \text{ K; } T_3 = 541 \text{ K}$$

$$\therefore p_3 = p_2 \times \left(\frac{T_3}{T_2}\right)^{\gamma/(\gamma-1)} = 20{\cdot}4 \times \left(\frac{541}{1273}\right)^{1{\cdot}4/0{\cdot}4} 1{\cdot}02 \text{ bar}$$

$$\frac{p_4v_4}{T_4} = \frac{p_1v_1}{T_1},\quad \text{also}\quad p_1v_1^{\gamma} = p_4v_4^{\gamma}$$

$$\therefore \frac{T_4}{T_1} = \left(\frac{v_1}{v_4}\right)^{\gamma-1} = \left(\frac{p_4}{p_1}\right)^{(\gamma-1)/\gamma};\ \frac{541}{1273} = \left(\frac{0{\cdot}032}{v_4}\right)^{\gamma-1};\ v_4 = 0{\cdot}256 \text{ m}^3$$

$$p_4 = \left(\frac{T_4}{T_1}\right)^{\gamma/(\gamma-1)} \times p_1 = \left(\frac{541}{1273}\right)^{1{\cdot}4/0{\cdot}4} \times 51 = 2{\cdot}55 \text{ bar}$$

(*a*) The heat rejected per cycle is rejected during the isothermal compression 3 to 4 and is equivalent to the work done in that process.

$$\therefore \text{heat rejected} = mRT_3 \ln\left(\frac{p_4}{p_3}\right)$$

$$\therefore \text{heat rejected/cycle} = 0{\cdot}457 \times 0{\cdot}287 \times 541 \ln \frac{2{\cdot}55}{1{\cdot}02}$$

$$= 63{\cdot}45 \text{ kJ} \quad \textit{Ans.}$$

(*b*) The heat added per cycle is the heat equivalent of the work done in isothermal expansion

$$= mRT_2 \ln\left(\frac{p_1}{p_2}\right)$$

$$= 0{\cdot}457 \times 0{\cdot}287 \times 1273 \ln \frac{51}{20{\cdot}4}$$

$$= 152{\cdot}90 \text{ kJ} \quad \textit{Ans.}$$

(*c*) The heat equivalent of the work done per cycle = 152·90 − 63·45 = 89·45 kJ

$$\therefore \text{ thermal efficiency} = \frac{89{\cdot}45 \times 100}{152{\cdot}90} = 57{\cdot}59\% \quad \textit{Ans.}$$

A check on this calculation is that the efficiency is also given by $\dfrac{T_1 - T_2}{T_1}$

namely, (1273 − 541)/1273 = 57·50% *Ans.*

7.4 Representation of Reversible Processes in Temperature–Entropy Diagram

This section is intended to be read in conjunction with section 4.5 dealing with the application of the first law to closed system processes. Here we are concerned not with the evaluation of heat and work transfers, but merely the evaluation of entropy changes for these various processes. Entropy being a property (unlike heat and work) and hence being uniquely determined by the end states, the entropy change will be independent of whether the process is performed in open or closed system, so that the restriction of closed systems used in section 4.5 does not apply here. As in the case of section 4.5, the perfect gas equations lead to particularly simple analytical relationships.

From equation (4.4b), section 4.4, the non-flow energy equation for the closed system, neglecting kinetic and potential energy effects, becomes (for unit mass)

$$dq - dw = du$$

and from equation (2.17), section 2.10, for the enthalpy h

$$dh = du + p\,dv + v\,dp$$

From equations (7.6) and (7.7) of section 7.3

$$dq_{rev} = T\,ds \quad \text{and} \quad dw_{rev} = p\,dv$$

Hence, for a reversible process we have

$$T\,ds = dw + du$$

i.e.

$$\boxed{T\,ds = p\,dv + du} \tag{7.8a}$$

or

$$T\,ds = p\,dv + dh - p\,dv - v\,dp$$

i.e.

$$\boxed{T\,ds = dh - v\,dp} \tag{7.8b}$$

But equations (7.8) are equations connecting properties only and must therefore apply quite generally, i.e. not only to reversible but also to irreversible processes. Equations (7.8) are of great importance in the analysis of both closed and open system processes, particularly when irreversibilities occur. They are known as the ENTROPY EQUATIONS (ref. 7.1).

In the case of the perfect gas only, equation (7.8a) leads to certain additional entropy relationships. Substituting in equation (7.8a) from the perfect gas equations (3.2) and (3.9a),

$$ds = \frac{R\,dv}{v} + c_v \frac{dT}{T} \qquad (7.9)$$

or

$$s_2 - s_1 = R \ln\left(\frac{v_2}{v_1}\right) + c_v \ln\left(\frac{T_2}{T_1}\right) \qquad (7.9a)$$

Alternatively, making use of equations (3.2) and (3.11a), equation (7.8b) may also be written in the forms

$$s_2 - s_1 = c_p \ln\left(\frac{v_2}{v_1}\right) + c_v \ln\left(\frac{p_2}{p_1}\right) \qquad (7.9b)$$

$$s_2 - s_1 = c_p \ln\left(\frac{T_2}{T_1}\right) - R \ln\left(\frac{p_2}{p_1}\right) \qquad (7.9c)$$

Equations (7.8) and (7.9) may now be applied to the evaluation of particular processes (see also section 4.5). All quantities refer to unit mass of working fluid.

(*a*) CONSTANT VOLUME PROCESS (*Fig.* 7.7)

General: $dq = du$ (equation 4.5)

i.e.

$$ds = \frac{dq}{T} = \frac{du}{T} \quad \text{or} \quad \Delta s_{12} = \int_1^2 \frac{du}{T} \qquad (7.10)$$

Perfect gas: $dq = du = c_v\,dT$

i.e.

$$ds = \frac{c_v\,dT}{T} \quad \text{or} \quad \Delta s_{12} = c_v \ln\left(\frac{T_2}{T_1}\right) \qquad (7.10a)$$

The equation of a constant volume line may be derived from equation (7.9a) for any arbitrary datum T_0, p_0, $v_0 = RT_0/p_0$, where $T_0 = 0°\text{C}$, $p_0 = 0{\cdot}006112$ bar abs.

i.e.

$$s = R \ln\left(\frac{v}{v_0}\right) + c_v \ln\left(\frac{T}{T_0}\right) = c_v \ln\left(\frac{T}{T_0}\right) \qquad (7.10b)$$

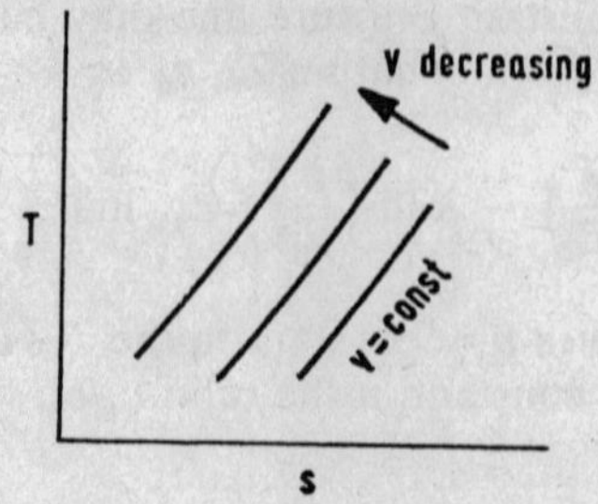

For perfect gas: $s = \text{constant} + c_v \ln\left(\dfrac{T}{T_0}\right)$

$v = \text{constant for all lines.}$

Fig. 7.7 Constant volume process in T–s diagram

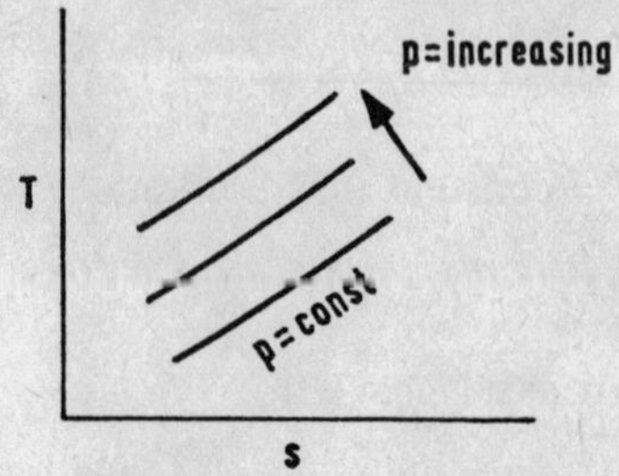

For perfect gas: $s = \text{constant} + c_p \ln\left(\dfrac{T}{T_0}\right)$

$p = \text{constant for all lines.}$

Fig. 7.8 Constant pressure process in T–s diagram

(*b*) CONSTANT PRESSURE PROCESS (*Fig.* 7.8)

General: $dq = dh$ (equation 4.6)

i.e. $$ds = \frac{dh}{T} \quad \text{or} \quad \Delta s_{12} = \frac{dh}{T} \tag{7.11}$$

Perfect gas: $dq = dh = c_p\,dT$

i.e. $$ds = \frac{c_p\,dT}{T}$$

or $$\Delta s_{12} = c_p \ln\left(\frac{T_2}{T_1}\right) \tag{7.11a}$$

The equation of a constant pressure line may be derived from equation (7.9c) again for any datum T_0, p_0, v_0

$$s = c_p \ln\left(\frac{T}{T_0}\right) - R \ln\left(\frac{p}{p_0}\right) = c_p \ln\left(\frac{T}{T_0}\right) \qquad (7.11b)$$

Note: The slope of the lines p = constant in the T–s diagram is less than that of the line v = constant in the ratio c_v/c_p.

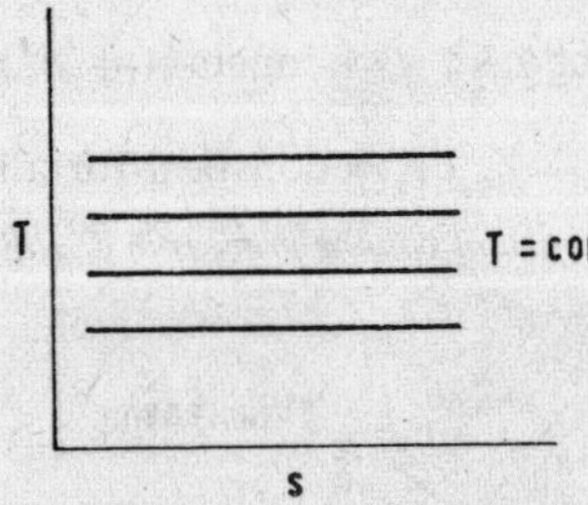

T = constant for all lines.

Fig. 7.9 Isothermal process in T–s diagram

(*c*) Isothermal process (*Fig.* 7.9)

General: $dq = dw + du$ (equation (4.7).)

i.e. $$ds = \frac{dw + du}{T}$$

or $$\Delta s_{12} = \int_1^2 \frac{dw + du}{T} \qquad (7.12)$$

Perfect gas: $dw = p\,dv = \dfrac{RT\,dv}{v}$ and $du = 0$

$$dq = \frac{RT\,dv}{v} + 0$$

i.e. $$ds = \frac{R\,dv}{v}$$

$$\Delta s_{12} = R \ln\left(\frac{v_2}{v_1}\right) \qquad (7.12a)$$

Isothermals are horizontal lines in the T–s diagram.

(*d*) ADIABATIC (REVERSIBLE) PROCESS (*Fig.* 7.10)

Since, in the adiabatic reversible process the entropy remains constant (see section 7.3), such processes are also known as ISENTROPICS.

General: $dq = 0, \quad ds = 0$ (7.13)

Perfect gas: $dq = 0, \quad ds = 0$ (7.13a)

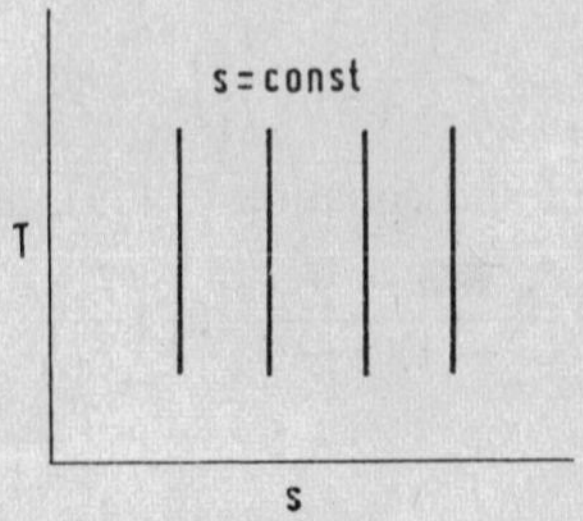

s = constant for all lines.

Fig. 7.10 Adiabatic reversible process in T–s diagram

Substitution in equation (7.9b) for the perfect gas only,

$$s_2 - s_1 = 0 = c_p \ln\left(\frac{v_2}{v_1}\right) + c_v \ln\left(\frac{p_2}{p_1}\right)$$

i.e. $$\frac{p_2}{p_1}\left(\frac{v_2}{v_1}\right)^{c_p/c_v} = \frac{p_2}{p_1}\left(\frac{v_2}{v_1}\right)^{\gamma} = \text{constant}$$

or $$pv^{\gamma} = \text{constant}$$

This is the isentropic state equation, as derived in section 4.5d.

Isentropics are vertical lines in the T–s diagram.

(*e*) POLYTROPIC (REVERSIBLE) PROCESS

General: $$dq = dw + du$$

i.e. $$ds = \frac{dw + du}{T}$$

or $$\Delta s_{12} = \int_1^2 \frac{dw + du}{T} \tag{7.14}$$

Perfect gas: $dw = p\,dv = \dfrac{RT\,dv}{v}$ where pv^n = constant

$$du = c_v\,dT$$

i.e.
$$ds = \frac{R\,dv}{v} + c_v\frac{dT}{T}$$

or
$$\Delta s_{12} = R\ln\left(\frac{v_2}{v_1}\right) + c_v\ln\left(\frac{T_2}{T_1}\right)$$

Since $\ln\left(\dfrac{v_2}{v_1}\right) = -\dfrac{1}{n-1}\ln\left(\dfrac{T_2}{T_1}\right)$ and $c_v = \dfrac{R}{\gamma - 1}$

from pv^n = constant and equation (3.12a), then

$$\Delta s_{12} = \frac{R(n-\gamma)}{(n-1)(\gamma-1)}\ln\left(\frac{T_2}{T_1}\right)$$

EXAMPLE 7.3

(*a*) *Constant volume heating*

One kg of fluid at pressure 1·5 bar and temperature 111·4°C is heated in a vessel with rigid walls of negligible coefficient of expansion, to a temperature of 300°C. Calculate the final pressure and the change in entropy (i) when the fluid is air which is treated as a perfect gas and (ii) when the fluid is steam.

SOLUTION

(*i*) $p_1v_1/T_1 = p_2v_2/T_2$ and $v_1 = v_2$, therefore

$p_2/p_1 = T_2/T_1 = 573/384{\cdot}4 = 1{\cdot}49$

$p_2 = 1{\cdot}49 \times 1{\cdot}5 = 2{\cdot}235$ bar *Ans.*

$$\Delta s_{12} = c_v\ln\left(\frac{T_2}{T_1}\right) = \frac{R}{\gamma-1}\ln\left(\frac{T_2}{T_1}\right) = \frac{0{\cdot}287}{0{\cdot}4}\ln 1{\cdot}49$$

$= 0{\cdot}286$ kJ/kg-K gain *Ans.*

(*ii*) From saturated steam tables, the specific volume of the steam is 1·159 m³/kg. From superheated steam tables at 300°C, the specific volume is 1·159 at a pressure which lies between 2 and 3 bar using abridged tables.

Interpolating as has been shown in Example 3.4, the final pressure is

$$2 + \frac{0{\cdot}157}{0{\cdot}4406} = 2{\cdot}357 \text{ bar} \quad \textit{Ans.}$$

$$\text{Change of entropy} = \left(7{\cdot}892 - \frac{0{\cdot}357}{1}(7{\cdot}892 - 7{\cdot}702)\right) - 7{\cdot}223$$

$$= 7{\cdot}824 - 7{\cdot}223 = +0{\cdot}601 \text{ kJ/kg-K} \quad \textit{Ans.}$$

(*b*) *Constant pressure process*

By placing a weight on top of a frictionless piston, free to move in a cylinder, it is possible to heat fluid in the cylinder under constant pressure conditions.

One kg of air at 1 bar pressure and 15°C is heated under such conditions to 149°C. Find the change in volume, the work done and the changes in internal energy, enthalpy and entropy.

SOLUTION

c_p for air $= 0{\cdot}992$ kJ/kg-K

$$\Delta s_{12} = c_p \ln\left(\frac{T_2}{T_1}\right) = 0{\cdot}992 \times \ln\left(\frac{422}{288}\right) = 0{\cdot}992 \times 0{\cdot}38184$$

$$= 0{\cdot}3788 \text{ kJ/kg-K} \quad \textit{Ans.}$$

$$V_1 = \frac{mRT_1}{p} = \frac{1 \times 0{\cdot}287 \times 288}{100 \times 1} = 0{\cdot}827 \text{ m}^3$$

$$V_2 = 0{\cdot}827 \times 422/288 = 1{\cdot}212 \text{ m}^3 \quad \textit{Ans.}$$

$$\text{Work done} = p(V_2 - V_1) = 100 \times 1(1{\cdot}212 - 0{\cdot}827) = 38{\cdot}5 \text{ kJ} \quad \textit{Ans.}$$

$$Q - W = \Delta U$$

$$1 \times 0{\cdot}992 \times (422 - 288) - 38{\cdot}5 = \Delta U_{12}$$

$$= 94{\cdot}4 \text{ kJ gain} \quad \textit{Ans.}$$

$$\Delta H = U_2 - U_1 + (pV_2 - pV_1) = H_2 - H_1$$

$$= \Delta U + W$$

$$= mc_p(T_2 - T_1) = 132{\cdot}9 \text{ kJ} \quad \textit{Ans.}$$

(*c*) *Isothermal process*, and (*d*) *Polytropic process*

Ideally, air (in a compressor) is compressed isothermally since then the work required is a minimum. Actually, in spite of jacket water cooling of the cylinders, the compression is polytropic with an index of compression, that is a value for n in the compression law pv^n = constant, between 1·2 and 1·35.

Consider a compressor with air at 1 bar and 15°C, compressing (*c*) isothermally to 27·59 bar (*d*) polytropically, the index being 1·30, to the same pressure. Compare the work done, the heat exchange with the surroundings, the final temperature, and the changes in internal energy and entropy due to the compression per kilogramme of air.

SOLUTION

(*c*) Isothermal compression:

$$v_1 = \frac{RT_1}{p_1} = \frac{0{\cdot}287 \times 288}{100 \times 1} = 0{\cdot}827 \text{ m}^3/\text{kg}$$

$$v_2 = RT_2/p_2$$

$$= \frac{0{\cdot}287 \times 288}{100 \times 27{\cdot}59} = 0{\cdot}030 \text{ m}^3/\text{kg} \quad \textit{Ans.}$$

$$T_2 = T_1 = 288 \text{ K}$$

Since the temperature is constant there is no change in internal energy, i.e.

$$q - w = \Delta u = 0 \quad \textit{Ans.}$$

Therefore the heat lost to the surrounding is the equivalent of the work done in compression:

$$q = w = RT \ln\left(\frac{v_1}{v_2}\right)$$

$$= -0{\cdot}287 \times 288 \ln\left(\frac{0{\cdot}827}{0{\cdot}030}\right)$$

$$= -263{\cdot}47 \text{ kJ/kg} \quad \textit{Ans.}$$

The negative sign indicates that work was done on the air and heat was lost to the surroundings.

$$\Delta s_{12} = R \ln\left(\frac{v_2}{v_1}\right) + c_v \ln\left(\frac{T_2}{T_1}\right)$$

$$= 0{\cdot}287 \ln\left(\frac{0{\cdot}03}{0{\cdot}827}\right) + 0 = -0{\cdot}287 \ln\left(\frac{0{\cdot}827}{0{\cdot}03}\right) + 0$$

$$= -0{\cdot}915 \text{ kJ/kg-K loss} \quad \textit{Ans.}$$

(*d*) Polytropic compression:

$$\frac{T_2}{T_1} = \left(\frac{p_2}{p_1}\right)^{(n-1)/n} = \left(\frac{27{\cdot}59}{1}\right)^{0{\cdot}3/1{\cdot}3} = 2{\cdot}15$$

Therefore

$$T_2 = 2{\cdot}15 \times 288 = 619{\cdot}2 \text{ K} \quad \textit{Ans.}$$

Change in internal energy $\Delta u_{12} = c_v\,(T_2 - T_1)$

$$= \frac{R}{\gamma - 1}(T_2 - T_1)$$

$$= \frac{0{\cdot}287}{0{\cdot}402}(619{\cdot}2 - 288)$$

$$= 237{\cdot}61 \text{ kJ/kg gain} \quad \textit{Ans.}$$

$$\Delta s_{12} = c_p \ln\left(\frac{T_2}{T_1}\right) - R \ln\left(\frac{p_2}{p_1}\right)$$

$$= \frac{\gamma}{\gamma - 1} R \ln\left(\frac{T_2}{T_1}\right) - R \ln\left(\frac{p_2}{p_1}\right)$$

$$= \frac{1{\cdot}402}{0{\cdot}402} \times 0{\cdot}287 \ln(619{\cdot}2/288) - 0{\cdot}287 \ln(27{\cdot}59/1)$$

$$= 0{\cdot}76617 - 0{\cdot}95211$$

$$= -0{\cdot}18594 \text{ kJ/kg} \quad \textit{Ans.}$$

The negative sign indicates a loss of entropy.

$$\text{Work done} = \frac{p_2v_2 - p_1v_1}{1 - n} = \frac{R(T_2 - T_1)}{1 - n}$$

$$= -\frac{0{\cdot}287(619{\cdot}2 - 288)}{0{\cdot}3}$$

$$= -316{\cdot}84 \text{ kJ/kg} \quad \textit{Ans.}$$

The negative sign indicates that work is done on the air.

Now, applying the first law,

$$q - w = \Delta u$$

$$q = w + \Delta u$$

$$= -316{\cdot}84 + 237{\cdot}61 = -79{\cdot}23 \text{ kJ/kg} \quad \textit{Ans.}$$

The negative sign indicates that heat is lost to the surroundings.

7.5 Reversible Steady Flow Processes

(See also section 5.8)

The S.F.E.E. (4.13) developed in section 4.6 may be written in differential form and for unit mass as follows—

$$dq - dw = dh + \tfrac{1}{2}\, d(c^2) + d(gZ) \tag{7.15}$$

Combining this with the entropy equation (7.8b)

$$T\, ds = dq = dh - v\, dp \text{ (for reversible processes only)}$$

the following equation is obtained

$$-\, v\, dp = dw + \tfrac{1}{2}\, d(c^2) + d(gZ) \tag{7.16}$$

(See also section 5.8, equation (5.13) which is the integrated form of equation (7.16).)

Equation (7.16) corresponds to the expression

$$p\, dv = dw$$

for the reversible non-flow process in the absence of kinetic and potential energy effects.

By integration of equation (7.16) we obtain, for unit mass,

$$\int_2^1 v\, dp = w_{12} + \tfrac{1}{2}(c_2^2 - c_1^2) + g(Z_2 - Z_1) \tag{7.17a}$$

and

$$\int_2^1 V\,dp = W_{12} + \tfrac{1}{2}m(c_2^2 - c_1^2) + g(Z_2 - Z_1) \qquad (7.17b)$$

for mass m.

Thus, while work in the reversible CLOSED SYSTEM PROCESS is represented by the area $\int_1^2 p\,dv$ (*Fig.* 7.11a), i.e. the area under the curve describing the process in the p–v diagram, between the lower and upper VOLUME LIMITS v_1 and v_2, respectively, the sum of work transfer and kinetic and potential energy changes for the reversible FLOW SYSTEM PROCESS is represented by the area under the curve describing the process path in the p–v diagram between the upper and lower PRESSURE LIMITS (*Fig.* 7.11b).

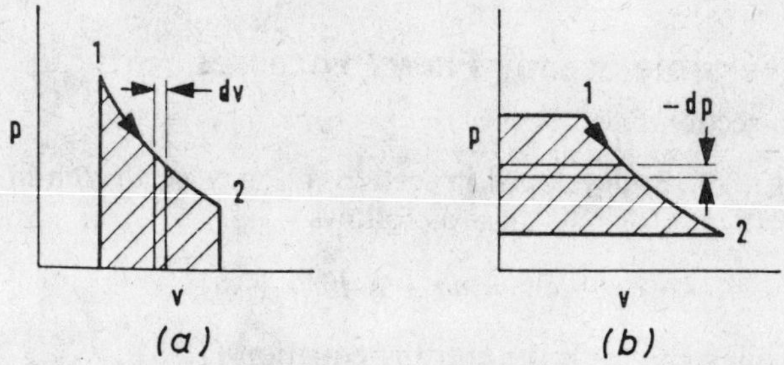

Fig. 7.11 Reversible steady flow process

For the compressor and turbine flow processes described in section 4.7, in which kinetic and potential energy effects were neglected, and described respectively by equations (4.15a) and (4.16b)

$$W_{12} = \int_2^1 v\,dp \qquad (7.18)$$

while for the nozzle, described by equation (4.17)

$$\tfrac{1}{2}(c_2^2 - c_1^2) = \int_2^1 v\,dp \qquad (7.19)$$

7.6 Entropy Changes in Irreversible Processes

As already indicated, entropy is often a useful measure of the degree of irreversibility of a particular process.

Consider an adiabatic irreversible closed system compression 1–2 applied to unit mass of working fluid (*Fig.* 7.12).

For the purpose of evaluating the entropy change $s_2 - s_1$, the process may be replaced by isentropic compression 1–2′ followed by constant volume heating 2′–2 since the expression $ds = dq/T$ only holds along a reversible path.

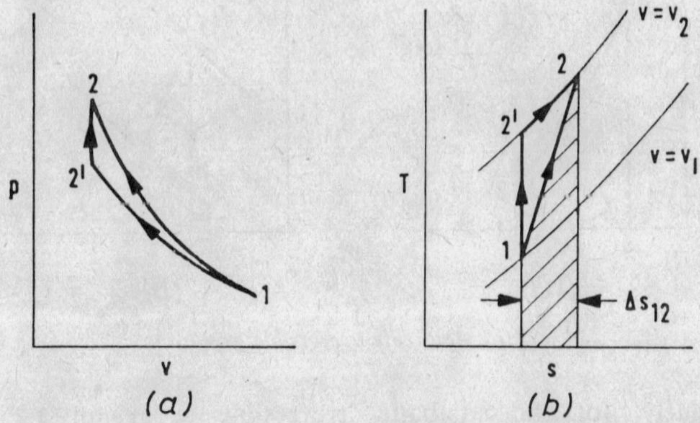

Fig. 7.12 Adiabatic irreversible compression

Thus, for a perfect gas

$$s_2 - s_1 = s_2 - s_{2'} = c_v \ln \left(\frac{T_2}{T_{2'}}\right)$$

where

$$\frac{T_{2'}}{T_1} = \left(\frac{v_1}{v_{2'}}\right)^{\gamma-1} \text{ (equation (4.9a)) and } \left(\frac{T_2}{T_1}\right) = \left(\frac{v_1}{v_2}\right)^{n-1} \quad (7.20a)$$

where $n > \gamma$ (equation (4.10)).

Since the process is irreversible (see also section 5.6)

$$dw_{irrev} \neq p\,dv \quad \text{i.e.} \quad w_{12} \neq \int_1^2 p\,dv$$

At the same time

$$dq_{irrev} \neq T\,ds \quad \text{i.e.} \quad q_{12} \neq \int_1^2 T\,ds \quad (7.20b)$$

This is evidently true since the process 1–2 is stipulated to be adiabatic, i.e. $q_{12} = 0$.

Thus areas under the temperature–entropy curve for an irreversible process do *not* represent heat added to the system, just as areas under the pressure–volume curve do *not* represent work done by the system.

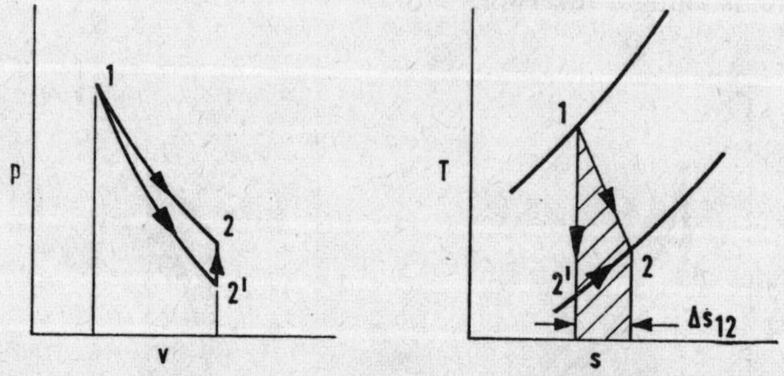

Fig. 7.13 Adiabatic irreversible expansion

Similarly, for the adiabatic irreversible expansion 1–2 (see *Fig.* 7.13) the entropy change is evaluated by dividing the process into an isentropic expansion 1–2′ followed by constant volume heating 2′–2,

$$s_2 - s_1 = s_2 - s_{2'} = c_v \ln \left(\frac{T_2}{T_{2'}}\right), \text{ as before}$$

These cases may be related directly to those discussed in section 5.6. The area under the curve 1–2 in the T–s diagram represents heat added if the process 1–2 is performed reversibly, i.e. $(q_{rev})_{1-2}$. But according to equation (5.6) of that section

$$(q_{rev})_{1-2} = (w_{rev})_{1-2} - (w_{irrev})_{1-2}$$

i.e. the shaded area represents the difference between the maximum work available for the process 1–2 and the work actually obtained.

In both cases, the entropy change $(s_2 - s_1)$ is a measure of *internal* irreversibility effects. An external irreversibility would not lead to an entropy change of the working fluid during an adiabatic change, since the dissipative effects occur wholly outside the working fluid.

THUS ENTROPY CHANGES OBSERVED DURING ADIABATIC PROCESSES ARE A MEASURE OF INTERNAL IRREVERSIBILITY.

This may be generalized in the form of yet another corollary of the second law—

Corollary 7. THE ENTROPY OF AN ISOLATED SYSTEM EITHER INCREASES OR IN THE LIMIT REMAINS CONSTANT.

EXAMPLE 7.4

Air in a piston–cylinder arrangement is initially at 1 bar and 15°C. The piston movement reduces the volume to one quarter of the original size. The compression is adiabatic and (*a*) reversible, the law of compression being $pv^{1\cdot4}$ = constant; and (*b*) irreversible, the final temperature being 6·6°C higher than in case (*a*).

Compare the work done and the change of entropy in the two cases.

SOLUTION

(*a*)

$$\frac{T_{2'}}{T_1} = \left(\frac{v_1}{v_2}\right)^{\gamma-1} = 4^{0\cdot4} = 1\cdot74$$

$$T_{2'} = 1\cdot74 \times 288 = 501\cdot1\text{K}$$

$$w' = \frac{p_1v_1 - p_2v_2}{\gamma - 1} = \frac{R(T_1 - T_{2'})}{0\cdot4}$$

$$= -\frac{0\cdot287}{0\cdot4}(501\cdot1 - 288)$$

$$= -152\cdot9 \text{ kJ/kg} \quad \textit{Ans.}$$

$$\Delta s = 0 \quad \textit{Ans.}$$

(*b*) $T_2 = 507\cdot7$ K (*Fig*. 7.12)

$$w = \Delta u_{12} = -c_v(T_2 - T_1) = -\frac{R}{\gamma - 1}(507\cdot7 - 288)$$

$$= -\frac{0\cdot287}{0\cdot4} \times 219\cdot7$$

$$= -157\cdot6 \text{ kJ/kg} \quad \textit{Ans.}$$

$$\Delta s = s_2 - s_{2'} = c_v \ln\left(\frac{T_2}{T_{2'}}\right)$$

$$= \frac{R}{\gamma - 1}\ln\left(\frac{T_2}{T_{2'}}\right) = \frac{0\cdot287}{0\cdot4}\ln\left(\frac{507\cdot7}{501\cdot1}\right)$$

$$= 0\cdot009\,259 \text{ kJ/kg-K gain} \quad \textit{Ans.}$$

If the system comprises not only the working fluid, but also that part of the enclosure affected by irreversibility effects, Corollary 7 applies not only to internal but also to external irreversibility effects.

Thus, taking the case of section 5.3, i.e. heat transfer through a finite temperature gradient, and including in the system not only the working fluid but also the high temperature reservoir supplying Q heat units at $(T + \Delta T)$ and the low temperature reservoir receiving Q heat units at $(T - \Delta T)$, the net entropy change of the system is that of the two reservoirs, since the entropy of the working fluid is returned to its original value.

Thus $$(\Delta S)_{system} = \frac{-Q}{T + \Delta T} + \frac{Q}{T - \Delta T} > 0 \qquad (7.21)$$

Hence the entropy of the system as a whole has increased in this case of external irreversibility, while the entropy of the working fluid has remained constant.

Thus, entropy changes may be regarded as indicating irreversibilities, both external and internal, when a system undergoes an adiabatic change.

When a system undergoes a non-adiabatic change, i.e. involving heat exchange with the surroundings, the resultant entropy change is due partly to the heat exchange process and partly to internal and external irreversibilities, if these are present, i.e.

$$\Delta S = (\Delta S)_{heat\ addition} + (\Delta S)_{irreversible} \qquad (7.22)$$

or $$\Delta S_{12} = \int_1^2 \left(\frac{dQ}{T}\right) + (\Delta S_{irrev})_{1-2} \qquad (7.22a)$$

which may be expressed in the form of the inequality

$$\Delta S_{12} > \int_1^2 \frac{dQ}{T} \qquad (7.22b)$$

The inequality (7.22b), which may be compared with the inequality (7.3), is an alternative form of the Clausius inequality for non-cyclic as opposed to cyclic processes and may be expressed in the form of a final corollary of the second law—

Corollary 8. WHEN A SYSTEM UNDERGOES A REAL, I.E. IRREVERSIBLE, NON-ADIABATIC PROCESS, THE GAIN IN ENTROPY OF THE SYSTEM EXCEEDS THAT WHICH WOULD RESULT FROM REVERSIBLE HEAT TRANSFER ONLY.

Reverting to differentials,

$$dQ \leqslant T\,dS \qquad (7.23)$$

Hence in any infinitesimal process, other than a fully reversible one, the heat transferred to the system is less than the product $T\,dS$, and therefore less than the area under the temperature–entropy curve.

This situation is analogous to that expressed by equation (7.20b) for the adiabatic irreversible process and, as already stated in that connexion, with the situation discussed in section 5.6 in which it was shown that areas under the pressure–volume curve do not represent work done, i.e.

$$dW \leqslant p\,dV$$

When a system undergoes an irreversible process between an initial equilibrium state and a final equilibrium state the change of entropy may nevertheless be calculated from

$$\Delta S = S_2 - S_1 = \int_1^2 \frac{dQ}{T}$$

along any reversible path between the initial and final points.

The truth of this statement is best realized by taking any ideal gas with constant specific heats and calculating $\int_1^2 dQ/T$ along three different reasonable paths, each consisting of two consecutive processes, e.g. constant pressure–isentropic, constant volume–isentropic, or constant temperature–isentropic, and showing that the result is the same for all three paths.

EXAMPLE 7.5

One mole of oxygen undergoes a free expansion from a volume of 4 litres to one of 8 litres, the temperature being 20°C. What is the entropy change (*a*) of the oxygen, (*b*) of the Universe, (*c*) of the oxygen if the expansion were performed reversibly and isothermally?

SOLUTION

$$G = 8{\cdot}3143 \text{ kJ/kg-mole-K}$$

Δs is calculated along a reversible isothermal path.

$$\Delta s = s_2 - s_1 = nR \ln\left(\frac{v_2}{v_1}\right)$$

$$= 1 \times 8{\cdot}3143 \times \ln 2$$

$$= 5{\cdot}645 \text{ kJ/K}$$

and will be the same for (*a*), and (*c*).

REFERENCES

7.1 Weir, C. D., *Irreversible Adiabatic State Changes with Special Reference to Steam Turbines*, Thermodynamics and Fluid Mechanics Convention, Cambridge, 1964, Paper 30 (Inst.Mech.Eng., London).

7.2 MAYHEW, Y. R. and ROGERS, G. F. C., *Thermodynamic Properties of Fluids and Other Data*, SI units (Oxford, Basil Blackwell) 3rd edition.
7.3 KEENAN, J. H. and KEYES, J. G., *Thermodynamic Properties of Steam* (New York, John Wiley, 1948).

EXERCISES

7.1 A closed system undergoes a reversible process from 220 K to 440 K with the heat transferred per degree in temperature being given by $dQ/dT = 0{\cdot}45$ kJ/K and the work by $dW/dT = 0{\cdot}12$ kJ/K. Find the change in internal energy and entropy for the process.
(*Ans:* 72·5 kJ, 0·311 kJ/K)

7.2 A tank having a volume of 0·14 m³ filled with a perfect gas is equipped with a stirrer. The state of the gas before rotation of the stirrer is $p = 1{\cdot}37$ bar, $t = 15$°C, after stirring $p = 2{\cdot}1$ bar. During the process 5·3 kJ of heat was irreversibly transferred to the surroundings at 5°C. Calculate the change of entropy of the gas if $c_p = 1{\cdot}05$ kJ/kg-K and $c_v = 0{\cdot}80$ kJ/kg-K, as well as the entropy change of the surroundings.
(*Ans:* 0·0913 kJ/kg; 0·0191 kJ/kg)

7.3 An ideal gas at 330 K with $\gamma = 1{\cdot}4$ in a cylinder is isentropically compressed from 0·7 bar to 4·2 bar. 2500 kJ/kg of heat are then added reversibly at constant temperature, followed by adiabatic expansion to the original temperature. The cycle is closed by reversible isothermal rejection of heat. The cycle efficiency is 0·30. Draw the cycle on a T–s diagram.
(*Ans:* $\Delta s_{12} = 0$, $\Delta s_{23} = 4{\cdot}54$ kJ/kg-K, $\Delta s_{34} = 0{\cdot}76$ kJ/kg-K)

7.4 Atmospheric air at 1 bar and 15°C is compressed isothermally to 14 bar, and then expanded adiabatically and reversibly back to atmospheric pressure. Given that $R = 0{\cdot}287$ kJ/kg-K, $c_p = 0{\cdot}992$ kJ/kg-K, deduce the final temperature and specific volume.
(*Ans:* 135·5 K, 0·387 m³/kg)

7.5 One kg of gas expands adiabatically and reversibly, its temperature falling from 240°C to 115°C while the volume is doubled. During the process, the gas does 90 kJ/kg of work. Determine the values of the specific heats c_p and c_v.
(*Ans:* 1·010 kJ/kg-K; 0·720 kJ/kg-K)

7.6 An air compressor draws in air at ambient conditions of 1 bar and 15°C, and delivers it at 7 bar. Determine the work done per kg of air pumped if the compression process is:

(*a*) isothermal,
(*b*) adiabatic,
(*c*) polytropic according to $pv^{1{\cdot}2}$ = constant.

Neglect any change in K.E. and P.E.
(*Ans:* (*a*) 161·0 kJ; (*b*) 215·5 kJ; (*c*) 190·2 kJ)

7.7 The index n to the expansion curve of an oil engine is 1·19 between the temperatures 1200°C and 1070°C. The mass of working fluid per cycle is 0·09 kg, and the properties may be assumed to be as for air ($\gamma = 1{\cdot}4$; $R = 0{\cdot}287$ kJ/kg-K). If the oil is supplied per cycle adds 144 Btu to the working fluid, find the percentage of this heat added to the fluid between the above temperatures.
(*Ans:* 6·44%)

7.8 1·8 kg of a gas (not air) having a volume of 0·51 m³ at 2·05 bar and 27°C are compressed adiabatically (reversibly) until the volume is reduced to one-third of the original, at which point the pressure is 8·56 bar. Find the value of the specific heats at constant pressure and constant volume.
Hint: Start with $s_2 - s_1 = 0$
(*Ans:* $c_v = 0{\cdot}0642$ kJ/kg-K; $c_p = 0{\cdot}835$ kJ/kg-K)

7.9 A perfectly lagged vessel is filled with air at a pressure of 0·14 bar and at a temperature equal to the atmospheric value of 15°C. Atmospheric air is allowed to enter the vessel until the pressure inside has risen to the atmospheric value of 1 bar. Determine the final temperature inside the vessel. For air take $R = 0{\cdot}287$ kJ/kg-K, $\gamma = 1{\cdot}4$, $c_p = 1{\cdot}009$ kJ/kg-K, $c_v = 0{\cdot}717$ kJ/kg-K.
(*Ans:* 380 K, 107°C)

7.10 An ideal reversible engine cycle is made up as follows: isothermal compression from P_1 to P_2, followed by expansion at constant pressure, V_2 to V_3, thereafter isentropic expansion back to P_1 at V_4, and finally a contraction, at P_1 from V_4 to V_1. Determine an expression for the efficiency in terms of the overall pressure ratio P_2/P_1 and the overall temperature ratio T_3/T_1: assume that the working fluid is a perfect gas. Proceed to show that, for a given value of T_3/T_1 the maximum efficiency occurs when $P_2/P_1 = (T_3/T_1)^{\gamma/\gamma-1}$. Illustrate the original cycle and the cycle of maximum efficiency on the T–S field and show that the maximum efficiency is given by

$$\eta = 1 - \frac{\ln T_3/T_1}{(T_3/T_1 - 1)}$$

7.11 In a Carnot cycle for 1 kg of air the volumes at the beginning and end of the upper isothermal stage and at the end of the succeeding adiabatic stage are in the proportion 3·7:10; the highest pressure is 21 bar and the highest temperature 556 K. Find (*a*) the values of pressure, volume, and temperature at the end points of the four stages and (*b*) the efficiency of the cycle.

Sketch the pressure–volume and temperature–entropy diagrams of the cycle.

(*Ans:*	21	9·0	5·44	13·77	bar
	556	556	483	483	K
	0·076	0·177	0·253	0·108	m³

$\eta = 0{\cdot}131$ or 13·1%)

7.12 Discuss briefly the significance of entropy as a thermodynamic property.

Show that, for 1 kg of a perfect gas, the entropy change for any process connecting states p_1, v_1, T_1 and p_2, v_2, T_2 is given by

$$s_2 - s_1 = c_v \ln \left(\frac{p_2}{p_1}\right) + c_p \ln \left(\frac{v_2}{v_1}\right)$$

From the above relationship deduce relationships between pressure, volume, and temperature for isentropic processes.

Determine the external heat addition as well as the entropy, enthalpy, and internal energy change for a diatomic perfect gas of molecular weight 28 undergoing heating from 35 bar and 315°C to 1100°C:

(*a*) at constant volume,
(*b*) at constant pressure.

(*Ans:* (*a*) $\Delta s = 0{\cdot}598$ kJ/kg-K $\quad \Delta u = \Delta q = 553$ kJ/kg
(*b*) $\Delta s = 0{\cdot}836$ kJ/kg-K $\quad \Delta h = \Delta q = 771$ kJ/kg
$\Delta u = 553$ kJ/kg)

Chapter 8

Properties of the Pure Substance

8.1 General Remarks

The pure substance has been defined in section 2.7 as a substance having one chemical structure but capable of existing in more than one phase.

It has also been noted that the state of a pure substance is defined by the values of two properties, and that the most widely used properties for this purpose are pressure p and specific volume v. All other properties are then uniquely determined in terms of the given properties. They may be found either by careful experiment or, in certain cases, by the use of suitable equations of state. The best known of these equations of state is that for the perfect gas, equation (3.2)

$$pv = RT \tag{8.1}$$

but this is an approximation to the actual relationship between the three properties pressure p, specific volume v, and temperature T, applicable only under certain clearly defined conditions to the gaseous phase of pure substances. In general the relationship between these three properties is represented in a state or phase diagram, as already briefly discussed in section 3.5. Such a representation requires a three-dimensional surface, known as a p–v–T surface, although in general phase diagrams are represented in two-dimensional form, with superimposed lines of constant values of properties other than those represented by the two axes. Thus in section 3.5 the phase diagram (*Fig.* 3.3) had lines of constant temperature, the axes representing pressure p and specific volume v respectively.

8.2 The p–v–T Surface for a Pure Substance (*Fig.* 8.1)

Water being the best known pure substance existing in the various possible forms of aggregation at or near normal temperatures, and

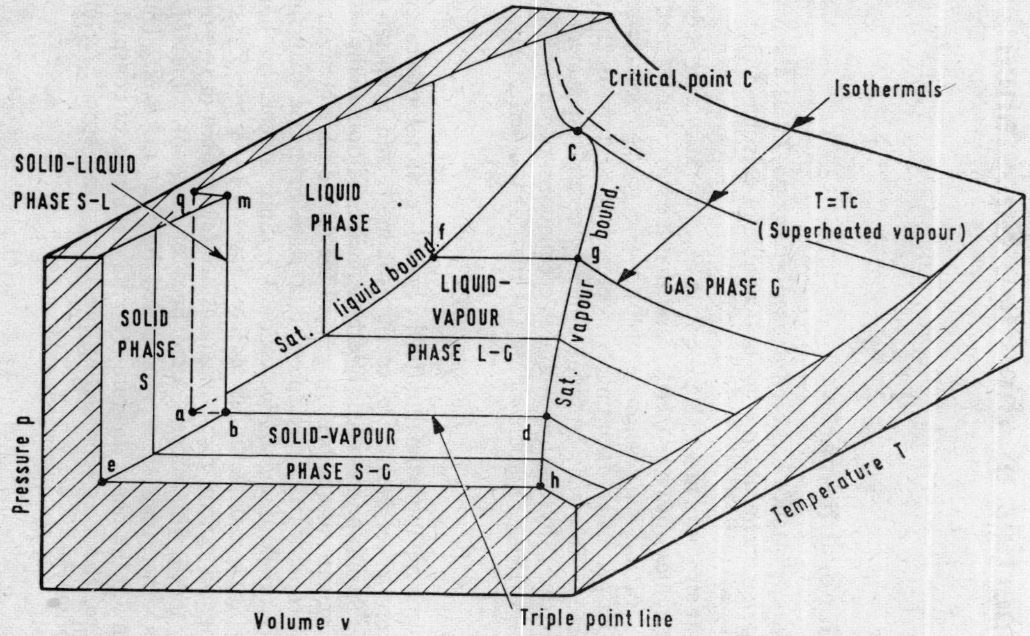

Fig. 8.1 p–v–T surface for H_2O

being very widely used in thermodynamic processes, illustrates the general behaviour of pure substances in the most readily intelligible manner. *Fig*. 8.1 illustrates the p–v–T surface for water.

In three regions, S, L, and G, water exists in a single phase, viz. solid (ice), liquid (water) and gas (steam) respectively. In each of these regions the state is uniquely defined by any two properties, e.g. p and v, p and T, v and T.

Between the single-phase regions are the two-phase regions, where two phases exist in equilibrium, viz. liquid–vapour, solid–vapour and solid–liquid. In practice, in the case of water, only the first of these is of interest, the latter two regions being omitted from most phase diagrams. Within the two-phase regions pressure and temperature are interdependent. Thus in the liquid–vapour phase any one value of pressure is associated with a unique value of temperature, these pairs of values being referred to as saturation pressure p_s and saturation temperature T_s. (The use of the term saturation derives from the fact that these values are associated with vapour existing in equilibrium with liquid, i.e. saturated with liquid.) Instead of pressure and temperature, specific volume with *either* pressure *or* temperature are the appropriate pairs of properties defining the state of a pure substance in the liquid–vapour or any other two-phase state.

Three phases exist in equilibrium along the line *abd*, the locus of the so-called triple point. Whilst, in the two phase regions, a fraction x of one phase implies the existence of a fraction $(1 - x)$ of the remaining phase, the fraction x being uniquely fixed by the relevant pair of property values, this is not the case along the triple point line where, for a fraction x of one phase, the fractions x' and x'' of the remaining two phases are not individually fixed by the pairs of property values v and T or v and p. (It may be shown that in this special case the state is uniquely determined by any pair of the properties entropy s, internal energy u, and specific volume v.)

In the important liquid–vapour phase, the liquid is in equilibrium with vapour and is known as saturated liquid. Along the phase boundary *ac* the vapour fraction is zero, this boundary being known as the SATURATED LIQUID LINE. Similarly the phase boundary *dc* is known as the SATURATED VAPOUR LINE.

The transitions from one phase to another, i.e. solid to liquid, solid to vapour, or liquid to vapour (known respectively as FUSION, SUBLIMATION and EVAPORATION) when the pressure is held fixed, are accompanied by significant volume changes and, in addition, by the absorption or release of heat, the relevant heat quantities, on a basis of unit mass of pure substance, being referred to as the LATENT HEATS of fusion, sublimation and evaporation, respectively. For fusion the volume change is indicated by the step between the lines *qa* and *mb*, for evaporation by the step from *ac* to *cd*, and for sublimation by the step between *eb* and *dh*, in each case at constant pressure.

8.3 Determination and Tabulation of Properties

It has already been shown in the previous section that the state of the pure substance is defined by the values of any two intensive properties for the homogeneous, i.e. single-phase system, and of two selected independent properties for the heterogeneous, i.e. two-phase system. Thus for pure substances in the liquid, solid or gaseous phases, it is usual to take pressure and temperature as the two properties defining the state, though either of these could be coupled with specific volume as the second defining property. In the various two-phase states, however, pressure and temperature are no longer independent and the relevant defining properties become *either* specific volume and pressure *or* specific volume and temperature.

However, although the *state* is thus determined by the values of only two properties, the *values* of the remaining properties cannot be deduced from the given two properties by purely mathematical reasoning, except in those cases where a simple equation of state can be formulated, e.g. the perfect gas. Thus, in general, it is necessary, in the case of single-phase states, to determine experimentally the relationships between pressure, specific volume and temperature, the resultant p–v–T data being tabulated. It remains to establish corresponding values of the extensive properties internal energy u, enthalpy h and entropy s, from these p–v–T data by mathematical analysis and/or experiment. In fact, a combination of the two methods is required except in the case of certain equations of state applicable only under certain clearly defined conditions, which permit the exclusive use of mathematical analysis (see Appendix II).

The experimental determination of u–h–s or related data involves the use of calorimetry. The determination of internal energy is made possible by the application of the energy equation (4.5), section 4.5, for the constant volume closed system,

$$m\Delta u_{12} = Q_{12} \tag{8.2}$$

i.e. the normal calorimeter system. Since the internal energy is always measured relative to an arbitrary datum level, p_0, T_0 (usually $p_0 = 0{\cdot}006112$ bar, $T_0 = 273{\cdot}16$ K or, in certain cases, 1 bar, 0°C), equation (8.2) may be applied to a constant volume process commencing from this datum level, so that for any state 1, defined by p_1, T_1, and for unit mass of substance:

$$u_1 = u_1 - u_0 = \frac{Q_{01}}{m} = q_{01} \tag{8.2a}$$

Equation (8.2a) applies only if there is no volume change between the datum state p_0, T_0 and the required state p_1, T_1. If a change in specific volume occurs, as is usually the case, the heating process required for the evaluation of internal energy has to be performed in two stages. Thus, considering the point 2 (*Fig.* 8.2) identified by

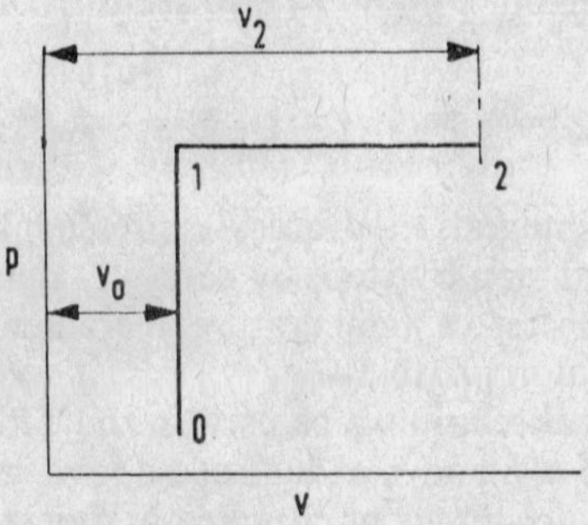

Fig. 8.2 Evaluation of internal energy

$p = p_2 = p_1$, $v = v_2$, the internal energy at 2 is obtained by first performing a constant volume heating process from datum state 0 to state 1, followed by a constant pressure heating process from state 1 to 2. For the former, equation (8.2a) is clearly applicable, while for the latter, applying equation (4.6) of section 4.5 for constant pressure heating,

$$Q_{12} = m\Delta u_{12} + p_1 \Delta V_{12} \tag{8.3}$$

or for unit mass of substance

$$\Delta u_{12} = q_{12} - p_1 \Delta v_{12} \tag{8.3a}$$

so that, combining equations (8.2a) and (8.3a)

$$u_2 = q_{01} + q_{12} - p_1 \Delta v_{12} = q_{01} + q_{12} - p_1(v_2 - v_0) \tag{8.4}$$

Equation (8.4) thus represents the most general relationship for the evaluation of internal energy of any pure substance, provided the datum level p_0, T_0 is associated with a single-phase state, as is usually the case. If the datum level corresponds to a two-phase state, it is necessary to specify the specific volume v_0 together with p_0 and T_0.

Having determined the internal energy for any state, such as state 2 (*Fig.* 8.2) experimentally, using equation (8.4), enthalpy h and entropy s may be derived mathematically from the equations defining these properties.

Thus, for enthalpy, using equation (2.17), section 2.10

$$h = u + pv$$

or for point 2

$$h_2 = u_2 + p_2 v_2 \tag{8.5}$$

while, for entropy, from equation (7.4), section 7.1

$$s_2 = s_2 - s_0 = \int_1^2 \left(\frac{dq}{T}\right)_{rev} = \int_0^1 \frac{dq}{T} + \int_1^2 \frac{dq}{T} \tag{8.6}$$

In general, the integration process expressed by equation (8.6) has to be performed step by step, by carefully determining heat increments dq for successive temperature increments dT, first along path 0–1 and then along path 1–2.

The only analytical relationships used in the determination of the extensive properties u, h and s, as outlined above, are equations (8.5) and (8.6) based on *definitions* of enthalpy and entropy, respectively. Hence the amount of experimental work required to obtain these u–h–s data is very great. In actual determination of u–h–s data the amount of experimental work is drastically reduced by making use of certain further analytical relationships, obtained from the so-called THERMODYNAMIC RELATIONS. These are treated in detail in Appendix II, being considered too advanced for incorporation in the text. As an example, however, the determination of entropy is greatly simplified by confining the calorimetric experiments, as opposed to the experiments associated with the determination of p–v–T data, to a constant pressure process at the reference pressure p_0, by the use of the following equation, derived from these thermodynamic relations—

$$s = s_0 + \int_{T_0}^{T} \left(\frac{c_p\, dT}{T}\right)_{p_0} - \left[\int_{p_0}^{p} \left(\frac{\partial v}{\partial T}\right)_p dp\right]_T \tag{8.7}$$

The second term on the right-hand side of equation (8.7) implies a series of calorimetric experiments at the reference pressure p_0, the third term a series of p–v–T determinations at the constant temperature T over the pressure range p_0 to p.

Further simplifications are possible by the use of other analytical relationships (see Appendix II).

(*i*) Property relationships in the liquid phase

Many liquids are virtually incompressible, except at very high pressures. Hence a single value of the specific volume, namely that on the saturated liquid boundary, v_f, may be used for any given temperature T, irrespective of the actual pressure p.

Again, the internal energy u is found to be essentially a function of temperature only, so that for tabulation purposes it is usually sufficient to give the internal energy value on the saturated liquid line, u_f, for the given temperature T, irrespective of pressure p. These remarks apply also to the enthalpy h and entropy s, for which the appropriate values of the saturated liquid line, i.e. h_f and s_f, are usually taken.

Summarizing, the quantities usually listed in saturated liquid tables are (ref. 8.1)

(*a*) Saturation pressure p
(*b*) Saturation temperature T
(*c*) Specific volume v_f
(*d*) Saturated liquid internal energy u_f
(*e*) Saturated liquid enthalpy h_f
(*f*) Saturated liquid entropy s_f

EXAMPLE 8.1

Read the saturated liquid values of internal energy, enthalpy and entropy for steam at (*a*) 1 bar, (*b*) 32 bar, and (*c*) 80°C.

(*a*) 417 kJ/kg, 417 kJ/kg, 1·303 kJ/kg-K
(*b*) 1 021 kJ/kg, 1 025 kJ/kg, 2·679 kJ/kg-K
(*c*) 334·7 kJ/kg, 334 kJ/kg, 1·075 kJ/kg-K.

(*ii*) Property relationships for saturated vapours

Properties of saturated vapours are derived from those of saturated liquids, by considering the constant temperature and pressure evaporation process connecting corresponding points on the saturated liquid line *ac* (*Fig.* 8.1) and on the saturated vapour line *cd* (*Fig.* 8.1). However, since this process takes place in a single p–v plane through the p–v–T surface represented by *Fig.* 8.1, it is simpler to show this process in a two-dimensional phase diagram. In fact two diagrams are conveniently used, viz. the p–v diagram (*Fig.* 8.3*a*) and the temperature–entropy or T–s diagram (*Fig.* 8.3*b*).

For a fixed saturation pressure and temperature, p and T, the following relationships hold, for the saturated vapour state,

$$\text{Specific volume: } v_g = v_f + v_{fg} \qquad (8.8)$$

$$\text{Internal energy: } u_g = u_f + u_{fg} \qquad (8.9)$$

$$\text{Enthalpy: } h_g = h_f + h_{fg} \qquad (8.10)$$

$$\text{Entropy: } s_g = s_f + s_{fg} \qquad (8.11)$$

where the suffix fg refers to the transition from the saturated liquid state f on the saturated liquid line ac (*Figs*. 8.3*a* and *b*) to the saturated vapour state g on the saturated vapour line cd (*Figs*. 8.3*a* and *b*).

The above properties are found in saturated vapour tables (ref. 8.1).

EXAMPLE 8.2

(*a*) Evaluate the enthalpy of saturated vapour at 25 bar.

At 24 bar, $h_g = 2802$ kJ/kg
At 26 bar, $h_g = 2803$ kJ/kg
At 25 bar, $h_g = 2802 + \frac{1}{2} \times 1 = 2802{\cdot}5$ kJ/kg

(*b*) Evaluate the enthalpy of saturated vapour at 21·7°C.

At 21°C, $h_g = 2539{\cdot}4$ kJ/kg
At 22°C, $h_g = 2541{\cdot}2$ kJ/kg
At 21·7°C, $h_g = 2539{\cdot}4 + 0{\cdot}7 \times 1{\cdot}8 = 2540{\cdot}66$ kJ/kg

(*c*) Evaluate the entropy of saturated vapour at 25 bar and at 21·7°C.

At 24 bar, $s_g = 6{\cdot}272$ kJ/kg-K
At 26 bar, $s_g = 6{\cdot}242$ kJ/kg-K
At 25 bar, $s_g = 6{\cdot}272 - \frac{1}{2} \times 0{\cdot}03 = 6{\cdot}257$ kJ/kg-K

At 21°C, $s_g = 8{\cdot}334$ kJ/kg-K
At 22°C, $s_g = 8{\cdot}297$ kJ/kg-K
At 21·7°C, $s_g = 8{\cdot}334 - 0{\cdot}7 \times 0{\cdot}037$
$= 8{\cdot}3081$ kJ/kg-K

The use of calorimetry may be avoided altogether in this case by making use of an analytical relationship valid only for phase changes such as the liquid–vapour transition, namely CLAPEYRON'S EQUATION, which in turn is derived from considerations based on the Carnot cycle (see section 7.3). Thus, considering a Carnot cycle operating between the saturated liquid and vapour lines, over a pressure and temperature range dp and dT (*Figs.* 8.3*a* and *b*) the following relationships hold—

In the p–v diagram (*Fig.* 8.3*a*) $\oint dw = \oint p\,dv = v_{fg}\,dp$ (i)

In the T–s diagram (*Fig.* 8.3*b*) $\oint dq = \oint T\,ds = s_{fg}\,dT = \dfrac{h_{fg}}{T}\,dT$ (ii)

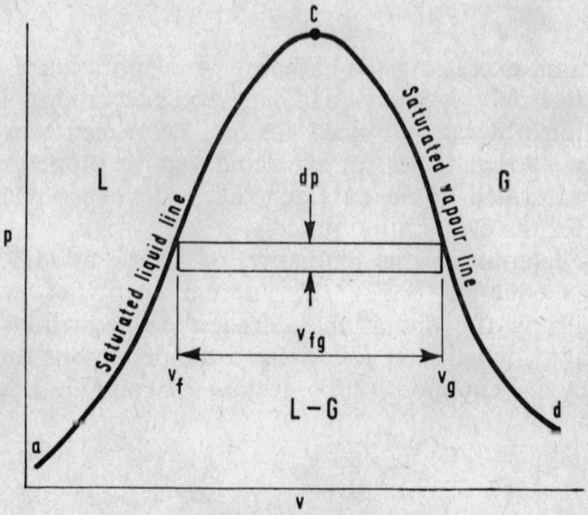

(a)

Fig. 8.3a p–v diagram for steam

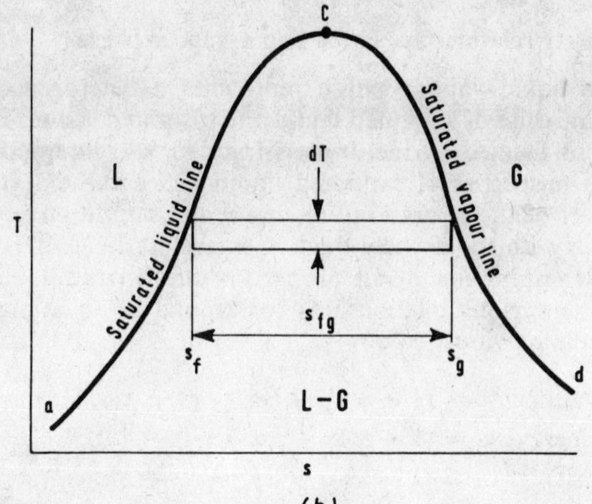

(b)

Fig. 8.3b T–s diagram for steam

But for a cyclic process, from the first law equation (4.2a), section 4.3

$$\oint dw = \oint dq \tag{iii}$$

so that, substituting from equations (i) and (ii) in (iii)

$$h_{fg} = v_{fg}T\frac{dp}{dT} \tag{8.12}$$

This equation, expressing the enthalpy of evaporation h_{fg} in terms of p–v–T data only, is known as Clapeyron's equation. The use of this equation obviates the need for any calorimetric experiments whatsoever, so that tables for saturated vapour properties may be based on saturated liquid data together with experimental p–v–T data only for the evaporation process.

Having determined the enthalpy of evaporation, h_{fg}, from Clapeyron's equation (8.12), the internal energy of evaporation u_{fg}, and entropy of evaporation s_{fg}, required in equations (8.9) and (8.11), above, immediately follow from the definitions for enthalpy and entropy, i.e. equations (2.17), section 2.10 and (7.4), section 7.1, giving

$$u_{fg} = h_{fg} - pv_{fg} \tag{iv}$$

$$s_{fg} = \frac{q_{fg}}{T} = \frac{h_{fg}}{T} \tag{v}$$

(*iii*) Property relationships in the liquid–vapour phase

Since the liquid–vapour phase represents a heterogeneous state, namely a mixture of saturated liquid and saturated vapour in varying proportions defined by the dryness fraction x, a knowledge of the respective properties of saturated liquid and saturated vapour for given values of the saturation pressure p and saturation temperature T, together with the dryness fraction x (see section 3.6), is sufficient for the determination of all property values in the liquid–vapour phase. The various relationships correspond to equations (8.8) to (8.11) in the previous section—

Specific volume: $v = (1 - x)v_f + xv_g = v_f + xv_{fg}$ (8.13)

Internal energy: $u = (1 - x)u_f + xu_g = u_f + xu_{fg}$ (8.14)

Enthalpy: $h = (1 - x)h_f + xh_g = h_f + xh_{fg}$ (8.15)

Entropy: $s = (1 - x)s_f + xs_g = s_f + xs_{fg}$ (8.16)

EXAMPLE 8.3

Evaluate the enthalpy, entropy and internal energy of steam **at 28°C**, 0·99 dry.

SOLUTION

(*a*) Enthalpy $h = 117{\cdot}3 + 0{\cdot}99 \times 2434{\cdot}8$

$$= 2522{\cdot}1 - 0{\cdot}01 \times 2434{\cdot}8 = 2497{\cdot}752 \text{ kJ/kg}$$
$$\simeq 2497{\cdot}8 \text{ kJ/kg}$$

(*b*) Entropy $s = 0{\cdot}409 + 0{\cdot}99 \times 8{\cdot}085$

$$= 8{\cdot}494 - 0{\cdot}01 \times 8{\cdot}085 = 8{\cdot}41315$$
$$\simeq 8{\cdot}4132 \text{ kJ/kg-K}$$

(*c*) Internal energy $u_{fg} = u_g - u_f$

$$= 2393{\cdot}2 - 116{\cdot}9 = 2276{\cdot}3 \text{ kJ/kg}$$
$$u = 2393{\cdot}2 - 0{\cdot}01 \times 2276{\cdot}3$$
$$= 2370{\cdot}4 \text{ kJ/kg}$$

(*iv*) **Property relationships in the vapour phase**

To avoid confusion with saturated vapours, this phase is usually referred to as the superheated vapour phase, or occasionally, as in the introduction to this chapter, the gaseous phase, i.e. region *G* in *Fig.* 8.1.

The vapour phase, unlike the liquid–vapour phase, represents a homogeneous state. However, whereas in the first homogeneous state considered, i.e. the liquid phase, the substance was virtually incompressible, and the extensive properties internal energy *u*, enthalpy *h*, and entropy *s*, were essentially functions of temperature only, neither of these conditions applies in the vapour phase.

Vapours are highly compressible, and the properties *u*, *h* and *s* are all strongly dependent not only on temperature but also on pressure. The evaluation of property values in the vapour phase necessitates both calorimetric experiments and experiments for the determination of *p–v–T* data.

Some reduction in the amount of experimental work required is again possible by the use of the thermodynamic relations (Appendix II).

The following property values are usually listed in (superheated) vapour tables—

For a given pressure *p*, a series of temperatures *T* above the saturation temperature, and for each of the latter, the following:

Specific volume *v* Enthalpy *h* Entropy *s*

Internal energy values are derived from the relationship

$$u = h - pv$$

EXAMPLE 8.4

Evaluate the specific volume, enthalpy, and entropy of steam vapour at 16 bar and 290°C.

At 250°C and 15 bar

$$v = 0{\cdot}1520 \text{ m}^3/\text{kg}$$
$$h = 2925 \text{ kJ/kg}$$
$$s = 6{\cdot}711 \text{ kJ/kg-K}$$

At 250°C and 20 bar

$$v = 0{\cdot}1115 \text{ m}^3/\text{kg}$$
$$h = 2904 \text{ kJ/kg}$$
$$s = 6{\cdot}547 \text{ kJ/kg-K}$$

At 250°C and 16 bar

$$v = 0{\cdot}1520 - \tfrac{1}{5} \times 0{\cdot}0405 = 0{\cdot}1439 \text{ m}^3/\text{kg}$$
$$h = 2925 - \tfrac{1}{5} \times 21 = 2920{\cdot}8 \text{ kJ/kg}$$
$$s = 6{\cdot}711 - \tfrac{1}{5} \times 0{\cdot}164 = 6{\cdot}6782 \text{ kJ/kg-K}$$

At 300°C and 15 bar

$$v = 0{\cdot}1697 \text{ m}^3/\text{kg}$$
$$h = 3039 \text{ kJ/kg}$$
$$s = 6{\cdot}919 \text{ kJ/kg-K}$$

At 300°C and 20 bar

$$v = 0{\cdot}1255 \text{ m}^3/\text{kg}$$
$$h = 3025 \text{ kJ/kg}$$
$$s = 6{\cdot}768 \text{ kJ/kg-K}$$

At 300°C and 16 bar

$$v = 0{\cdot}1679 - \tfrac{1}{5} \times 0{\cdot}0442 = 0{\cdot}16086 \text{ m}^3/\text{kg}$$
$$h = 3039 - \tfrac{1}{5} \times 14 = 3036{\cdot}2 \text{ kJ/kg}$$
$$s = 6{\cdot}919 - \tfrac{1}{5} = 6{\cdot}8888 \text{ kJ/kg-K}$$

Finally interpolating between 250°C and 300°C at 16 bar:

At 290°C and 16 bar we have

$$v = 0{\cdot}1439 + \tfrac{40}{50} \times (0{\cdot}16086 - 0{\cdot}1439) = 0{\cdot}15746 \text{ m}^3/\text{kg}$$
$$h = 2920{\cdot}8 + \tfrac{4}{5}(3036{\cdot}2 - 2920{\cdot}8) = 2943{\cdot}88 \text{ kJ/kg}$$
$$s = 6{\cdot}6782 + \tfrac{4}{5}(6{\cdot}8888 - 6{\cdot}6782) = 6{\cdot}7203 \text{ kJ/kg-K}$$

8.4 The Pressure–Volume (p–v) Phase Diagram (*Fig.* 8.4)

This is the diagram (*Fig.* 3.3) already discussed in sections 3.5 and 3.6.

The solid, liquid, and gaseous phases are denoted by *S*, *L*, and *G*. Temperature, instead of being represented by the third co-ordinate axis as in p–v–T surface (*Fig.* 8.1) is shown by a series of lines of constant temperature or isothermals. In addition, in accordance with equation (8.13), above, for the specific volume in the liquid–vapour (or WET VAPOUR) phase, lines of constant dryness fraction x are superimposed on the liquid–vapour region.

The critical isothermal is that temperature above which the substance can exist only as a (superheated) vapour, whereas below the critical isothermal, combined compression and abstraction of

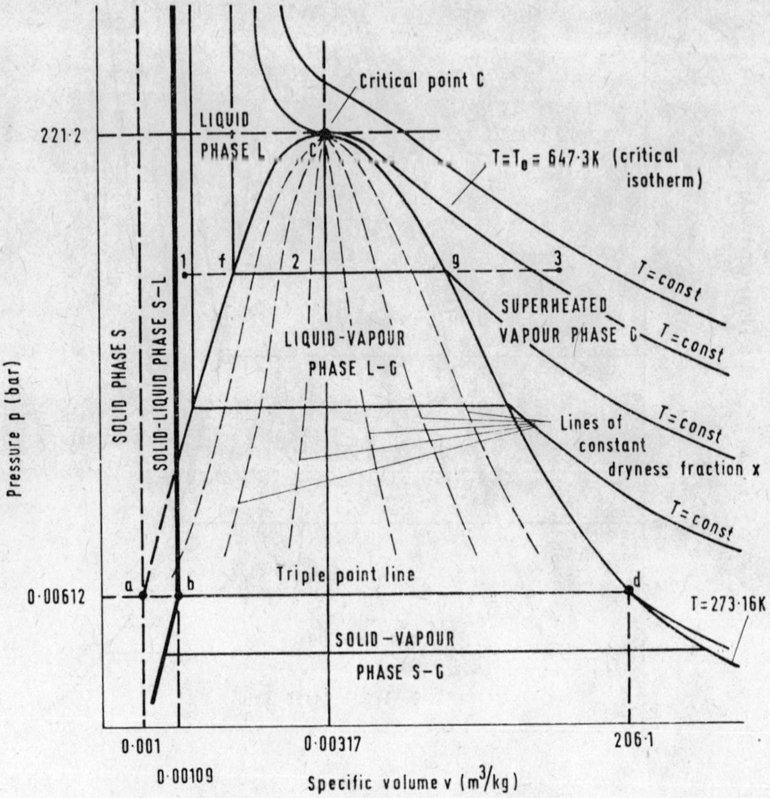

Fig. 8.4 p–v phase diagram for H_2O

heat results in transformation of the vapour first into a wet vapour, and ultimately into a liquid.

Superimposed on the diagram is the typical constant pressure heating process 1–*f*–2–*g*–3, already discussed in section 3.6.

8.5 The Temperature–Entropy (*T*–*s*) Diagram

(*Fig.* 8.5)

Since any two suitably chosen properties define the state of the pure substance, it is possible to employ temperature *T* and entropy *s*

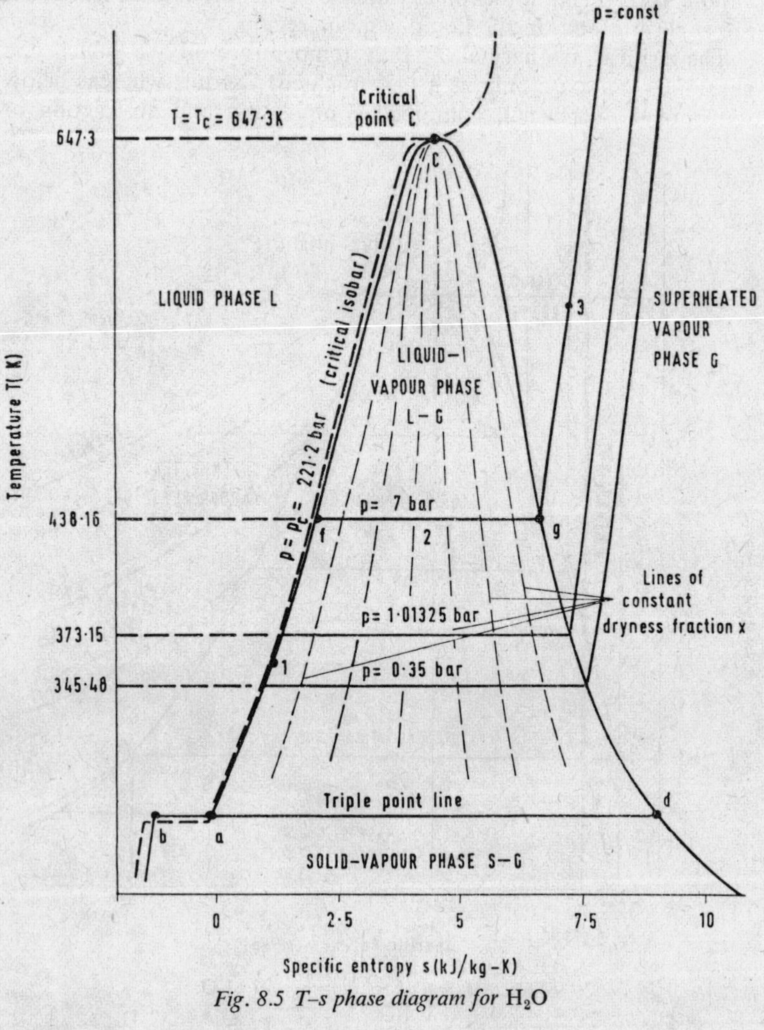

Fig. 8.5 T–s phase diagram for H_2O

as alternative axes for the phase diagram in place of pressure p and specific volume v.

This form of phase diagram has already been discussed in section 7.3, and has been shown to simplify the representation of many processes and, in particular, to identify irreversibilities in a more direct fashion than is the case with the pressure–volume diagram.

As with the p–v diagram, the various phases are clearly indicated, the same symbols being used in *Fig.* 8.5, both for phase regions and phase boundaries, as in *Fig.* 8.4. Superimposed on the temperature–entropy diagram are lines of constant pressure or ISOBARS and, in the liquid–vapour region, lines of constant dryness fraction x.

The typical constant pressure heating process discussed in section 3.6, i.e. 1–*f*–2–*g*–3, is also superimposed on *Fig.* 8.5.

8.6 The Enthalpy–Entropy (h–s) Diagram
(*Fig.* 8.6)

This is a very widely used form of phase diagram, particularly in the case of cycle calculations involving flow processes. Its usefulness is immediately apparent from the form of the steady flow energy equation, i.e. equation (4.13b), section 4.6

$$q_{net} - w_{net} = h_{net} - h_1 + g(Z_2 - Z_1) + \tfrac{1}{2}(c_2^2 - c_1^2) \quad (8.17)$$

Since such cycles usually incorporate, in their ideal form, isentropic processes, e.g. expansion in nozzles or turbines, and since gravitational effects are usually negligible, equation (8.17) reduces to the following simpler forms, the entropy being constant in each case—

(*a*) Nozzle: $0 - 0 = h_2 - h_1 + \tfrac{1}{2}(c_2^2 - c_1^2)$

or $$\Delta KE = -\Delta h \quad (8.18)$$

(*b*) Turbine (neglecting KE changes):

$$-w_{net} = h_2 - h_1$$

or $$w_{net} = -\Delta h \quad (8.19)$$

(*c*) Compressor (neglecting KE changes):

$$w_{net} = \Delta h \quad (8.20)$$

Constant pressure heating or cooling processes are equally common (e.g. boilers and condensers in steam plant). Applying equation (8.17) in the absence of kinetic energy and gravitational effects, as well as external work,

$$q_{net} = h_2 - h_1 = \Delta h \quad (8.21)$$

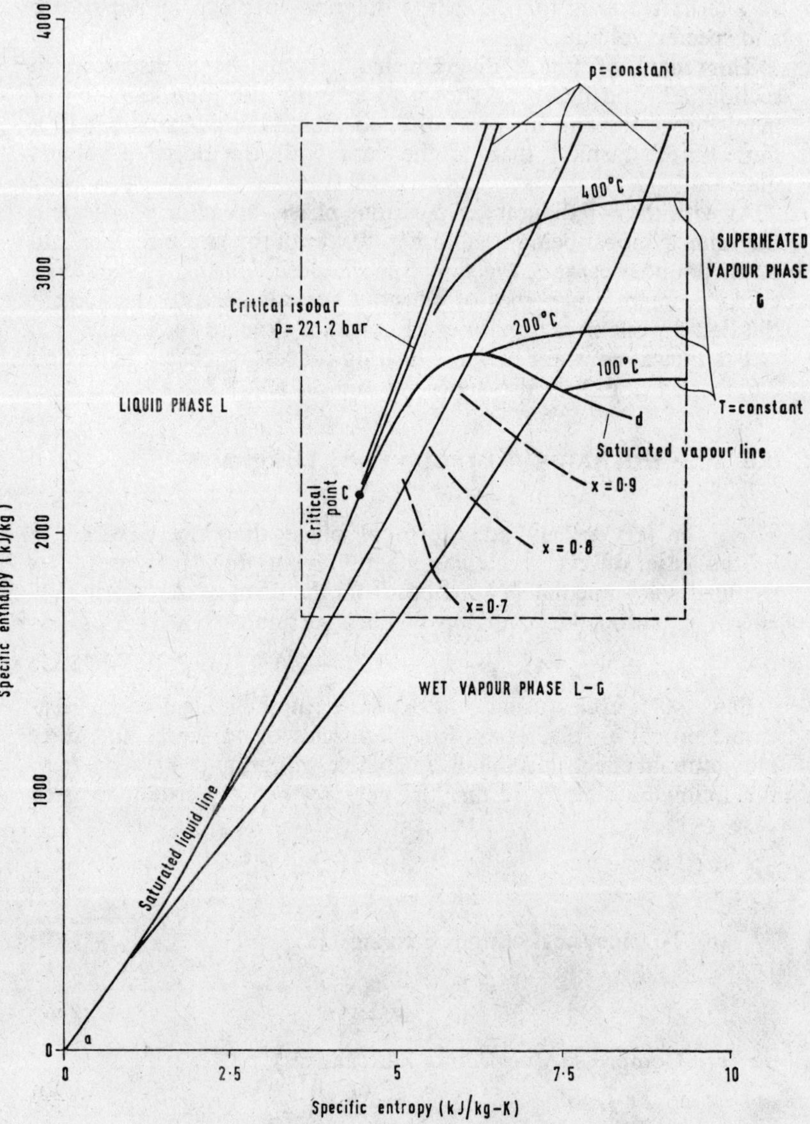

Fig. 8.6 h–s phase diagram for H_2O

Enthalpy–entropy (h–s) or MOLLIER DIAGRAMS for steam provide very full information concerning property values, namely

(*a*) Lines of constant pressure
(*b*) Lines of constant temperature
(*c*) Lines of constant dryness fraction x in liquid–vapour phase
(*d*) Lines of constant specific volume (not shown in *Fig*. 8.6).

It should be noted that, in general, Mollier charts for steam cover only the liquid–vapour and superheated vapour phases. Enthalpy and entropy values for the liquid phase must therefore be obtained from tables.

EXAMPLE 8.5

A throttling process is one of constant enthalpy. By Mollier or h–s chart determine the condition of the steam after throttling which is initially at 7 bar and 0·92 dry and finally at 0·05 bar.

SOLUTION

Locate the point 7 bar and 0·92 dry on the chart. From there travel along the constant enthalpy (a horizontal) line to the pressure line of 0·05 bar. It is seen to be in the superheat region at a temperature of about 53°C.

To evaluate the ideal amount of work obtainable from a steam turbine per pound of steam it is only necessary to evaluate the isentropic enthalpy drop between the supply conditions and the exhaust conditions.

For example, a turbine receives steam at 10 bar and 350°C and exhausts it at 0·06 bar. Evaluate the isentropic heat drop by means of the Mollier diagram of h–s chart (see Fig. 8.7).

$$h_1 = 3158 \text{ kJ/kg}$$
$$h_2 = 2248 \text{ kJ/kg}$$
$$\Delta h_{12} = 910 \text{ kJ/kg}$$

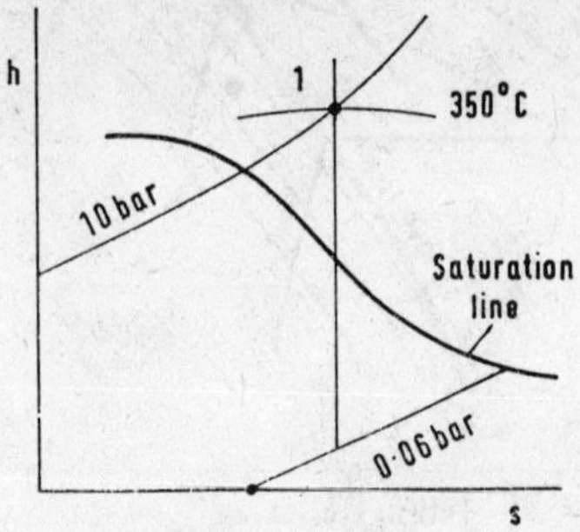

Fig. 8.7

If the amount of steam used by the turbine is 250 kg/min the horsepower output equals ideally

$$(250 \times 910)/60 = 3800 \text{ kW}$$

It will be obvious that an isentropic process can be treated in a similar way on the temperature–entropy chart as the throttling process was treated in the enthalpy–entropy chart.

8.7 The Pressure–Enthalpy (p–h) Diagram
(*Fig.* 8.8)

This is the last of the common forms of representation of the phase diagram and is particularly appropriate to refrigeration or heat pump cycles. Superimposed on the diagram are

(*a*) Lines of constant temperature

(*b*) Lines of constant entropy

(*c*) Lines of constant dryness fraction x in liquid–vapour phase

(*d*) Lines of constant specific volume.

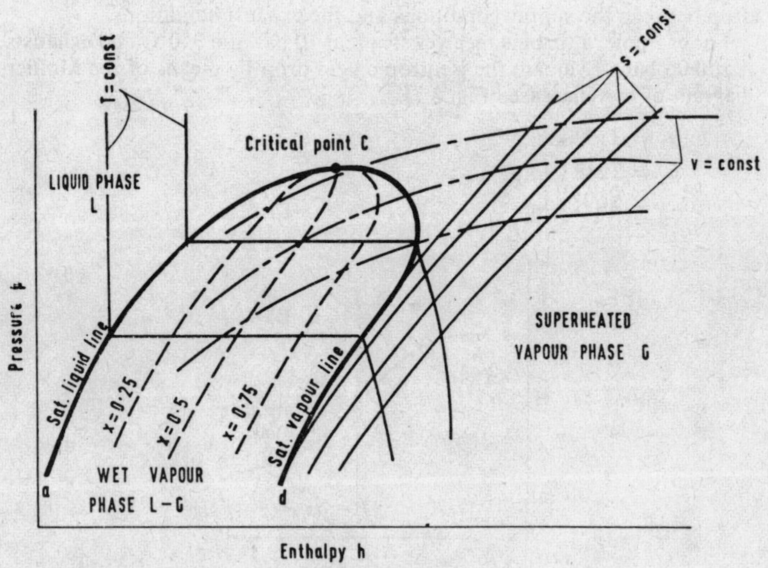

Fig. 8.8 p–h phase diagram

8.8 The Temperature–Pressure (T–p) Diagram
(*Fig.* 8.9)

This diagram is used mainly to show pressure–temperature relationships for the three types of phase change already discussed in section 8.2, namely fusion, sublimation and evaporation. Since, during these processes, pressure and temperature are uniquely

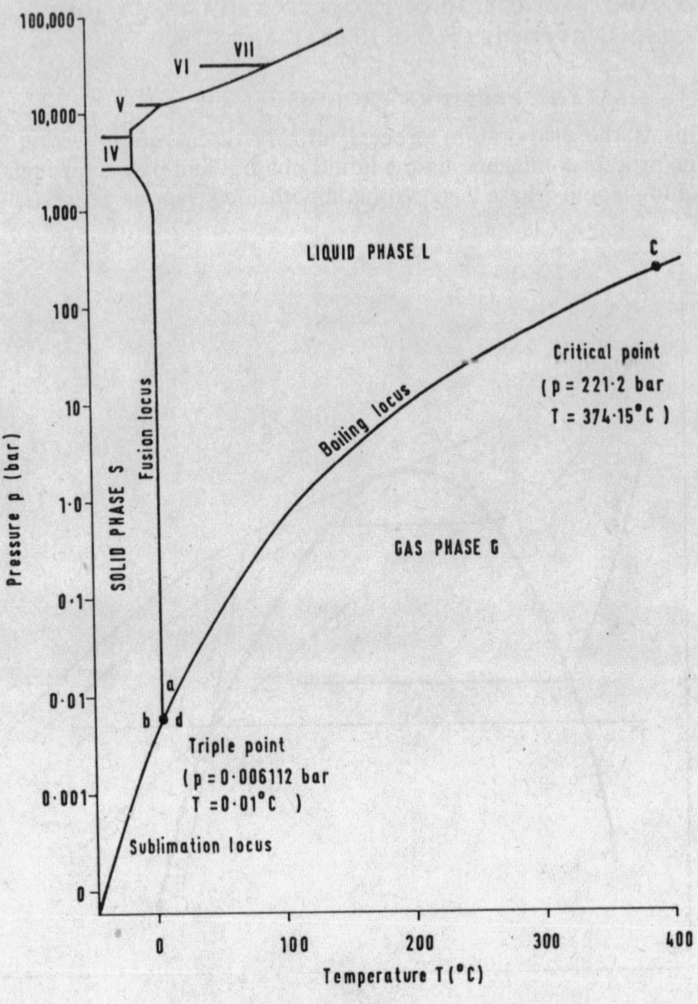

Fig. 8.9 T–p phase diagram for H_2O

related, they are represented by lines rather than areas, as in the case of the phase diagrams discussed earlier. Furthermore, the temperature–pressure diagram clearly indicates the existence of sub-phases, or allotropic forms. It will be noted that the triple point now appears as a single point.

8.9 Representation of Processes in the Various Phase Diagrams (*Figs.* 8.10, 8.11, and 8.12)

(*a*) Constant pressure process 1–*f*–2–*g*–3

This is the evaporation process already discussed in section 3.6. The process commences in the liquid phase *L* and passes through the liquid–vapour phase *L–G* to the superheated vapour phase *G*.

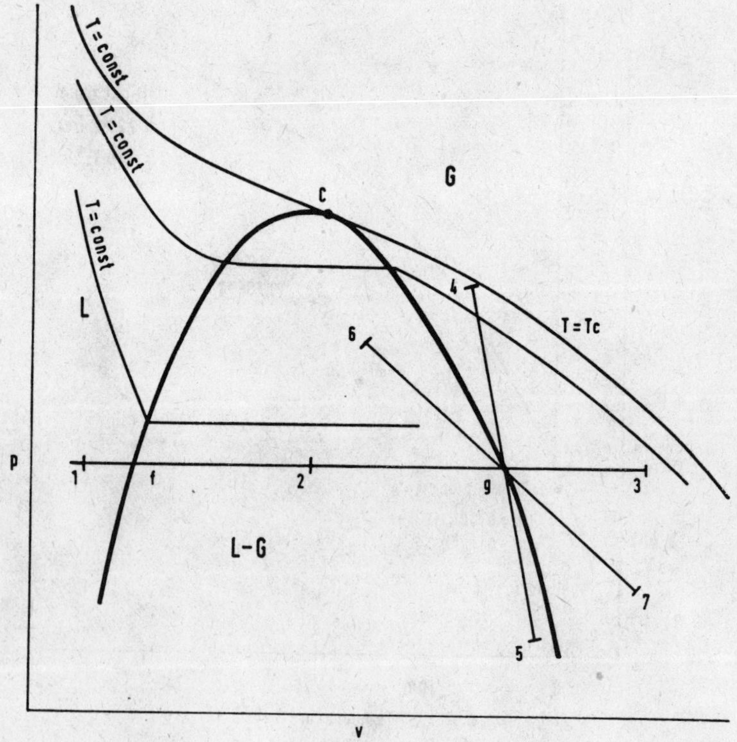

Fig. 8.10 p–v diagram

(*b*) CONSTANT ENTROPY PROCESS 4–*g*–5

This is the process which, ideally, occurs in nozzles or turbines (equations (8.18) and (8.19), section 8.6) using initially superheated vapour. Provided the pressure ratio p_4/p_5 is sufficiently large, the

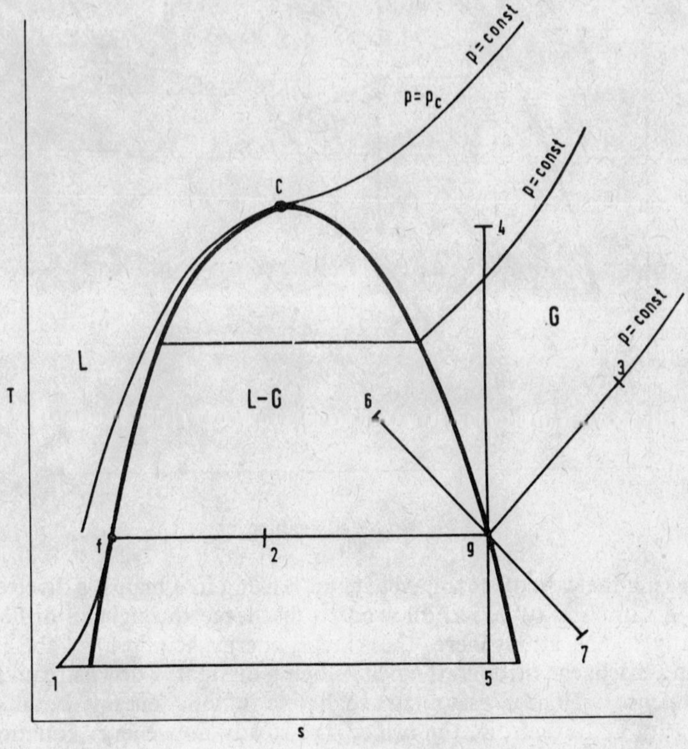

Fig. 8.11 T–s diagram

terminal point 5 lies in the liquid–vapour phase, the final dryness fraction being x_5 and being determined by the condition of constant entropy*: $s_5 = s_4 = (s_f)_5 + x_5(s_{fg})_5$ (i)

(*c*) CONSTANT ENTHALPY PROCESS 6–*g*–7

This process forms the basis for experiments for the determination of the dryness fraction x (in the case shown x_6) of wet vapour, i.e. vapour in the liquid–vapour phase using the throttling calorimeter.

* In the notation used in the following sections, viz. $(s_f)_5$, $(s_{fg})_5$, $(h_f)_6$, $(h_{fg})_6$, etc., the number suffix refers to pressure p_5, p_6, etc. as shown in Fig. 8.10. Thus, $(s_f)_5$ is the saturated liquid entropy at p_5; $(h_{fg})_6$ is the enthalpy of evaporation at p_6.

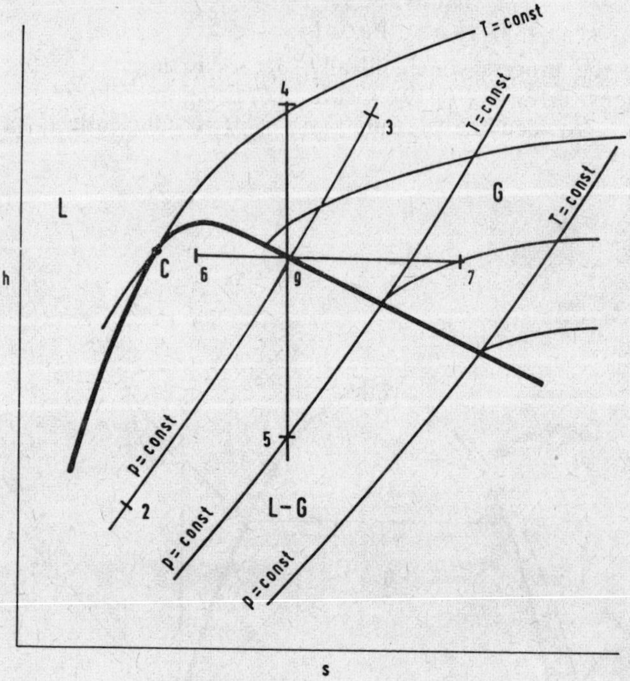

Fig. 8.12 h–s diagram

Taking the case of steam, wet steam flowing in a mains is diverted into a sampling tube and allowed to discharge through an orifice, usually to the atmosphere, the kinetic energy acquired by the jet during discharge through the orifice being dissipated downstream in turbulence. Hence, assuming negligible kinetic energy both at entry to, and exit from, the orifice, the steady flow energy equation (8.17) becomes

$$q_{net} - w_{net} = \Delta h + \Delta \text{ KE} + \Delta \text{ PE}$$

$$0 - 0 = \Delta h + 0 + 0$$

or
$$\Delta h = 0 \tag{8.22}$$

Applying equation (8.22) to the throttling process 6–*g*–7 commencing in the liquid–vapour phase and terminating in the superheated vapour phase,

$$h_7 = (h_f)_6 + x_6(h_{fg})_6$$

or
$$x_6 = \frac{h_7 - (h_f)_6}{(h_{fg})_6} \tag{8.22a}$$

The quantities to be determined experimentally are the initial temperature T_6 (or pressure p_6) and the final temperature T_7 and pressure p_7. Saturated vapour tables for the condition 6 provide values for both the liquid enthalpy $(h_f)_6$ and the enthalpy of evaporation $(h_{fg})_6$ while superheated vapour tables provide the enthalpy value h_7.

Note: It is essential that the vapour leaving the orifice be superheated. This limits the use of the throttling calorimeter to relatively high initial dryness fractions.

(*d*) OTHER PROCESSES

Clearly an infinite number of processes is possible. All can be represented in each of the various phase diagrams listed, but depending on the particular process under consideration some diagrams are more suitable than others.

EXAMPLE 8.6

(1) Solve the problem in Example 8.5 by means of steam tables.

At 7 bar and 0·92 dry

$$h = 2764 - 0{\cdot}08 \times 2067 = 2598{\cdot}6 \text{ kJ/kg}$$

At 0·05 bar, $h = 2598{\cdot}6$ kJ/kg which is greater than h_g at 0·05 bar so the steam is superheated.

We now have to find in the superheated steam tables at 0·05 bar the temperature at which $h = 2\,598{\cdot}6$ kJ/kg.

It falls between 50°C and 100°C.

At 50°C, $h = 2594$ kJ/kg
At 100°C, $h = 2688$ kJ/kg

So the temperature required is

$$50 + \frac{2598{\cdot}6 - 2594}{2688 - 2594} \times 50 = 52{\cdot}5°\text{C}$$

(2) Using steam tables evaluate the ideal horsepower of the turbine in Example 8.5.

At 10 bar and 350°C

$h = 3158$ kJ/kg
$s = 7{\cdot}301$ kJ/kg-K

At 0·06 bar

$$s = 7{\cdot}301$$
$$= 0{\cdot}521 + x \times 7{\cdot}808$$
$$x = (7{\cdot}301 - 0{\cdot}521)/7{\cdot}808 = 0{\cdot}869$$
$$h = 152 + 0{\cdot}869 \times 2\,415 = 2\,252 \text{ kJ/kg}$$
$$\Delta h = 3158 - 2\,252 = 906 \text{ kJ/kg}$$

This compares with 910 kJ/kg by the charts, so the horsepower is approximately of the same value.

(3) Steam at 3·5 bar saturated is expanded isothermally to 0·05 bar. Find the condition of the steam after expansion, using the steam tables, and evaluate the property changes resulting.

At 3·5 bar saturated

$$h_g = 2732 \text{ kJ/kg}$$
$$v_g = 0{\cdot}5241 \text{ m}^3\text{/kg}$$
$$s_g = 6{\cdot}941 \text{ kJ/kg-K}$$
$$t = 138{\cdot}9°\text{C}$$

At 0·05 bar and 138·9°C

$$h = 2688 + \frac{38{\cdot}9}{50} \times 96 = 2762{\cdot}6 \text{ kJ/kg}$$

h increases by 30·6 kJ/kg

$$v = 34{\cdot}42 + \frac{38{\cdot}9}{50} \times 4{\cdot}62 = 38{\cdot}01 \text{ m}^3\text{/kg}$$

v increases by 37·486 m³/kg

$$s = 8{\cdot}768 + \frac{38{\cdot}9}{50} \times 0{\cdot}24 = 8{\cdot}9547 \text{ kJ/kg-K}$$

s increases by 0·0137 kJ/kg-K

REFERENCE

8.1 MAYHEW, Y. R. and ROGERS, G. F. C., *Thermodynamic Properties of Fluids and Other Data*, SI units (Oxford, Blackwell) 3rd edition.

EXERCISES

8.1 1 kg of steam at 7 bar occupies a volume of 0·24 m³. Determine the condition, enthalpy, and internal energy of the contents.
(*Ans:* 0·869 dry; 2517 kJ; 2346 kJ)

8.2 Determine the density of 1 kg of steam initially at 5·5 bar having a dryness fraction of 0·78.

If 100 kJ are added at constant pressure, determine the condition, density, and enthalpy for the final state.

(*Ans:* 3·73 kg/m³; 0·828 dry; 3·52 kg/m³; 2391 kJ/kg)

8.3 For the initial conditions of the previous question, determine the final condition, density, and enthalpy for the addition of 563 kJ at constant pressure.

(*Ans:* 44·5°C superheat; 2·57 kg/m³; 2851·5 kJ/kg)

8.4 Steam, initially at 10 bar and 0·85 dryness fraction, expands down to 3·5 bar according to the hyperbolic law pv = constant. Determine the final condition and change in temperature of the steam. What external work per kilogramme is done during the change? Find also the change in internal energy and the amount of heat supplied or rejected.

(*Ans:* 0·9 dry; −41°C; 173·6 kJ/kg; 42 kJ/kg; 215·6 kJ/kg)

8.5 In extension of Exercise 8.4, use the approximate condition law for saturated steam that $pv^{16/15} = 1{\cdot}745$ (p in bar, v in m³/kg) to determine the pressure to which the steam would have to expand, still following the hyperbolic expansion law, in order to attain complete dryness.

Use steam tables to check the result so obtained, and estimate what will now be the overall temperature change.

(*Ans:* 0·70 bar, −89·9°C)

8.6 Determine the enthalpy, entropy, and internal energy of 1 kg of steam at:

(*a*) 13 bar and saturated
(*b*) 80°C and saturated
(*c*) 16 bar and 296°C

(*Ans:* (*a*) 2787 kJ/kg; 6·495 kJ/kg-K; 2590 kJ/kg
(*b*) 2643·2 kJ/kg; 7·611 kJ/kg-K; 2481·7 kJ/kg
(*c*) 3027 kJ/kg; 6·8792 kJ/kg-K; 2774·9 kJ/kg)

8.7 Determine the density of steam initially at 17 bar having a dryness fraction of 0·8. If 300 kJ/kg of heat are supplied at constant pressure, determine the condition, density and enthalpy.

(*Ans:* 0·956 dry, 8·97 kg/m³, 2712 kJ/kg)

8.8 Steam enters a radiator at 1·5 bar and 150°C, is condensed and leaves the radiator at the atmospheric pressure of 1 bar. What is the heat given out by the radiator per kg of steam condensed?

(*Ans:* 2356 kJ)

8.9 A steam main, closed at both ends has an internal volume of 1·5 m³ and is filled with saturated steam at 138·9°C. Determine the pressure in the main and the heat lost by steam when its quality is 0·7.

(*Ans:* 2·39 bar, 1760 kJ)

8.10 Steam at 14 bar and dryness 0·98 flows through a valve in a steam main. The valve is partially closed and as a result the downstream pressure is reduced to 4 bar. Determine the condition of the steam downstream of the valve and the change of entropy for the process. Take steam velocities and heat loss as negligible.

(*Ans:* 5·4°C superheat; $\Delta s = 0{\cdot}539$ kJ/kg-K)

8.11 A steam main of internal diameter 10 cm carries wet steam at 8 bar. A separator and throttling calorimeter are fitted in order to determine the dryness fraction. During a 10 min test 0·14 kg of water are separated and 0·66 kg of steam are condensed in the calorimeter. The pressure and temperature in the calorimeter are 0·9976 bar and 102·2°C. Calculate the dryness fraction of the steam in the main and the main flow in kg/h when the main flow velocity is 2·44 m/s.

(*Ans:* 0·795 dry, 285 kg/s)

8.12 A boiler supplies steam at 11 bar and 0·96 dry to a reciprocating steam engine. Before entering the engine steam is throttled so that the actual pressure at the start of expansion is 1·5 bar. Expansion follows the law pv = constant and the exhaust pressure is 0·06 bar.

Use the steam tables to find the condition of the steam at the start of expansion and calculate the heat transfer during the expansion process. Check the result you obtain, using the Mollier diagram for steam, and make a sketch of the whole process in the enthalpy–entropy plane.

(*Ans:* Superheated $T = 134{\cdot}6$°C; 547 kJ/kg)

8.13 Represent the Carnot cycle for a condensable vapour, operating between temperature and pressure limits of T and $T + dT$, and p and $p + dp$, respectively, in suitable pressure–volume and temperature–entropy diagrams. From considerations of cycle work, and using the two diagrams, show that

$$\frac{dp}{dT} = \frac{h_{fg}}{Tv_{fg}}$$

where p = pressure T = temperature

h_{fg} = latent heat of evaporation

v_{fg} = specific volume change during evaporation.

Hence, using temperature, pressure, and volume data only, deduce the latent heat of evaporation of steam at 2 bar and compare the value thus found with the tabulated value.

Comment on the general importance of equations giving thermal properties in terms of temperature, pressure, volume data.

(*Ans:* 2180 kJ/kg, cf. 2202 kJ/kg)

8.14 Discuss briefly the behaviour of wet and superheated vapours and comment on the statement that the perfect gas is a limiting case of superheated vapour.

Superheated steam at a pressure and temperature of 0·75 bar and 313°C expands isentropically to a pressure of 0·05 bar. Find the final condition of the steam:

(*a*) using tables for superheated steam,
(*b*) treating steam as a perfect gas,

and comment on differences between the results obtained.

For triatomic gases: $\gamma = 1{\cdot}33$
atomic weights: H = 1
O = 16

(*Ans:* (*a*) dry saturated
(*b*) 0·982 dry)

8.15 Steam escapes to atmosphere through a leak in a main where the pressure is 14 bar; ambient pressure is 1 bar and the observed steam temperature in the immediate vicinity of the leak is 150°C. Use the steam tables provided to find the initial dryness fraction of steam in the main and sketch your solution in the *h–s* field.

Steam from the main expands adiabatically and reversibly in a small turbine the exhaust pressure of which is 1·1 bar. Use the Mollier chart to calculate the work done per kilogramme of steam and used find also the final dryness fraction.
(*Ans:* 0·993; 449 kJ; 0·844)

8.16 A fluid expands reversibly in a turbine from a pressure p_1 and specific volume v_1 to $p_1/3$ and $2v_1$, according to the law $(p + a)v = b$, a and b being constants. The fluid enters with a velocity $c = \frac{1}{2}(p_1v_1)^{\frac{1}{2}}$ and the inlet and outlet pipes have the same cross-sectional area. Calculate the work done per kilogramme of fluid passing through the turbine and the heat transfer, assuming that the internal energy is constant during expansion. (*Hint:* the equation defining the process path has two constants and the properties p and v are given for two points on the process path.)
(*Ans:* $0{\cdot}55p_1v_1$; $+\,0{\cdot}59p_1v_1$)

8.17 State the first law of thermodynamics for open systems with unsteady flow.

Steam from a supply of 15 bar and 250°C enters a drum having a volume of 1·4 m³ containing steam at 3·5 bar having a dryness fraction of 0·9. If the final pressure in the drum is 11·5 bar show that, assuming adiabatic conditions throughout, the final temperature of the trapped steam is approximately 265°C. Hence find the change of mass content in the drum.
(*Ans:* Mass increased by 2·86 kg)

8.18 Derive an expression for the first law of thermodynamics in its most general form for a system with simultaneous inflow and outflow, and comment on the significance of the various terms. From this derive simplified expressions:

(*a*) for an open system with steady flow, and gas as the working fluid,
(*b*) for the closed system.

Initially dry saturated steam at 14 bar is expanded adiabatically and reversibly to 1 bar. Find the work done per kg if the expansion takes place:

(*a*) in an engine cylinder,
(*b*) in a turbine.

Find also the total work per kg performed in the steam engine cylinder, neglecting valve and other losses, and compare it with that for the steam turbine.
(*Ans:* (*a*) 401 kJ/kg; (*b*) 453 kJ/kg)

8.19 With the aid of typical pressure–volume and temperature–entropy diagrams, describe qualitatively the main differences between permanent gases and wet vapours.
In terms of the appropriate quantities for the saturated liquid and saturated vapour, respectively, write expressions for:

(*a*) the specific volume,
(*b*) the internal energy,
(*c*) the enthalpy,
(*d*) the entropy,

of a wet vapour of dryness fraction x.
Dry saturated steam at 28 bar expands isentropically to 0·07 bar. Find the values of specific volume, internal energy, enthalpy, and entropy at the end of expansion, as well as the work done, assuming that the expansion takes place in a steam turbine.
(*Ans:* (*a*) 15·02 m^3/kg; 1823 kJ/kg; 1923 kJ/kg; 6·209 kJ/kg-K
(*b*) 16·45 m^3/kg; 1983 kJ/kg; 2099 kJ/kg; 6·759 kJ/kg-K)

8.20 What are the differences between open or flow systems and closed systems?
State the first law of thermodynamics in its most general form for open systems with unsteady flow and derive from this expression the corresponding expression for closed systems.
The clearance volume of a steam engine contains 0·1 kg of steam at 4 bar and 200°C when admission begins; steam from a supply at 15 bar and 250°C enters until the pressure is 10 bar. Assuming that the volume remains constant and that the process is adiabatic, find the mass which enters and show that the resultant temperature is approximately 317°C.
(*Ans:* 0·1 kg)

8.21 A Carnot cycle operates between 36 bar and 1 bar, the working fluid being steam which is saturated at the commencement of expansion. Find the efficiency, work ratio, and specific consumption of this cycle, and compare it with the performance of a cycle operating between the same temperature and pressure limits but using air as the working fluid, assuming it to have the properties of a perfect gas.

N.B. $\text{Work ratio} = \dfrac{\text{net work per cycle}}{\text{positive work per cycle}}$

(*Ans:* Steam: 28·4%; 0·82; 7·55 kg/kWh
Air: 27·96%; 0·680; 38·01 kg/kWh)

Chapter 9

The Kinetic Theory of Gases

9.1 The Perfect Gas

It has already been pointed out (section 3.5) that under certain conditions the behaviour of pure substances in the vapour phase approaches that for the perfect gas as defined by the state equation (3.2)

$$pv = RT \tag{9.1}$$

In particular this applies at temperatures more than twice the critical temperature, or alternatively at very low pressures compared with the critical pressure, provided that the state in question is sufficiently far removed from the saturated vapour condition.

Because of the existence of the above simple state equation and the further experimental fact that the internal energy of pure substances approaching the perfect gas condition expressed by equation (9.1) is a function of temperature only, other property relationships may be established for the perfect gas in analytical form, thus eliminating the need for extensive tables of properties or state diagrams, as discussed in the previous section.

The so-called permanent gases, i.e. air, hydrogen H_2, nitrogen N_2, oxygen O_2, carbon dioxide CO_2, etc., conform very closely to perfect gas behaviour under normal conditions of pressure and temperature, since they then satisfy the condition $T > 2T_c$ stated above, where T_c = critical temperature. Nevertheless, at sufficiently low temperatures or high pressures they depart radically from perfect gas behaviour, just as steam was seen to do under normal conditions of pressure and temperature.

A better insight into the behaviour of perfect gases, and reasons for departures from perfect gas behaviour in the case of real substances, may be gained from the kinetic theory of gases, here presented in simplified form.

9.2 Kinetic Theory of Gases—General Discussion

The simplified kinetic theory of gases is a reminder that, although we have treated the thermodynamic fluid throughout as a continuum, it is, in fact, composed of individual atoms or molecules, i.e. discrete particles. The behaviour displayed by the fluid in bulk is ultimately dependent on the properties of these individual particles.

The assumptions made in the formulation of the simple kinetic theory are as follows—

(*a*) The molecules are identical (i.e. the existence of isotopes is ignored).
(*b*) The molecules are negligibly small compared with the mean free path, i.e. the statistical mean distance travelled by molecules between successive encounters.
(*c*) There are no forces of attraction or repulsion between the molecules, i.e. all energy is in the form of kinetic energy, as opposed to potential energy.
(*d*) The molecules are perfectly elastic, i.e. there is no loss of kinetic energy during collisions. (The term molecule has been used to cover, at the same time, atoms in the case of monatomic gases.)

These assumptions will be shown to be appropriate to perfect gas behaviour and cannot apply strictly to any real gas.

It is instructive, at this stage, to introduce some numerical values. In Chapter 3 brief mention was made of Avogadro's law, according to which equal volumes of gases under identical conditions of pressure and temperature contain equal numbers of molecules.

It has been shown experimentally that the number of individual molecules of any gas in a volume of 1 m^3 under standard conditions of pressure and temperature is $2{\cdot}68 \times 10^{25}$. Similarly, 1 kg-mol (i.e. one kilogramme molecule) of any gas contains $5{\cdot}81 \times 10^{26}$ individual molecules, which is the Avogadro number. Under standard conditions 1 kg-mol occupies 22·41 m^3. The diameter of a molecule is of the order of $2\text{–}3 \times 10^{-10}$ m so that, using the above information on numbers of molecules per unit volume, the mean free path is of the order of ten molecular diameters, i.e. $2\text{–}3 \times 10^{-9}$ m.

Suppose the gas to be in equilibrium in a vessel at rest and let the velocity c of any molecule be resolved into the components c_x, c_y, and c_z along three mutually perpendicular co-ordinate axes. Consider an area A of a plane surface in the y–z plane, i.e. perpendicular to the x-direction forming part of the enclosure. The surface will experience a force, measurable in terms of pressure, i.e. force per unit area, due to the impact of successive molecules and to the resultant reversal of the velocity component c_x under the assumed conditions of perfectly elastic impact. The change of momentum due to a single impact thus becomes

$$F_x = 2Mc_x \tag{i}$$

where M is the mass of an individual molecule. To consider the total effect, the molecules are divided into groups according to their velocities. Let n be the number of molecules in unit volume of

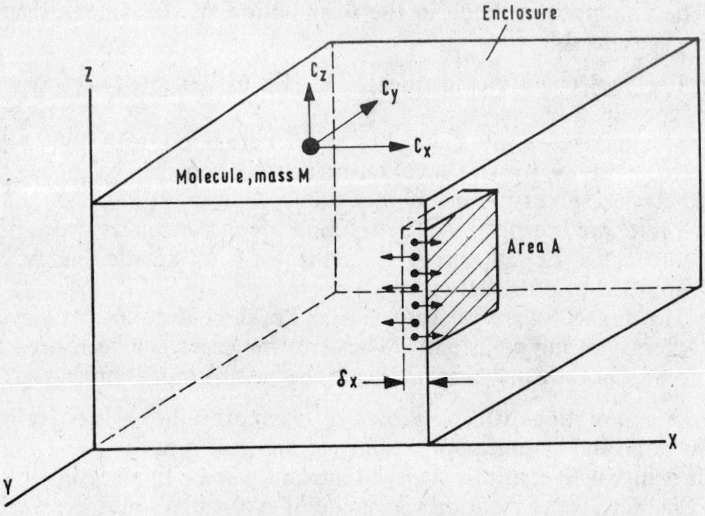

Fig. 9.1 Illustrating molecular motion

gas having an x-component of velocity of c_x, or differing from c_x by an amount not exceeding $\pm\frac{1}{2}\delta c_x$. There will be many such groups with different values of c_x. However, considering the particular group n, at any instant half the molecules within the group, i.e. $\frac{1}{2}n$, will be travelling towards the surface element A and half away from it. Within a small distance δx of the surface, the number of molecules approaching would be $\frac{1}{2}nA\,\delta x$, while a molecule travelling towards the surface, but initially separated by distance δx from it, would reach it in time

$$\delta t = \frac{\delta x}{c_x} \tag{ii}$$

provided that it does not encounter other molecules during the journey. However, such encounters have no effect on the pressure experienced by the surface since, under conditions of fully elastic impact, momentum is simply transferred from one molecule to another without loss. Hence the change in momentum at the surface A during time δt is given using equations (i) and (ii) by

$$\delta\,(\text{momentum}) = 2Mc_x\,(\tfrac{1}{2}nA\,\delta x) = nMc_x^2 A\,\delta t$$

Hence the change of momentum in unit time, i.e. the rate of change of momentum which may be equated to the force F_x on the surface, becomes

$$F_x = nMc_x^2 A \qquad \text{(iii)}$$

Since other groups, with different velocities, are involved, the pressure, i.e. force per unit area, due to all these groups becomes

$$p_x = \sum_{i=1}^{i=N} nMc_x^2 = M\sum_{i=1}^{i=N} nc_x^2 = MN\bar{c}_x^2 \qquad \text{(iv)}$$

where N = total number of molecules per unit volume and $\bar{c}_x$ = root mean square velocity for all molecules in the x-direction.

Now for any one molecule the resultant velocity becomes

$$c = \sqrt{(c_x^2 + c_y^2 + c_z^2)} \qquad \text{(v)}$$

the same reasoning applying also to the root mean square velocity—

$$\bar{c} = \sqrt{(\bar{c}_x^2 + \bar{c}_y^2 + \bar{c}_z^2)} = \sqrt{(3\bar{c}_x^2)} \qquad \text{(vi)}$$

since, statistically, the velocities are evenly distributed in all directions.

Substituting from equation (vi) in equation (iv), and since pressure is exerted uniformly in all directions

$$p_x = p = MN\bar{c}_x^2 = \tfrac{1}{3}MN\bar{c}^2 \qquad (9.2)$$

Now the product $M \times N$ represents the mass per unit volume, or the reciprocal of the specific volume, whence

$$pv = \tfrac{1}{3}\bar{c}^2 \qquad (9.3)$$

Equation (9.3) closely resembles equation (9.1), the perfect gas equation and suggests that the term $\frac{1}{3}\bar{c}^2 = \frac{2}{3} \times$ (kinetic energy per unit mass) in the former, assuming the motion of the molecules to be purely translational, corresponds to the term RT in the latter.

Since the internal energy u of the perfect gas is evidently related to the molecular kinetic energy level, and in the absence of intermolecular forces of any kind is equal to it, we may write

$$pv = \tfrac{1}{3}\bar{c}^2 = \tfrac{2}{3}\text{KE} = RT = \text{constant} \times u \qquad (9.4)$$

Thus absolute zero temperature is that temperature at which the molecular kinetic energy level is zero. The simple kinetic theory has thus given us a most valuable insight into the fundamental nature of temperature.*

The very interesting results obtained in this manner by applying the momentum equation derived from Newton's second law of motion illustrate the usefulness of statistical methods. The root mean square (r.m.s.) velocity $\bar{c}$ which appears in equations (9.2), (9.3) and (9.4) is derived from the statistical distribution of molecular velocities, assumed to follow a definite law such that the number of molecules n_0 and n_∞ (in a total number N) having respectively zero and infinite velocity is zero, while there is a maximum number n_{max} whose velocity is either equal to a particular value c or differs from it by a defined amount $\pm \frac{1}{2}\delta c$. By dividing the total number N into individual groups n_i whose velocity lies within the band $c_i \pm \frac{1}{2}\delta c$, we may draw a distribution curve, such as that shown in *Fig*. 9.2.

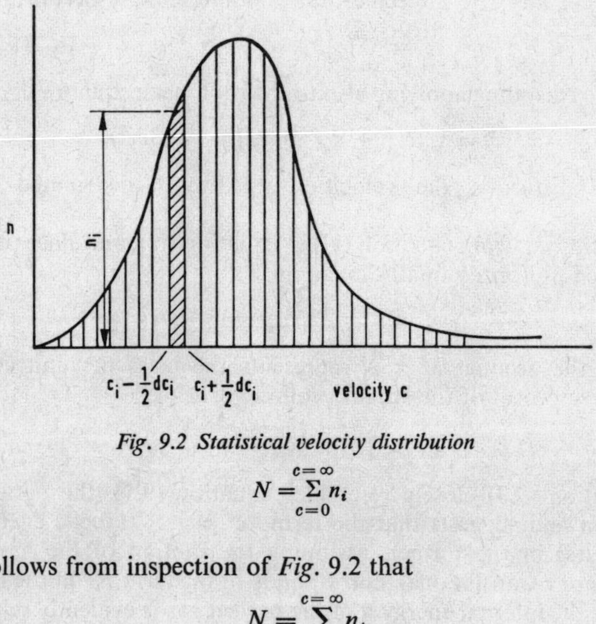

Fig. 9.2 Statistical velocity distribution

$$N = \sum_{c=0}^{c=\infty} n_i$$

It follows from inspection of *Fig*. 9.2 that

$$N = \sum_{c=0}^{c=\infty} n_i$$

whilst the root mean square velocity is clearly defined as

$$\bar{c}^2 = \sum_{c=0}^{c=\infty} \left(\frac{n_i c_i^2}{N}\right)$$

* Compare with the significance of absolute zero on the perfect gas scale of temperature (section 3.3) and the thermodynamic scale of temperature (section 6.5).

It was shown by Maxwell and confirmed by Boltzmann that the form of the distribution represented by *Fig.* 9.2 is of the exponential form

$$dn_i = \frac{4N}{\sqrt{\pi}}\beta^3 c^2 \exp(-\beta^2 c^2)dc$$

or

$$\frac{dn_i}{dc} = \frac{4N}{\sqrt{\pi}}\beta^3 c^2 \exp(-\beta^2 c^2) \qquad \text{(vii)}$$

where β is a distribution constant which, by methods beyond the scope of the present treatment but discussed in reference 9.1, may be shown to be inversely proportional to the square root of the absolute temperature T.

The form of the Maxwell–Boltzmann distribution function suggests an alternative method of plotting the distribution curve (*Fig.* 9.2) shown in *Fig.* 9.3. The area under the curve represents the total

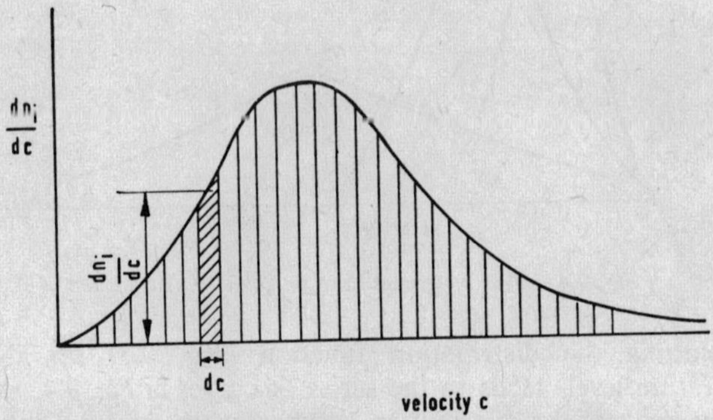

Fig. 9.3 Alternative presentation of statistical velocity distribution

$$N = \int_{c=0}^{c=\infty} \frac{dn_i}{dc}\,dc = \int dn_i = N$$

number of molecules whilst the root mean square velocity $\bar{c}$ is clearly obtained from the expression

$$\bar{c}^2 = \int_0^N \frac{c^2\,dn_i}{N} \qquad \text{(viii)}$$

Combining equations (vii) and (viii),

$$\bar{c} = \left[\frac{4\beta^3}{\sqrt{\pi}}\int_0^\infty c^4 \exp(-\beta^2 c^2)\,dc\right]^{1/2} \qquad \text{(ix)}$$

and making use of standard integration procedures

$$\bar{c}^2 = \frac{4\beta^3}{\sqrt{\pi}} \frac{3\sqrt{\pi}}{8\beta^5} = \frac{3}{2} \frac{1}{\beta^2} \tag{x}$$

Since β has already been stated to be inversely proportional to the square root of the absolute temperature, the above result, derived from more detailed statistical analysis merely confirms the deduction from equation (9.4), namely that the root mean square velocity itself varies as the square root of the absolute temperature.

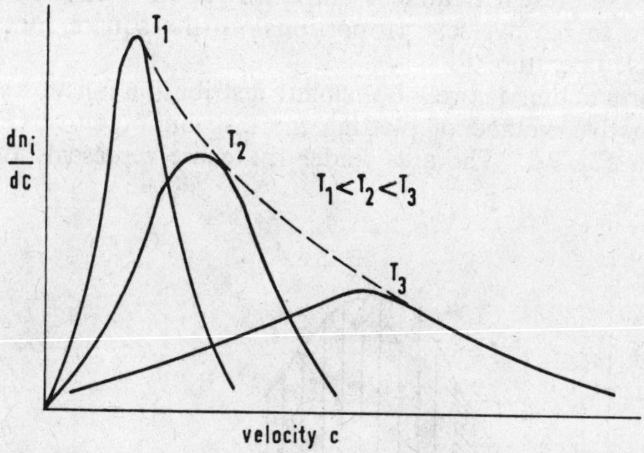

Fig. 9.4 Effect of temperature on velocity distribution

Replotting the distribution function (*Fig.* 9.3) for different temperature levels leads to the series of curves in *Fig.* 9.4, showing that the peak of the distribution curve is progressively displaced in the direction of higher velocities as the temperature increases.

9.3 Kinetic Energy and Specific Heats

From equation (9.4) above

$$RT = \tfrac{2}{3}\text{KE} = \tfrac{2}{3}u \tag{9.4a}$$

or

$$u = \tfrac{3}{2}RT \tag{9.4b}$$

In general terms,

$$u = f(T) \tag{9.4c}$$

It has already been shown in section 3.5 that, for the perfect gas,

$$du = c_v \, dT \tag{i}$$

and $$(c_p - c_v) = \frac{\gamma - 1}{\gamma} c_p = (\gamma - 1)\, c_v = R \tag{ii}$$

so that, differentiating equation (9.4b) and substituting for *du* and *R* from equations (i) and (ii)

$$du = c_v\, dT = \tfrac{3}{2}(\gamma - 1)c_v\, dT$$

or $$\gamma = c_p/c_v = \tfrac{2}{3} + 1 = 1{\cdot}667 \tag{9.5a}$$

This value of the ratio of specific heats has, in fact, been confirmed experimentally for monatomic gases, at moderate temperatures, indicating that for such gases the assumptions underlying the simple kinetic theory are substantially correct.

For gases with more complex molecular structures, however, the assumption that the internal energy of the gas derives solely from the translational kinetic energy of the molecules is no longer tenable. Thus, the diatomic molecule, in addition to the three degrees of freedom associated with translational motion in three perpendicular directions, has two further degrees of freedom effective for the storage of kinetic energy, namely those of rotation about two axes, both perpendicular to the axis of the molecule, the latter being considered as two concentrated masses separated by a distance, in the manner of a dumbbell. The energy associated with rotation about the molecule's own axis is insignificant due to the very much smaller moment of inertia about that axis. It follows from the principle of equipartition of energy (ref. 9.1) (the proof is beyond the scope of this book) and from equation (9.4b) that the kinetic energy associated with five degrees of freedom, as with diatomic molecules, becomes

$$u_2 = \frac{5}{3} u_1 = \frac{5}{3}\left(\frac{3}{2}\right) RT = \frac{5}{2} RT \tag{9.4d}$$

where suffixes 1 and 2 refer to monatomic and diatomic molecules, respectively.

Applying auxiliary equations (i) and (ii), above, the specific heat ratio γ for the diatomic molecules is given by

$$du_2 = (c_v)_2\, dT = \tfrac{5}{2}\,(\gamma_2 - 1)(c_v)_2\, dT$$

i.e. $$\gamma_2 = \left(\frac{c_p}{c_v}\right)_2 = \frac{7}{5} = 1{\cdot}4 \tag{9.5b}$$

At normal temperatures the value $\gamma = 1{\cdot}4$ has been confirmed experimentally for representative diatomic gases such as nitrogen, hydrogen and oxygen. At higher temperatures, however, the specific heat ratio decreases fairly rapidly due to the fact that the molecules now possess, in addition to translational and rotational kinetic energy, vibrational kinetic energy. The latter behaves in an altogether more complex manner than the first two forms of kinetic energy and the principle of equipartition of energy no longer applies.

For triatomic molecules, e.g. carbon dioxide and water vapour, there are at least six degrees of freedom, three of translation and three of rotation, and it may be shown that, for such molecules, at normal temperature

$$\gamma_3 = \left(\frac{c_p}{c_v}\right)_3 = \frac{4}{3} = 1{\cdot}333 \tag{9.5c}$$

Table 9.1

	GAS	γ	c_p/R	c_v/R	$(c_p - c_v)/R$
Monatomic gases	He	1·659	2·52	1·519	1·001
	Ne	1·640			
	A	1·670	2·52	1·509	1·008
	Xe	1·660			
Diatomic gases	H_2	1·410	3·42	2·438	0·9995
	O_2	1·401	3·52	2·504	1·004
	N_2	1·404	3·50	2·448	1·005
	CO	1·404	3·50	2·488	1·005
	NO	1·400	3·52	2·512	1·005
Triatomic gases	CO_2	1·304	4·40	3·38	1·027

In Table 9.1, values of γ are obtained experimentally from measurements of the acoustic velocity, i.e. the velocity of propagation of sound waves.

It may be shown that, for a perfect gas, the acoustic velocity is given by

$$a = (10^3 \gamma RT)^{1/2}$$

where γ = ratio of specific heats,
R = gas constant (kJ/kg-K),
T = absolute temperature (K),
a = acoustic velocity (m/s).

whilst values of the specific heat at constant pressure and constant volume, c_p and c_v, are obtained by calorimetry.

Agreement of γ with the values predicted by the simple kinetic theory is remarkably good. The values of c_p/R and c_v/R when compared with those predicted by application of equation (ii)

$$\frac{c_p}{R} = \frac{\gamma}{\gamma - 1} = 2{\cdot}50 \text{ for monatomic gases}$$
$$= 3{\cdot}50 \text{ for diatomic gases}$$
$$= 4{\cdot}0 \text{ for triatomic gases}$$

and

$$\frac{c_v}{R} = \frac{1}{\gamma - 1} = 1{\cdot}50 \text{ for monatomic gases}$$
$$= 2{\cdot}50 \text{ for diatomic gases}$$
$$= 3{\cdot}0 \text{ for triatomic gases}$$

again show remarkably good agreement, at least in the case of monatomic and diatomic gases. Agreement for triatomic gases is rather less good, and experiment shows that for gases having a more complex molecular structure, still greater divergences occur. This is due to the fact that the principle of equipartition of energy, whilst very adequate for monatomic and diatomic molecules having respectively three and five clearly defined degrees of freedom, cannot be applied with the same degree of confidence to more complex molecules.

9.4 Other Results Derived from the Kinetic Theory of Gases

(*a*) AVOGADRO'S LAW

For two gases at the same pressure, it follows from equation (9.2) that for equal volumes

$$p = \tfrac{1}{3} M_1 N_1 \bar{c}_1^{\,2} = \tfrac{1}{3} M_2 N_2 \bar{c}_2^{\,2} \qquad \text{(i)}$$

where N_1 and N_2 are the respective total number of molecules occupying the given volume.

It was shown by Maxwell that, at a given temperature, the average kinetic energy of different molecules is constant—

$$\tfrac{1}{2} M_1 \bar{c}_1^{\,2} = \tfrac{1}{2} M_2 \bar{c}_2^{\,2} \quad \text{or} \quad M_1 \bar{c}_1^{\,2} = M_2 \bar{c}_2^{\,2} \qquad \text{(ii)}$$

Substituting from equation (ii) in equation (i)

$$N_1 = N_2 \qquad (9.6)$$

Equation (9.6) is expressed verbally in the form of Avogadro's law—

EQUAL VOLUMES OF DIFFERENT GASES, AT THE SAME PRESSURE AND TEMPERATURE, CONTAIN EQUAL NUMBERS OF MOLECULES.

(*b*) THE GIBBS–DALTON LAW

It follows from the discussion in section 9.2 that for a vessel filled with a mixture of different gases, the pressure exerted on the containing walls is due to the combined momentum changes experienced by molecules of the different gases at the wall. Hence, from equation (9.2)

$$p = \tfrac{1}{3} M_1 N_1 \bar{c}_1^{\,2} + \tfrac{1}{3} M_2 N_2 \bar{c}_2^{\,2} + \dots \qquad (9.7)$$

But the individual terms on the right hand side of equation (9.7) represent, by comparison with equation (9.2), the pressure which each constituent gas would exert if it alone occupied the full volume of the vessel. This leads to the following statement, usually known as Dalton's law—

THE PRESSURE OF A GASEOUS MIXTURE IS THE SUM OF THE PRESSURES WHICH EACH COMPONENT WOULD EXERT IF IT ALONE OCCUPIED THE VOLUME OF THE MIXTURE AT THE MIXTURE TEMPERATURE.

9.5 Departures from Perfect Gas Behaviour

Real gases differ considerably from the hypothetical or perfect gas having the molecular properties listed in section 9.2. In particular, intermolecular forces, due both to gravitational and electrical field effects, are considerable while, at the same time, the volume occupied by the molecules cannot be completely neglected. These two effects result in a modification of the perfect gas equation (9.1) to include correction terms for both pressure and volume, resulting in van der Waals' equation

$$\left(p + \frac{a}{v^2}\right)(v - b) = RT \qquad (9.8)$$

Van der Waals (1873) made the premise that pressure would be increased by the reduced mean free path resulting from finite size of the molecules in the ratio total volume/actual volume, i.e.

$$\frac{p_{actual}}{p_{ideal}} = \frac{p_{actual}}{RT/v} = \frac{v}{v - b} \quad \text{whence} \quad p = \frac{RT}{v - b} \qquad \text{(i)}$$

while intermolecular attractive forces would lead to a reduction in pressure given by

$$p = \frac{RT}{v - b} - \frac{a}{v^2} \qquad \text{(ii)}$$

which transposes to the form of equation (9.8).

This equation, a cubic in v, can readily be represented in the p–v phase diagram (see section 8.4) (see Fig. 9.5).

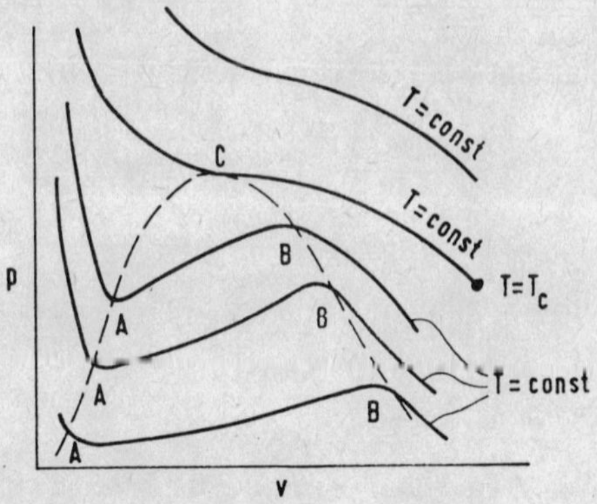

Fig. 9.5 Illustrating van der Waals' equation

The constants a and b may be determined in accordance with known conditions at the "critical point" C, i.e. by substitution of $p = p_c$, $v = v_c$ in equation (9.8) and in the differentiated form of equation (9.8), since experimentally $(dp/dv)_c = 0$.

The irregularities in behaviour demonstrated by van der Waals' equation are most pronounced at low specific volume, i.e. at high densities, and in this region isothermals plotted with the aid of equation (9.8) display distinct minima A and maxima B. These suggest at least a rough approximation to the sharp discontinuities in isothermals observed for real substances at the saturated liquid and saturated vapour boundaries (*Fig.* 8.4, section 8.4). Hence not only does the kinetic theory provide us with a physical understanding of perfect gas behaviour but indirectly also with that of non-perfect gases and vapour.

The perfect gas equation (9.1) is approached most closely by van der Waals' equation when the specific volume v is large, i.e. at very low pressures, as is borne out by experimental results.

It must be emphasized that van der Waals' equation is only one of many equations of state which have been proposed, but none of which completely expresses the behaviour of real substances, throughout the possible field of states.

EXAMPLE 9.1

The constants in van der Waals' equation (9.8) may be expressed in terms of the properties at the critical point as follows (see appendix III, section III.3)—

$$a = \frac{27}{64}\frac{R^2T_c^{\,2}}{p_c} \quad \text{and} \quad b = \frac{1}{8}\frac{RT_c}{p_c}$$

where R = gas constant $= \frac{G}{M}$ (see equation (3.7))

$$= \frac{8{\cdot}3143}{18} = 0{\cdot}462 \text{ kJ/kg-K}$$

$$= 0{\cdot}462 \times 10^3 \text{ J/kg-K}$$

For steam the critical conditions are

$$p_c = 221{\cdot}2 \text{ bar} = 2{\cdot}212 \times 10^7 \text{ N/m}^2$$
$$T_c = 374{\cdot}15^\circ\text{C} = 647{\cdot}15 \text{ K}$$

Using van der Waals' equation, determine the specific volume of steam under the following conditions:

(*a*) 80·0 bar, 700°C (973 K)

(*b*) 0·75 bar, 300°C (573 K)

SOLUTION

$$a = \frac{27}{64}\frac{(0{\cdot}462)^2(647{\cdot}15)^2 \times 10^6}{2{\cdot}212 \times 10^7} = 1{\cdot}701 \times 10^3$$

$$b = \frac{1}{8}\frac{0{\cdot}462 \times 10^3 \times 647{\cdot}15}{2{\cdot}212 \times 10^7} = 1{\cdot}69 \times 10^{-3}$$

Substitution in equation (9.8) gives

$$\left(p + \frac{1{\cdot}701 \times 10^3}{v^2}\right)(v - 1{\cdot}69 \times 10^{-3}) = 0{\cdot}462 \times 10^3 T$$

Case (*a*)

$$\left(80 \times 10^5 + \frac{1{\cdot}701 \times 10^3}{v^2}\right)(v - 1{\cdot}69 \times 10^{-3}) = 0{\cdot}462 \times 10^3 \times 973$$

By trial and error:

$$v = 5{\cdot}38 \times 10^{-2}\ \text{m}^3/\text{kg}$$

(cf. $5{\cdot}476 \times 10^{-2}$ m³/kg from tables)

Case (*b*)

$$\left(0{\cdot}75 \times 10^5 + \frac{1{\cdot}701 \times 10^3}{v^2}\right)(v - 1{\cdot}69 \times 10^{-3}) = 0{\cdot}462 \times 10^3 \times 573$$

By trial and error:

$$v = 3{\cdot}50\ \text{m}^3/\text{kg}$$

(cf. 3·521 m^3/kg from tables)

Since the constants a and b involve the gas constant R, usually expressed in kJ/kg-K, and pressure, usually expressed in bar, confusion can be avoided by reducing them to basic SI units. Hence R is expressed in J/kg-K and p_c in N/m^2. All other pressures must likewise be expressed in N/m^2.

The agreement with experimentally determined values is thus very good, and would be even better either at higher temperatures than for condition (*a*), or lower pressures than for conditions (*b*).

9.6 Statistical Treatment of Entropy

It was stated in the introduction to Chapter 7 that a completely satisfactory treatment of entropy was possible only in statistical terms. Since the statistical treatment of entropy involves certain concepts not dealt with in the present chapter, the matter is dealt with separately in Appendix I.

REFERENCE

9.1 LEE, J. F., SEARS, F. W. and TURCOTTE, D. C., *Statistical Thermodynamics* (Addison-Wesley).

EXERCISES

9.1 The kinetic theory of gases indicates that the mean kinetic energy of a monatomic gas molecule is related to temperature by the following equation:

$$\tfrac{1}{2}mc^2 = \tfrac{3}{2}kT$$

k is Boltzmann's constant and the other symbols have the conventional meaning. Show how it is possible, given the above information, to calculate the adiabatic index (γ) for a monatomic gas.

State what is meant by "equi-partition of energy" and use this principle to evaluate γ for a triatomic gas. How will the ratio of specific heats of real gases deviate from the predicted values of the kinetic theory at elevated temperatures?

9.2 The Maxwell-Boltzmann distribution function of molecular speeds can be written

$$\frac{dn_x}{dx} = F(x^2)$$

$$\text{where } x = \frac{c}{c_m}, \quad c_m = \sqrt{\frac{2kT}{m}}$$

$$F(x^2) = \left(\frac{4N}{\sqrt{\pi}}\right) x^2 c^{-x^2}$$

Compute the function $F(x^2)$ for values of x:

$$x = 0, 0{\cdot}2, 0{\cdot}4, 0{\cdot}6, 0{\cdot}8, 1{\cdot}0, 1{\cdot}5, 2{\cdot}0, 3{\cdot}0$$

and hence plot $F(x^2)$ against x.

Hence deduce the root mean square velocity $\bar{c}$ in terms of the most probable speed c_m.

(*Ans:* $\bar{c} = 1{\cdot}13c_m$)

9.3 Find the fractional number of molecules of a gas having velocities with x-components between $\bar{c}_m$ and $1{\cdot}01c_m$.

(*Ans:* $2{\cdot}08 \times 10^{-3}$)

9.4 Given the result, from kinetic theory,

$$\bar{c}^2 = \frac{3kT}{m}$$

where k = Boltzmann's constant ($1{\cdot}3803 \times 10^{-23}$ J/mol-K)
m = molecular mass (kg)
T = absolute temperature (K)
$\bar{c}$ = r.m.s. speed of molecules (m/s)

find the r.m.s. speed at 300 K of (*a*) oxygen and (*b*) hydrogen.

m for hydrogen = $3{\cdot}32 \times 10^{-27}$ kg/molecule

(*Ans:* (*a*) 482 m/s; (*b*) 1928 m/s)

9.5 Given the Boltzmann constant $k = 1{\cdot}3803 \times 10^{-23}$ J/mol-K in conjunction with the expression:

$$R = k/m$$

where R = specific gas constant
m = molecular mass ($3{\cdot}32 \times 10^{-27}$ kg/molecule for hydrogen)

find the specific heats at constant volume and constant pressure (*a*) for hydrogen; (*b*) for oxygen, assuming perfect gas behaviour and $\gamma = 1{\cdot}4$ for both gases.

(*Ans:* (*a*) $c_v = 2{\cdot}49$ kcal/kg-K; (*b*) $c_v = 0{\cdot}155$ kcal/kg-K;
$c_p = 3{\cdot}48$ kcal/kg-K; $c_p = 0{\cdot}217$ kcal/kg-K)

9.6 Compute the number of molecules per m^3 in a gas at 40°C:

(*a*) at 10^{-5} bar
(*b*) at 10^{-1} bar
(*c*) at 10^{3} bar

given the Avogadro number $n = 2{\cdot}68 \times 10^{25}$ molecules/m^3 under standard conditions (1·01325 bar, 0°C).

(*Ans:* (*a*) $2{\cdot}31 \times 10^{20}$; (*b*) $2{\cdot}31 \times 10^{24}$; (*c*) $2{\cdot}31 \times 10^{28}$)

9.7 A manned space vehicle enters a region of the stratosphere where the atmospheric pressure and temperature are 10^{-2} mm of mercury and 320 K respectively. Dry nitrogen is supplied to a high-speed auxiliary turbine, which is required to operate during the re-entry phase of the flight, from a high-pressure storage vessel mounted in the external frame of the vehicle. The vessel has a volume of 3×10^3 cm^3 and is charged initially to a pressure of 20 atmospheres. For successful operation the turbine requires an absolute minimum starting pressure of 15 atmospheres.

The material from which the vessel was manufactured is porous; the effect of the porosity is equivalent to a large number of microscopic holes piercing the wall of the vessel and totalling in area 10^{-6} cm^2. What will be the flight time before the pressure failure warning light is activated in the cabin? Make a list of the assumptions you consider necessary in order to obtain a solution. The following data may be assumed:

$$k = 1{\cdot}38 \times 10^{-23} \text{ J/mol-K}$$
$$N_0 = 6{\cdot}0251 \times 10^{26}$$
$$\text{mol. wt. } N_2 = 28$$
$$\bar{v} = 0{\cdot}925\sqrt{v^2}$$

(*Ans:* $\frac{p_t}{p_0} = -\frac{A}{V}\frac{1}{4}\bar{v}t$ where V = volume of container
$\bar{v} = 533{\cdot}9$ m/s
$t = 69{\cdot}4 \times 10^3$ sec $\simeq$ 19 h 16 min)

Chapter 10

The Perfect Gas

Note: The student who is new to the subject should concentrate, in the first instance, on sections 10.1 to 10.4 inclusive, treating the basic properties of perfect and semi-perfect gases.

The remaining sections of this chapter deal with details of particular processes and may be omitted at a first reading.

It was stated in Chapter 3, section 3.5 that under certain conditions the behaviour of pure substances in the gaseous phase approximates to that given by the equation of state for the perfect gas

$$pv = RT = f_1(T) \tag{10.1}$$

when *coherent* units are employed, or

$$pv = 10^{-2}RT = f_1(T) \tag{10.1a}$$

with p in bar (10^5 N/m^2), v in m^3/kg, R in kJ/kg-K (10^3 J/kg-K), T in K.

The conditions under which this approximation obtains were elucidated by the kinetic theory of gases in Chapter 9.

It is fortunate that the second of the alternative essential conditions for perfect gas behaviour, namely pressures well below the critical pressure p_c, or temperatures well above the critical temperatures T_c, should be satisfied by several important substances widely considered in engineering thermodynamics: the so-called permanent gases oxygen, hydrogen, nitrogen, air, carbon dioxide and several others, under normal conditions of pressure and temperature.

The use of the perfect gas equation (10.1) under these circumstances greatly simplifies the evaluation of property changes, and of heat and work transfers, during processes undergone by substances to which the equation may be applied.

10.1 Joule's Law—Internal Energy of the Perfect Gas

The kinetic theory of gases demonstrates two essential attributes of the perfect gas—

(*a*) The fact that it satisfies the state equation $pv = RT$.
(*b*) The fact that the internal energy is a function of temperature only, since no energy effects other than kinetic energy effects are possible in the absence of intermolecular forces of any kind

$$u = f_2(T) \tag{10.2}$$

(see also equation (9.4c), section 9.3).

Experimental confirmation of equation (10.1) is readily provided, whilst confirmation of equation (10.2), i.e. of the dependence of internal energy of "permanent gases" on temperature only, was furnished by Joule in a classical experiment (*Fig.* 10.1).

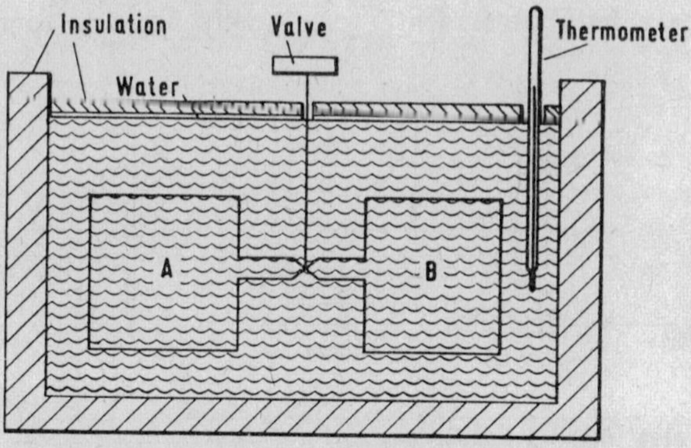

Fig. 10.1 Joule's experiment

Two vessels *A* and *B* were immersed in a water bath and interconnected through a cock initially closed. Vessel *B* was initially evacuated, only vessel *A* being filled with gas. Sudden opening of the interconnecting cock resulted in pressure equalization between the two vessels. Had the temperature of the gas changed significantly in the process, heat transfer between the surrounding water bath and the gas would have taken place, resulting in a change of water temperature. In fact no such change was observed.

Application of the N.F.E.E. (equation (4.4b)) to the process undergone by the gas, in which neither heat nor work transfer occurs, indicates that the internal energy remains constant, i.e.

$$Q - W = \Delta U$$

or

$$\Delta U = 0 \tag{10.3}$$

Joule therefore concluded that, since no measurable temperature change accompanied this constant internal energy process, in spite of the very large changes in pressure and specific volume, internal energy itself must be independent of the latter properties and therefore a function of temperature only, as expressed by equation (10.2).

The statement that the internal energy of perfect and semi-perfect gases is a function of temperature only is known as Joule's law. Real gases depart from the behaviour (Appendix II, section II.6).

10.2 The Joule–Thomson Experiment (*Fig.* 10.2)

Additional experimental proof of the independence of internal energy from properties other than temperature was obtained by Joule and Thomson in 1852, using a simple steady flow apparatus in which a hot "permanent" gas, usually air, was forced under

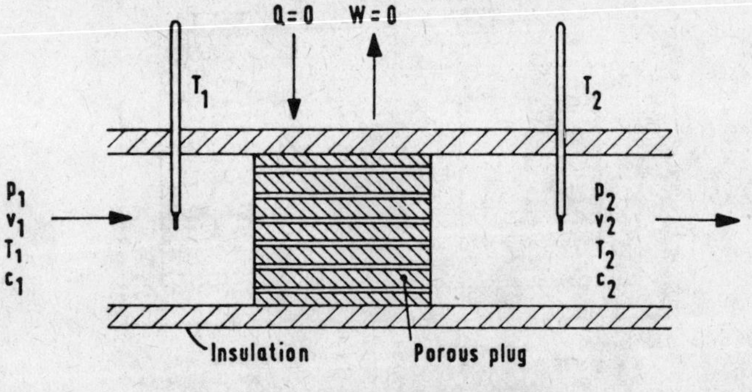

Fig. 10.2 Joule–Thomson experiment

pressure through a porous plug in a thermally insulated pipe of uniform bore, the bore being large enough to make kinetic effects negligible. As in the case of Joule's constant internal energy experiment, major changes in pressure and specific volume occurred, but applying the steady flow energy equation (S.F.E.E., equation (4.13b)) in the absence of heat, work, kinetic and potential energy effects,

$$Q - W = 0 = \Delta H \tag{10.4}$$

i.e. the process could be shown to be one of constant enthalpy.

Since, for such permanent gases, the state equation (10.1) is very closely obeyed, and since by the definition of enthalpy (2.17),

$$h = u + pv = u + RT = f_2(T) + f_1(T) \tag{10.5}$$

using equations (10.1) and (10.2)

$$h = f_3(T) \tag{10.6}$$

Hence, on the basis of the *experimental* relationships expressed by equations (10.1) and (10.2), resulting in equation (10.6), Joule and Thomson concluded that enthalpy, like internal energy, was a function of temperature only. However, with the much more accurate measurement possible with the "porous plug" apparatus, as opposed to the earlier Joule experiment, small temperature differences were, in fact, observed (approximately 0·3°C decrease per atmosphere pressure difference for air and 0·03°C increase per atmosphere pressure difference for hydrogen, with both gases at 0°C).

This led to the conclusion that, even in the absence of enthalpy change as expressed by equation (10.4), small temperature changes were possible as a result of changes in pressure and specific volume. Hence equation (10.6) is no longer strictly accurate and should be replaced by

$$h = f_4(T, p) \tag{10.7}$$

and using equation (10.5)

$$u = f_5(T, p) \tag{10.8}$$

It will be noted that the specific volume v is not included in the property relationships (10.7) and (10.8) since in fact any property of a pure substance is defined by any other two independent properties.

The dependence of enthalpy on pressure, as well as temperature, for real gases is illustrated in general terms in *Fig.* 10.3, and is expressed by the Joule–Thomson coefficient, defined as

$$\mu = \left(\frac{\partial T}{\partial p}\right)_h \tag{10.9}$$

This can be expressed in terms of various thermodynamic relations (see Appendix II) and is a quantity of great importance in connexion with the liquefaction of gases.

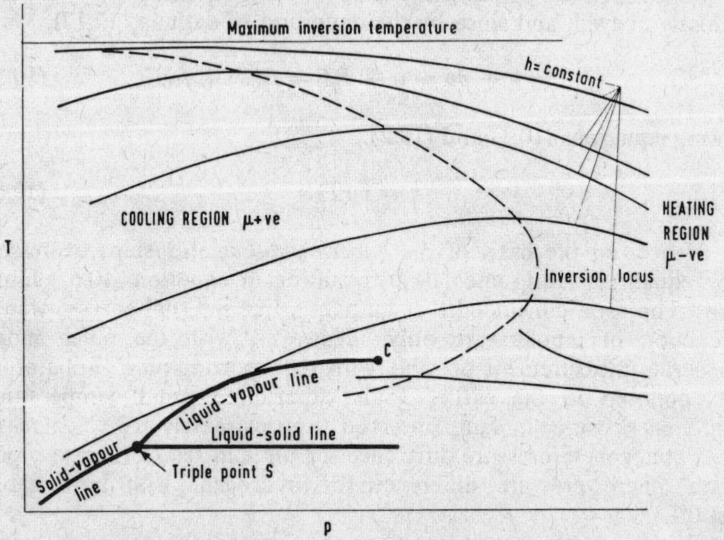

Fig. 10.3 Temperature–pressure diagram for a gas

10.3 Semi-perfect Gases

Although the Joule–Thomson effect is of great importance it is numerically small, i.e. the influence of the pressure term in equation (10.7), and hence also in equation (10.8), is sufficiently small for equation (10.2) to remain substantially correct. Similarly, observed deviations from the state equation are usually sufficiently small for equation (10.1) to remain substantially correct. Equations (10.1) and (10.2) together define the SEMI-PERFECT GAS, i.e. a semi-perfect gas obeys the equations

$$pv = f_1(T) = RT \tag{10.1}$$

$$u = f_2(T) \tag{10.2}$$

A perfect gas is one in which, in addition to equation (10.2), the specific heat at constant volume, defined by equation (3.9), is itself constant—

$$c_v = \left(\frac{du}{dT}\right)_{v=const.} = \text{constant} \tag{10.10}$$

which is not the case for the semi-perfect gas where c_v is itself a function of temperature.

Gas constants

It has already been shown in Chapter 3, section 3.5, that the state equation $pv = RT$ may be generalized to apply to *any* semi-perfect (or perfect) gas by the use of Avogadro's hypothesis, according to which the number of molecules of any gas contained in a given volume at a given pressure and temperature is constant.

From equation (3.7), the particular gas constant R, the molecular weight M and the universal gas constant G are related by

$$R = \frac{G}{M}$$

so that, writing v_m for the molar volume, i.e. the volume occupied by one kilogramme molecule or kg-mol (see equation (3.6)) of the gas under given conditions of temperature T and pressure p, equation (10.1) becomes

$$pv_m = 10^{-2}MRT = 10^{-2}GT \tag{10.11}$$

whilst, alternatively, for any mol number n of a particular gas

$$pV = 10^{-2}nGT = 10^{-2}mRT \tag{10.12}$$

where V = volume occupied by m kg of gas.

Equations (10.1), (10.11) and (10.12) are all closely related, the latter two being merely more general forms of equation (10.1).

The universal gas constant G has already been shown (section 3.5) to have the invariant value

$$G = 8{\cdot}3143 \text{ kJ/kg-mol-K}$$

Hence, the particular gas constant for any semi-perfect or perfect gas of known molecular weight M becomes

$$R = \frac{G}{M} = \frac{8{\cdot}3143}{M} \quad \text{kJ/kg-K}$$

Specific heats

The specific heat at constant volume is defined as (equation (3.9))

$$c_v = \left(\frac{du}{dT}\right)_{v=const.} \tag{10.13}$$

Whilst for the perfect gas the specific heat at constant volume is constant, it follows from equation (10.2) that, for the semi-perfect gas

$$c_v = f_6(T) \tag{10.13a}$$

i.e. the specific heat at constant volume, like internal energy, is a function of the temperature only, so that the derivative (du/dT) in equation (10.13) may be used without the restriction $v =$ constant,

$$c_v = \frac{du}{dT} \tag{10.13b}$$

From the definition of the specific heat at constant pressure (equation (3.11))

$$c_p = \left(\frac{dh}{dT}\right)_{p=const.} \tag{10.14}$$

and since, from equations (10.5) and (10.6), the enthalpy h is given by

$$h = u + pv = u + RT = f_2(T) + f_1(T) = f_3(T)$$

it follows from the above two equations that

$$c_p = f_7(T) \tag{10.14a}$$

i.e. the specific heat at constant pressure, like that at constant volume, is a function of temperature only. Hence the derivative (dh/dT) in equation (10.14) may be used without the restriction $p =$ constant,

$$c_p = \frac{dh}{dT} \tag{10.14b}$$

Differentiating equation (10.5) with respect to the temperature T and substituting from equations (10.13b) and (10.14b)

$$\frac{dh}{dT} = \frac{du}{dT} + R$$

i.e.
$$c_p = c_v + R$$
or
$$c_p - c_v = R \tag{10.15}$$

Equation (10.15) for semi-perfect gases is thus identical with equation (3.12) for perfect gases.

The units normally used for c_p, c_v and R in equation (10.15) are kJ/kg-K.

EXAMPLE 10.1

(1) Determine the volume, and specific heat at constant volume, of 1·5 kg of air at 7 bar and 60°C (molecular weight 28·97), given that the specific heat at constant pressure is 1·0049 kJ/kg-K. The universal gas constant $G = 8{\cdot}3143$ kJ/kg-mol-K.

SOLUTION

$$R = G/M = 8{\cdot}3143/28{\cdot}97 = 0{\cdot}2871 \text{ kJ/kg-K}$$

From equation (10.1a)

$$pV = 10^{-2}mRT$$
$$7V = 10^{-2} \times 1{\cdot}5 \times 0{\cdot}2871 \times 333$$
$$V = 0{\cdot}205 \text{ m}^3 \quad \textit{Ans.}$$

Since $c_p - c_v = R$,

$$1{\cdot}0049 - c_v = 0{\cdot}2871$$
$$c_v = 0{\cdot}7178 \text{ kJ/kg-K} \quad \textit{Ans.}$$

(2) The specific heats of a gas are given by $c_p = a + kT$ and $c_v = b + kT$, where a, b and k are constants and T is in K. Show that, for isentropic expansion of this gas $T^b v^{a-b} e^{kT} =$ a constant.

It is necessary to note that since the specific heats are functions of temperature the ratio c_p/c_v will not now be a constant and so pv^γ is not now a constant. However $c_p - c_v = (a - b) =$ constant, so we have here a semi-perfect gas and the gas law still holds. For other equations we must go to the first law of thermodynamics.

SOLUTION

From the first law,

$$0 = c_v\,dT + \frac{p\,dv}{J} = (b + kT)\,dT + \frac{RT}{Jv}\,dv$$

but

$$\frac{R}{J} = c_p - c_v = (a - b)$$

Substituting this and dividing throughout by T

$$0 = b\frac{dT}{T} + k\,dT + (a - b)\frac{dv}{v}$$

Integrating we have

$$\text{Constant} = b \ln T + kT + (a - b) \ln v$$

or

$$T^b v^{a-b} e^{kT} = \text{constant} \quad \textit{Ans.}$$

(3) 1·36 kg of gas occupying 0·0568 m³ at 55·1 bar expand isentropically until the temperature is 240°C. If $a = 0{\cdot}945$, $b = 0{\cdot}660$, and $k = 0{\cdot}00018$ calculate the work done in the expansion, assuming semi-perfect gas behaviour.

SOLUTION

$$\text{Work done} = -\Delta u = u_1 - u_2$$
$$R = a - b = 0{\cdot}285 \text{ kJ/kg-K}$$
$$10^2 \times 55{\cdot}1 \times 0{\cdot}0568 = 1{\cdot}36 \times 0{\cdot}285 \times T_1$$
$$T_1 = 806 \text{ K}$$

Also

$$T_2 = 240°\text{C} = 513 \text{ K}$$
$$\begin{aligned}\text{Work done} &= +\int_{513}^{806} 1{\cdot}36 c_v dT \\ &= 1{\cdot}36 \int_{513}^{806} (0{\cdot}660 + 0{\cdot}00018T)dT \\ &= 1{\cdot}36 \Big[0{\cdot}660T + 0{\cdot}00009T^2\Big]_{513}^{806} \\ &= 314{\cdot}5 \text{ kJ}\end{aligned}$$

The relationship between the specific heats expressed by equation (10.15) applies to a particular gas. Like the state equation (10.1), the specific heat relationship (10.15) can be generalized to apply to any semi-perfect (or perfect gas) by using the universal gas constant (equation (10.11)) and applying the equation to 1 kg-mol

$$C_p - C_v = G \tag{10.16}$$

where

$$G = MR \tag{10.17}$$
$$C_p = Mc_p \tag{10.18}$$
$$C_v = Mc_v \tag{10.19}$$

the latter two quantities being the molar specific heats at constant pressure and constant volume respectively, measured in kJ/kg-mol-K.

The specific heat ratio $\gamma = c_p/c_v$

The fundamental significance of the specific heat ratio γ has already been discussed in Chapter 9, section 9.3 in connexion with the kinetic theory of gases, where it was shown that, using the simple molecular model for the perfect gas (section 9.2)

$\gamma = 1{\cdot}667$ for monatomic gases
$\gamma = 1{\cdot}400$ for diatomic gases
$\gamma = 1{\cdot}333$ for triatomic gases.

As briefly discussed in section 9.3, the specific heat ratio of real gases departs progressively from the above theoretical values as the temperature is raised due to the acquisition, by the individual molecules, of forms of kinetic energy other than that associated with translational and rotational motion. Thus, the semi-perfect gas defined by equations (10.1) and (10.2) possesses specific heats

c_p and c_v which increase progressively with increasing temperature in accordance with this modified kinetic theory, but whose difference remains constant, according to equation (10.15). In this case, the specific heat ratio γ gradually decreases with temperature.

The variation of the specific heat at constant volume with temperature is conveniently expressed as a polynomial

$$c_v = k_0 + k_1T + k_2T^2 + k_3T^3 + \ldots \qquad (10.20)$$

where k_0, k_1, k_2, etc., are empirical constants.

Hence, using equations (10.15) and (10.20)

$$c_p = R + k_0 + k_1T + k_2T^2 + k_3T^3 + \ldots \qquad (10.21)$$

so that the specific heat ratio γ becomes

$$\gamma = \frac{c_p}{c_v} = \left(\frac{R}{k_0 + k_1T + k_2T^2 + k_3T^3}\right) + 1 \qquad (10.22)$$

If the empirical constants k_0, k_1, k_2, k_3, etc., are positive, as they generally are, it follows from equation (10.22) that γ diminishes with increasing temperature.

Mean specific heats, internal energy and enthalpy contents

It is evident from equations (10.20) and (10.21) that when a semi-perfect gas is heated (or cooled) through a finite temperature range, the mean specific heats over the temperature range in question are given by

$$(\bar{c}_v)_{12} = \frac{\int_1^2 c_v\,dT}{(T_2 - T_1)} = \frac{u_2 - u_1}{T_2 - T_1} \qquad (10.23)$$

and

$$(\bar{c}_p)_{12} = \frac{\int_1^2 c_p\,dT}{(T_2 - T_1)} = \frac{h_2 - h_1}{T_2 - T_1} \qquad (10.24)$$

the last terms in the above equations being derived from equation (10.13b) and (10.14b).

Internal energy values for particular semi-perfect gases are usually given in tabulated form in relation to an arbitrary reference temperature T_0. Substituting in equations (10.23) and (10.24) the reference internal energy u_0 and enthalpy h_0 respectively, we have

$$u_2 - u_1 = (u_2 - u_0) - (u_1 - u_0)$$
$$= (\bar{c}_v)_{02}(T_2 - T_0) - (\bar{c}_v)_{01}(T_1 - T_0) \qquad (10.23a)$$

and

$$h_2 - h_1 = (h_2 - h_0) - (h_1 - h_0)$$
$$= (\bar{c}_p)_{02}(T_2 - T_0) - (\bar{c}_p)_{01}(T_1 - T_0) \qquad (10.24a)$$

Typical curves for various semi-perfect gases, showing the variation of the *molar* specific heat at constant volume, are shown in *Fig*. 10.4.

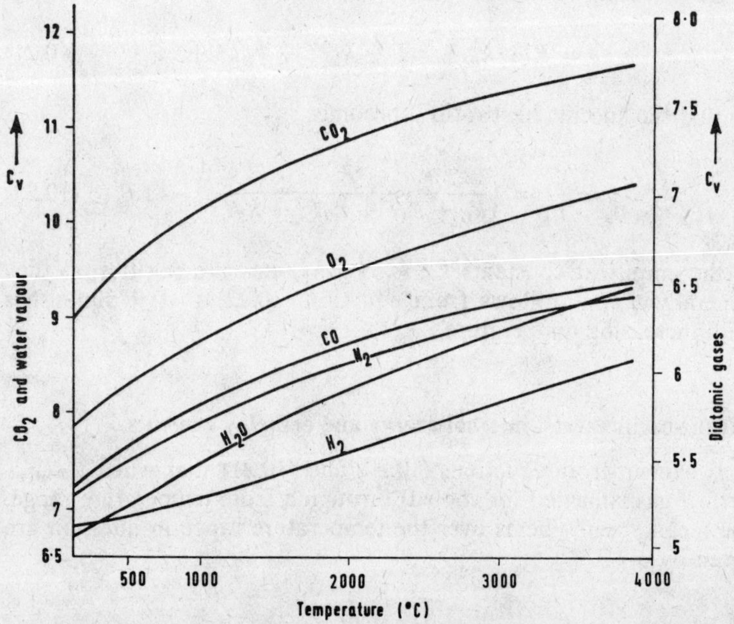

Fig. 10.4 Molar specific heats at constant volume
(kcal/kg-mol-K between 100°C and t°C)

Mean specific heat values between any two temperatures T_1 and T_2 may be derived from charts such as *Fig*. 10.4 or tables using equation (10.23a) in conjunction with equation (10.23). The difference formula, equation (10.15), applies equally to the mean specific heats $\bar{c}_p$ and $\bar{c}_v$ as to the original point values, i.e. for any temperature range T_1 to T_2

$$(\bar{c}_p)_{12} - (\bar{c}_v)_{12} = R \qquad (10.15a)$$

It is interesting to note that the specific heat at constant pressure for air increases from 1·0023 kJ/kg-K at 175 K to 1·2949 kJ/kg-K at 3000 K, but that it remains sensibly constant ($1{\cdot}0023 < c_p < 1{\cdot}0511$) between 175 K and 600 K. Hence, over this temperature range, air may be treated as a perfect gas according to equations (10.1) and (10.10).

EXAMPLE 10.2

Using tables of properties of gases, evaluate the mean specific heats and mean values of their ratio, for air in the temperature range 270 K to 450 K, at a pressure of 1 bar.

SOLUTION

Using *Thermodynamic Functions of Gases*, vol. 2, (ref. 10.1):

at 270 K $\quad h_M = 11\,515$ kJ/kg-mol

at 450 K $\quad h_M = 16775$ kJ/kg-mol

Mean specific heat at constant pressure is given by

$$(\bar{c}_p)_M = \frac{\Delta h_M}{\Delta T} = \frac{16775 - 11\,515}{450 - 270}$$

$$= 29{\cdot}25 \text{ kJ/kg-mol-K}$$

or $$\bar{c}_p = \frac{(\bar{c}_p)_M}{M} = \frac{29{\cdot}25}{28{\cdot}96} = 1{\cdot}011 \text{ kJ/kg-K}$$

But $$\bar{c}_v = \bar{c}_p - R = (1{\cdot}011 - 0{\cdot}287)$$
$$= 0{\cdot}724 \text{ kJ/kg-K}$$

$$\bar{\gamma} = \frac{\bar{c}_p}{\bar{c}_v} = \frac{1{\cdot}011}{0{\cdot}724} = 1{\cdot}400 \quad \textit{Ans.}$$

Entropy changes

Expressions for entropy changes for the perfect gas have been derived in Chapter 7, section 7.4. Using the entropy equation (7.8a) of that section as the basis for entropy determinations for the semi-perfect gas we have

$$T\,ds = p\,dv + du \tag{10.25}$$

and substituting for $p = RT/v$ from equation (10.1) and $du = c_v\,dT$ from equation (10.13b) in equation (10.25)

$$ds = \frac{R\,dv}{v} + c_v \frac{dT}{T} \tag{10.25a}$$

However, this expression cannot be integrated directly, as in the case of the perfect gas since, from equations (10.13) and (10.20), the specific heat at constant volume c_v is a function of temperature, i.e. $c_v = f_6(T)$

$$s_2 - s_1 = R\int_1^2 \frac{dv}{v} + \int_1^2 c_v \frac{dT}{T}$$

$$= R \ln\left(\frac{v_2}{v_1}\right) + \int_1^2 c_v \frac{dT}{T} \tag{10.25b}$$

With the aid of tabulated values of c_v for particular semi-perfect gases, it is possible to evaluate the integral

$$\int_1^2 c_v \frac{dT}{T}$$

and hence the entropy change $(s_2 - s_1)$.

Entropy values, based on equation (10.25b), for semi-perfect gases may be computed from tabulated values of the integral

$$\int_{T_0}^{T} c_v \frac{dT}{T}$$

in equation (10.25b), T_0 being the same arbitrary reference temperature as is employed in the evaluation of internal energy and enthalpy (ref. 10.1).

10.4 Perfect Gases

Whilst the semi-perfect gas is defined by equations (10.1) and (10.2), the perfect gas obeys equations (10.1) and (10.10), i.e. its state equation is the same as that of the semi-perfect gas, but its specific heat at constant volume, instead of being temperature-dependent, is now constant. This results in further simplifications of the various analytical relationships established for the semi-perfect gas and completely eliminates the need for tabulated values.

In practice, perfect gas behaviour may be assumed for permanent gases when the temperature range in question is limited, although in many theoretical studies, perfect gas behaviour is assumed over a much wider range of temperatures in order to take advantage of the analytical simplifications which are possible only on the basis of perfect gas behaviour.

DEFINITION

A perfect gas obeys the equations

$$pv = RT \tag{10.1}$$

$$c_v = \frac{du}{dT} = \text{constant} \tag{10.13c}$$

the restriction of constant volume in equation (10.13) having been removed in accordance with equation (10.13b).

It follows from equation (10.15), which holds equally for semi-perfect and perfect gases, in conjunction with equation (10.13) that

$$c_p = \frac{dh}{dT} = \text{constant} \tag{10.14c}$$

so that the perfect gas may be defined either by equations (10.1) and (10.13c) or (10.1) and (10.14c).

GAS CONSTANTS

All relationships established under this heading in section 10.3 for the semi-perfect gas apply equally to the perfect gas, and are therefore not repeated here.

PROPERTY RELATIONSHIPS

(*a*) *Specific heats* (equations (10.13c) and (10.14c))

$$\left.\begin{aligned} c_v &= \text{constant} \\ c_p &= \text{constant} \end{aligned}\right\} \tag{10.26}$$

$$\frac{c_p}{c_v} = \gamma = \text{constant} \tag{10.27}$$

$$c_p - c_v = (\gamma - 1)c_v = \frac{\gamma - 1}{\gamma}\, c_p = R = \frac{G}{M} \tag{10.28}$$

(*b*) *Internal energy and enthalpy*

$$u - u_0 = c_v(T - T_0) \tag{10.29}$$

$$h - h_0 = c_p(T - T_0) \tag{10.30}$$

cf. equations (10.23a) and (10.24a).

(*c*) *Entropy*

Substituting in equation (10.25a) for c_v = constant, from equation (10.26)

$$s_2 - s_1 = R \ln\left(\frac{v_2}{v_1}\right) + c_v \ln\left(\frac{T_2}{T_1}\right) \tag{10.31}$$

See equation (7.9a). By substitution from the state equation (10.1) either the volume ratio v_2/v_1 or the temperature ratio T_2/T_1 may be eliminated from equation (10.31)

$$s_2 - s_1 = R \ln\left(\frac{v_2}{v_1}\right) + c_v \ln\left(\frac{p_2 v_2}{p_1 v_1}\right)$$

$$= (R + c_v) \ln\left(\frac{v_2}{v_1}\right) + c_v \ln\left(\frac{p_2}{p_1}\right)$$

and again substituting from equation (10.15)

$$s_2 - s_1 = c_p \ln\left(\frac{v_2}{v_1}\right) + c_v \ln\left(\frac{p_2}{p_1}\right) \tag{10.32}$$

See equation (7.9b).

Alternatively

$$s_2 - s_1 = R \ln\left(\frac{T_2 p_1}{T_1 p_2}\right) + c_v \ln\left(\frac{T_2}{T_1}\right)$$

$$= (R + c_v) \ln\left(\frac{T_2}{T_1}\right) + R \ln\left(\frac{p_1}{p_2}\right)$$

and substituting from equation (10.15)

$$s_2 - s_1 = c_p \ln\left(\frac{T_2}{T_1}\right) - R \ln\left(\frac{p_2}{p_1}\right) \tag{10.33}$$

See equation (7.9c.)

In view of the extreme importance of the above perfect gas property relationships they are summarized in Table 10.1 below.

Table 10.1

Specific heats	*Internal energy and enthalpy*	*Entropy*
$c_v = \text{constant}$	$u - u_0 = c_v(T - T_0)$	$s_2 - s_1 = R \ln\left(\frac{v_2}{v_1}\right) + c_v \ln\left(\frac{T_2}{T_1}\right)$
		or
$c_p = \text{constant}$	$h - h_0 = c_p(T - T_0)$	$s_2 - s_1 = c_p \ln\left(\frac{v_2}{v_1}\right) + c_v \ln\left(\frac{p_2}{p_1}\right)$
		or
$c_p - c_v = R = \frac{G}{M}$		
$c_p/c_v = \gamma$		$s_2 - s_1 = c_p \ln\left(\frac{T_2}{T_1}\right) - R \ln\left(\frac{p_2}{p_1}\right)$

EXAMPLE 10.3

$0{\cdot}1\ m^3$ of a perfect gas at 2·8 bar and 120°C is allowed to expand isentropically to 1 bar. Then 20 kJ of heat is supplied at constant pressure. If the ratio of specific heats is 1·4 and the specific heat at constant pressure is 1·0049 kJ/kg-K, calculate the changes in entropy of the process.

If the two processes above are replaced by a reversible polytropic expansion find the index of expansion between the original and final states. What is then the change of entropy?

SOLUTION

(*a*) The law of isentropic expansion is $pV^{1\cdot4}$ = constant. Therefore

$$(V_2/V_1) = (p_1/p_2)^{1/1\cdot4} = (2\cdot8/1)^{1/1\cdot4}$$

giving $V_2 = 0{\cdot}1 \times 2{\cdot}8^{1/1\cdot4} = 0{\cdot}209\ m^3$.

$$T_2/T_1 = (p_2/p_1)^{(\gamma-1)/\gamma} = (1/2\cdot8)^{0\cdot4/1\cdot4} = 1\cdot343$$

giving $T_2 = (273 + 120)/1{\cdot}343 = 292\ K = 19°C$.

Now

$$s_2 - s_1 = R \ln\left(\frac{V_2}{V_1}\right) + c_v \ln\left(\frac{T_2}{T_1}\right)$$

$$c_v = 1\cdot0049/1\cdot4 = 0\cdot7170\ kJ/kg\text{-}K$$

$$R = c_p - c_v = 1\cdot0049 - 0\cdot7170 = 0\cdot2879\ kJ/kg\text{-}K$$

Therefore

$$s_2 - s_1 = 0\cdot2879 \ln\left(\frac{0\cdot209}{0\cdot100}\right) + 0\cdot7170 \ln\left(\frac{292}{393}\right)$$

$$= 0\cdot2125 - 0\cdot2125 = 0 \quad \textit{Ans.}$$

in accordance with the definition of an isentropic expansion.

(*b*) Constant pressure heating:

$$\frac{V_3}{V_2} = \frac{T_3}{T_2} \text{ and } mc_p(T_3 - T_2) = 20$$

$$m = \frac{pV}{RT} \times 10^2 = \frac{2\cdot8 \times 0\cdot1 \times 10^2}{0\cdot2879 \times 393} = 0\cdot2542\ kg$$

Therefore

$$0\cdot2542 \times 1\cdot0049(T_3 - T_2) = 20$$

$$T_3 - T_2 = 78\cdot1°C$$

$$T_3 = 292 + 78\cdot1 = 370\cdot1K$$

$$V_3 = 0\cdot209 \times 370\cdot1/292 = 0\cdot256\ m^3$$

$$s_3 - s_2 = c_p \ln(V_3/V_2) + c_v \ln(p_3/p_2)$$

$$= 1\cdot0049 \ln(0\cdot265/0\cdot209)$$

$$= 0\cdot2365\ kJ/kg\text{-}K$$

Total change of entropy

$$s_3 - s_1 = (s_2 - s_1) + (s_3 - s_2) = (0 + 0{\cdot}2365)\ \text{kJ/kg-K}$$

or for 0·2542 kg

$$\Delta s_{23} = 0{\cdot}2542 \times 0{\cdot}2365 = 0{\cdot}0601\ \text{kJ/K} \quad \textit{Ans.}$$

(*c*) Reversible polytropic expansion:

Let the index be *n*.

$$\frac{T_1}{T_3} = \left(\frac{V_3}{V_1}\right)^{n-1} = \left(\frac{p_1}{p_3}\right)^{(n-1)/n}$$

$$\frac{493{\cdot}0}{370{\cdot}1} = \left(\frac{0{\cdot}265}{0{\cdot}1}\right)^{n-1}$$

whence $n = 1{\cdot}0615$ *Ans.*

The overall change of entropy is the same as (*a*) + (*b*) since the state points are identical.

Note: The remainder of this chapter, dealing with particular processes, may be omitted at a first reading.

10.5 Particular Processes Undergone by Perfect Gases

Note: This section should be read in conjunction with sections 4.5, and 4.7 in Chapter 4 and sections 7.4, 7.5 and 7.6 in Chapter 7.

(*a*) CLOSED SYSTEMS

Reversible processes undergone by closed systems have been analysed in section 4.5 in terms of the pressure–volume diagram, and in section 7.4 in terms of the temperature–entropy diagram.

Expressions for heat added, work done, and changes in property values were derived in terms of the N.F.E.E. for unit mass (equation (4.4b))

$$q_{12} - w_{12} = \Delta u_{12} \qquad (10.34)$$

(neglecting kinetic and potential energy changes)

where

$$w_{12} = \int_1^2 p\,dv \qquad (10.35)$$

and

$$q_{12} = \int_1^2 T\,ds \qquad (10.36)$$

in conjunction with the property equations for the perfect gas outlined in section 3.5 and developed in detail in the present chapter.

(*b*) OPEN SYSTEMS

Reversible processes undergone by open systems have been analysed in section 4.7, for a number of steady flow processes.

Expressions for heat added and work done were derived in terms of the S.F.E.E. for unit mass (equation (4.13b))

$$q_{12} - w_{12} = [h_2 + \tfrac{1}{2}c_2{}^2 + gZ_2] - [h_1 + \tfrac{1}{2}c_1{}^2 + gZ_1] \quad (10.37)$$

in which, generally, kinetic and potential energy effects cannot be neglected. Furthermore, it was shown in section 7.5 that for *reversible* processes

$$w_{12} + \tfrac{1}{2}(c_2{}^2 - c_1{}^2) + g(Z_2 - Z_1) = \int_2^1 v\,dp \quad (10.38)$$

and $$q_{12} = \int_1^2 T\,ds \quad (10.39)$$

Irreversibilities were discussed in section 7.6, with particular reference to the fact that areas in the pressure–volume diagrams do not represent work, and areas in the temperature–entropy diagram do not constitute heat added, in the case of irreversible processes.

10.5.1 Reversible closed system processes (see sections 4.5 and 7.4)

(*a*) **Constant volume process** (see *Fig.* 10.5)

p–v–T relationships

$$\frac{v_2}{v_1} = 1{\cdot}0 \qquad \frac{p_2}{p_1} = \frac{T_2}{T_1} \quad (10.40)$$

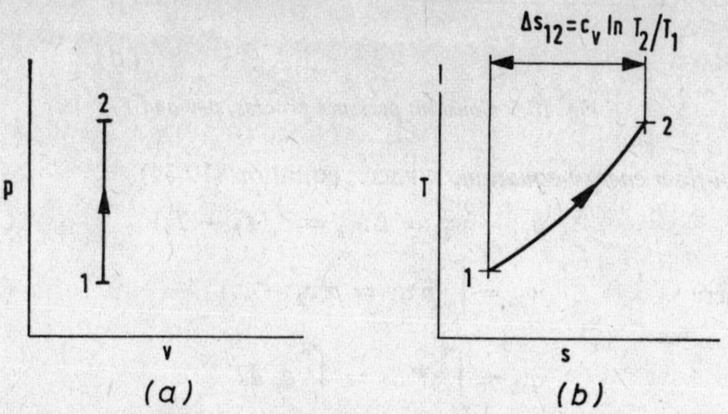

Fig. 10.5 Constant volume process, p–v and T–s

Non-flow energy equation, N.F.E.E., equation (10.34)

$$q_{12} - w_{12} = \Delta u_{12} = c_v(T_2 - T_1) \tag{10.34}$$

where

$$w_{12} = \int_1^2 p\,dv = 0 \tag{10.41}$$

$$q_{12} = \int_1^2 T\,ds = \int_1^2 c_v\,dT = c_v(T_2 - T_1) \tag{10.42}$$

Entropy change, equation (10.31)

$$\Delta s_{12} = \int_1^2 \frac{dq}{T} = \int_1^2 c_v \frac{dT}{T} = c_v \ln\left(\frac{T_2}{T_1}\right) \tag{10.43}$$

Note: Constant volume processes are sometimes described as ISOCHORIC PROCESSES.

(*b*) **Constant pressure process** (see *Fig.* 10.6)

p–v–T relationships

$$\frac{p_2}{p_1} = 1{\cdot}0 \qquad \frac{v_2}{v_1} = \frac{T_2}{T_1} \tag{10.44}$$

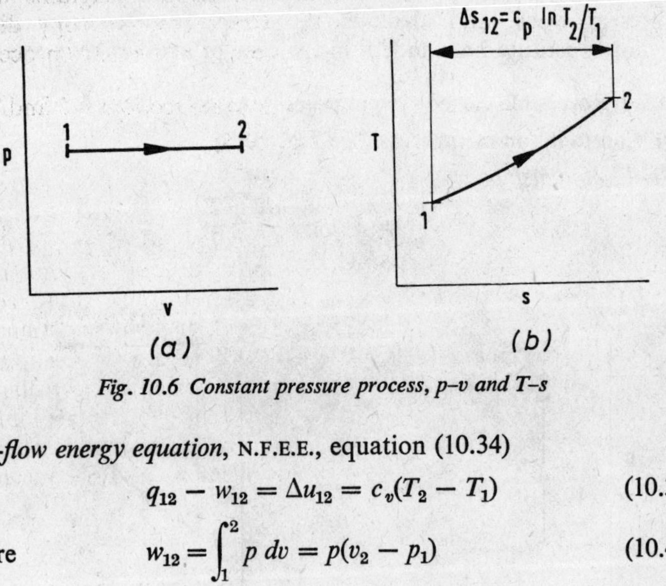

Fig. 10.6 Constant pressure process, p–v and T–s

Non-flow energy equation, N.F.E.E., equation (10.34)

$$q_{12} - w_{12} = \Delta u_{12} = c_v(T_2 - T_1) \tag{10.34}$$

where

$$w_{12} = \int_1^2 p\,dv = p(v_2 - p_1) \tag{10.45}$$

$$q_{12} = \int_1^2 T\,ds = \int_1^2 c_p\,dT$$

$$= c_p(T_2 - T_1) \tag{10.46}$$

Entropy change, equation (10.33)

$$\Delta s_{12} = \int_1^2 \frac{dq}{T} = \int_1^2 c_p \frac{dT}{T} = c_p \ln\left(\frac{T_2}{T_1}\right) \tag{10.47}$$

Note: Constant pressure processes are sometimes described as ISOBARIC PROCESSES.

(*c*) **Isothermal process** (see *Fig.* 10.7)

p–v–T relationships

$$\frac{T_2}{T_1} = 1{\cdot}0 \qquad \frac{v_2}{v_1} = \frac{p_1}{p_2} \tag{10.48}$$

or

$$\boxed{pv = \text{constant}} \tag{10.48a}$$

Equation (10.48a) is usually referred to as BOYLE'S LAW.

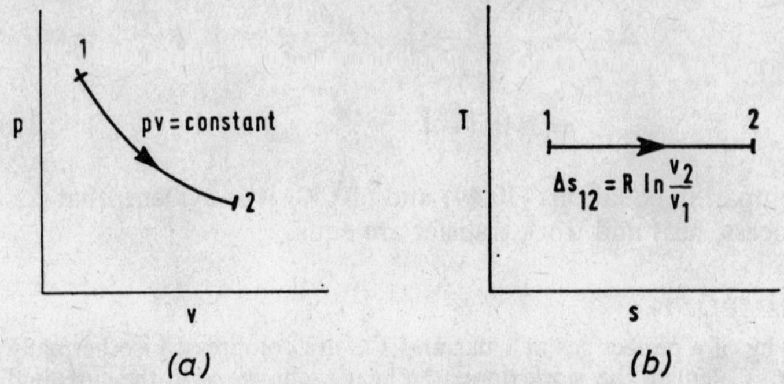

Fig. 10.7 Isothermal process, p–v and T–s

Non-flow energy equation N.F.E.E., equation (10.34)

$$q_{12} - w_{12} = \Delta u_{12} = c_v(T_2 - T_1) = 0 \tag{10.34}$$

where

$$w_{12} = \int_1^2 p\, dv = \int_1^2 \frac{RT}{v}\, dv = RT \ln\left(\frac{v_2}{v_1}\right) \tag{10.49}$$

$$= p_1 v_1 \ln\left(\frac{v_2}{v_1}\right) \tag{10.49a}$$

$$= p_2 v_2 \ln\left(\frac{v_2}{v_1}\right) \tag{10.49b}$$

The ratio v_2/v_1 in equations (10.49), (10.49a) and (10.49b) may also be replaced by p_1/p_2.

$$q_{12} = \int_1^2 T\,ds$$

where, from equation (10.34),

$$q_{12} = w_{12}$$

i.e.

$$q_{12} = RT \ln\left(\frac{v_2}{v_1}\right) \tag{10.50}$$

$$= p_1 v_1 \ln\left(\frac{v_2}{v_1}\right) \tag{10.50a}$$

$$= p_2 v_2 \ln\left(\frac{v_2}{v_1}\right) \tag{10.50b}$$

Entropy change, equation (10.31)

$$\Delta s_{12} = \int_1^2 \frac{dq}{T} = \int_1^2 RT \frac{dv}{v} \frac{1}{T} = \int_1^2 R \frac{dv}{v}$$

$$= R \ln\left(\frac{v_2}{v_1}\right) \tag{10.51}$$

Comparing equations (10.49) and (10.50) it is evident, that for this process, heat and work transfer are equal.

EXAMPLE 10.4

1·5 kg of a perfect gas at 1 bar and 7°C are compressed isothermally to 6 bar. Calculate the work done, the heat exchange with the surroundings, the state change, and the entropy change for the process.

Take $R = 0{\cdot}296$ kJ/kg-K and $c_p = 1{\cdot}035$ kJ/kg-K.

SOLUTION

$$p_1 V_1 = 10^{-2} mRT$$

$$V_1 = (10^{-2} \times 1{\cdot}5 \times 0{\cdot}296 \times 280)/1 = 1{\cdot}245 \text{ m}^3$$

$$V_2 = V_1 \times \frac{p_1}{p_2} = 1{\cdot}245 \times \tfrac{1}{6} = 0{\cdot}2076 \text{ m}^3$$

$$\text{Work done} = mRT \ln\left(\frac{V_2}{V_1}\right)$$

$$= -1{\cdot}5 \times 0{\cdot}296 \times 280 \times \ln 6$$

$$= -222{\cdot}5 \text{ kJ} \quad \textit{Ans.}$$

From the first law

$$Q - W = \Delta U = 0$$
$$Q = W = -222{\cdot}5 \text{ kJ} \quad \textit{Ans.}$$

The negative sign indicates that work is done on the gas and heat is lost to the surroundings.

Entropy change is given by

$$S_2 - S_1 = mR \ln\left(\frac{V_2}{V_1}\right)$$
$$= 1{\cdot}5 \times 0{\cdot}296 \ln(\tfrac{1}{6})$$
$$= -0{\cdot}796 \text{ kJ/K, a loss} \quad \textit{Ans.}$$

(*d*) **Adiabatic reversible or isentropic processes*** (*Fig.* 10.8)

p–v–T relationships. (Derived from equations (10.31), (10.32) or (10.33) from the condition of constant entropy $\Delta s = 0$.)

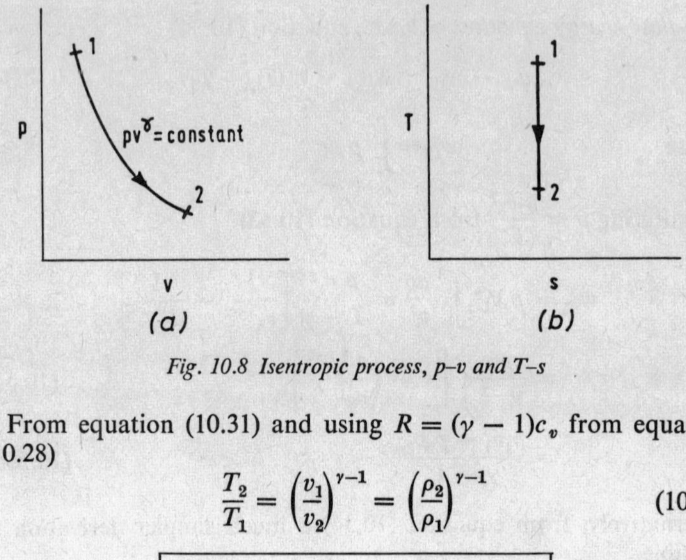

Fig. 10.8 Isentropic process, p–v and T–s

From equation (10.31) and using $R = (\gamma - 1)c_v$ from equation (10.28)

$$\frac{T_2}{T_1} = \left(\frac{v_1}{v_2}\right)^{\gamma-1} = \left(\frac{\rho_2}{\rho_1}\right)^{\gamma-1} \qquad (10.52)$$

or

$$\boxed{Tv^{\gamma-1} = T\rho^{-(\gamma-1)} = \text{constant}} \qquad (10.52a)$$

where ρ = density = $1/v$.

* The terms *adiabatic reversible* and *isentropic* are synonymous. The former term (see also section 4.5d) implies a process in which heat transfer is absent, whilst all irreversibilities are eliminated. For such a process, since $dQ = 0$, and from the definition of entropy as $S_2 - S_1 = (dQ/T)_{rev}$ (equation (7.4),) the entropy change is zero, i.e. the process is one of constant entropy or isentropic.

From equation (10.32) and using $c_p/c_v = \gamma$ from equation (10.27)

$$\frac{p_2}{p_1} = \left(\frac{v_1}{v_2}\right)^\gamma = \left(\frac{\rho_2}{\rho_1}\right)^\gamma \tag{10.53}$$

or

$$\boxed{pv^\gamma = p\rho^{-\gamma} = \text{constant}} \tag{10.53a}$$

From equation (10.33)

$$\frac{T_2}{T_1} = \left(\frac{p_2}{p_1}\right)^{(\gamma-1)/\gamma} \tag{10.54}$$

or

$$\boxed{pT^{-(\gamma-1)/\gamma} = \text{constant}} \tag{10.54a}$$

EQUATIONS (10.52) TO (10.54) ARE KNOWN AS THE ISENTROPIC STATE EQUATIONS FOR THE PERFECT GAS.*

Non-flow energy equation, N.F.E.E., equation (10.34)

$$q_{12} - w_{12} = \Delta u_{12} = c_v(T_2 - T_1) \tag{10.34}$$

where

$$w_{12} = \int_1^2 p\, dv$$

Substituting $p = \dfrac{p_1 v_1^\gamma}{v^\gamma}$ from equation (10.53)

$$w_{12} = p_1 v_1^\gamma \int_1^2 \frac{dv}{v^\gamma} = \frac{p_1 v_1^\gamma}{1-\gamma}\left[\frac{1}{v_2^{1-\gamma}} - \frac{1}{v_1^{1-\gamma}}\right]$$

$$= \frac{p_1 v_1}{\gamma - 1}\left[1 - \left(\frac{v_1}{v_2}\right)^{\gamma-1}\right] \tag{10.55}$$

$$= \frac{p_1 v_1 - p_2 v_2}{\gamma - 1} \tag{10.55a}$$

Alternatively, from equation (10.34) a much simpler derivation is possible.

Since $$w_{12} = -\Delta u_{12} = c_v(T_1 - T_2)$$

* The isentropic state equations (10.52) to (10.54) have already been discussed briefly in Chapter 4, section 4.5, equations (4.9a, b and c).

It should be noted, however, that the derivation given in that chapter was based solely on first law considerations. It is only with the introduction of entropy as a measure of irreversibility in equations (10.31) to (10.33) that the derivation of the isentropic state equations may be said to have been put on a rigorous basis.

and substituting from equation (10.28), for c_v in terms of R,

$$w_{12} = \frac{R(T_1 - T_2)}{\gamma - 1}$$

$$= \frac{p_1v_1 - p_2v_2}{\gamma - 1} \quad (10.56)$$

$$q_{12} = 0 \text{ (by definition)} \quad (10.57)$$

Entropy change

$$\Delta s_{12} = 0 \quad (10.58)$$

(*e*) **Polytropic reversible processes** (*Fig*. 10.9)

p–v–T relationships

As stated in section 4.5, the polytropic reversible process for the perfect gas may be defined by the equation

$$\boxed{pv^n = p\rho^{-n} = \text{constant}} \quad (10.59)$$

or

$$\frac{p_2}{p_1} = \left(\frac{v_1}{v_2}\right)^n = \left(\frac{\rho_2}{\rho_1}\right)^n \quad (10.59a)$$

This leads to the further forms

$$\left(\frac{T_2}{T_1}\right) = \left(\frac{v_1}{v_2}\right)^{n-1} \quad (10.60)$$

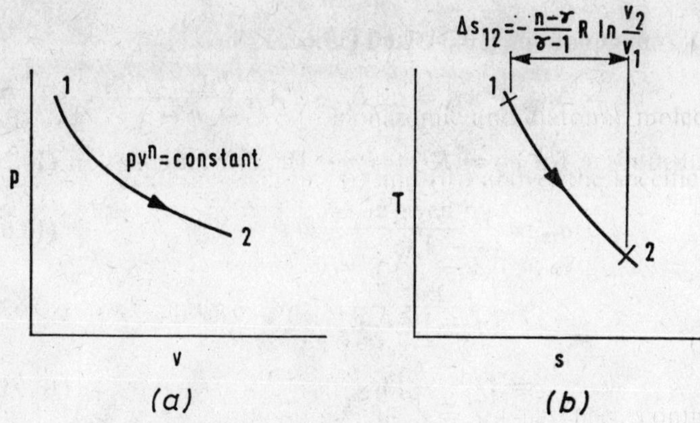

Fig. 10.9 Polytropic reversible process, p–v and T–s

or $$Tv^{n-1} = T\rho^{-(n-1)} = \text{constant} \tag{10.60a}$$

and $$\left(\frac{T_2}{T_1}\right) = \left(\frac{p_2}{p_1}\right)^{(n-1)/n} \tag{10.61}$$

or $$pT^{-(n-1)/n} = \text{constant} \tag{10.61a}$$

Equations (10.59), (10.60) and (10.61) may be compared respectively with equations (10.53), (10.52) and (10.54) for the isentropic process.

Non-flow energy equation, N.F.E.E., equation (10.34)

$$q_{12} - w_{12} = \Delta u_{12} = c_v(T_2 - T_1) \tag{10.34}$$

where $$w_{12} = \int_1^2 p\, dv$$

Substituting for $p = \dfrac{p_1 v_1^{\,n}}{v^n}$ from equation (10.55)

$$w_{12} = p_1 v_1^{\,n} \int_1^2 \frac{dv}{v} = \frac{p_1 v_1^{\,n}}{1-n}\left[\frac{1}{v_2^{\,1-n}} - \frac{1}{v_1^{\,1-n}}\right]$$

$$= \frac{p_1 v_1}{n-1}\left[1 - \left(\frac{v_1}{v_2}\right)^{n-1}\right] \tag{10.62}$$

$$= \frac{p_1 v_1 - p_2 v_2}{n-1} \tag{10.62a}$$

Equations (10.62) and (10.62a) may be compared respectively with equations (10.55) and (10.55a) for the isentropic process.

$$q_{12} = \int_1^2 T\, ds$$

where, from equations (10.34) and (10.62a)

$$q_{12} = \Delta u_{12} + w_{12} = c_v(T_2 - T_1) + \frac{p_1 v_1 - p_2 v_2}{n-1}$$

and substituting for $pv = RT = (\gamma - 1)c_v T$ from equation (10.28)

$$q_{12} = \frac{\gamma - n}{(n-1)(\gamma - 1)} R(T_1 - T_2) \tag{10.63}$$

$$= \frac{\gamma - n}{(n-1)(\gamma - 1)}(p_1 v_1 - p_2 v_2) \tag{10.63a}$$

$$= \left(\frac{\gamma - n}{\gamma - 1}\right) w_{12} \tag{10.63b}$$

(Compare with equations (4.11a) and (4.11b).)

Note that in an expansion, heat is added if $n < \gamma$ whilst for a compression, heat is added if $n > \gamma$

Entropy change

$$\Delta s_{12} = \int_1^2 \frac{dq}{T} = \int_1^2 \frac{\gamma - n}{(n-1)(\gamma-1)} \left(-\frac{R\,dT}{T}\right)$$

$$= \frac{(n-\gamma)}{(n-1)(\gamma-1)} R \ln\left(\frac{T_2}{T_1}\right) \quad (10.64)$$

or substituting from equation (10.59) to (10.61)

$$\Delta s_{12} = \frac{(n-\gamma)}{n(\gamma-1)} R \ln\left(\frac{p_2}{p_1}\right) \quad (10.64a)$$

$$\Delta s_{12} = -\frac{(n-\gamma)}{(\gamma-1)} R \ln\left(\frac{v_2}{v_1}\right) \quad (10.64b)$$

Note: It should be noted that the above process is reversible and diabatic and in the limiting case with zero heat addition, reduces to the adiabatic reversible process described in the previous section, i.e. $n = \gamma$. All other processes dealt with above may be regarded as special cases of the polytropic reversible process. Thus

for the constant volume process: $n = \infty$
for the constant pressure process: $n = 0$
for the isothermal process: $n = 1$
for the reversible adiabatic process: $n = \gamma$.

An adiabatic irreversible process obeying the same law pv^n = constant (equation (10.59)) as the above diabatic reversible process is also possible. This will give rise to an entropy gain numerically equal to that given by equation (10.64), (10.64a), or (10.64b). However, this entropy gain is now the result of an internal irreversibility, not of external heat addition (see section 7.6).

EXAMPLE 10.5

Take the same gas as in Example 10.4 (p. 210) and compress it reversibly and polytropically between the same two volumes, the index of compression being 1·3. Calculate the final state, the work done and the heat exchange and change of entropy.

SOLUTION

$$\frac{T_2}{T_1} = \left(\frac{V_1}{V_2}\right)^{n-1} = \left(\frac{p_2}{p_1}\right)^{(n-1)/n}$$

$$T_2 = 280 \times 6^{0\cdot3} = 479\cdot1 \text{ K}$$

$$p_2 = 1 \times 6^{1\cdot3} = 10\cdot25 \text{ bar}$$

$$\text{Work done} = \frac{p_1V_1 - p_2V_2}{n-1} = \frac{mR(T_1 - T_2)}{n-1}$$

$$= \frac{1\cdot5 \times 0\cdot296(280 - 479\cdot1)}{0\cdot3}$$

$$= -295 \text{ kJ} \quad \textit{Ans.}$$

$$\text{Heat exchange} = \frac{\gamma - n}{(n-1)(\gamma - 1)} R(T_1 - T_2)$$

$$= \frac{\gamma - n}{n-1} \times -295$$

$$c_p - c_v = 0\cdot296 \text{ kJ/kg-K and } c_p = 1\cdot035 \text{ kJ/kg-K}$$

$$\gamma = c_p/c_v = 1\cdot035/0\cdot739 = 1\cdot402$$

$$Q_{12} = \frac{1\cdot10^2}{0\cdot402} \times -295 = -810 \text{ kJ} \quad \textit{Ans.}$$

The change of entropy Δs_{12}

$$= -\frac{n-\gamma}{\gamma - 1} mR \ln\left(\frac{V_2}{V_1}\right)$$

$$= -\frac{1\cdot3 - 1\cdot402}{0\cdot402} \times 1\cdot5 \times 0\cdot296 \ln \tfrac{1}{6}$$

$$= -0\cdot202 \text{ kJ/K} \quad \textit{Ans.}$$

10.5.2 Reversible open system processes (see sections 4.7 and 7.5)

(*a*) **Constant pressure process** (see *Fig.* 10.6)

This case is analogous to case (*a*) in section 4.7, a perfect gas replacing steam as the working fluid, e.g. a heat exchanger, with air at moderate temperatures.

p–v–T relationship, equation (10.44)

$$\frac{p_2}{p_1} = 1 \qquad \frac{v_2}{v_1} = \frac{T_2}{T_1} \tag{10.65}$$

Steady flow energy equation, S.F.E.E., equation (10.37)

$$q_{12} - w_{12} = (h_2 + \tfrac{1}{2}c_2^2 + gZ_2) - (h_1 + \tfrac{1}{2}c_1^2 + gZ_1) \quad (10.37)$$

In this application KE and PE effects are generally negligible, and external work effects are zero, i.e. $w_{12} = 0$.

$$\therefore q_{12} = h_2 - h_1 = c_p(T_2 - T_1) \quad (10.66)$$

See also equation (4.14).

Entropy change, equations (10.33) and (10.47)

$$\Delta s_{12} = c_p \ln\left(\frac{T_2}{T_1}\right) \quad (10.67)$$

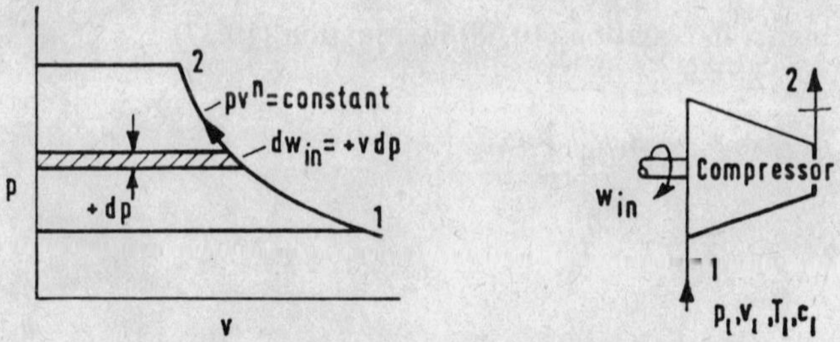

Fig. 10.10 Compression flow process

(*b*) **Compressor process** (*Fig.* 10.10)

p–v–T relationships

For reversible polytropic compression, pressure, volume and temperature are related by equations (10.59), (10.60) and (10.61), i.e.

$$\frac{p_2}{p_1} = \left(\frac{v_1}{v_2}\right)^n \quad (10.59)$$

$$\frac{T_2}{T_1} = \left(\frac{v_1}{v_2}\right)^{n-1} \quad (10.60)$$

$$\frac{T_2}{T_1} = \left(\frac{p_2}{p_1}\right)^{(n-1)/n} \quad (10.61)$$

Steady flow energy equation, S.F.E.E., equation (10.37)

$$q_{12} - w_{12} = (h_2 + \tfrac{1}{2}c_2^2 + gZ_2) - (h_1 + \tfrac{1}{2}c_1^2 + gZ_1) \quad (10.37)$$

where, from equation (10.38), and neglecting PE effects

$$w_{12} + \tfrac{1}{2}(c_2{}^2 - c_1{}^2) = \int_2^1 v\,dp = +\frac{n(p_1v_1 - p_2v_2)}{(n-1)} \qquad (10.68)$$

Note that $\int_2^1$ is a negative quantity. Since KE effects are usually small

$$w_{12} = \int_2^1 v\,dp \qquad (10.68a)$$

The fact that $\int_2^1 v\,dp$ is negative implies that work is done *on* the system.

Substituting from equation (10.68) in equation (10.37)

$$q_{12} = h_2 - h_1 + \int_2^1 v\,dp$$

$$= c_p(T_2 - T_1) + \frac{n}{n-1}(p_1v_1 - p_2v_2)$$

$$= \frac{(\gamma - n)}{(n-1)(\gamma - 1)} R\,(T_1 - T_2) \qquad (10.69)$$

See also equation (10.63).

Note that since, in a compression, generally $T_2 > T_1$, heat is added when $n > \gamma$, and abstracted when $n < \gamma$. In the "ideal" rotary compressor $n = \gamma$, i.e. the compression is isentropic.*

Entropy change

This is given, as in the case of closed system polytropic reversible processes, by equations (10.64), (10.64a) and (10.64b).

* The definition of an "ideal" compression process is a somewhat arbitrary matter. In the case of uncooled rotary, aerodynamic compressors employing aerofoil blades, as in the axial flow compressor, or alternatively radial vanes, as in the centrifugal compressor, the work input is a minimum under isentropic conditions, i.e. when $n = \gamma$.

On the other hand, in reciprocating compressors, where extensive cooling of the cylinder by suitably designed water jackets with the further possibility of intercooling between stages can readily be effected, the work input may be reduced still further by approaching the "ideal" of isothermal or constant temperature compression, i.e. $n = 1$.

(*c*) **Turbine or expander process** (see *Fig.* 10.11)

p–v–T relationships

These are given by equations (10.59), (10.60) and (10.61), above.

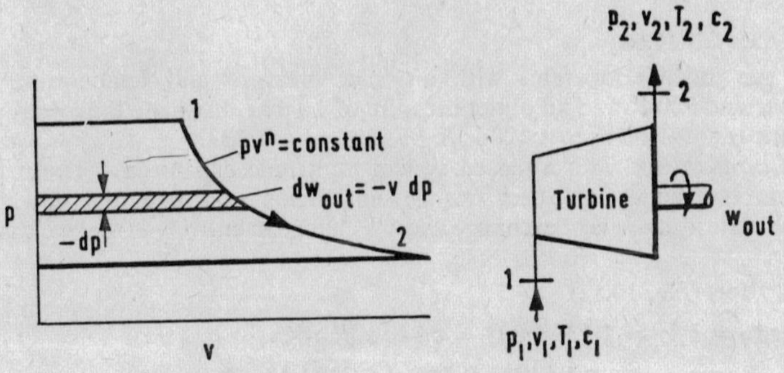

Fig. 10.11 Expansion flow process

Steady flow energy equation, S.F.E.E., equation (10.37)

From equation (10.38), and neglecting PE effects

$$w_{12} + \tfrac{1}{2}(c_2^2 - c_1^2) = \int_2^1 v\,dp = \frac{n}{n-1}(p_1v_1 - p_2v_2) \qquad (10.70)$$

Note that $\int_2^1 v\,dp$ is now a positive quantity, implying that work is done by the system, when KE effects are negligible. Substituting in the S.F.E.E., equation (10.37)

$$q_{12} = h_2 - h_1 + \int_2^1 v\,dp$$

$$= c_p(T_2 - T_1) + \frac{n}{n-1}(p_1v_1 - p_2v_2)$$

$$= \frac{\gamma - n}{(n-1)(\gamma - 1)} R(T_1 - T_2) \qquad (10.71)$$

Equations (10.70) and (10.71) for the expander are evidently identical with equations (10.68) and (10.69) for the compressor. However, since in expanders generally $T_2 < T_1$, work is done *by* the system whilst heat is *added* when $n < \gamma$, and *abstracted* when $n > \gamma$, the reverse of the compressor situation. In the ideal turbine $n = \gamma$, i.e. expansion is isentropic.

Entropy change

This is given, as in the case of compressors, by equations (10.64), (10.64a) and (10.64b).

In the "ideal" turbine, as in the "ideal" compressor, $h = \gamma$, i.e. the expansion is isentropic.

EXAMPLE 10.6a

A gas turbine operates with an inlet pressure and temperature of 7 bar and 920 K and an outlet pressure of 1·1 bar. Estimate the isentropic work output for $c_p = 1{\cdot}120$ kJ/kg-K and $\gamma = 1{\cdot}333$.

Compare this with a closed system isentropic expansion between the same two states. Neglect changes in kinetic and potential energies. Calculate also the heat exchange and the change in entropy in the two cases.

SOLUTION (*Fig.* 10.12)

$$c_v = c_p/\gamma = 1{\cdot}120/1{\cdot}333 = 0{\cdot}840 \text{ kJ/kg-K}$$

$$R = c_p - c_v = 1{\cdot}120 - 0{\cdot}840 = 0{\cdot}280 \text{ kJ/kg-K}$$

$$W_{turb} = \frac{\gamma}{\gamma - 1} R(T_1 - T_2)$$

$$= \frac{1{\cdot}333}{0{\cdot}333} \times 0{\cdot}280(920 - T_2) \text{ kJ/kg}$$

Now $\dfrac{T_2}{T_1} = \left(\dfrac{p_2}{p_1}\right)^{(\gamma-1)/\gamma}$ and therefore

$$T_2 = 920\left(\frac{1{\cdot}1}{7}\right)^{0{\cdot}333/1{\cdot}333} = 580 \text{ K}$$

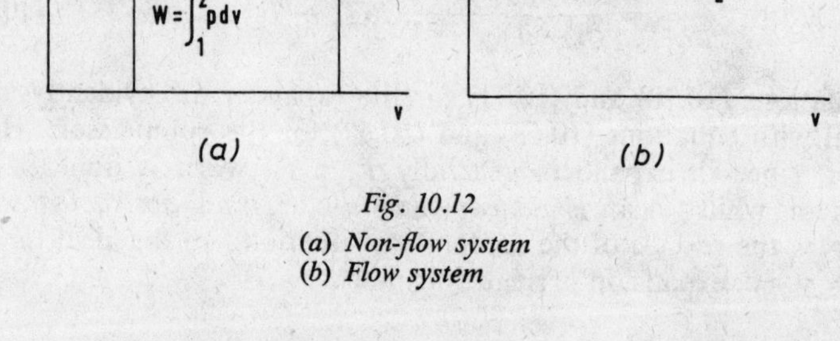

Fig. 10.12
(a) Non-flow system
(b) Flow system

so that

$$W_{turb} = \frac{1{\cdot}333}{0{\cdot}333} \times 0{\cdot}280(920 - 580)$$

$$= 382 \text{ kJ/kg} \quad \textit{Ans.}$$

For the closed system isentropic expansion (equation (10·55a)),

$$W_{isen} = \frac{1}{\gamma - 1} R(T_1 - T_2)$$

$$= 382/1{\cdot}333 = 286 \text{ kJ/kg} \quad \textit{Ans.}$$

The difference between the non-flow and the flow processes is seen very clearly on the p–v field (*Fig.* 10.12).

For turbine

$$q_{12} = 0 \text{ since isentropic} \quad \textit{Ans.}$$

or from (10.71), $q_{12} = c_p(T_1 - T_2) + \frac{\gamma}{\gamma - 1} R(T_2 - T_1)$

but since $c_p = \frac{\gamma}{\gamma - 1} R$

$$q_{12} = 0$$

likewise $\Delta s = s_2 - s_1 = -\frac{n - \gamma}{\gamma - 1} R \ln\left(\frac{v_2}{v_1}\right)$ (equation (10·64b))

and since $n = \gamma \quad \therefore \Delta s = 0$ *Ans.*

This is also the case in the non-flow process.

EXAMPLE 10.6b

Take the same example (10.6*a*) and assume a polytropic reversible expansion with the index 1·20 and compare the results with those for the isentropic expansion through the turbine.

SOLUTION

The work done $= \frac{n}{n - 1} R(T_1 - T_2)$ (equation (10.70))

$$T_2 = 920 \left(\frac{1{\cdot}1}{7}\right)^{0{\cdot}2/1{\cdot}2} = 675 \text{ K}$$

$$W_{turb} = \frac{1{\cdot}2}{0{\cdot}2} \times 0{\cdot}280(920 - 675)$$

$$= 411 \text{ kJ/kg} \quad \textit{Ans.}$$

Heat transferred (equation (10.71)):

$$q_{12} = c_p(T_2 - T_1) + \frac{n}{n-1} R(T_1 - T_2)$$

$$= 1{\cdot}120(675 - 920) + \frac{1{\cdot}2}{0{\cdot}2} \times 0{\cdot}280(920 - 675)$$

$$= 137 \text{ kJ/kg} \quad \textit{Ans.}$$

Entropy change (equation (10.64)):

$$\Delta s_{12} = \frac{n - \gamma}{(n-1)(\gamma - 1)} R \ln \left(\frac{T_2}{T_1}\right)$$

$$= \frac{1{\cdot}2 - 1{\cdot}333}{0{\cdot}2 \times 0{\cdot}333} \times 2{\cdot}80 \ln \left(\frac{675}{920}\right)$$

$$= 0{\cdot}01745 \text{ kJ/kg-K increase} \quad \textit{Ans.}$$

(*d*) **Nozzle process** (*Fig*. 10.13)

p–v–T relationships

This is again given by equations (10.59), (10.60) and (10.61) above.

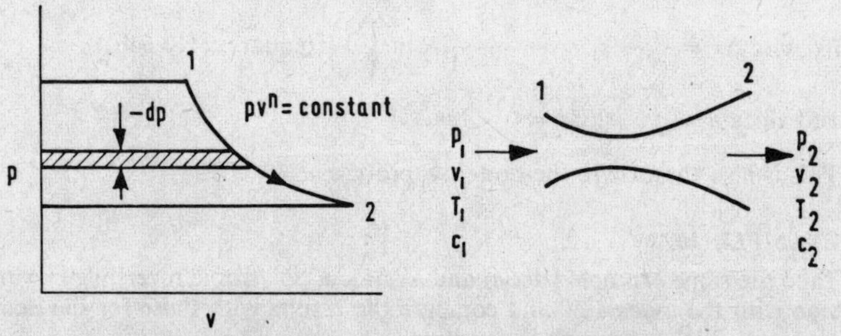

Fig. 10.13 Expansion process for nozzle, p–v and T–s diagrams

Steady flow energy equation

From equation (10.38), and neglecting PE as well as external work effects

$$\tfrac{1}{2}(c_2{}^2 - c_1{}^2) = \int_2^1 v\, dp = \frac{n}{n-1}(p_1 v_1 - p_2 v_2) \qquad (10.72)$$

$$= \frac{nR}{n-1}(T_1 - T_2) \qquad (10.72a)$$

and substituting in the S.F.E.E., equation (10.37)

$$q_{12} = h_2 - h_1 + \int_2^1 v\,dp$$

$$= c_p(T_2 - T_1) + \frac{n}{n-1}(p_1v_1 - p_2v_2)$$

$$= \frac{(\gamma - n)}{(n-1)(\gamma-1)}R(T_1 - T_2) \qquad (10.73)$$

This equation is again identical with equations (10.69) and (10.71) for the compressor and expander.

Since $\int_2^1 v\,dp$ is clearly a positive quantity (see *Fig.* 10.12), equation (10.72) indicates a gain in kinetic energy.

The position concerning heat addition or rejection is identical with that for the turbine.

10.5.3 Irreversible processes (see section 7.6)

Irreversible processes have been discussed in section 7.6. Equations (10.35) and (10.36) for closed systems and equations (10.38) and (10.39) for open systems which have been used to derive expressions for heat and work transfer in sections 10.5.1 and 10.5.2 above, for reversible processes, no longer apply and have to be replaced by more complicated relationships.

THE IRREVERSIBLE ADIABATIC PROCESS, pv^n = CONSTANT.

(*a*) **Closed systems** (*Fig.* 10.14, also section 7.6)

As mentioned at the end of section 10.5.1, the relationship pv^n = constant may describe not only a polytropic reversible process

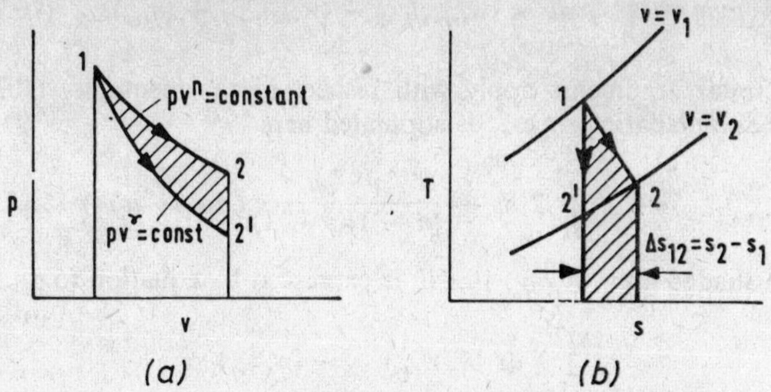

Fig. 10.14 Irreversible process, p–v and T–s diagrams

obeying equations (10.59) to (10.64), but also an adiabatic irreversible process.

While the N.F.E.E. (10.34) is still applicable in the latter case (see also section 5.7), work is no longer represented by equation (10.35), i.e.

$$(w_{irrev})_{1-2} \neq \int_1^2 p\,dv \tag{10.74}$$

nor heat transferred by equation (10.36), i.e.

$$(q_{irrev})_{1-2} \neq \int_1^2 T\,ds \tag{10.75}$$

From equation (5.6a)

$$\int_1^2 p\,dv = (w_{irrev})_{1-2} + (q_{rev})_{1-2}$$

(see also *Fig.* 10.14) and substituting for $(q_{rev})_{1-2}$ from equation (10.63a),

$$(w_{irrev})_{1-2} = \int_1^2 p\,dv - \frac{\gamma - n}{(n-1)(\gamma - 1)}(p_1v_1 - p_2v_2) \tag{10.74a}$$

Equation (10.74a) replaces the inequality (10.74). Since for an irreversible adiabatic expansion $n < \gamma$ (see *Fig.* 7.13), $(w_{irrev})_{1-2}$ is always less than the area under the curve describing the process path, i.e. $\int_1^2 p\,dv$. The shaded area in *Fig.* 10.14*a* is expressed by equation (5.8), i.e.

$$\int_1^2 p\,dv - \int_1^{2'} p\,dv = (w_{irrev})_{1-2} - (w_{rev})_{1-2'} + (q_{rev})_{1-2} \tag{10.74b}$$

Similar arguments apply with respect to the inequality (10.75). For an adiabatic process, as stipulated here

$$(q_{irrev})_{1-2} = 0 = \int_1^2 T\,ds - \frac{\gamma - n}{(n-1)(\gamma - 1)}(p_1v_1 - p_2v_2) \tag{10.75a}$$

The shaded area in *Fig.* 10.14*b* is expressed by equation (5.6), i.e.

$$\begin{aligned}\int_1^2 T\,ds &= (w_{rev})_{1-2} - (w_{irrev})_{1-2}\\ &= (q_{rev})_{1-2}\end{aligned} \tag{10.75b}$$

For adiabatic irreversible compression processes, $n > \gamma$ and substitution in equation (10.74a) shows that for such processes, $(w_{irrev})_{1-2}$ always exceeds the area under the curve describing the process path, i.e. $\int_1^2 p\,dv$.

Since the difference between the exponent n and the "ideal" exponent γ is a measure of the degree of irreversibility of the process, it should be possible to establish a relationship between n, γ and entropy gain.

Thus, considering any adiabatic irreversible process of the form pv^n = constant, the entropy gain is expressed by the property equation (10.32) which applies equally to reversible and irreversible processes.

Rearranging this equation (with $c_p = \gamma c_v$)

$$\frac{p_2}{p_1}\left(\frac{v_2}{v_1}\right)^{\gamma} = \exp\left(\frac{s_2 - s_1}{c_v}\right) \tag{10.32a}$$

but from the assumed process equation pv^n = constant

$$\frac{p_2}{p_1}\left(\frac{v_2}{v_1}\right)^{n} = 1{\cdot}0 \tag{10.59a}$$

so that, combining equations (10.32a) and (10.59a)

$$(n - \gamma)\ln\left(\frac{v_2}{v_1}\right) = -\left(\frac{s_2 - s_1}{c_v}\right) \tag{10.76}$$

Since any irreversibility leads to an increase in entropy (Corollary 8 of the second law, section 7.6), if follows that

$$-\left(\frac{s_2 - s_1}{c_v}\right) < 0$$

so that for $(v_2/v_1) > 1$, i.e. an expansion process

$$n < \gamma \tag{10.77a}$$

whilst for $(v_2/v_1) < 1$, i.e. a compression process

$$n > \gamma \tag{10.77b}$$

These relationships, have already been referred to in connexion with equation (10.74a) (see also *Figs*. 7.11 and 7.12).

For the more general case of diabatic irreversible processes, in which some heat transfer $(q_{irrev})_{1-2}$ occurs, the inequalities (10.74) and (10.75) still obtain, but equations (10.74a) and (10.74b) are modified as follows—

$$(w_{irrev})_{1-2} = \int_1^2 p\,dv - \frac{\gamma - n}{(n-1)(\gamma-1)}(p_1v_1 - p_2v_2) + (q_{irrev})_{1-2} \tag{10.74c}$$

$$\int_1^2 p\,dv - \int_1^{2'} p\,dv = (w_{irrev})_{1-2} - (w_{rev})_{1-2} + (q_{rev})_{1-2} - (q_{irrev})_{1-2} \tag{10.74d}$$

while equations (10.75a) and (10.75b) are replaced by the following equation

$$(q_{irrev})_{1-2} = \int_1^2 T\,ds - \frac{1}{(n-1)}(p_1v_1 - p_2v_2) + (w_{irrev})_{1-2} \tag{10.75c}$$

and

$$\int_1^2 T\,ds - \int_1^{2'} T\,ds = (w_{rev})_{1-2} - (w_{irrev})_{1-2} + (q_{irrev})_{1-2} \tag{10.75d}$$

(*b*) **Open systems**

As in the case of closed systems, the equation pv^n = constant may describe adiabatic irreversible, as well as diabatic reversible processes.

While the S.F.E.E. (10.37) is still applicable in the former case (see also section 5.8), work and heat transfer are no longer correctly represented by equations (10.38) and (10.39)

i.e. $$(w_{irrev})_{1-2} + \tfrac{1}{2}(c_2^2 - c_1^2) + g(Z_2 - Z_1) \neq \int_2^1 v\,dp \tag{10.78}$$

and $$(q_{irrev})_{1-2} \neq \int_1^2 T\,ds \tag{10.79}$$

Adapting the arguments of section 5.6 it may readily be shown that for adiabatic processes and open systems

$$\int_2^1 v\,dp = (w_{irrev})_{1-2} + [\tfrac{1}{2}(c_2^2 - c_1^2) + g(Z_2 - Z_1)]_{irrev} + (q_{rev})_{1-2}$$

so that, substituting for $(q_{rev})_{1-2}$ from equations (10.69) or (10.71) for open systems (which are identical with equation (10.63) for closed systems)

$$(w_{irrev})_{1-2} + [\tfrac{1}{2}(c_2^2 - c_1^2) + g(Z_2 - Z_1)]_{irrev} = \int_2^1 v\,dp - \frac{\gamma - n}{(n-1)(\gamma-1)}(p_1v_1 - p_2v_2) \tag{10.78a}$$

which replaces the inequality (10.78).

As in the case of closed systems, equation (10.75a), for adiabatic processes

$$(q_{irrev})_{1-2} = 0 = \int_1^2 T\,ds - \frac{\gamma - n}{(n-1)(\gamma-1)}(p_1v_1 - p_2v_2) \quad (10.79a)$$

The argument leading to the formulation of equations (10.76), (10.77a) and (10.77b) giving the relationship between the exponents n and γ in the case of closed systems apply equally to open systems.

For the more general case of adiabatic irreversible processes, equations (10.78a) and (10.79a) are modified as follows (cf. equations (10.74c) and (10.75c))—

$$(w_{irrev})_{1-2} + [\tfrac{1}{2}(c_2^2 - c_1^2) + g(Z_2 - Z_1)]_{irrev}$$

$$= \int_2^1 v\,dp - \frac{\gamma - n}{(n-1)(\gamma-1)}(p_1v_1 - p_2v_2) + (q_{irrev})_{1-2} \quad (10.78b)$$

$$(q_{irrev})_{1-2} = \int_1^2 T\,ds - \frac{\gamma - n}{(n-1)(\gamma-1)}(p_1v_1 - p_2v_2) \quad (10.79b)$$

Since, in the case of flow processes, the path limits are expressed more conveniently as a pressure ratio p_2/p_1 rather than a volume ratio v_2/v_1 as in the case of closed systems, equation (10.76) may be expressed for open systems in the form

$$\left(\frac{n-\gamma}{n}\right)\ln\left(\frac{p_1}{p_2}\right) = -\left(\frac{s_2 - s_1}{c_v}\right) \quad (10.80)$$

EXAMPLE 10.7

Oxygen at 17 bar and 500°C is expanded irreversibly to 1·5 bar with a loss of heat to the surroundings of 45 kJ/kg during the process. Calculate the work done and the change of entropy for (*a*) a non-flow process, and (*b*) a flow process.

Take the index for the expansion as $n = 1{\cdot}28$; $RO_2 = 0{\cdot}260$ kJ/kg-K.

SOLUTION

From the gas equation $10^2 p_1v_1 = RT$,

$$v_1 = \frac{0{\cdot}260 \times 773{\cdot}15}{17 \times 10^2} = 0{\cdot}118 \text{ m}^3/\text{kg}$$

$$p_1v_1^{1{\cdot}28} = p_2v_2^{1{\cdot}28}$$

$$v_2 = 0{\cdot}118\left(\frac{17}{1{\cdot}5}\right)^{1/1{\cdot}28} = 0{\cdot}668 \text{ m}^3/\text{kg}$$

$$T_2 = \frac{10^2 \times 1{\cdot}5 \times 0{\cdot}668}{0{\cdot}260} = 386 \text{ K}$$

Using Table A2, reference 10.4,

At 773·16 K $\quad h = 23\,600$ kJ/kg-mol

At 386 K $\quad h = 11\,150$ kJ/kg-mol

$$\text{Mean } C_p = \frac{23\,600 - 11\,150}{773{\cdot}16 - 386} = 32{\cdot}20 \text{ kJ/kg-mole-K}$$

$$\text{or mean } c_p = 32{\cdot}20/32{\cdot}0 = 1{\cdot}005 \text{ kJ/kg-K}$$

$$c_p - c_v = R \qquad 1{\cdot}005 - c_v = 0{\cdot}260$$

$$c_v = 0{\cdot}745 \text{ kJ/kg-K}$$

$$\gamma = \frac{1{\cdot}005}{0{\cdot}745} = 1{\cdot}35$$

n is given as 1·28.

Work done by the expanding gas

(*a*) Non-flow process (equation (10.74*c*))

$$(w_{irrev})_{1-2} = \int_1^2 p\,dv - \frac{\gamma - n}{(\gamma - 1)(n - 1)}(p_1v_1 - p_2v_2) + (q_{irrev})_{1-2}$$

$$= \frac{(p_1v_1 - p_2v_2)10^2}{0{\cdot}28}$$

$$- \frac{0{\cdot}07}{0{\cdot}35 \times 0{\cdot}28}(p_1v_1 - p_2v_2)10^2 + (q_{irrev})_{1-2}$$

$$= \frac{(p_1v_1 - p_2v_2)10^2}{0{\cdot}28}\left(1 - \frac{0{\cdot}07}{0{\cdot}35}\right) + (q_{irrev})_{1-2}$$

$$= \frac{(17 \times 0{\cdot}118 - 1{\cdot}5 \times 0{\cdot}668)10^2}{0{\cdot}28}(0{\cdot}8) + (q_{irrev})_{1-2}$$

$$= 287{\cdot}2 \text{ kJ/kg} + (q_{irrev})_{1-2}$$

$$= (287{\cdot}2 - 45{\cdot}0) \text{ kJ/kg}$$

$$= 242{\cdot}2 \text{ kJ/kg} \quad \textit{Ans.}$$

(*b*) Flow process (equation (10.78*b*))

Note that it is only the integral that is different in the two cases. The $\int_1^2 p\,dv$ is replaced by $\int_2^1 v\,dp$ in this case (*b*).

$$(w_{irrev})_{1-2}$$

$$= \frac{1{\cdot}28(17 \times 0{\cdot}118 - 1{\cdot}5 \times 0{\cdot}668)10^2}{0{\cdot}28}\left(1 - \frac{0{\cdot}07}{0{\cdot}35 \times 1{\cdot}28}\right) + (q_{irrev})_{1-2}$$

$$= 388{\cdot}0 \text{ kJ/kg} + (q_{irrev})_{1-2}$$

$$= (388{\cdot}0 - 45{\cdot}0) = 343{\cdot}0 \text{ kJ/kg} \quad \textit{Ans.}$$

Entropy change

(*a*) Non-flow process. From equation (10.76),

$$(n - \gamma) \ln\left(\frac{v_2}{v_1}\right) = -\left(\frac{s_2 - s_1}{c_v}\right)$$

$$-0{\cdot}07 \ln\left(\frac{0{\cdot}668}{0{\cdot}118}\right) = -\frac{(s_1 - s_2)}{0{\cdot}745}$$

$$(s_2 - s_1) = 0{\cdot}09035 \text{ kJ/kg-K}$$

(*b*) Flow system. From equation (10.80),

$$\frac{n - \gamma}{n} \ln\left(\frac{p_1}{p_2}\right) = \left(\frac{s_2 - s_1}{c_v}\right)$$

$$-\frac{0{\cdot}07}{1{\cdot}28} \ln\left(\frac{17}{1{\cdot}5}\right) = \frac{s_2 - s_1}{0{\cdot}745}$$

$$(s_2 - s_1) = 0{\cdot}09035 \text{ kJ/kg-K} \quad \textit{Ans.}$$

Both cases are associated with an increase in entropy, and since the end states are identical so is the change of entropy.

REFERENCES

10.1 DIN, F., *Thermodynamic Functions of Gases*, Vol. 2 (Butterworth, 1962).

10.2 KEENAN, J. H., and KAYE, J., *Gas Tables* (John Wiley, 1953).

10.3 GEYER, E. W., and BRUGES, E. A., *Tables of Properties of Gases with Dissociation Theory and its Applications* (Longmans, 1948).

10.4 BENSON, R. S., *Advanced Engineering Thermodynamics* (Pergamon, 1967).

EXERCISES

10.1 Calculate the mass of 0·0375 m³ of a perfect gas at 0°C and 1 bar if its molecular weight is 16 and the universal gas constant is 8·3143 kJ/kg-mol-K. Calculate also the specific heat at constant pressure and constant volume, assuming the gas to be diatomic.
(*Ans:* 0·00264 kg, 1·82 kJ/kg-K, 1·30 kJ/kg-K)

10.2 Find the change in internal energy of 0·15 kg of air when the temperature increases from $-10°C$ to 220°C. Treat as a perfect gas taking $\gamma = 1{\cdot}4$, $R = 0{\cdot}2871$ kJ/kg-K. If 30 kJ of heat are supplied to the gas during the process, how much external work is performed?

(*Ans:* 24·78 kJ; 5·22 kJ)

10.3 In an air standard cycle, the pressure at the end of compression is 30 bar and the volume 0·00023 m³. Heat is then supplied at constant pressure until the volume of the working fluid is doubled. Find the heat supplied and the change of internal energy, taking $\gamma = 1{\cdot}4$.

(*Ans:* 24·1 kJ; 17·2 kJ)

10.4 Atmospheric air at 1·015 bar and 15°C is compressed isothermally to 13·80 bar and is thereafter expanded adiabatically and without friction back to 1·015 bar. Given that $R = 0{\cdot}2871$ kJ/kg-K and $c_p = 1{\cdot}0049$ kJ/kg-K deduce the final temperature and specific volume.

(*Ans:* $-136°C$, 0·0385 m³/kg)

10.5 A balloon envelope of 170 m³ capacity is filled from gas storage cylinders charged initially to a pressure of 50 bar. Filling takes place very slowly, and the pressure in the storage cylinders is reduced to 1·8 bar when the envelope is just filled. Taking atmospheric pressure as 1 bar, determine the volume of the cylinders and the heat absorbed into them from the atmosphere.

(*Hint:* Treat cylinder volume constant and process as isothermal.)

(*Ans:* 3·52 m³; 17000 kJ)

10.6 Find how far the balloon of Exercise 10.5 is inflated if the storage cylinder volume is 3·52 m³ and the cylinder pressure drop is the same as before (i.e. 50 — 1·8), assuming that the bottles and balloon are made of perfect non-conductors. Take $\gamma = 1{\cdot}4$.

(*Ans:* 71·5% filled)

10.7 An air compressor takes in air at 0·965 bar and 20°C, compresses it reversibly according to the law $pv^{1\cdot 2}$ = constant, and delivers to a receiver at constant pressure of 9·65 bar. What is the temperature at the end of compression? Calculate, per kilogramme of air,

(*a*) the work done and the heat rejected during compression,
(*b*) the work done during delivery, and
(*c*) the work done in the whole cycle of suction, compression and delivery.

(*Ans:* $-89{\cdot}2$ kJ, $-44{\cdot}5$ kJ, -56 kJ, $-107{\cdot}5$ kJ)

10.8 A perfect gas at temperature T_1 is heated at constant pressure to T_2 and then expanded reversibly, according to the law pv^n = constant, until the temperature is once more T_1. What is the required value of n if the changes of entropy during the separate processes are equal?

(*Ans:* $n = 2$)

10.9 One kilogramme of a perfect gas is compressed through a pressure ratio of 4:1 according to the law $p = av + b$, and a and b being constant. The molecular weight of the gas is 44·01 and the ratio of the specific heats, γ, is 1·285. The initial pressure and temperature are 3·5 bar and 450 K respectively while the final volume is 0·01 m^3.

Calculate the work done, the heat transfer, and the change of entropy during the process and sketch the path in the T–s plane.

(*Ans:* −124·5 kJ; +68·5 kJ; +0·171 kJ/kg-K)

10.10 Define what is meant by the term "isentropic." A real gas is observed to have specific heats which depend on temperature according to the following relationships:

$$c_v = a + \alpha T$$
$$c_p = b + \alpha T$$

a, b and α being constants and T being the absolute temperature. Using the first law of thermodynamics derive an expression relating the absolute temperature T and the volume v of the gas during an isentropic process.

One kilogramme of gas at 1 bar occupies 0·875 m^3; after isentropic compression the temperature is 650°C. Calculate the work done if the constants a, b and α are respectively 0·157, 0·227 and 0·000025.

(*Ans:* $T^a v^{(b-a)} e^{\alpha T}$ = constant, 263·3 kJ)

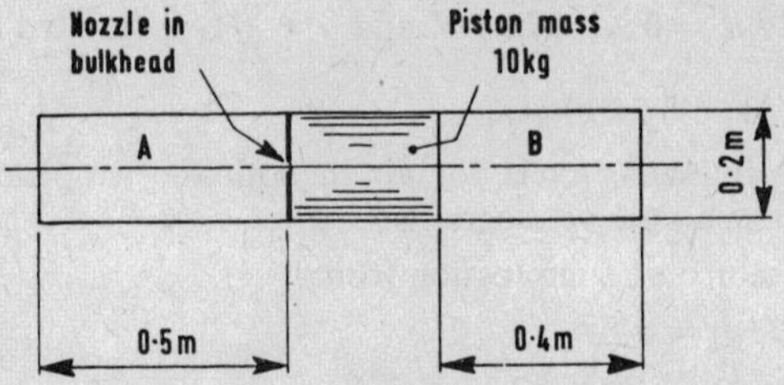

Fig. 10.15

10.11 *Fig.* 10.15 shows a schematic arrangement of a gun which is operated by compressed air and employed in a high-speed extrusion process. When primed the pressure in A (p_A) is 106 bar, while that in B (p_B) is 15 bar. The piston stays on the nozzle by virtue of the effective area difference. Triggering is achieved by a device which has the effect of breaking the seal between piston and nozzle so that the pressure p_A becomes effective over the whole of the piston area.

Assuming that the processes in A and B are reversible and adiabatic, calculate the distance moved by the piston at the instant when the piston velocity is a maximum, and find the value of the latter.

(*Ans:* 0·24 m, 87 m/s)

10.12 A rocket nozzle test rig requires a continuous and steady flow of Kerosene (specific gravity 0·82) at 24·2 bar over a test period of one and a half minutes. Compressed nitrogen is supplied from high-pressure bottles through a reducing valve to the Kerosene storage tank and so expels the Kerosene which then passes to the rocket motor.

Calculate the required bottle capacity if 1·82 kg of Kerosene are consumed per second and the maximum pressure in the nitrogen bottles is 207 bar. Assume that the starting temperature is 15·5°C, the minimum allowable pressure is 25 bar and that the bottles and storage tanks are:

(*a*) perfect conductors,
(*b*) perfect insulators.

(*Ans:* (*a*) 0·0372 m^3, (*b*) 0·0279 m^3)

10.13 Show from the First Law of Thermodynamics that the equation pv^γ = constant describes the adiabatic expansion of a perfect gas, and the change in entropy of unit mass of a perfect gas expanding from pressure p_1 to pressure p_2 is given by:

$$s_2 - s_1 = c_p \ln\left(\frac{T_2}{T_1}\right) - c_v(\gamma - 1)\ln\left(\frac{p_2}{p_1}\right)$$

1 kg of gas, initially at a pressure of 9·5 bar and a temperature of 500°C, expands in a cylinder according to the law $pv^{1 \cdot 333}$ = constant, to a pressure of 1 bar. Calculate the overall entropy change, assuming $R = 0{\cdot}248$ kJ/kg-K and $\gamma = 1{\cdot}4$. Comment briefly on the result.

(*Ans:* 0·06975 kJ/kg-K)

10.14 Plot the following cycle for air in both the temperature–entropy and pressure–volume diagrams:

(i) Isentropic compression from
$p_1 = 1$ bar
$T_1 = 15$°C
to $p_2 = 7$ bar

(ii) Isothermal heat addition, such as to give a volume ratio
$V_3/V_2 = 10$

(iii) Adiabatic irreversible expansion to the original temperature $T_1 = 15$°C with an efficiency of 80%.

(iv) Isothermal heat rejection to the original conditions p_1, T_1.

10.15 Formulate the energy equation for steady flow and hence determine the conditions governing throttling in the case of a perfect gas.

Air at 7 bar and 25°C is passed through a throttle valve, where its pressure is reduced to 1·4 bar. Neglecting kinetic energy effects, find the changes of internal energy and entropy resulting from the process.

Plot the process approximately in the pressure–volume and temperature–entropy diagrams, and discuss the significance of areas under the process curves in these two cases.

10.16 Discuss briefly the significance of entropy as a thermodynamic property.

Show that, for 1 kg of a perfect gas, the entropy change for any process connecting states p_1, u_1, T_1 and p_2, u_2, T_2 is given by

$$s_2 - s_1 = c_v \ln (p_2/p_1) + c_p \ln (V_2/V_1)$$

From the above relationship deduce relationships between pressure, volume, and temperature for isentropic processes.

Determine the external heat addition as well as the entropy, enthalpy, and internal energy change for a diatomic perfect gas of molecular weight 28 undergoing heating from 34·5 bar and 315°C to 1090°C:

(*a*) at constant volume,
(*b*) at constant pressure.

(*Ans:* (*a*) 577·5 kJ/kg, 0·6225 kJ/kg-K, 806 kJ/kg, 577·5 kJ/kg
(*b*) 806 kJ/kg, 0·8695 kJ/kg-K, 806 kJ/kg, 577·5 kJ/kg)

10.17 A cylinder contains 1 kg of air at a pressure and temperature of 8·5 bar and 95°C. The air then expands against a piston, the temperature being maintained constant at 95°C by adding heat from an external source through a temperature gradient of 60°C, and the work done being 30 kJ. Subsequently the air is compressed isothermally, at 95°C, the heat being abstracted by an external sink through a temperature gradient of 60°C, until the original state is reached.

What is the change of entropy between the beginning and end of the complete process:

(*a*) for the air in the cylinder,
(*b*) for the complete system consisting of air, cylinder, source, and sink.

Is this process reversible?
(*Ans:* (*a*) 0, (*b*) 0·0274 kJ/kg-K)

10.18 Describe clearly the nature of thermodynamic irreversibility in the following cases:

(*a*) unresisted expansion of a gas when suddenly released from a containing vessel to atmosphere,
(*b*) expansion of gas in a cylinder in the presence of mechanical friction, viscous shearing effects, and inertia effects.

An engine cylinder contains 0·0227 kg of air at an initial pressure and temperature of 34·5 bar and 815°C, respectively. The air then expands according to the law:

$$pv^{1 \cdot 25} = \text{constant}$$

to 5·17 bar. Find the work done and heat added during expansion assuming the process to take place:

(*a*) polytropically and reversibly,
(*b*) adiabatically and irreversibly.

What is the expansion efficiency in the second case?

(*Ans:* (*a*) $W = 8{\cdot}56$ kJ, $Q = 3{\cdot}14$ kJ, (*b*) $W = 5{\cdot}41$ kJ, $Q = 0$)

10.19 Discuss briefly the significance of the ratio of specific heats in thermodynamic calculations involving gases.

A closed vessel contains nitrogen at 15°C and 17·2 bar. A valve is opened momentarily and the pressure falls rapidly and isentropically to 6·9 bar. Some time later the temperature is again 15°C, and the pressure is observed to be 8·94 bar. Find γ for nitrogen.

Universal gas constant $G = 8{\cdot}3143$ kJ/kg-K.

Molecular weight of nitrogen = 28.

(*Ans:* $\gamma = 1{\cdot}4$)

10.20 Explain the use of the temperature–entropy diagrams for the representation of thermodynamic processes and its advantages over the pressure–volume diagram.

Air expands irreversibly from 6·9 bar and 555 K to 1·38 bar the entropy gain being 0·1675 kJ/kg-K. Find the equivalent law of expansion in the form pv^n = constant and calculate the work done, comparing it with the equivalent isentropic work.

(*Ans:* $n = 1{\cdot}225$, $w = 142$ kJ/kg, cf. 206 kJ/kg)

10.21 Explain clearly the difference between the properties of a substance, and the quantities heat and work. Hence show qualitatively that entropy is a property.

With the aid of a temperature–entropy diagram and using the relationship between entropy s, volume v and pressure p,

$$s_2 - s_1 = c_p \ln\left(\frac{v_2}{v_1}\right) + c_v \ln\left(\frac{p_2}{p_1}\right)$$

find the entropy change for a flow type compression according to the law $pv^{1\cdot16}$ = constant, for air at 1·38 bar and 26·5°C to 5·5 bar.

Assuming that the above process is replaced by isentropic compression between the same pressure limits followed by constant pressure heating to the same state, compare work done and heat added in the two cases.

(*Ans:* 0·1246 kJ/kg-K, $W = 207$ kJ/kg, cf. 148 kJ/kg and $Q =$ 59·5 kJ/kg)

10.22 What are the characteristics of the perfect gas?

Deduce expressions for the entropy change undergone by a perfect gas between any two state points in terms of pressure and specific volume ratio.

With the aid of this expression evaluate the entropy change undergone by a perfect gas when expanding adiabatically and irreversibly through a pressure ratio of 4 from an initial pressure of 6·9 bar and temperature of 65·5°C according to the relationship $pv^{1\cdot 2}$ = constant.

Find also the useful work done during expansion and compare it with that obtained for a polytropic reversible process obeying the same pressure–volume relationship.

Take $c_p = 1\cdot 005$ kJ/kg-K, $\gamma = 1\cdot 4$, and evaluate the various quantities for 1 lb of gas.

(*Ans:* $\Delta s = 0\cdot 1655$ kJ/kg-K, $W = 48\cdot 6$ kJ/kg, cf. 100·6 kJ/kg)

10.23 A certain mass of sulphur dioxide is contained in a vessel of 0·142 m³ capacity, at a pressure and temperature of 23·1 bar and 18°C, respectively. A valve is opened momentarily and the pressure falls immediately to 6·9 bar. Some time later the temperature is again 18°C, and the pressure is observed to be 9·1 bar. Estimate the value of γ for the gas.

Universal gas constant $G = 8\cdot 3143$ kJ/kg-mol-K.

(*Ans:* $\gamma = 1\cdot 28$)

10.24 Two vessels, each of 1 m³ capacity, are connected by a throttle valve. One vessel is filled with air at 6·9 bar and 49°C and the other with air at 1·38 bar and 15°C. The valve is suddenly opened and pressure and temperature equilibrium are eventually established.

Assuming that the process occurs in two phases, viz. rapid isentropic expansion of the air in the high-pressure vessel and simultaneous charging of the low-pressure vessel, until pressure balance is established, followed by gradual temperature equalization, accompanied by secondary pressure equalization, calculate the energy loss from the high-pressure vessel during the first phase, as well as the equalization pressure.

For air: $R = 0\cdot 2871$ kJ/kg-K

$\gamma = 1\cdot 40.$

(*Ans:* 686 kJ, 4·14 bar)

Chapter 11

Basic Flow Equations

In Chapter 4 we discussed several forms of the energy equation derived from the first law of thermodynamics. The two most important were that for the closed system in which only energy transfers take place across the boundaries of the system, but in which mass transfers are absent, the so-called non-flow energy equation N.F.E.E., equation (4.4), and on the other hand that for the open or flow system in which both energy and mass transfer occur simultaneously and which, in the absence of time variable effects, reduces to the steady flow energy equation, S.F.E.E., equation (4.13).

These equations were derived without reference to any laws other than the first law of thermodynamics. However, any flow process such as that described by the steady flow energy equation (S.F.E.E.) is subject not only to the first law, but, as shown in Chapter 7, also to the second law of thermodynamics as expressed most conveniently by the ENTROPY EQUATIONS, equations (7.8a) and (7.8b). In addition to the above two laws of thermodynamics, the flow of fluids is subject also to the laws of mechanics, in particular Newton's second law as expressed by the MOMENTUM EQUATION and the law of conservation of mass, expressed as the so-called CONTINUITY EQUATION. Finally, the behaviour of the particular fluid under consideration is governed by the STATE EQUATION, e.g. that for the perfect gas discussed in Chapter 10.

Since consideration of the basic flow equations thus involves not only the laws of thermodynamics, but also those of fluid mechanics, it is impossible to separate the equations derived from these laws and for this reason a chapter on the basic flow equation is not considered inappropriate in a text on basic engineering thermodynamics.

In the particular case where the working fluid may be treated as a perfect gas, certain particularly convenient and elegant formulations of the basic flow equations result; these form the basis of the science of GAS DYNAMICS.

In the present chapter certain restrictions will be placed on the flows considered, namely

(*a*) The flow is treated as essentially *one-dimensional*, i.e. variations in fluid properties normal to the direction of flow are considered negligible.
(*b*) The flow is assumed to be steady, i.e. variations in the properties of the fluid or in flow velocities occur only with respect to *distance*, not with respect to *time*.
(*c*) The working fluid is a pure substance occurring in a single phase only.

11.1 Formulation of the General Equations Governing One-dimensional Flow of a Compressible Fluid with Heat Addition, Friction and External Work

As shown in section 4.6 the flow is studied within a system contained partly by physical and partly by artificial boundaries, i.e. partly by the walls of the duct or vessel under consideration, partly by sections through the entering and leaving flow.

The problem to be solved may be formulated as follows.

Given the fluid properties (pressure p, density ρ, enthalpy h, entropy s, etc.) any two of which completely define the state of the fluid under condition (*c*) above, as well as the velocity c at the entry section of the system, and given further the numerical values of the rate of external heat addition, wall friction, and external work done by or on the fluid between the sections concerned, as well as the geometry of the duct, to find the changes in the above quantities, i.e. Δp, $\Delta \rho$, Δh, Δs and Δc, between the two sections and hence the conditions at the exit section.

As already discussed in the introduction to this chapter, five equations are available for the solution of these five unknowns, each based on a fundamental physical law, namely

(1) Steady flow energy equations (first law of thermodynamics)
(2) Entropy equation (second law of thermodynamics)
(3) Momentum equation (Newton's second law)
(4) Continuity equation (law of conservation of mass)
(5) State equation (either from observation or in limit from fundamental structure of matter).

These will, for convenience, be treated in the order (4), (3), (1), (2), (5).

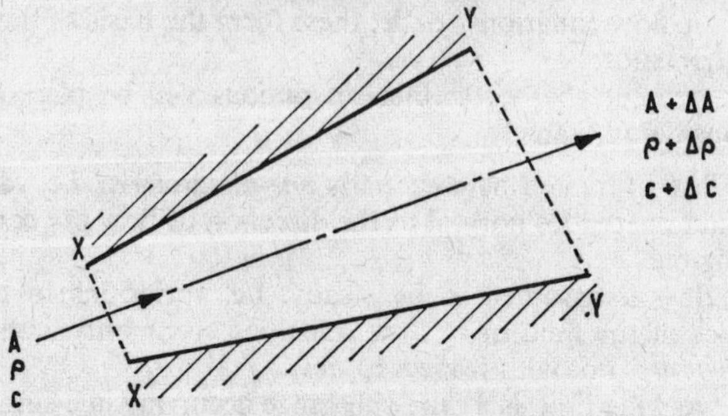

Fig. 11.1 Illustrating continuity equation

CONTINUITY EQUATION (see *Fig.* 11.1)

This is the statement of the law of conversation of mass. Since the flow is assumed to be steady, there is no possibility of accumulation of mass within the system and hence the continuity equation may be formulated as follows—

Rate of inflow of mass = rate of outflow of mass

i.e. $$A\rho c = (A + \Delta A)(\rho + \Delta\rho)(c + \Delta c)$$

or $$\frac{\Delta A}{A} + \frac{\Delta\rho}{\rho} + \frac{\Delta c}{c} = 0 \qquad (11.1)$$

In differential form, equation (11.1) becomes

$$\frac{dA}{A} + \frac{d\rho}{\rho} + \frac{dc}{c} = 0 \qquad (11.1a)$$

MOMENTUM EQUATION (*Fig.* 11.2)

The fluid element within the system shown in *Fig.* 11.2 is in equilibrium under the action of five external forces and the inertia force experienced by the fluid element. Resolving in the direction of motion the five force components are

(*a*) *Pressure force at section XX*

$$(F_p)_1 = pA \qquad (i)$$

(*b*) *Pressure force at section YY*

$$\begin{aligned}(F_p)_2 &= -(p + \Delta p)(A + \Delta A)\\ &= -pA - A\Delta p - p\Delta A \qquad (ii)\end{aligned}$$

(neglecting the second order term $\Delta p\Delta A$).

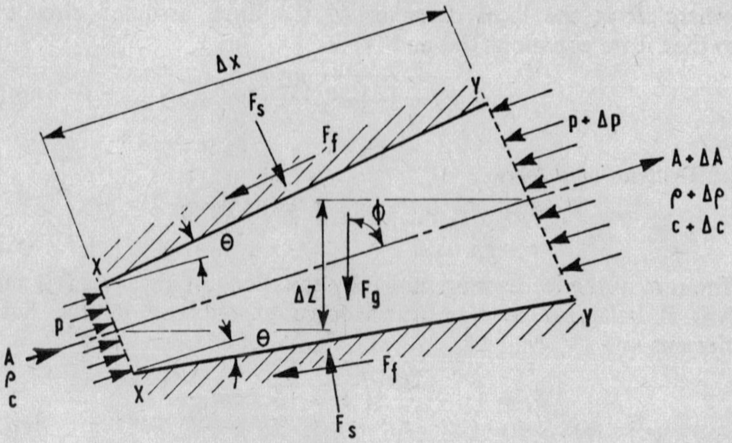

Fig. 11.2 Illustrating momentum equation

(*c*) *Side wall pressure force*

$$F_s = (p + \tfrac{1}{2}\Delta p)\Delta A_w \sin\theta$$
$$= (p + \tfrac{1}{2}\Delta p)\Delta A \simeq p\Delta A \qquad \text{(iii)}$$

(neglecting the second order term $\frac{1}{2}\Delta p\Delta A$), where $\Delta A_w \sin\theta$ is the wall surface ΔA_w projected on to section YY.

(*d*) *Side wall friction force*

$$F_f = \tau\Delta A_w \cos\theta \qquad \text{(iv)}$$

where τ is the wall shear stress which, according to the usual definition of a friction factor f, may be written in the form

$$\tau = \frac{f\rho c^2}{8} \qquad \text{(v)}$$

so that, substituting from (v) in (iv) and writing

$$\Delta A_w \cos\theta = \frac{\Delta A \cos\theta}{\sin\theta} = \Delta A \cot\theta$$

and $$\Delta A = \frac{(D + 2\Delta x \tan\theta)^2 - D^2\pi}{4} \simeq \frac{4A \tan\theta\Delta x}{D}$$

$$\Delta A_w \cos\theta = \frac{4A}{D}\Delta x$$

(where D is the local diameter of the duct, assumed circular) so that from equations (iv) and (v)

$$F_f = -\tfrac{1}{2}\rho c^2 f\frac{A\Delta x}{D} \qquad \text{(vi)}$$

(*e*) *Gravitational force*

$$F_g = -g(\rho + \tfrac{1}{2}\Delta\rho)(A + \tfrac{1}{2}\Delta A)\Delta x \cos\phi$$
$$\simeq -g\rho A\Delta Z \qquad \text{(vii)}$$

The sum of the forces represented by equations (i), (ii), (iii), (vi) and (vii) is balanced by the inertia force experienced by the fluid element—

$$F_i = \left(\rho + \frac{\Delta\rho}{2}\right)\left(A + \frac{\Delta A}{2}\right)\Delta x\,\frac{\Delta c}{\Delta t}$$
$$\simeq \rho A c\Delta c \qquad \text{(viii)}$$

so that equating the sum of equations (i), (ii), (iii), (vi) and (vii) and equation (viii) we have, after slight simplification

$$\Delta p + \tfrac{1}{2}\rho c^2 f\frac{\Delta x}{D} + g\rho\Delta Z + \rho c\Delta c = 0 \qquad (11.2)$$

or, alternatively, in differential form

$$dp + \tfrac{1}{2}\rho c^2 f\frac{dx}{D} + g\rho\, dZ + \rho c\, dc = 0 \qquad (11.2a)$$

STEADY FLOW ENERGY EQUATION

This has already been established in section 4.6 in the form of equations (4.13a) and (4.13b), and may be written

$$Q - W = m(\Delta h + \tfrac{1}{2}\Delta c^2 + g\Delta Z) \qquad \text{(i)}$$

Putting $m = Ac\rho$ and $Q = q\,\Delta x$ where q = rate of heat addition per unit length of duct, we obtain

$$q\Delta x - \Delta W = (Ac\rho)(\Delta h + c\Delta c + g\Delta Z) \qquad (11.3)$$

or written in differential form

$$q\,dx - dW = Ac\rho(dh + c\,dc + g\,dZ) \qquad (11.3a)$$

ENTROPY EQUATION

Inasmuch as the property of entropy arose out of considerations based on the second law of thermodynamics, we may regard the entropy equations (7.8a) and (7.8b) of Chapter 7 as mathematical expressions of the second law, applicable quite generally to reversible and irreversible processes, as well as to closed and open systems.

Since in flow problems enthalpy is a more appropriate property function than internal energy, equation (7.8b) will be used, i.e.

$$T\Delta s = \Delta h - \frac{1}{\rho}\Delta p \tag{11.4}$$

or in differential form

$$T\,ds = dh - \frac{1}{\rho}\,dp \tag{11.4a}$$

STATE EQUATION

The state of the pure substance is defined by the values of two of its properties (see Chapter 8), the most convenient combination being pressure and specific volume. In certain cases, these property relations may be expressed by an EQUATION OF STATE, connecting any one "dependent" property, e.g. temperature, enthalpy, entropy, with the two "independent" properties of pressure (p) and specific volume (v), the latter being replaced for convenience by its reciprocal, the density ρ. These property relationships have been discussed in some detail in Chapter 8, while methods of deriving new property relationships from existing ones are discussed in detail in Appendixes II and III.

Equations of state, as such, are formulated most readily for gases, e.g.

the perfect gas equation $pv = RT$ (i)

or van der Waals' equation $\left(p + \frac{a}{v^2}\right)(v - b) = RT$ (ii)

but irrespective of whether a suitable state equation exists or not, any three of the four thermodynamic properties used in the formulation of the present problem—

(*a*) pressure p
(*b*) density ρ
(*c*) enthalpy h
(*d*) entropy s

may be considered to be functionally related to each other by equations of the form

$$h = f_1(p, \rho) \tag{11.5a}$$

or

$$s = f_2(p, \rho) \tag{11.5b}$$

so that, for an incremental change of state between sections XX and YY (*Fig.* 11.2) suitable differential or finite difference expressions may be derived.

In the case of the perfect gas as the working fluid we may write, using the relationships of Chapter 10

$$dh = c_p\,dT = \frac{c_p\,d(p/\rho)}{R}$$

But
$$d\left(\frac{p}{\rho}\right) = \frac{\rho\,dp - p\,d\rho}{\rho^2}$$

and
$$\frac{c_p}{R} = \frac{\gamma R}{(\gamma - 1)R} = \frac{\gamma}{\gamma - 1}$$

whence
$$dh = \frac{\gamma}{\gamma - 1}\left(\frac{\rho\,dp - p\,d\rho}{\rho^2}\right) \tag{11.6a}$$

or in finite difference form

$$\Delta h = \frac{\gamma}{\gamma - 1}\,\frac{\rho\Delta p - p\Delta\rho}{\rho^2} \tag{11.6b}$$

Equations (11.1), (11.2), (11.3), (11.4), and (11.5) completely define the given problem and may be solved by the general methods applicable to simultaneous first order algebraic equations for the five dependent variables Δp, $\Delta\rho$, Δh, Δs and Δc.

Generally, it is possible to simplify these equations considerably since, of the "disturbing influences" acting on the flow and considered in the present treatment, namely

(1) area change
(2) fluid friction
(3) external heat addition
(4) external work

all but one can in many cases be neglected without serious loss of accuracy.

Furthermore, for most gases, solutions of adequate accuracy may be obtained by using the perfect gas state equation $pv = RT$. The combination of the perfect gas equation with *one* of the above "disturbing influences" leads to solutions of considerable elegance having wide application in different types of duct flow. These flows will be discussed briefly in a later section, but are listed here for convenience—

(i) Perfect gas flow with area change only (nozzles, diffusers).
(ii) Perfect gas flow with fluid friction only (pipes and ducts). This flow is known as FANNO FLOW.
(iii) Perfect gas flow with external heat addition only (pipes and ducts). This flow is known as RAYLEIGH FLOW.

11.2 Relationships between Steady Flow Energy Equation and the Integrated Momentum or Bernoulli Equation

The relationship between the S.F.E.E. and the Bernoulli equation, the first being derived from the first law of thermodynamics and the second from Newton's second law, has already been discussed briefly in section 4.9. It was shown in that section that in the special case of an adiabatic reversible process only, and in the absence of external work, these two equations, although derived from two quite distinct physical principles, become identical. It was also shown that the S.F.E.E. has a more general significance than the Bernoulli equation, since it is not restricted to reversible processes, and since work transfer may be included.

Using the extended form of the momentum equation, i.e. equation (11.2), in which dissipation effects and hence irreversibilities are taken into account, and introducing the internal entropy change Δs_i as a quantitative measure of irreversibility (see Chapter 7, section 7.6, equation (7.22)) it is possible to extend the comparison between the S.F.E.E. and the momentum equation in section 4.9 and to find interesting new relationships.

In the absence of external work, the S.F.E.E. equation (11.3) becomes

$$q\Delta x = Ac\rho(\Delta h + c\,\Delta c + g\Delta Z) \tag{11.7}$$

Substituting for Δh from the entropy equation (11.4)

$$\Delta h = T\,\Delta s + \frac{\Delta p}{\rho} \tag{i}$$

$$q\Delta x = Ac\rho\left[\frac{\Delta p}{\rho} + c\,\Delta c + g\Delta Z + T\Delta s\right] \tag{11.7a}$$

where, from equation (7.22), the entropy change Δs is given as the sum of the *external* and *internal* entropy changes, the former being due to external heat transfer, and the latter to dissipation effects or internal irreversibilities. Hence, writing

$$\Delta s = \Delta s_e + \Delta s_i = \frac{q\Delta x}{(Ac\rho)T} + \Delta s_i \qquad \text{(ii)}$$

where $q\Delta x$ = heat transferred in unit time to the system,

$Ac\rho$ = mass of fluid receiving heat in unit time within the system,

and substituting from equation (ii) in equation (11.7a)

$$0 = Ac\rho\left(\frac{\Delta p}{\rho} + c\,\Delta c + g\,\Delta Z + T\Delta s_i\right) \qquad (11.8)$$

or

$$\Delta p + \rho c\,\Delta c + g\rho\,\Delta Z + \rho T\,\Delta s_i = 0 \qquad (11.8a)$$

Equation (11.8a) is derived directly from the S.F.E.E. from purely thermodynamic considerations. However, it is seen to be strikingly similar in form to the momentum equation (11.2) and since it contains pressure as an explicit variable, it is known as the PRESSURE–ENERGY EQUATION.

Rearranging the momentum equation (11.2) very slightly

$$\Delta p + \rho c\,\Delta c + g\rho\,\Delta Z + \tfrac{1}{2}\rho c^2 f\,\frac{\Delta x}{D} = 0 \qquad (11.9)$$

the similarity between the two equations becomes immediately obvious. Since all terms other than the last are identical, the latter must correspond, i.e.

$$\rho T\,\Delta s_i = \tfrac{1}{2}\rho c^2 f \frac{\Delta x}{D} \qquad (11.10)$$

The significance of equation (11.10) lies in the fact that it provides a direct quantitative correlation between the intensity of dissipative effects, measured by the friction factor f, and the internal entropy gain Δs_i as the thermodynamic expression of the resultant internal irreversibility.

In section 4.9, the comparison between the momentum and energy equations was extended also to their integrated forms.

The integration of the S.F.E.E., equation (11.7), in the absence of heat as well as work transfer effects may be performed without difficulty, i.e.

$$(h_2 - h_1) + \tfrac{1}{2}(c_2^2 - c_1^2) + g(Z_2 - Z_1) = 0 \qquad (11.7b)$$

The momentum equation (11.9) can be integrated directly only when

(1) the flow is incompressible, i.e. ρ = constant,
(2) the flow is non-dissipative, i.e. $f = 0$.

In that particular case, the momentum equation reduces to the simple form of the Bernoulli equation (4.23)

$$\frac{p_2 - p_1}{\rho} + \tfrac{1}{2}(c_2^2 - c_1^2) + g(Z_2 - Z_1) = 0 \qquad (11.9a)$$

Furthermore, since from the entropy equation (11.4), for the special case of an *adiabatic reversible* process only, in which $\Delta s_i = \Delta s_e = 0$,

$$\Delta h = \frac{1}{\rho}\Delta p$$

or for an *incompressible* fluid, by integration of the above equation

$$h_2 - h_1 = \frac{1}{\rho}(p_2 - p_1) \qquad \text{(iii)}$$

Substitution from equation (iii) in equation (11.9a) reduces the latter to a form identical with the S.F.E.E. as expressed by equation (11.7b) for the more general case of compressible, adiabatic flow, both non-dissipative and dissipative. Thus, once again, the energy equation is more general than the momentum equation.

Removing next the restriction of incompressible flow, the integrated momentum equation (11.9) becomes

$$\int_1^2 \frac{1}{\rho}\,dp + \tfrac{1}{2}(c_2^2 - c_1^2) + g(Z_2 - Z_1) = 0 \qquad (11.9b)$$

Reverting once again to the entropy equation (11.4), for an adiabatic reversible process, but removing the restriction of incompressible flow, we may write

$$h_2 - h_1 = \int_1^2 \frac{1}{\rho}\,dp \qquad \text{(iv)}$$

so that, substituting from equation (iv) in equation (11.9b), the latter reduces once again to the form of the S.F.E.E. (11.7b).

Summarizing the above two results—

The integrated momentum equation and the S.F.E.E. reduce to identical forms provided heat and work transfers are absent, and the flow is free from dissipative effects.

Finally, removing the second restriction of non-dissipative flow, the momentum equation (11.9), unlike the S.F.E.E. (equation (11.7)) is now no longer capable of direct integration.

Thus, dividing equation (11.9) by the density ρ, and rewriting in differential form, the second and third terms may be integrated directly to yield $\frac{1}{2}(c_2^2 - c_1^2)$ and $g(Z_2 - Z_1)$, respectively.

However, the first term, i.e. (dp/ρ), upon integration after substituting from the entropy equation (11.4a), is no longer equal to the enthalpy difference $(h_2 - h_1)$ but must be written

$$\int_1^2 \frac{dp}{\rho} = (h_2 - h_1) - \int_1^2 T\,ds$$

where, from equation (11.10) and putting $ds = ds_i$ for adiabatic flow (equation (7.22)),

$$T\,ds = \tfrac{1}{2}c^2 f \frac{dx}{D}$$

i.e.

$$\int_1^2 \frac{dp}{\rho} = (h_2 - h_1) - \int_1^2 \tfrac{1}{2}c^2 f \frac{dx}{D} \qquad \text{(v)}$$

Since the last term of equation (11.9) upon integration cancels the last term of auxiliary equation (v), above, we have from equation (11.9)

$$\int_1^2 \frac{dp}{\rho} = -\int_1^2 c\,dc - \int_1^2 g\,dZ - \int_1^2 \tfrac{1}{2}c^2 f \frac{dx}{D}$$

so that by substitution in equation (v)

$$(h_2 - h_1) - \int_1^2 \tfrac{1}{2}c^2 f \frac{dx}{D} + \tfrac{1}{2}(c_2^2 - c_1^2) + g(Z_2 - Z_1) + \int_1^2 \tfrac{1}{2}c^2 f \frac{dx}{D} = 0$$

or

$$(h_2 - h_1) + \tfrac{1}{2}(c_2^2 - c_1^2) + g(Z_2 - Z_1) = 0 \qquad \text{(11.9c)}$$

Hence, ultimately, the momentum equation (11.9) once again reduces to the form of the S.F.E.E., but *direct integration* is possible only in the first case, i.e. when the flow is both reversible and non-dissipative.

EXAMPLE 11.1

An incompressible fluid of density 1·2 kg/m³ and at 15°C flows through a horizontal pipe of 0·15 m diameter with a velocity of 76 m/s. Find (*a*) the local pressure gradient, (*b*) the local entropy gradient, when $f = 0{\cdot}005$.

SOLUTION

From equation (11.9), eliminating the terms $\rho c\Delta c$ (flow is incompressible, $\Delta c = 0$) and $g\rho\Delta Z$ (pipe is horizontal),

$$\frac{dp}{dx} = \frac{\frac{1}{2}\rho c^2 f}{D} = \frac{\frac{1}{2} \times 1{\cdot}2 \times (76)^2 \times 0{\cdot}005}{0{\cdot}15}$$

$$= 115{\cdot}5 \text{ (N/m}^2\text{)/m run} \quad \textit{Ans.}$$

From equation (11.10), converting into consistent units,

$$\frac{ds}{dx} = \frac{\frac{1}{2}c^2 f}{TD} = \frac{\frac{1}{2} \times (76)^2 \times 0{\cdot}005}{288 \times 0{\cdot}15}$$

$$= 0{\cdot}333 \text{ (J/kg-K)/m run} \quad \textit{Ans.}$$

11.3 Basic Concepts of One-dimensional Gas Dynamics

As already stated in section 11.1, simple and elegant solutions of the flow equations (11.1) to (11.5) are possible when the number of "disturbing influences" on the flow is reduced to one, and when the perfect gas state equation is assumed.

11.3.1 Mach number

Such analyses are further simplified by introducing a new parameter, the Mach number M, into the equations. The Mach number, which is a measure of the importance of compressibility effects, may be defined as

$$\text{Mach number } M = \frac{\text{local particle velocity } c}{\text{local acoustic velocity } a} \qquad (11.11)$$

When $M < 1$, the flow is described as SUBSONIC.
When $M = 1$, the flow is described as SONIC.
When $M > 1$, the flow is described as SUPERSONIC.

The acoustic or sound velocity is the propagation velocity of an infinitely small disturbance relative to the fluid ahead of it, and may be derived from simple momentum considerations, applied to a duct of uniform cross-section A.

Let AA represent a plane acoustic wave of amplitude Δp propagating to the left with velocity a' into the gas just ahead of the wave front which, itself, has a particle velocity c. Since the wave front AA is a propagating one, the system is not a steady flow system, as was stipulated in the introduction to the present chapter.

However, it may readily be converted to a steady flow system by superimposing on the entire flow field the propagation velocity a' from left to right. The resultant flow system may then be analysed using equations (11.1) to (11.5).

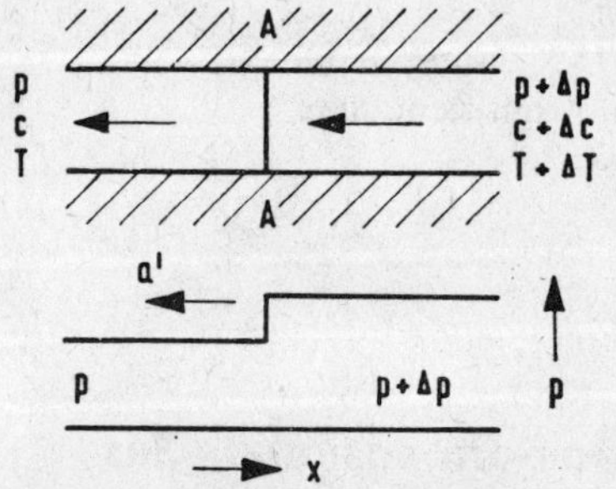

Fig. 11.3 Illustrating sound waves

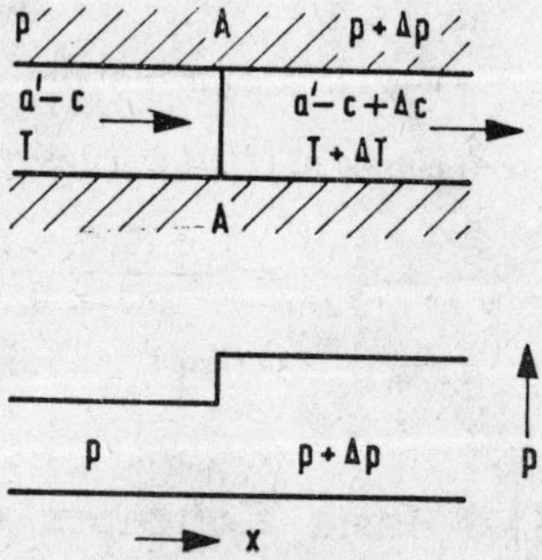

Fig. 11.4 Illustrating sound waves

Referring to *Fig.* 11.4, from the continuity equation (11.1)

$$\frac{\Delta\rho}{\rho} + \frac{\Delta c}{a' - c} = 0 \tag{i}$$

From the momentum equation (11.2), neglecting dissipative and gravitational effects

$$\Delta p + \rho(a' - c)\Delta c = 0 \tag{ii}$$

Hence, substituting for Δc from equation (i) in equation (ii)

$$\frac{\Delta p}{\Delta\rho} = (a' - c)^2 \tag{11.12}$$

But since the flow is non-dissipative, and the propagation velocity is high, leading to negligible heat transfer to the gas in the wake of wave AA, we may regard state changes as isentropic, i.e.

$$\left(\frac{\Delta p}{\Delta\rho}\right) \rightarrow \left(\frac{\partial p}{\partial\rho}\right)_{s=const} \tag{iii}$$

and using the perfect gas isentropic state equation (10.53a)

$$pv^\gamma = \text{constant} \quad \text{or} \quad p\rho^{-\gamma} = \text{constant} \tag{iv}$$

and combining it with the perfect gas state equation (10.1)

$$\frac{p}{\rho} = RT \tag{v}$$

substitution from the auxiliary equations (iii), (iv) and (v) in equation (11.12) leads to

$$(a' - c)^2 = \gamma RT \tag{11.12a}$$

Finally, since $(a' - c)$ represents the velocity of the disturbance relative to the gas ahead of it, it is by definition equal to the acoustic velocity a.

Hence we may write
For the general case: acoustic velocity

$$a = \left[\left(\frac{\partial p}{\partial \rho}\right)_{s=const}\right]^{1/2} \tag{11.13a}$$

For the perfect gas: acoustic velocity

$$a = (\gamma RT)^{\frac{1}{2}} = \left(\gamma \frac{p}{\rho}\right)^{1/2} \tag{11.13b}$$

whilst the Mach number from equation (11.11),

$$M = \frac{c}{a} \tag{11.11a}$$

11.3.2 Stagnation conditions

A further concept which greatly simplifies the analysis of various types of one-dimensional flow is that of STAGNATION ENTHALPY which, in the case of perfect gases, is associated with a readily defined STAGNATION TEMPERATURE. The stagnation enthalpy is the value of enthalpy attained by a group of particles having a velocity c and specific enthalpy h, when that group of particles is brought to rest in a fully reversible manner, and without heat or work transfer, i.e. isentropically.

Applying the steady flow energy equation (11.3) to this case, and using the suffix t to denote stagnation or TOTAL HEAD conditions, we have

$$h_t = h + \tfrac{1}{2}c^2 \tag{11.14}$$

or making the perfect gas substitution, from equation (10.30)

$$h = c_p T$$

we have

$$c_p T_t = c_p T + \tfrac{1}{2}c^2$$

i.e.
$$T_t = T\left[1 + \tfrac{1}{2}\frac{c^2}{c_pT}\right] \tag{11.14a}$$

But from equation (11.13b), in conjunction with the perfect gas specific heat relationships (10.15)

$$c_p - c_v = \left(\frac{\gamma - 1}{\gamma}\right) c_p = R$$

so that substitution in equation (11.14a) gives

$$T_t = T\left[1 + \tfrac{1}{2}(\gamma - 1)\frac{c^2}{\gamma RT}\right]$$

$$= T\left[1 + \tfrac{1}{2}(\gamma - 1)\frac{c^2}{a^2}\right] \tag{11.14b}$$

Finally, since that ratio (c/a) represents, by definition, the Mach number M, we may write

$$T_t = T[1 + \tfrac{1}{2}(\gamma - 1)M^2] \tag{11.15}$$

The stagnation temperature for a perfect gas stream is thus simply related to the so-called static temperature through the Mach number M.

Since, in fact, pressure is measured more readily than temperature, it is helpful to have a relationship, corresponding to equation (11.14b), but involving the stagnation and static pressures instead of temperatures.

Making use of the isentropic state relationship implied by the definition of stagnation conditions in conjunction with the mathematical statement of that relationship, i.e. equation (10.54), equation (11.15) expressed in terms of pressure, becomes

$$p_t = p[1 + \tfrac{1}{2}(\gamma - 1)M^2]^{\gamma/\gamma - 1} \tag{11.16}$$

The stagnation pressure has a further and important significance in that, for adiabatic flow, changes in stagnation pressure are a measure of internal irreversibility, see also equation (11.10), but that, unlike the internal entropy change Δs_i, changes in stagnation pressure are readily measurable.

Reverting to the steady flow energy equation (11.3) in conjunction with the entropy equation (11.4a) the latter may be written, substituting stagnation for static conditions and putting

$$ds_t = ds$$

(since the stagnation and static conditions are isentropically related to each other)

$$T_t\,ds_t = dh_t - \frac{1}{\rho_t}dp_t \tag{i}$$

Also, from the S.F.E.E., equation (11.3a), in the absence of heat and work transfer and substituting for dh_t from equation (11.14)

$$dh_t = dh + c\,dc \tag{ii}$$

and neglecting gravitational effects

$$dh_t = 0 \tag{iii}$$

so that combining auxiliary equations (i) and (iii)

$$T_t\,ds_t = -\frac{1}{\rho_t}dp_t$$

or

$$\frac{dp_t}{\rho_t T_t} = -\,ds_t \tag{11.17}$$

Finally from the perfect gas state equation

$$\rho_t T_t = \frac{p_t}{R}$$

i.e.

$$\frac{dp_t}{p_t} = -\frac{ds_t}{R}$$

or

$$\frac{(p_t)_2}{(p_t)_1} = \exp\{-\,[(s_t)_2 - (s_t)_1]/R\} \tag{11.17a}$$

Equation (11.17a) provides a most convenient and important relationship between a physically measurable quantity, stagnation pressure, and internal irreversibility—

$$\Delta s = (s_t)_2 - (s_t)_1 = s_2 - s_1$$

EXAMPLE 11.2

Air flows in a duct at a pressure of 2·1 bar and a velocity of 215 m/s. The temperature of the air is 28°C at the point of measurement. What are the isentropic stagnation temperature and pressure?

Assume air as a perfect gas with the specific heat at constant pressure $c_p = 1{\cdot}004$ kJ/kg-K.

SOLUTION

From equation (11.14) and converting to thermal units (see section 2.3),

$$\frac{c^2}{2J} = h_t - h = c_p(T_t - T)$$

$$\tfrac{1}{2} \times 215^2 \times 10^{-3} = 1{\cdot}004[T_t - (273 + 28)]$$

$$T_t = (273 + 28 + 23) = 324\ \text{K} = 41°\text{C} \quad \textit{Ans.}$$

$$\frac{T_t}{T} = \left(\frac{p_t}{p}\right)^{(\gamma-1)/\gamma}$$

$$= \left(\frac{324}{301}\right)^{0{\cdot}4/1{\cdot}4} = 1{\cdot}0214$$

$$p_t = 2{\cdot}1 \times 1{\cdot}0214 = 2{\cdot}145\ \text{bar} \quad \textit{Ans.}$$

11.4 One-dimensional Flow Subject to a Single Disturbing Influence

11.4.1 Influence of area change (*Fig.* 11.5) (isentropic flow through nozzles and diffusers)

Since area change is the only disturbing influence, the flow is

(1) reversible,
(2) free from gravitational effects,
(3) free from work and heat transfer effects,

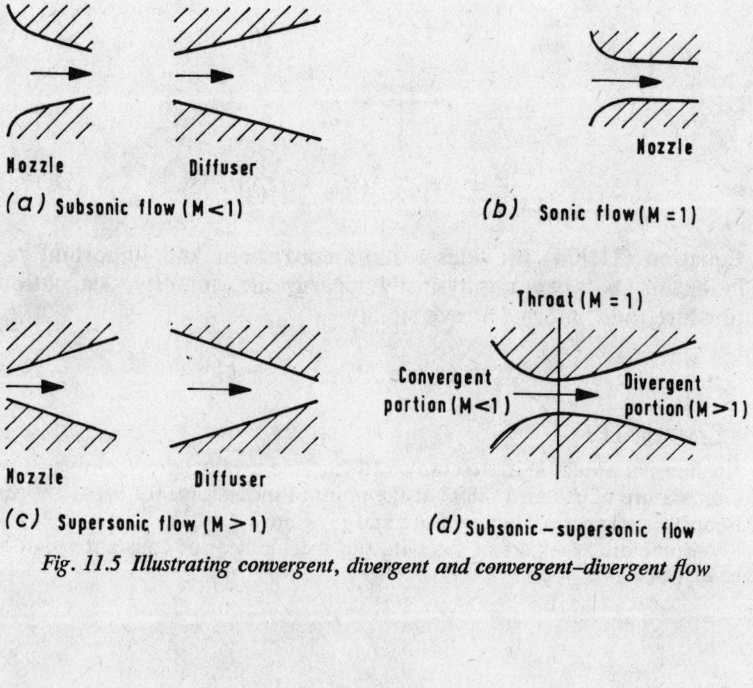

Fig. 11.5 Illustrating convergent, divergent and convergent–divergent flow

and in a purely thermodynamic sense may be described as adiabatic reversible. Condition (1) leads to constant entropy (see equation (11.4)),

$$s = s_t \tag{i}$$

Condition (2) leads to elimination of terms containing the height Z above an arbitrary datum in the relevant equations of section 11.1 and 11.2. Conditions (2) and (3) lead to constant stagnation enthalpy (see equations (11.3) and (11.14)),

$$h_t = \text{constant} \tag{ii}$$

(*a*) **General equations**

For the general case of any compressible fluid, as opposed to the perfect gas, solutions may be obtained very readily as follows—

Energy equation, equation (ii)

The energy equation (11.3a) in the absence of heat and work transfer may be stated as

$$dh + c\,dc = 0 \tag{iii}$$

Entropy equation, equation (i)

The entropy equation (11.4a) for isentropic, i.e. constant entropy, flow becomes

$$dh - \frac{dp}{\rho} = 0 \tag{iv}$$

Continuity equation, equation (11.1a)

This may be stated as

$$\frac{dA}{A} + \frac{d\rho}{\rho} + \frac{dc}{c} = 0 \tag{v}$$

Eliminating dh from equations (iii) and (iv)

$$c\,dc + \frac{dp}{\rho} = 0$$

and substituting for $c\,dc = -\left(\frac{dA}{A} + \frac{d\rho}{\rho}\right)c^2$ from equation (v) we obtain

$$\frac{dp}{\rho} - c^2\left(\frac{d\rho}{\rho} + \frac{dA}{A}\right) = 0 \tag{vi}$$

But from equation (11.13a) the acoustic velocity becomes, with (i) above,

$$a^2 = \left(\frac{\partial p}{\partial \rho}\right)_s = \frac{dp}{d\rho} \tag{vii}$$

so that substituting from equation (vii) in equation (vi)

$$\frac{dp}{\rho}\left(1 - \frac{c^2}{a^2}\right) = c^2 \frac{dA}{A}$$

or writing $(c/a) = M$ from equation (11.11a), where M is the local Mach number

$$\frac{dA}{dp} = \frac{A}{\rho c^2}(1 - M^2) \qquad (11.18)$$

The following results may be inferred from equation (11.18) (see *Fig.* 11.5)

(i) When $M < 1$, (dA/dp) is positive, i.e. for SUBSONIC FLOW a convergent duct is required for decreasing pressure, associated with accelerating flow, see equation (iii) above, i.e. nozzles; and a divergent duct for decelerating flow, i.e. diffusers.
(ii) When $M = 1$, $(dA/dp) = 0$, i.e. the attainment of SONIC CONDITIONS coincides with a minimum cross-sectional flow area.
(iii) When $M > 1$, (dA/dp) is negative, i.e. for SUPERSONIC FLOW a divergent duct is required for decreasing pressure or accelerating flow (nozzles) and a convergent duct for decelerating flow (diffusers).

Equation (11.18) does not, in itself, provide a direct relationship between the local area and the corresponding pressure and velocity. However, for most compressible fluids other than perfect gases, in particular condensible vapours, it is possible to formulate isentropic state relationships of the form

$$p\rho^{-n} = \text{constant} \qquad (11.19)$$

where the index n may be regarded as constant, at least over a limited pressure range.

Substituting from equation (11.19) in the energy equation (iii), above, between limits 1 and 2, and assuming that, at the entry section, the particle velocity is zero, i.e. $c_1 = 0$, we obtain (see *Fig.* 11.6)

$$\int_1^2 \rho\, dp = -\int_1^2 c\, dc$$

or after substituting for ρ from equation (11.19)

$$\frac{2n}{n-1}\left[1 - \left(\frac{p_2}{p_1}\right)^{(n-1)/n}\right]\frac{p_1}{\rho_1} = c_2{}^2 \qquad (11.20)$$

Equation (11.20) provides a relationship between pressure and velocity, but not between either of these quantities and area. However, by using the continuity equation (11.1a) in its integrated form and putting

$$A\rho c = \text{constant} = m \tag{viii}$$

we have, by substituting for

$$c_2 = \frac{m}{A_2\rho_2} = \frac{m}{A_2\rho_1}\left(\frac{p_1}{p_2}\right)^{1/n}$$

(where m is the mass flow rate) in equation (11.20)

$$\frac{m}{A_2} = \left\{\frac{2n}{n-1}\,p_1\rho_1\left[\left(\frac{p_2}{p_1}\right)^{2/n} - \left(\frac{p_2}{p_1}\right)^{(n+1)/n}\right]\right\}^{1/2} \tag{11.21}$$

Clearly equation (11.21) may be used to establish the pressure ratio at any other section, whilst equations (11.19) and (11.20) then provide the corresponding values of density and velocity.

The resultant distribution of pressure, density and velocity for the convergent-divergent nozzle of *Fig.* 11.6 is shown in *Fig.* 11.7.

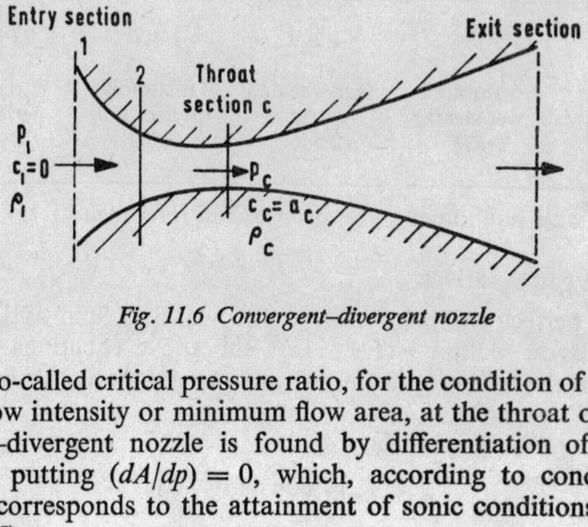

Fig. 11.6 Convergent–divergent nozzle

The so-called critical pressure ratio, for the condition of maximum mass flow intensity or minimum flow area, at the throat of the convergent–divergent nozzle is found by differentiation of equation (11.21), putting $(dA/dp) = 0$, which, according to condition (*b*), above, corresponds to the attainment of sonic conditions (see also *Fig.* 11.5).

Denoting the critical pressure ratio by $p_c/p_1 = r_c$ and substituting in equation (11.20) we have, putting $a_2 = a_c$ and $\gamma = n$ in equation (11.13b),

$$r_c = \frac{p_c}{p_1} = \left(\frac{2}{n+1}\right)^{n/(n-1)} \tag{11.22}$$

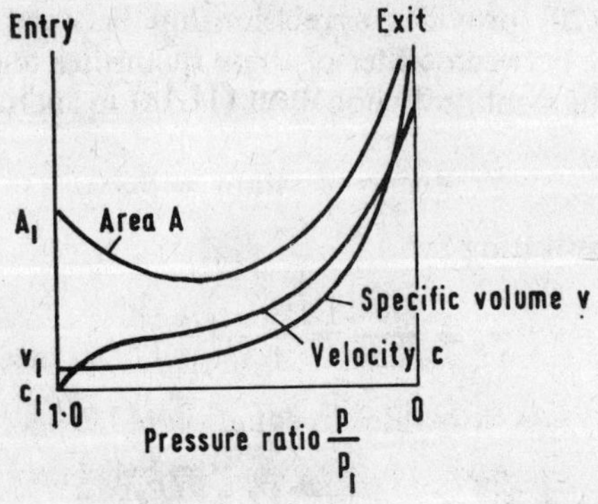

Fig. 11.7 Pressure, density and velocity distribution in convergent–divergent nozzle

The condition of sonic flow at the throat corresponding to equation (11.22) is also referred to as CHOKING FLOW.

The following table gives the critical pressure ratio for four values of n.

Table 11.1

n	1·135	1·300	1·400	1·667
	(saturated wet steam)	(superheated steam)	(diatomic perfect gas)	(triatomic perfect gas)
$\frac{p_c}{p_1}$	0·577	0·546	0·528	0·487

Other nozzle phenomena are discussed in reference 11.1.

(*b*) Perfect gas equations

Assuming perfect gas behaviour of the working fluid and, again, zero velocity at section 1 (*Fig.* 11.6), the above equations may be rewritten in the following form, for any general section 2—

Energy equation, equation (ii)

$$c_p T_t = \text{constant}$$

i.e. substituting from equation (11.15)

$$c_p T[1 + \tfrac{1}{2}(\gamma - 1)M^2] = \text{constant}$$

or with $c_p = \text{constant}$ and $M_1 = 0$

$$\frac{T_2}{T_1} = \frac{1}{1 + \frac{1}{2}(\gamma - 1)M_2^{\,2}} \tag{11.23}$$

Entropy equation, equation (i)

$$s = s_t = \text{constant}$$

so that from equation (11.17a) the stagnation pressure p_t is constant, and substituting finally for stagnation pressure from equation (11.16).

$$\frac{(p_t)_2}{(p_t)_1} = 1 = \frac{p_2}{p_1}\left[\frac{1 + \frac{1}{2}(\gamma - 1)M_2^2}{1 + \frac{1}{2}(\gamma - 1)M_1^2}\right]^{\gamma/(\gamma-1)}$$

or

$$\frac{p_2}{p_1} = \left[\frac{1}{1 + \frac{1}{2}(\gamma - 1)M_2^2}\right]^{\gamma/(\gamma-1} \qquad (11.24)$$

Continuity equation, equation (11.1a)

$$A\rho c = m$$

so that, substituting for $c = Ma$ from equation (11.11a) and for $a = (\gamma RT)^{1/2}$ from equation (11.13b), together with the isentropic state relationship (10.53),

$$A_2\rho_2c_2 = A_2\rho_2M_2a_2 = A_2\rho_1\left(\frac{p_2}{p_1}\right)^{1/\gamma} M_2\sqrt{(\gamma RT_2)}$$

$$= A_2\rho_1\left(\frac{p_2}{p_1}\right)^{1/\gamma} M_2\sqrt{(\gamma RT_1)}\left(\frac{T_2}{T_1}\right)^{1/2} = m$$

and finally substituting for T_2/T_1 from equation (11.23) and for p_2/p_1 from equation (11.24) we ultimately obtain

$$\frac{m}{A_2} = \sqrt{\left(\frac{\gamma}{R}\right)}\frac{p_1}{\sqrt{T_1}}\frac{M_2}{[1 + \frac{1}{2}(\gamma - 1)M_2^2]^{(\gamma+1)/2(\gamma-1)}} \qquad (11.25)$$

Equation (11.25) provides a convenient relationship between the area of the duct and the local Mach number from which, using equations (11.23) and (11.24), the local temperature and pressure may immediately be derived.

The various properties of the flow in relation to the duct shape already discussed for the general case, may readily be deduced from equation (11.25).

Thus, by logarithmic differentiation of equation (11.25)

$$\frac{dA}{A} = \frac{dM(M^2 - 1)}{[1 + \frac{1}{2}(\gamma - 1)M^2]M} \qquad (11.25a)$$

From this the following results, similar to those inferred from the general case, may be inferred (see *Fig.* 11.5)

(i) When $M < 1$, (dA/dM) is negative, i.e. for *subsonic flow*, a convergent duct is required for accelerating flow (nozzles) and a divergent duct for decelerating flow (diffusers).
(ii) When $M = 1$, $(dA/dM) = 0$, and (d^2A/dM^2) is positive, i.e. the attainment of *sonic conditions* coincides with a minimum cross-sectional flow area, usually referred to as the throat.
(iii) When $M > 1$, (dA/dM) is positive, i.e. for *supersonic flow*, a divergent duct is required for accelerating flow (nozzles) and a convergent duct for decelerating flow (diffusers).

It is thus evident that the Mach number M has an over-riding influence on the general behaviour of the flow.

EXAMPLE 11.3a

A convergent nozzle has an exit area of 6·5 cm². Air enters the nozzle with a stagnation pressure of 8 bar and a stagnation temperature of 120°C. Determine the mass rate of flow for back pressures of 5·67 bar, 4·23 bar and 2·42 bar. Take $\gamma = 1{\cdot}4$. See *Fig.* 11.6.

SOLUTION

The critical pressure ratio (equation (11.22))

$$\frac{p_c}{p_1} = \left(\frac{2}{\gamma+1}\right)^{\gamma/(\gamma-1)} = 0{\cdot}528$$

$$\therefore\ p_c = 0{\cdot}528 \times 8 = 4{\cdot}23 \text{ bar}$$

∴ back pressure of 4·23 bar represents the critical conditions. At this pressure the Mach number is unity at exit and the velocity, the acoustic velocity.

$$\frac{T_c}{T_1} = \frac{2}{\gamma+1} = 0{\cdot}833$$

$$\therefore\ T_c = 0{\cdot}833 \times 393{\cdot}2 = 327{\cdot}0 \text{ K}$$

$$a_c = \sqrt{(\gamma R T_c)} = \sqrt{(1{\cdot}4 \times 0{\cdot}2871 \times 10^3 \times 327{\cdot}0)}$$

$$= 362{\cdot}0 \text{ m/s}$$

$$\rho_c = \frac{p_c}{RT_c} = \frac{4{\cdot}23 \times 10^2}{0{\cdot}2871 \times 327{\cdot}0}$$

$$= 4{\cdot}50 \text{ kg/m}^3$$

$$\therefore \text{ Max. mass flow} = \rho_c A_c c_c$$

$$= 4{\cdot}50 \times 6{\cdot}5 \times 10^{-4} \times 362{\cdot}0$$

$$= 1{\cdot}059 \text{ kg/s} \quad \textit{Ans.}$$

For the back pressure of 2·42 bar since this is below the critical pressure of 4·23 bar the mass flow must be 1·059 kg/s and the nozzle is choked. If the flow down to 2·42 bar is to be controlled, it will be necessary to have a divergent portion added to the nozzle after the throat. Let the subscript e denote the exit from this divergent length.

$$\frac{T_1}{T_e} = \left(\frac{p_1}{p_e}\right)^{(\gamma-1)/\gamma} = \left(\frac{\rho_1}{\rho_e}\right)^{\gamma-1} = 1 + \frac{\gamma-1}{2} M_e^2$$

$$\frac{393{\cdot}2}{T_e} = \left(\frac{8}{2{\cdot}42}\right)^{2/7} \qquad T_e = 279{\cdot}2 \text{ K}$$

$$M_e = 1{\cdot}453 = \frac{c_e}{\sqrt{(\gamma R T_e)}} = \frac{c_e}{\sqrt{(1{\cdot}4 \times 0{\cdot}2871 \times 10^3 \times 279{\cdot}2)}}$$

$$c_e = 506 \text{ m/s}$$

$$\rho_e = \frac{p_e}{RT_e} = \frac{10^2 \times 2{\cdot}42}{0{\cdot}2871 \times 279{\cdot}2} = 3{\cdot}02 \text{ kg/m}^3$$

But $\dot{m} = 1{\cdot}059 \text{ kg/s} = 3{\cdot}02 A_e \times 506$

$$A_e = 6{\cdot}93 \times 10^{-4} \text{ m}^2 = 6{\cdot}93 \text{ cm}^2$$

For the back pressure 5·67 bar we have from equation (11.15)

$$\frac{T_t}{T} = \frac{T_1}{T} = 1 + \frac{\gamma-1}{2} M^2$$

and for isentropic flow

$$\frac{T_1}{T} = \left(\frac{p_1}{p}\right)^{(\gamma-1)/\gamma} = \left(\frac{\rho_1}{\rho}\right)^{\gamma-1} \tag{1}$$

$$\therefore \frac{p_1}{p} = \left(1 + \frac{\gamma-1}{2} M^2\right)^{\gamma/(\gamma-1)} \tag{2}$$

and

$$\frac{\rho_1}{\rho} = \left(1 + \frac{\gamma-1}{2} M^2\right)^{1/(\gamma-1)} \tag{3}$$

We know $p = 5{\cdot}67$ bar, $p_1 = 8$ bar so we can find M from equaion (2); equation (1) discloses T; equation (3) discloses ρ; and $m = \rho A c$ where $c = M \times \sqrt{(\gamma RT)}$ as before.

Gas tables are usually used for such solutions. These contain the following items for isentropic flow with $\gamma = 1{\cdot}4$,

$$M \qquad \frac{p}{p_1} \qquad \frac{\rho}{\rho_1} \qquad \frac{T}{T_1} \qquad \frac{A}{A_c} \qquad \frac{F}{F_c} \qquad \frac{A}{A_c} \times \frac{p}{p_1}$$

and are used as follows.

$$\frac{p}{p_1} = \frac{5{\cdot}67}{8{\cdot}00} = 0{\cdot}708; \text{ at this point from tables } M = 0{\cdot}72;\ T/T_1 = 0{\cdot}906$$

$$\therefore\ T = 0{\cdot}906 \times 393{\cdot}2 = 356{\cdot}1 \text{ K}$$

$$a = \sqrt{(\gamma RT)} = \sqrt{(1{\cdot}4 \times 0{\cdot}2871 \times 10{\cdot}3 \times 356{\cdot}1)}$$
$$= 379{\cdot}0 \text{ m/s}$$

$$c = 0{\cdot}72 \times 379{\cdot}0 = 273{\cdot}0 \text{ m/s}$$

$$\rho = \frac{p}{RT} = \frac{5{\cdot}67 \times 10^2}{0{\cdot}2871 \times 356{\cdot}1} = 5{\cdot}55 \text{ kg/m}^3$$

$$\dot{m} = \rho Ac = 5{\cdot}55 \times 6{\cdot}5 \times 10^{-4} \times 273{\cdot}0$$
$$= 0{\cdot}985 \text{ kg/s} \quad \textit{Ans.}$$

Note: F in the gas tables is called the impulse function and is defined by

$$F = pA + \rho Ac^2$$

For a perfect gas $\rho c^2 = \dfrac{p}{RT} c^2 = \dfrac{p}{\gamma RT} \cdot \gamma c^2 = \gamma p M^2$

so $$F = pA(1 + \gamma M^2)$$

We can further establish, expressing for isentropic flow, a dimensionless impulse function as

$$\frac{F}{p_1 A_c} = \frac{pA}{p_t A_c}(1 + \gamma M^2)$$

from which we derive

$$\frac{F}{F_c} = \frac{1 + \gamma M^2}{M\sqrt{[2(\gamma + 1)(1 + \frac{1}{2}(\gamma - 1)M^2)]}}$$

This function is used in problems connected with jet propulsion to determine the reaction of the flow on the duct.

EXAMPLE 11.3b

Steam at 10 bar and 200°C is expanded in a nozzle to 0·34 bar. The throat area is 6·5 cm² and the inlet velocity is negligible. Estimate the mass flow and the exit velocity. See *Fig*. 11·6.

SOLUTION

Assuming isentropic flow and taking the index $n = 1{\cdot}3$ for steam,

$$\frac{p_c}{p_1} = \left(\frac{2}{n+1}\right)^{n/n-1} = \left(\frac{2}{2{\cdot}3}\right)^{1{\cdot}3/0{\cdot}3} = 0{\cdot}546$$

$$p_c = 10 \times 0{\cdot}55 = 5{\cdot}5 \text{ bar}$$

At 10 bar and 200°C, from steam tables and equations of section 8.3,

$$v_1 = 0{\cdot}2061 \text{ m}^3\text{/kg},\ h_1 = 2829 \text{ kJ/kg},\ s_1 = 6{\cdot}695 \text{ kJ/kg-K}$$

At the throat, $s_c = s_1$ and

$$s_c = s_f + x_c s_{fg} \quad \text{at 5·5 bar}$$
$$1{\cdot}897 + 4{\cdot}893x_c = 6{\cdot}695 \qquad x_c = 0{\cdot}981$$
$$h_c = h_f + x_c h_{fg} \quad \text{at 5·46 bar}$$
$$h_c = 656 + 0{\cdot}981 \times 2097 = 2711 \text{ kJ/kg}$$
$$(\Delta h)_{1-c} = 2829 - 2711 = 118 \text{ kJ/kg} = 118 \times 10^3 \text{ J/kg}$$
$$c_c = \sqrt{[2(\Delta h)_{1-c}]}$$
$$= \sqrt{[2 \times 118 \times 10^3]} = 485 \text{ m/s}$$

Since $\dot{m} = \rho A_c = A_c c_c / v_c$ and $v_c = 0{\cdot}981(v_g)_c$

$$\dot{m} = 0{\cdot}981 \times 0{\cdot}3427 = 0{\cdot}336 \text{ m}^3\text{/kg}$$
$$= \frac{6{\cdot}5 \times 10^{-4} \times 485}{0{\cdot}336} = 0{\cdot}9395 \text{ kg/s} \quad \textit{Ans.}$$

At the exit

$$s_e = s_c = s_1$$

and

$$s_e = s_f + x_e s_{fg} \text{ at 0·34 bar}$$

Therefore

$$0{\cdot}980 + 6{\cdot}745x_e = 6{\cdot}695 \qquad x_e = 0{\cdot}849$$
$$h_e = h_f + x_e h_{fg} = 302 + 2328 \times 0{\cdot}849 = 2249{\cdot}5 \text{ kJ/kg}$$
$$c_e = \sqrt{[2(\Delta h)_{1-e}]} = \sqrt{[2 \times 579{\cdot}5 \times 10^3]} = 1075 \text{ m/s} \quad \textit{Ans.}$$
$$v_e = 0{\cdot}849 \times 4{\cdot}649 = 3{\cdot}94 \text{ m}^3\text{/kg}$$
$$A_e = \dot{m}v_e/c_c = (0{\cdot}9395 \times 3{\cdot}94)/1075$$
$$= 3{\cdot}44 \times 10^{-3} \text{ m}^2 = 34{\cdot}4 \text{ cm}^2$$

If this convergent–divergent nozzle has an efficiency of 85 per cent. By definition this means that

$$\frac{\text{actual enthalpy drop}}{\text{isentropic enthalpy drop}} = 0{\cdot}85$$

i.e. $$\frac{\Delta h}{(\Delta h)_s} = 0{\cdot}85 \quad \text{or} \quad h = 0{\cdot}85 \times 579{\cdot}5 = 493 \text{ kJ/kg}$$

c_e now becomes $\sqrt{(2 \times 493 \times 10^3)} = 989{\cdot}5$ m/s

h_e now becomes $2829 - 493 = 2336$ kJ/kg

x_e now becomes $\dfrac{2336 - 302}{2328} = 0{\cdot}875$

v_e now becomes $0{\cdot}874 \times 4{\cdot}649 = 4{\cdot}06$ m³/kg

A_e now becomes $\dfrac{0{\cdot}9395 \times 4{\cdot}06}{989{\cdot}5} = 3{\cdot}86 \times 10^{-3} \text{ m}^2 = 38{\cdot}6 \text{ cm}^2$

From these results it is seen that friction and eddy losses "improve" the condition of the exit steam at the expense of mass flow. Losses in convergent–divergent nozzles mostly occur in the divergent portion but these are usually expressed as a percentage of the isentropic enthalpy drop as in this example.

Two other terms might be mentioned here. The velocity coefficient which is the ratio of the actual exit velocity to the exit velocity calculated on the same pressure drop assuming isentropic flow:

$$k_c = \frac{c_e}{(c_e)_{isen.}}$$

The coefficient of discharge takes account of any constriction in the flow at the throat and influences the mass flow:

$$k_m = \frac{m}{(m)_{isen.}} = \frac{\text{actual mass flow}}{\text{mass flow "isentropically"}}$$

11.4.2 Influence of dissipative effects (pipe flow with friction)

Since dissipative effects in the form of fluid friction are the only disturbing influence, the flow is

(1) free from work and heat transfer,
(2) free from area change,
(3) free from gravitational effects,

and in a purely thermodynamic sense may be described as *adiabatic irreversible*.

Condition (1) leads to constant stagnation enthalpy, see equations (11.3) and (11.14), i.e.

$$h_t = \text{constant} \qquad \text{(i)}$$

Condition (1) with the internal entropy gain implied by the irreversible nature of the flow, but in the absence of external entropy gain due to zero heat transfer, leads to the special form of the entropy equation (11.4) derived from comparison of the pressure energy equation (11.8a) and the momentum equation (11.2), i.e. equation (11.10)

$$T\,ds_i = \tfrac{1}{2}c^2 f \frac{dx}{D} \qquad \text{(ii)}$$

Condition (2) leads to the elimination of terms containing the area change dA in equations (11.1) and (11.2).

Condition (3) again leads to the elimination of terms containing the height Z above an arbitrary datum, in the relevant equations of sections (11.1) and (11.2).

(*a*) **General equations**

For the general case of any compressible fluid, as opposed to the perfect gas, solutions may be obtained as follows—

Energy equation, equation (i)

The energy equation (11.3a) in the absence of heat and work transfer may be stated as

$$dh + c\,dc = 0 \qquad \text{(iii)}$$

Entropy equation, equation (ii)

The entropy equation (11.4a) for adiabatic irreversible flow becomes

$$T\,ds_i = dh - \frac{dp}{\rho} \qquad \text{(iv)}$$

since $ds = ds_i$ = internal entropy gain with $ds_e = 0$ = external entropy gain.

Continuity equation, equation (11.1a)

Since area changes are absent, this reduces to

$$\frac{d\rho}{\rho} + \frac{dc}{c} = 0 \qquad \text{(v)}$$

Eliminating *dh* from equations (iii) and (iv), and substituting for *dc* from equation (v) we obtain

$$T\,ds_i + \frac{dp}{\rho} - \frac{c^2\,d\rho}{\rho} = 0 \qquad \text{(vi)}$$

It can be shown that for this type of flow there exists a limiting condition, corresponding to the critical condition for convergent–divergent nozzles, where the flow intensity reaches a maximum, equation (11.22), which coincides with the attainment of a maximum value of entropy, i.e. $ds_i = 0$ in equation (vi) above. Furthermore, it can be shown that this condition of maximum entropy coincides with sonic flow in the pipe and can be approached either from initially subsonic or initially supersonic conditions. However, transition from subsonic to supersonic flow, or vice versa, would imply a violation of the second law since an entropy decrease is ruled out. These properties of adiabatic irreversible pipe flow are represented most conveniently in the enthalpy–entropy or MOLLIER DIAGRAMS (Chapter 8) the resultant curves being known as FANNO LINES (*Fig*. 11.8).

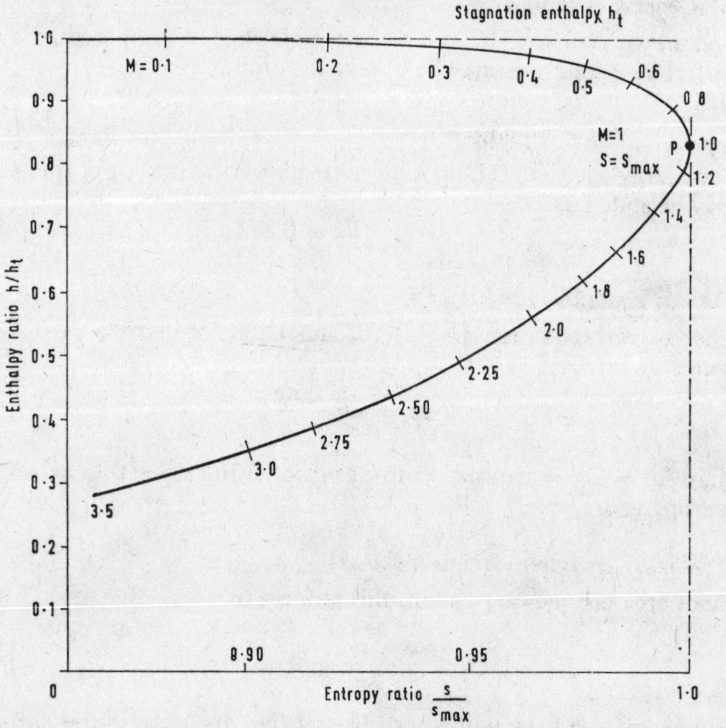

Fig. 11.8 Fanno line for air

The form of the Fanno lines may be determined with the aid of the condition of constant stagnation enthalpy, introducing the mass flow rate *m*, as in the case of variable area duct flow.

Thus, from equation (i), above, derived from the energy equation (11.3)

$$h_t = h + \tfrac{1}{2}c^2 = \text{constant} \tag{vii}$$

and substituting from the continuity equation (11.1)

$$m = A\rho c \quad \text{or} \quad c = \frac{m}{A\rho} \tag{viii}$$

$$h + \frac{m^2}{2A^2\rho^2} = \text{constant} \tag{11.26}$$

Equation (11.26) constitutes a relationship between the enthalpy *h* and the density *ρ*. The required relationship between enthalpy and entropy corresponding to equation (11.26) may readily be deduced from appropriate state diagrams (see Chapter 8) or equations of state where such are available, e.g. for the perfect gas.

Fig. 11.8 shows a generalized Fanno line, expressed as an enthalpy ratio h/h_t. Each curve exhibits a point of maximum entropy, P, the tangent at this point being vertical. But vertical lines in the enthalpy–entropy diagram are, by definition, ISENTROPICS, so that at the point of maximum entropy the various properties of the working fluid are connected by relations appropriate to isentropic changes. In particular, pressure and density changes may be represented by the partial derivative $(\partial p/\partial \rho)_{s=const}$ which, according to equation (11.13a), may be equated to the square of the acoustic velocity—

$$\left(\frac{\partial p}{\partial \rho}\right)_{s=const} = a^2 \tag{ix}$$

Substituting from equation (ix) in equation (vi), the latter reduces to

$$\left(\frac{\partial p}{\partial \rho}\right)_{s=const} = c^2$$

$$\therefore \qquad c^2 = a^2 \tag{11.27}$$

whence

$$\frac{c}{a} = M = 1 \tag{11.27a}$$

Hence the critical condition for adiabatic, irreversible pipe flow, as for nozzle flow, occurs at a Mach number of unity.

The following results may be inferred from equation (11.27) together with the previous equations (see *Fig.* 11.8)—

(i) When $M < 1$, i.e. for *subsonic flow*, the entropy increases, as does also the flow velocity c, until the above limiting condition of maximum entropy is reached. At the same time, the condition of constant stagnation enthalpy h_t, equation (i), together with the progressive increase in flow velocity c, leads to a progressive reduction in the enthalpy h (see equation (11.14)).

(ii) When $M = 1$, i.e. for *sonic conditions*, the entropy reaches a maximum value, and any further addition to the duct length would imply a transition to supersonic conditions with progressive *reduction* of entropy, contrary to the second law.

(iii) When $M > 1$, i.e. for *supersonic flow*, the entropy can increase only when the flow is decelerated from a Mach number in excess of unity to sonic conditions, i.e. $M = 1$.

These relations should be compared with those for isentropic variable area flow.

The physical significance of the condition of maximum entropy at Mach number unity is as follows.

For given values of subsonic entry Mach number and the duct characteristics, i.e. friction factor f and diameter D, the Mach number at exit will be less than unity, if the duct length is less than a certain critical value, but will be equal to unity when the critical value of duct length has been reached. Any increase in length beyond that value leads to an adjustment of the entry conditions so as to maintain the exit Mach number at the critical value of unity. Similar arguments apply to initially supersonic flow which again leads to an exit Mach number of unity when the critical duct length is reached. The condition of unity Mach number at the exit is known as choking, and corresponds closely to the condition of sonic flow at the throat of convergent–divergent nozzles.

(*b*) Perfect gas equations

Assuming perfect gas behaviour for the working fluid, considerable simplifications of the above equation become possible—

Energy equation, equation (i)

The condition of constant stagnation enthalpy may be expressed in the form

$$c_p T_t = \text{constant}$$

i.e. substituting from equation (11.15)

$$T[1 + \tfrac{1}{2}(\gamma - 1)M^2] = \text{constant}$$

or

$$\frac{T_2}{T_1} = \frac{1 + \frac{1}{2}(\gamma - 1)M_1^2}{1 + \frac{1}{2}(\gamma - 1)M_2^2} \tag{11.28}$$

Entropy equation, equation (ii)

The most convenient statement of the entropy equation is again provided by the stagnation pressure equation (11.17a) in conjunction with equation (11.16)

$$\frac{(p_t)_2}{(p_t)_1} = \frac{p_2}{p_1}\left[\frac{1 + \frac{1}{2}(\gamma - 1)M_2^2}{1 + \frac{1}{2}(\gamma - 1)M_1^2}\right]^{\gamma/(\gamma-1)} = \exp\,[-(s_2 - s_1)/R] \tag{11.29}$$

Continuity equation, equation (11.1a)

$$A\rho c = \text{constant}$$

i.e.

$$\frac{c_2}{c_1} = \frac{M_2 a_2}{M_1 a_1} = \frac{M_2\sqrt{T_2}}{M_1\sqrt{T_1}} = \frac{\rho_1}{\rho_2} = \frac{p_1}{p_2}\frac{T_2}{T_1}$$

whence with equation (11.28)

$$\frac{p_2}{p_1} = \frac{M_2}{M_1}\frac{\sqrt{T_2}}{\sqrt{T_1}} = \frac{M_1}{M_2}\left[\frac{1 + \frac{1}{2}(\gamma - 1)M_1^2}{1 + \frac{1}{2}(\gamma - 1)M_2^2}\right]^{1/2} \qquad (11.30)$$

Combining equations (11.29) and (11.30)

$$\frac{M_1}{M_2}\left[\frac{1 + \frac{1}{2}(\gamma - 1)M_2^2}{1 + \frac{1}{2}(\gamma - 1)M_1^2}\right]^{\frac{1}{2}(\gamma+1)/(\gamma-1)} = \exp\left[-(s_2 - s_1)/R\right] \qquad (11.31)$$

Equation (11.31) corresponds to equation (11.25) for variable area isentropic flow, and like the latter provides a relationship between Mach number and the external disturbing influence, in this case dissipative effects.

The internal entropy gain representing dissipative effects is directly related to frictional effects by equation (11.10) which, written in differential form, becomes

$$T\,ds_i = \tfrac{1}{2}c^2 f\frac{dx}{D}$$

or making the appropriate substitution for acoustic velocity and Mach number from equations (11.11) and (11.13b)

$$\frac{ds_i}{R} = \tfrac{1}{2}\gamma M^2 f\frac{dx}{D} \qquad (x)$$

After logarithmic differentiation of equation (11.31) and substitution from auxiliary equation (x) above, the following relationship, equivalent to equation (11.31), is obtained

$$\frac{1}{\gamma}\left(\frac{1}{M_1^2} - \frac{1}{M_2^2}\right) + \frac{\gamma + 1}{2\gamma}\ln\left[\frac{M_1^2(1 + \frac{1}{2}(\gamma - 1)M_2^2)}{M_2^2(1 + \frac{1}{2}(\gamma - 1)M_1^2)}\right] = \frac{fx}{D} \qquad (11.32)$$

Equation (11.32) makes possible the prediction of the local Mach number M_2 in terms of the known entry Mach number M_1 (at $x = 0$), the distance x separating the section under consideration from the entry section, and the pipe friction factor and diameter. The local temperature and pressure may be derived by substituting the Mach number M_2 from equation (11.32) in equations (11.28) and (11.30), assuming that the entry temperature T_1 and pressure p_1 are known.

With the aid of enthalpy values calculated from equation (11.28), using the perfect gas enthalpy relationship

$$h = c_p T \tag{xi}$$

and entropy values calculated from equation (11.31), it is possible to construct families of Fanno lines (see *Fig.* 11.8), each line being characterized by a particular value of the stagnation enthalpy $h_t = c_p T_t$. These curves demonstrate very clearly the various properties of Fanno flow already discussed in the previous section dealing with the general equations.

By logarithmic differentiation of equation (11.31) we obtain

$$\frac{ds_i}{R} = \frac{dM}{M} - \frac{\gamma + 1}{2} \frac{M\,dM}{1 + \frac{1}{2}(\gamma - 1)M^2}$$

$$= \frac{(1 - M^2)\,dM}{M[1 + \frac{1}{2}(\gamma - 1)M^2]} \tag{11.31a}$$

From this, the following results may be inferred—

(i) When $M < 1$, i.e. for *subsonic flow*, since ds_i must be positive, (ds_i/dM) is positive, so that there is a continuous increase of Mach number from the value obtaining at the entry section of the pipe or duct, up to the critical value.

(ii) When $M = 1$, $(ds_i/dM) = 0$, i.e. the attainment of *sonic conditions* coincides with maximum entropy of the flow. Any further increase in Mach number would lead to a loss of entropy, contrary to the second law.

(iii) When $M > 1$, i.e. for *supersonic flow*, since ds_i must still be positive, (ds_i/dM) is negative, so that there is a continuous decrease of Mach number from the value obtaining at the entry section of the pipe or duct, if that value exceeds unity.

The important temperature, pressure and entropy relationships represented by equations (11.28) to (11.32) have been evaluated for perfect gases having values of γ of 1·67 (monatomic), 1·40 (diatomic) and 1·33 (triatomic) and are summarized in so-called Fanno tables (ref. 11.2).

EXAMPLE 11.4

A convergent–divergent nozzle is connected to a tube and air is flowing through both. The pressure upstream of the nozzle is 7 bar and the temperature 42°C. The throat diameter of the nozzle is 0·65 cm and the diameter at exit and also of the tube is 1·30 cm.

Pressure readings are taken at points along the tube, two such points being at sections where $x/D = 0{\cdot}0$ and $x/D = 35$, the pressure reading at the latter position being 0·559 bar.

Use these data to calculate the average friction coefficient between the two points.

Assume isentropic flow to the exit of the nozzle and that the entire flow is adiabatic.

SOLUTION

Using tables, for the nozzle exit e, and denoting upstream conditions by u,

$$\left(\frac{A_e}{A_c}\right)_1 = \left(\frac{1{\cdot}30}{0{\cdot}65}\right)^2 = 4{\cdot}0$$

$$\frac{p_e}{p_u} = \frac{p_1}{p_u} = 0{\cdot}0298$$

$$p_e = p_1 = 0{\cdot}0298 \times 7 = 0{\cdot}208 \text{ bar}$$

$$M_e = M_1 = 2{\cdot}94$$

where 1 refers to the entry conditions to the pipe.

Taking this value of M_1 to the Fanno tables for frictional, adiabatic, constant area flow for a perfect gas, $\gamma = 1{\cdot}4$ we find*

$$\left(\frac{p}{p^*}\right)_1 = 0{\cdot}2256 \quad \text{and} \quad \left(\frac{fL}{D}\right)_1 = 0{\cdot}5129$$

But

$$\frac{p_2}{p_1} = \frac{(p_2/p^*)}{(p_1/p^*)}$$

$$\therefore \left(\frac{p_2}{p^*}\right) = \frac{0{\cdot}559}{0{\cdot}208} \times 0{\cdot}2256$$

$$= 0{\cdot}6055$$

where 2 refers to the exit conditions from the pipe.

This value for $(p_2/p^*)_2$ gives us $M_2 = 1{\cdot}59$, and

$$\left(\frac{fL}{D}\right)_2 = 0{\cdot}2702$$

$$\therefore \frac{f\Delta L}{D} = 0{\cdot}5129 - 0{\cdot}2702 = 0{\cdot}2427$$

and $$\frac{\Delta L}{D} = 35$$

* In Fanno tables starred quantities p^*, T^*, etc. refer to a hypothetical section distant L from the section in question, where choking flow occurs, i.e. $M^* = 1$.

so the average coefficient of friction is

$$\frac{0{\cdot}2427}{35} = 0{\cdot}00693 \quad \textit{Ans.}$$

11.4.3 Influence of external heat transfer (frictionless pipe flow with heat addition)

Since external heat transfer is the only disturbing influence, the flow is

(1) free from work transfer,
(2) free from area change,
(3) free from dissipative effects,
(4) free from gravitational effects,

and in a purely thermodynamic sense may be described as *diabatic reversible*.

Condition (1), in the presence of heat transfer only, leads to the following form of the energy equation (11.3), expressed in terms of the stagnation enthalpy h_t

$$q\,dx = Ac\rho\,dh_t \tag{i}$$

Condition (2) leads to elimination of terms containing the area change dA in equations (11.1) and (11.2), i.e. in the continuity and momentum equations.

Condition (3) leads to elimination of the internal entropy gain term in equation (11.8a), i.e. the pressure–energy equation

$$dp + \rho c\,dc = 0 \tag{ii}$$

Condition (4) has already been taken into account in equation (ii) above by omission of the gravitational term $g\rho\,dZ$. Similar terms would be omitted from all relevant equations of sections 11.1 and 11.2.

(*a*) General equations

For the general case of any compressible fluid, as opposed to the perfect gas, solutions may be obtained as follows—

Energy equation, equation (i)

The energy equation (11.3a) may be stated as

$$q\,dx = Ac\rho(dh + c\,dc) \tag{iii}$$

Entropy equation

This may be stated most conveniently in the form of equation (11.4a) putting

$$ds = ds_e + ds_i = ds_e = \frac{q\,dx}{A\rho c}$$

where ds_e = external entropy gain, ds_i = internal entropy gain = 0.

$$\therefore T\,ds_e = \frac{q\,dx}{A\rho c} = dh - \frac{1}{\rho}\,dp \qquad \text{(iv)}$$

Continuity equation, equation (11.1a)

Since area changes are absent, this reduces to

$$\frac{d\rho}{\rho} + \frac{dc}{c} = 0 \qquad \text{(v)}$$

Eliminating dh from equations (iii) and (iv), and substituting in (v),

$$\frac{1}{\rho}\,dp - \frac{c^2}{\rho}\,d\rho = 0 \qquad \text{(vi)}$$

Equation (vi) above is identical except for the term $T\,ds_i$ with equation (vi) of the previous subsection 11.4.2, dealing with adiabatic irreversible pipe flow. By analogy with that equation, it may readily be shown that there exists again a limiting condition for which the flow at the pipe exit section becomes sonic, and which coincides with a condition of maximum entropy.

Hence again putting

$$a^2 = \left(\frac{\partial p}{\partial \rho}\right)_{s=const} \qquad \text{(vii)}$$

from equation (11.13a), equation (vi) reduces to

$$\left(\frac{\partial p}{\partial \rho}\right)_{s=const} = c^2$$

or

$$c^2 = a^2 \qquad \text{(11.32)}$$

whence

$$\frac{c}{a} = M = 1 \qquad \text{(11.32a)}$$

The limiting condition for diabatic reversible flow, as for nozzle flow and adiabatic irreversible flow, thus occurs at a Mach number of unity.

The following results may be inferred from equation (11.32) together with the previous equations—

(i) When $M < 1$, i.e. for *subsonic flow*, the entropy increases, as does also the flow velocity c, when heat is transferred to the fluid, until the above limiting condition of maximum entropy is reached. At the same time the energy equation (iii) indicates that with heat transfer to the fluid, the sum $(dh + c\,dc)$ is positive, so that, up to a certain limiting value of the flow velocity c, the enthalpy increment dh is positive, i.e. there is a continuous increase in enthalpy. However, beyond this limiting value, the acceleration term $c\,dc$ becomes so large, that the enthalpy increment dh assumes negative values in order to ensure balance of equation (iii). This may be expressed qualitatively by stating that the rate of increase of kinetic energy exceeds the rate of external energy transfer. Hence the flow attains a condition of *maximum enthalpy* before the critical condition of *maximum entropy* is reached.

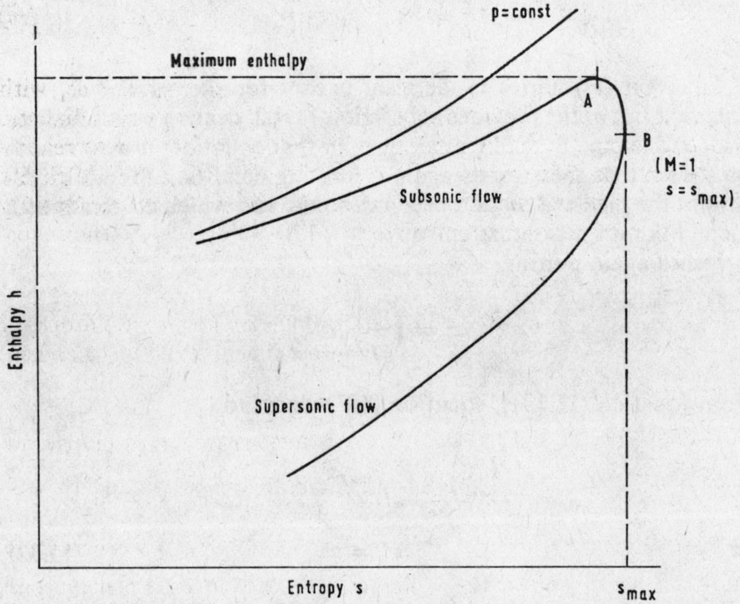

Fig. 11.9 Rayleigh line

The processes described above refer to heat transfer to the fluid. When heat is abstracted from the fluid, the situation is reversed. In this respect diabatic reversible flow differs radically from adiabatic irreversible flow where the process can proceed only in a direction of increasing entropy.

(ii) When $M = 1$, i.e. for *sonic conditions* the entropy reaches a maximum value and any further heat transfer to the fluid would imply a reduction of entropy which, from the entropy equation (iv), is clearly absurd, and would constitute a violation of the second law.

(iii) When $M > 1$, i.e. for *supersonic flow*, the entropy when heat is transferred to the fluid can increase only when the flow is decelerated from a Mach number in excess of unity towards the critical sonic condition, $M = 1$.

These relations should be compared with those for isentropic variable area flow and adiabatic irreversible pipe flow. They can be represented graphically in the enthalpy–entropy or Mollier diagram (Chapter 8) the resultant curves being known as RAYLEIGH LINES (*Fig.* 11.9).

The form of the Rayleigh lines may be determined by substituting from the continuity equation (11.1)

$$m = A\rho c \quad \text{or} \quad c = \frac{m}{A\rho}$$

in equation (vi)

$$dp - \left(\frac{m^2}{A^2 \rho^2}\right) d\rho = 0$$

which, upon integration, yields

$$p + \frac{m^2}{A^2\rho} = p + \rho c^2 = \text{constant} \qquad (11.33)$$

Equation (11.33) constitutes a relationship between two properties, the pressure p and the density ρ, and hence, since any two property values define the state of the pure substance in a single phase state (see Chapter 8), also between the two properties required for plotting Rayleigh lines in the Mollier diagram, viz. the enthalpy h and the entropy s. When equations of state are available, as for the perfect gas, the calculations are greatly simplified.

The physical significance of the condition of maximum entropy at Mach number unity is very similar to that for the Fanno lines in adiabatic irreversible flow. If heat transfer to the fluid occurs beyond the section where the flow first becomes sonic, an adjustment of the entry conditions takes place so as to render the flow wholly subsonic (or wholly supersonic) along the entire length of the duct. The condition of unity Mach number at exit is again known as choking.

(*b*) Perfect gas equations

Assuming perfect gas behaviour for the working fluid, the general equations may again be considerably simplified—

Energy equation, equation (i)

The energy equation (i), for a finite length x of duct becomes

$$qx = (Ac\rho)c_p[(T_t)_2 - (T_t)_1] = mc_p[(T_t)_2 - (T_t)_1] \qquad \text{(viii)}$$

or

$$\frac{(T_t)_2}{(T_t)_1} = \frac{T_2\,[1 + \frac{1}{2}(\gamma - 1)M_2^2]}{T_1\,[1 + \frac{1}{2}(\gamma - 1)M_2^2]} = 1 + \frac{qx}{mc_pT_t} \qquad (11.34)$$

Entropy equation

The most convenient form of the entropy equation, in this case, is provided by the derived equation (11.33), expressing the Rayleigh line condition

$$p + \rho c^2 = \text{constant} \qquad \text{(ix)}$$

or writing $\rho_1 c_1^2 = \dfrac{\gamma p_1 c_1^2}{a_1^2} = \gamma p_1 M_1^2$

and $\rho_2 c_2^2 = \dfrac{\gamma p_2 c_2^2}{a_2^2} = \gamma p_2 M_2^2$

from the acoustic velocity equation (11.13b), equation (ix) reduces to

$$\frac{p_2}{p_1} = \frac{1 + \gamma M_1^2}{1 + \gamma M_2^2} \qquad (11.35)$$

Continuity equation, equation (11.1a)

Since $A\rho c$ = constant and A = constant, then

$$\rho c = \text{constant}$$

It follows that

$$\frac{\rho_2}{\rho_1} = \frac{c_1}{c_2} \quad \text{or} \quad \frac{p_2}{p_1}\frac{T_1}{T_2} = \frac{M_1 a_1}{M_2 a_2} = \frac{M_1\sqrt{T_1}}{M_2\sqrt{T_2}}$$

i.e.

$$\frac{T_2}{T_1} = \frac{M_2^2 p_2^2}{M_1^2 p_1^2} = \frac{M_2^2}{M_1^2}\left[\frac{1+\gamma M_1^2}{1+\gamma M_2^2}\right]^2 \tag{11.36}$$

Combining equations (11.34) and (11.36)

$$\frac{M_2^2(1+\gamma M_1^2)^2[1+\frac{1}{2}(\gamma-1)M_2^2]}{M_1^2(1+\gamma M_2^2)^2[1+\frac{1}{2}(\gamma-1)M_1^2]} = 1 + \frac{qx}{mc_p(T_t)_1} \tag{11.37}$$

Equation (11.37) corresponds to equation (11.32) for adiabatic irreversible flow and equation (11.25) for variable area isentropic flow, and like the latter equations provides a relationship between Mach number and the external disturbing influence, in this case heat transfer.

The local pressure and temperature may be derived by substituting the Mach number M_2 from equation (11.37) in equations (11.35) and (11.36), respectively, assuming that the entry values of these quantities are known.

With the aid of enthalpy values calculated from equation (11.36) using the perfect gas enthalpy relationship

$$h = c_p T \tag{x}$$

and entropy values calculated from the perfect gas entropy equation (10.33), namely

$$s_2 - s_1 = c_p \ln\left(\frac{T_2}{T_1}\right) - R\ln\left(\frac{p_2}{p_1}\right)$$

$$= \left[\frac{\gamma}{\gamma-1}\ln\left(\frac{T_2}{T_1}\right) - \ln\left(\frac{p_2}{p_1}\right)\right]R \tag{11.38}$$

it is possible to construct families of Rayleigh lines, each line being characterized by a particular value of the constant in equation (11.33)

$$p + \rho c^2 = \text{constant} \tag{xi}$$

which is itself a function of the intensity of mass flow, and hence of entry Mach number M_1.

By logarithmic differentiation of equation (11.38), after substitution from equations (11.35) and (11.36) we obtain

$$\frac{ds_e}{R} = \frac{(1-M^2)\,dM}{[\frac{1}{2}(\gamma-1)M](1+\gamma M^2)} \tag{11.38a}$$

Equation (11.38a) corresponds to equation (11.31a) for adiabatic irreversible flow and leads to the establishment of properties for the

Rayleigh line very similar to those for the Fanno line (see *Figs.* 11.8 and 11.9)—

(i) When $M < 1$, i.e. for *subsonic flow*, since ds_e must be positive when heat is transferred to the fluid, there is a continuous increase of Mach number from the value obtaining at the entry section of the pipe or duct, up to the critical value. However, since in the present case, unlike that of adiabatic irreversible flow in the previous section, it is also possible to reduce the entropy of the working fluid by heat abstraction, the subsonic branch of the Rayleigh line may be traversed also in the reverse direction, i.e. from initially high (subsonic) Mach numbers to lower values.

Heating of the working fluid under subsonic conditions leads to a progressive increase in enthalpy up to a maximum value. This condition of *maximum enthalpy*, as opposed to the condition of *maximum entropy*, therefore occurs on the subsonic branch of the Rayleigh line. The value of the Mach number at this point may be determined from equation (11.36), using the perfect gas enthalpy relationship, (viii), above. Thus, by logarithmic differentiation of equation (11.36)

$$\frac{dh}{dM} = c_p \frac{dT}{dM} = \frac{2(1 - \gamma M^2)}{M(1 + \gamma M^2)} \tag{11.36a}$$

whence, for maximum enthalpy

$$\frac{dT}{dM} = 0 \quad \text{or} \quad M = \frac{1}{\sqrt{\gamma}} \tag{11.39}$$

Equation (11.39) locates the point of maximum enthalpy on the Rayleigh line.

(ii) When $M = 1$, $(ds_e/dM) = 0$, i.e. as in the case of the Fanno line, the attainment of *sonic conditions* coincides with maximum entropy of the flow.

(iii) When $M > 1$, i.e. for *supersonic flow*, the differential coefficient (ds_e/dM) in equation (11.38a) undergoes a change of sign. Hence, for heat transfer to the fluid, when the entropy increases continuously, there must be a reduction of Mach number from an initially supersonic condition, ultimately to the critical sonic condition $M = 1$, as shown by the supersonic or lower branch of the Rayleigh line. On the other hand, heat abstraction from the working fluid causes the supersonic branch of the Rayleigh line to be traversed in the opposite direction, in the direction of increasing Mach numbers. In this respect the supersonic branch of the Rayleigh line, like the subsonic branch, differs from that of the Fanno line, in that in the latter all processes occur in the direction of increasing entropy.

EXAMPLE 11.5

In a test of a combustion chamber for a jet engine, the gaseous mixture of air and paraffin enters the chamber with a velocity of 76 m/s at 52°C and a pressure of 0·35 bar. The heat of reaction of the mixture for one mixture strength amounted to 1115 kJ/kg of mixture.

Assuming friction is negligible, that the cross-sectional area of the chamber is constant, and that the mixture may be treated as air without serious error, determine the properties of the gases at exit from the combustion chamber. If the mixture strength is steadily increased so increasing the heat of reaction, what is the maximum amount of heat for the given conditions. Perfect gas conditions may be assumed with $\gamma = 1{\cdot}4$ and $R = 0{\cdot}2871$ kJ/kg-K.

SOLUTION

Denoting the inlet by subscript 1 and the exit by subscript 2, the sonic velocity $a_1 = \sqrt{(\gamma R T_1)} = \sqrt{(1{\cdot}4 \times 0{\cdot}2871 \times 10^3 \times 325)}$

$$= 368 \text{ m/s}$$

$$\therefore\ M_1 = \frac{c_1}{a_1} = \frac{76}{368} = 0{\cdot}207$$

Taking this value of M_1 to the isentropic gas tables, we find the stagnation pressure and the stagnation temperature at section 1, thus

$$M_1 = 0{\cdot}207 \qquad (p/p_t)_1 = 0{\cdot}9705 \qquad (T/T_t)_1 = 0{\cdot}99150$$

$$(p_t)_1 = 0{\cdot}35/0{\cdot}9705 = 0{\cdot}361 \text{ bar}$$

$$(T_t)_1 = 325/0{\cdot}9915 = 328 \text{ K}$$

Now $(T_t)_2 - (T_t)_1 = \Delta h/c_p$ and $\Delta h = 1115$ kJ/kg

$$c_p = \frac{\gamma}{\gamma - 1}\frac{R}{J} = \frac{1{\cdot}4}{0{\cdot}4} \times 0{\cdot}2871 = 1{\cdot}005 \text{ kJ/kg-K}$$

$$\therefore\ (T_t)_2 - (T_t)_1 = \frac{1115}{1{\cdot}005} = 1110 \text{ K}$$

$$(T_t)_2 = 328 + 1110 = 1438 \text{ K and } \frac{(T_t)_2}{(T_t)_1} = 4{\cdot}39$$

This value of $(T_t)_2/(T_t)_1$ is the link with section 2. Turning to the Rayleigh line tables for a perfect gas, $\gamma = 1{\cdot}4$ at $M_1 = 0{\cdot}207$ we find

$$\left(\frac{T_t}{T_t{}^*}\right)_1 = 0{\cdot}186 \qquad \left(\frac{p}{p^*}\right)_1 = 2{\cdot}264$$

$$\left(\frac{T}{T^*}\right)_1 = 0{\cdot}2193 \qquad \left(\frac{p_0}{p_t{}^*}\right)_1 = 1{\cdot}2323$$

Now

$$\left(\frac{T_t}{T_t^*}\right)_2 = \left(\frac{T_t}{T_t^*}\right)_1 \cdot \left(\frac{(T_t)_2}{(T_t)_1}\right) = 0{\cdot}186 \times 4{\cdot}39$$

$$= 0{\cdot}818$$

This value of $\left(\frac{T_t}{T_t^*}\right)_2$, with the same tables, gives us $M_2 = 0{\cdot}599$, also

$$\left(\frac{T}{T^*}\right)_2 = 0{\cdot}9167; \left(\frac{p}{p^*}\right)_2 = 1{\cdot}596; \left(\frac{p_t}{p_t^*}\right)_2 = 1{\cdot}075$$

We are now able to evaluate the remaining properties at exit 2.

$$T_2 = T_1 \times \frac{(T/T^*)_2}{(T/T^*)_1} = 325 \times \frac{0{\cdot}9167}{0{\cdot}2193} = 1360 \text{ K}$$

$$p_2 = p_1 \frac{(p/p^*)_2}{(p/p^*)_1} = 0{\cdot}35 \times \frac{1{\cdot}596}{2{\cdot}264} = 0{\cdot}2462 \text{ bar}$$

$$(p_t)_2 = (p_t)_1 \frac{(p_t/p_t^*)_2}{(p_t/p_t^*)_1} = 0{\cdot}361 \times \frac{1{\cdot}076}{1{\cdot}2323} = 0{\cdot}314 \text{ bar}$$

The limiting condition for the heat of reaction is reached when the Mach number at exit becomes unity. Any additional heat supply will reduce the mass flow rate and so reduce the original Mach number at entry.

Denoting this condition by 3,

$$M_3 = 1 \quad \text{and} \quad \left(\frac{T_t}{T_t^*}\right)_3 = 1$$

$$(T_t)_3 = (T_t)_1 \frac{(T_t/T_t^*)_3}{(T_t/T_t^*)_1} = 328 \times \frac{1}{0{\cdot}186} = 1770 \text{ K}$$

$$(\Delta h)_{max} = c_p[(T_t)_3 - (T_t)_1] = 1{\cdot}005(1770 - 328)$$

$$= 1450 \text{ kJ/kg}$$

REFERENCES

11.1 Shapiro, A. H., *The Dynamics and Thermodynamics of Compressible Fluid Flow*, Chapter 5 (New York, Ronald Press, 1953).

11.2 Keenan, J. H. and Kaye, J., *Gas Tables* (John Wiley, 1948).

EXERCISES

11.1 Air flows in a 5·08 cm diameter duct with a velocity of 117 m/s, the static pressure and temperature being 12·4 bar and 289 K respectively. Calculate the temperature and pressure to which the gas will rise if it is brought to rest isentropically in a Pitot tube. (See *Fig.* 11.10).

The air flows steadily from the duct through a throttle valve, and into a reservoir of 28·4 m³ capacity. Assuming that heat transfer through the walls of the reservoir is sufficient to render the process isothermal, estimate the time required to raise the vessel pressure firms 1·015 bar to 6·9 bar.

(*Ans:* 295 K, 13·3 bar, 62 s)

11.2 Steam expands through a convergent–divergent nozzle; the supply pressure and condition are 7 bar and 0·96 dry respectively while the back pressure is 1 bar. At the design condition, which assumes reversible flow, the throat pressure is $0{\cdot}58p_1$, p_1 being the supply pressure. Calculate the theoretical throat and exit plane velocities and determine the final condition of the steam. Find also the mass flow if the throat area is 6·5 cm².

In practice it is found that friction occurs in the diverging section of the nozzle and this effectively reduces the total enthalpy drop by 15% compared with the ideal. Find the actual exit plane velocity and condition and compare the ideal and irreversible flows by sketching the processes in the *h–s* field.

(*Ans:* 449 m/s, 800 m/s, 0·860, 0·684 kg/s; 736 m/s, 0·925)

11.3 *Fig.* 11.10 shows a pitot tube which is an instrument used for the measurement of fluid velocity. It is assumed that the outer tube will, by virtue of the wall slots, measure the existing static pressure

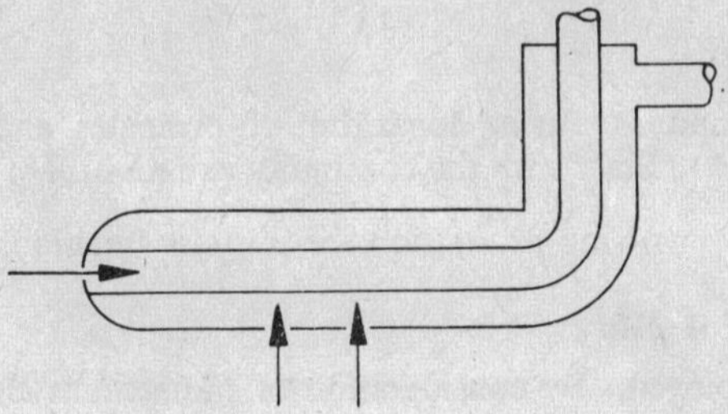

Fig. 11.10

in the fluid. In front of the pitot tube fluid is decelerated isentropically to zero velocity (i.e. the front of the tube is a stagnation point); as a consequence the pressure recorded by the centre tube will be larger than the static pressure by an amount dependent on the kinetic energy of the fluid. Derive an expression relating the square of the fluid velocity with the ratio of the absolute pressure readings.

It is frequently assumed that the flow is incompressible; in this case the first law reduces to the Bernoulli equation, i.e.

$$\frac{c_1^2}{2g_0} + \frac{p_1}{\rho} = \frac{c_2^2}{2g_0} + \frac{p_2}{\rho}$$

where the symbols have the usual meaning and changes of potential energy are ignored. Air flows in a pipe at 122 m/s, the static pressure and stream temperature being respectively 1·38 bar and 15°C. Is the assumption of incompressibility justified?

(*Ans:* $p_t = 1{\cdot}510$ bar, cf. 1·500 bar)

11.4 As a dimensionless parameter the Mach number is very convenient, but it has two obvious disadvantages:

1. It is not proportional to velocity alone,
2. At high speeds it tends towards infinity.

It is convenient therefore to divide the local velocity by one of three reference velocities. Generally the most useful of these is the critical acoustic velocity. The ratio is then

$$M^{*2} = c^2/a^{*2}$$

Derive an expression relating M with M^* and the ratio of the specific heats.

$$Ans: \frac{1}{M^{*2}} - \frac{2}{M^2(\gamma + 1)} - \frac{\gamma - 1}{\gamma + 1} = 0$$

11.5 Derive a relationship between the mass flow per unit area, stagnation temperature, static pressure and Mach number. Rearrange your result in terms of the molecular weight of the gas, m, and plot the mass flow parameter

$$\frac{\dot{m}}{A}\frac{\sqrt{T_t}}{p}\frac{1}{\sqrt{m}}$$

against M and M^* using logarithm co-ordinates and assuming $\gamma = 1{\cdot}4$. *Hint:* Start with the continuity equation, $Ac\rho = \dot{m}$.

11.6 In problems involving propulsion the impulse function, F, defined by

$$F = pA + \rho Ac^2$$

is often employed. By consideration of momentum changes in a duct between sections 1 and 2 which are perpendicular to the flow show that the net thrust on the duct is given by

$$F = F_2 - F_1$$

State any assumptions necessary.

During a test of a turbo jet engine measurements indicate a thrust of 8210 N when the flow rate is 13·6 kg/s. The temperature of the fluid at the nozzle entrance is 759°C and the velocity there is 91 m/s. The nozzle has no diverging section so that the fluid reaches atmospheric pressure somewhere outside the nozzle. Assuming isentropic flow in the nozzle and that air enters the engine perpendicular to the axis of thrust estimate the exit pressure.

Take atmospheric pressure as 1·015 bar.

(*Ans:* 0·8825 bar)

11.7 An air receiver, volume 0·284 m^3, is charged from an air main where the pressure and temperature are 6·89 bar and 21°C respectively. Air enters the receiver through a convergent nozzle having a diameter of 0·254 cm which is initially sealed by a metal diaphragm. Estimate the time required for the pressure in the receiver to rise to 1·725 bar once the diaphragm is broken. Assume that the vessel is initially evacuated and that the whole apparatus is insulated.

(*Ans:* $t = 70$ sec)

11.8 Air flows in a duct where the pressure and Mach number are 1·38 bar and 0·6 respectively, at a rate of 0·227 kg/s. The duct cross-sectional area is 6·44 cm^2.

Calculate:

(*a*) Stagnation temperature and a_t.
(*b*) The maximum percentage reduction in area which may be introduced without reducing the mass flow.
(*c*) For the reduction of (*b*) the velocity and pressure at the minimum area.

(*Ans:* (*a*) 292 K, 331 m/s)

11.9 Derive the following expressions for isentropic flow with the pressure ratio p/p_0 as parameter:

$$c = \sqrt{\left(\frac{2\gamma R}{\gamma - 1}\right)} \sqrt{T_t} \sqrt{\left[1 - \left(\frac{p}{p_t}\right)^{(\gamma-1)/\gamma}\right]}$$

$$M^2 = \frac{2}{\gamma - 1}\left[\left(\frac{p_t}{p}\right)^{(\gamma-1)/\gamma} - 1\right]$$

$$M^{*2} = \frac{\gamma + 1}{\gamma - 1}\left[1 - \left(\frac{p}{p_t}\right)^{(\gamma-1)/\gamma}\right]$$

$$\frac{A^*}{A} = \sqrt{\left[\frac{2}{\gamma - 1}\left(\frac{\gamma + 1}{2}\right)^{(\gamma+1)/(\gamma-1)}\right]} \left(\frac{p}{p_t}\right)^{1/\gamma} \sqrt{\left[1 - \left(\frac{p}{p_t}\right)^{(\gamma-1)/\gamma}\right]}$$

11.10 A convergent–divergent air nozzle used in an experimental investigation operates under the following upstream conditions:

Temperature $T_0 = 93$°C
Pressure $p_0 = 20{\cdot}7$ bar
Velocity $c_0 = 0$

If the ratio exit area/throat area $= A_e/A_t = 3{\cdot}0$, show that, under design conditions, the Mach number at exit, M_e, is approximately 2·63 and hence find the corresponding back pressure. Show that the nozzle, with the same upstream conditions, can also operate under conditions of isentropic flow for back pressures between 20·12 bar and 20·7 bar. Find the exit Mach number at 20·12 bar back pressure and comment on the conditions at the throat.

Outline briefly the phenomena likely to be observed during transition from a back pressure of 1 bar to 20·12 bar.
(*Ans:* 1 bar, 0·195)

11.11 For the special case of flow in a constant area duct with no friction the momentum equation can be written

$$p + \rho c^2 = F/A = \text{constant} \qquad (1)$$

where F is the momentum flux and A the area.

The equation of continuity is

$$A\rho c = \dot{m} = \text{constant} \qquad (2)$$

Combining (1) and (2) leads to

$$p + \dot{m}^2/A^2\rho = F/A \qquad (3)$$

For fixed values of the flow per unit area and the impulse function per unit area, equation (3) defines a unique relation between pressure and density which defines the Rayleigh line. Show that the equation of the Rayleigh line in the entropy–enthalpy plane is

$$\frac{s - s_b}{c_p} = \ln\left(\frac{T}{T_b}\right)\left[\frac{2}{(\gamma - 1) - \sqrt{[(\gamma + 1)^2 - 4\gamma T/T_b]}}\right]^{(\gamma-1)/\gamma}$$

where s_b and T_b correspond to the entropy and temperature at the point of maximum entropy. Show that a continuous change of properties along a Rayleigh line is possible through heat transfer.

11.12 A gaseous mixture of air and fuel enters a ram-jet combustion chamber with a velocity of 61 m/s, at a temperature of 49°C, and at a pressure of 0·344 bar. The heat reaction H of the mixture for the particular fuel/air ratio employed is 1 162 kJ/kg of mixture. It is desired to find the stream properties at the exit of the combustion chamber. It will be assumed that friction is negligible, that the cross-sectional area is constant, and that the properties of both the reactants and the products are equivalent to air in respect to molecular weight and specific heat.
(*Ans:* T_2 = 1 420 K, p_2 = 0·284 bar, c_2 = 327 m/s)

11.13 Air is brought into a tube through a converging–diverging nozzle designed for a Mach number of 2, and has initially a stagnation temperature of 278 K and a stagnation pressure of 6·89 bar. The air is to be cooled by some mechanism not involving friction.

(*a*) Assuming that there are no shocks, compute the maximum possible amount of heat rejection per pound of air passing through the duct. Find, for this limiting condition, the final temperature, pressure and stagnation pressure.

(*b*) Investigate the types of flow patterns which might exist in the nozzle-duct system if the amount of cooling is greater than that computed for part (*a*).

(*Ans:* 204·2 kJ/kg, 329 K, 0·88 bar)

11.14 The Fanno line describes the adiabatic flow in a constant area duct with wall friction. Show that the Fanno line is given, in the temperature entropy field, by

$$\frac{s - s^*}{c_p} = \ln \left(\frac{2}{\gamma + 1}\right)^{(\gamma-1)/2\gamma} \left(\frac{T}{T^*}\right)^{1/\gamma} \left[\frac{\gamma + 1}{2} - \frac{T}{T^*}\right]^{(\gamma-1)/2\gamma}$$

Demonstrate that movement along the Fanno line is achieved through the effect of friction.

11.15 A stream of air flows in an insulated tube of constant area having a cross-sectional area of 0·1 m². At a section 1 the pressure is 0·69 bar, the temperature is 4·5°C, and the mass velocity is 144·5 kg/m². The pressure in the space to which the tube exhausts is so low that a "choking" condition prevails, i.e. alterations in the pressure of the exhaust region have no effect on the flow through the tube.

(*a*) Calculate the Mach number at section 1.

(*b*) Calculate the Mach number, temperature, and pressure at the exit of the tube.

(*c*) Calculate the total force in the axial direction which must be exerted to hold stationary the section of the duct between section 1 and the exit.

(*Ans:* (*a*) 0·5, (*b*) 1·0, 243 K, 0·322 bar, (*c*) 1 585 N)

11.16 Air enters an insulated tube of 2·54 cm diameter through a converging–diverging nozzle with a throat diameter of 1·03 cm. The pressure and temperature in the low-velocity region at the nozzle entrance are 6·89 bar and 60°C, respectively. The flow through the nozzle may be assumed frictionless, and the average friction coefficient in the duct may be taken as 0·0025.

(*a*) Plot the pressure, temperature, Mach number, and isentropic stagnation pressure as a function of distance from the tube inlet, assuming shockless flow to the exit. Determine the maximum length of duct for these conditions.

(*b*) Suppose that the tube is 1 m long. Plot, versus the back pressure in the space to which the duct discharges, the following quantities: mass rate of flow, pressure, and Mach number at the duct inlet, and location of shock.

(*c*) Repeat part (*b*) for a tube 2·25 m long.

(*Ans:* (*a*) 1·46 m)

11.17 Flow in a duct of uniform cross-section under diabatic frictionless conditions is governed by the Rayleigh line condition,

$$p + \rho c^2 = \text{constant}$$

Explain the significance of this statement.

Hence show that for the perfect gas the following relationships hold

$$\frac{p_2}{p_1} = \frac{1 + \gamma M_1^2}{1 + \gamma M_2^2}$$

$$\frac{T_2}{T_1} = \frac{M_2^2(1 + \gamma M_1^2)^2}{M_1^2(1 + \gamma M_2^2)^2}$$

where p = pressure, T = temperature, M = Mach number.

Using these equations, describe the process of subsonic heating.

Air enters a duct of cross-sectional area 0·00925 m² with an initial Mach number $M_1 = 0{\cdot}5$ and temperature $T_1 = 334$ K and pressure $p_1 = 3{\cdot}45$ bar. Find the total heat addition required to produce the conditions of maximum entropy.

(*Ans:* 146·5 kJ/kg)

11.18 Given the stagnation pressure energy equation for adiabatic flow in a duct of constant cross-sectional area in the form

$$\frac{dp_t}{p_t} = -\frac{ds}{R/J}$$

and given further the relationship between entropy gain and frictional energy dissipation, viz.,

$$\frac{ds_i}{R/J} = \frac{\gamma}{2} M^2 f \frac{dx}{D}$$

show that the following equation connects Mach number change and energy dissipation:

$$f\frac{dx}{D} = \frac{2(1 - M^2)dM}{\gamma M^3[1 + \frac{1}{2}(\gamma - 1)M^2]}$$

Hence compare pipe lengths for the case of a pipeline of 1·22 m internal diameter, conveying methane, and operating with an entry velocity of 15·2 m/s and temperature of 21°C, when the exit Mach number is (*a*) unity, (*b*) 0·25.

The exit pressure in both cases is 6·89 bar, the pipe friction factor f is 0·02 and the gas constants for methane are as follows:

$R = 0{\cdot}518$ kJ/kg-K

$c_p = 2{\cdot}22$ kJ/kg-K

$\gamma = 1{\cdot}30$.

The integrated form of the Mach number equation may be assumed to be

$$\frac{1}{\gamma}\left[\frac{1}{M_1^2}-\frac{1}{M_2^2}\right]+\frac{\gamma+1}{2\gamma}\ln\frac{M_1^2[1+\frac{1}{2}(\gamma-1)M_2^2]}{M_2^2[1+\frac{1}{2}(\gamma-1)M_1^2]}=\frac{fL}{D}$$

(*Ans:* (*a*) 41·1 km; (*b*) 39·0 km)

11.19 Using the energy equation for a perfect gas together with the continuity equation show that the mass flow with isentropic and sub-critical expansion through a convergent nozzle for entry and exit pressure p_1 and p_2 respectively is given by

$$\frac{\dot{m}}{A_2}=p_1\left[\frac{2g\gamma}{\gamma-1}\frac{1}{RT_1}\left\{\left(\frac{p_2}{p_1}\right)^{2/\gamma}-\left(\frac{p_2}{p_1}\right)^{(\gamma+1)/\gamma}\right\}\right]^{1/2}$$

when the velocity of the gas entering the nozzle is zero. Hence show that for a maximum mass flow per unit area:

$$\frac{p_2}{p_1}=\left(\frac{2}{\gamma+1}\right)^{\gamma/(\gamma-1)}$$

What particular physical significance is to be attached to this result?

Air at 5·51 bar and 38°C flows from a reservoir through a convergent nozzle into a region of low pressure. Calculate the maximum rate of discharge per square inch of throat area and state the required downstream pressure to give this discharge.

(*Ans:* 0·42 kg/s, 2·92 bar)

11.20 Diabatic frictionless flow is characterized by constancy of the thrust function

$$F=p+\rho c^2$$

where p = pressure, ρ = density, c = velocity.

Comment on this statement, and discuss the representation of the resultant Rayleigh line in the temperature–entropy diagram.

Show that, when a perfect gas enters a frictionless duct of cross-sectional area A with stagnation temperature T_t, and Mach number M_1, and if the rate of heat addition along the duct is Q kJ/s, the relevant relationships for downstream stagnation temperature and Mach number are

$$1+\frac{Q}{M_1 a_1 \rho_1 c_p (T_t)_1}=\frac{(T_t)_2}{(T_t)_1}$$

$$\frac{(T_t)_2}{(T_t)_1}=\frac{M_2^2\,(1+\gamma M_1^2)^2\,[1+\frac{1}{2}(\gamma-1)M_2^2]}{M_1^2\,(1+\gamma M_2^2)^2\,[1+\frac{1}{2}(\gamma-1)M_1^2]}$$

where a_1 = upstream acoustic velocity,

ρ_1 = upstream density,

c_p = specific heat at constant pressure.

Air at 15°C and 6·89 bar enters a heated section of frictionless duct, having a cross-sectional area of 0·0462 m^2 with a velocity of 183 m/s. Heat is added at the rate of 715 kJ/s-m run. Find

(*a*) the length of heated ducting required to illustrate choked conditions,
(*b*) resultant change in pressure,
(*c*) resultant change in entropy.

(*Ans:* (*a*) 10·6 m, (*b*) −2·89 bar, (*c*) 0·329 kJ/kg-K)

11.21 Comment on the correspondence between the pressure–energy equation for one-dimensional, steady flow of a compressible fluid, in the form

$$\Delta p + \rho c \Delta c + \rho \Delta T + J \rho T \, \Delta s_i$$

(symbols having their usual meaning) and the momentum equation. Show that for flow without energy dissipation, the momentum and pressure–energy equations are identical and that Bernoulli's equation is the integrated form of those equations for incompressible flow.

Determine for adiabatic flow of air through a constant area duct the approximate entropy change associated with a frictional pressure drop of 0·0276 bar when the temperature and pressure are respectively 389 K and 1·72 bar. Density and temperature changes may be neglected for this purpose.

Given the expression

$$\frac{\Delta s}{R/J} = \frac{(1 - M^2)\Delta M}{M[1 + \frac{1}{2}(\gamma - 1)M^2]}$$

for adiabatic flow of a perfect gas in a constant area duct, find the actual density and temperature increments in the above case, when the local Mach number is 0·3.

(*Ans:* Δs = 0·00459 kJ/kg-K)

Chapter 12

Gaseous Mixtures

In Chapter 10 we discussed the properties of perfect and semi-perfect gases. The use of particular values of the various gas constants in the equations of Chapter 10 implies that a single chemical substance is involved. However, as already stated in connexion with the examples involving air as the working substance, mixtures of perfect or semi-perfect gases may be treated as if they were a single chemical substance, all the equations of Chapter 10 being applicable in full to such mixtures. It is necessary, however, first to establish values for the various gas constants used in these equations, the relevant relationships being discussed in the present chapter.

The use of the perfect gas equations may be extended from mixtures of gases (such as air) to mixtures of gases and vapours under certain conditions, particularly when the gas concerned is only partially "saturated" with vapour. The most commonly encountered gas–vapour mixture is, of course, air containing water vapour either in "unsaturated" or "saturated" forms: for example, in the atmosphere, in the exhaust gases of internal combustion engines burning hydrocarbon fuels, in the cooling towers of power stations in which condenser cooling water is recirculated after coming into direct contact with air for the purpose of "evaporative" cooling, and in air-conditioning plant. In addition to these and many other instances of mixtures of air and water vapour, there are many examples of gas–vapour mixtures in chemical engineering to which similar methods of analysis may be applied.

The basis of the analytical treatment of mixtures of gases or of gases and vapours is Dalton's law, or its refined form known as Gibbs' law. With the aid of this law and the relationships for perfect and semi-perfect gases of Chapter 10, it is possible to establish new relationships appropriate to mixtures of any given composition, by mass or volume.

Brief reference to Dalton's law has already been made in Chapter 9 dealing with the kinetic theory of gases (section 9.4). It was demonstrated in that chapter that not only does the kinetic theory permit the formulation of the perfect gas equation for a single chemical substance, but that the concept of pressure as the result of the impact of molecules on the walls of the containing vessel leads directly to the formulation of Dalton's law, according to which the pressure exerted by a mixture is equal to the sum of the pressures which each constituent would exert if it alone occupied the volume of the mixture at the same temperature.

12.1 Dalton's Law of Partial Pressures

In the previous eleven chapters we have been dealing, by implication, with pure substances, as defined in Chapter 2, section 2.7, i.e. with substances having a single chemical structure. Since the analysis of any process undergone by a pure substance involves the use of appropriate property functions, it is clear that the analysis of similar processes for mixtures of pure substances, whether gases or gases and vapours, would entail the experimental determination of property functions for an infinite number of combinations of two or more substances, were it not for the existence of Dalton's law. The resultant simplification is emphasized by the remark in Chapter 8, section 8.3, on the determination of the extensive properties (internal energy u, enthalpy h and entropy s) of the pure substance, by the use of calorimetry in addition to the accurate but nevertheless simpler techniques available for the determination of the intensive properties (pressure p, specific volume v and temperature T). Dalton's law, and its extension due to Gibbs, eliminates the necessity for such experimental property determinations of mixtures, Dalton's law applying essentially to the intensive property, pressure, while the Gibbs extension is relevant to the extensive properties.

Dalton's law, like the perfect gas equations, may be deduced from the kinetic theory of gases (Chapter 9). However, whereas the perfect gas equations are obeyed by real gases only under certain clearly defined conditions (Chapter 10), Dalton's law has been shown experimentally to apply not only to mixtures of gases but also, with negligible error, to mixtures of gases and vapours, over a very wide range of conditions.

DALTON'S LAW (see Chapter 9, section 9.4)

Dalton's law was enunciated in 1802 and was followed in 1875 by its extension due to Gibbs, which applies to the extensive thermodynamic properties.

THE PRESSURE OF A GASEOUS MIXTURE IS THE SUM OF THE PRESSURES WHICH EACH COMPONENT WOULD EXERT IF IT ALONE OCCUPIED THE VOLUME OF THE MIXTURE AT THE TEMPERATURE OF THE MIXTURE.

The Gibbs–Dalton law

THE INTERNAL ENERGY, ENTHALPY AND ENTROPY OF A GASEOUS MIXTURE ARE RESPECTIVELY EQUAL TO THE SUMS OF THE INTERNAL ENERGIES, ENTHALPIES AND ENTROPIES WHICH EACH COMPONENT OF THE MIXTURE WOULD HAVE IF EACH ALONE OCCUPIED THE VOLUME OF THE MIXTURE AT THE TEMPERATURE OF THE MIXTURE.

12.2 General Mathematical Formulation of Dalton's and Gibbs–Dalton Laws

Let a, b, . . . n be the constituents of a mixture of pure chemical substances. Then Dalton's law may be stated as follows—

$$p = p_a + p_b + \ldots + p_n \tag{12.1}$$

where p = mixture pressure, mixture occupying volume V at temperature T, and

p_a = partial pressure of constituent a,
p_b = partial pressure of constituent b,
p_n = partial pressure of constituent n.

The term "partial pressure" refers to the pressure exerted by each constituent when it alone occupies the volume V of the mixture at temperature T.

Certain additional relationships follow.

Volume. Each constituent may be thought of as occupying the full volume

$$V = V_a = V_b = \ldots = V_n \tag{12.2}$$

Temperature. Each constituent is at the mixture temperature

$$T = T_a = T_b = \ldots = T_n \tag{12.3}$$

Mass. The total mass equals the sum of the constituent masses

$$\sum_{i=1}^{n} m_i = m = m_a + m_b + \ldots + m_n \tag{12.4}$$

Specific volume. The composition by mass, i.e. proportions by mass of the individual constituents of the mixture, is known as the GRAVIMETRIC COMPOSITION—

$$m = V/v,\ m_a = V_a/v_a,\ m_b = V_b/v_b, \ldots, m_n = V_n/v_n \tag{12.5}$$

where V = total volume, v = specific volume.

Substituting from equations (12.2) and (12.5) in equation (12.4)

$$\frac{1}{v} = \frac{1}{v_a} + \frac{1}{v_b} + \ldots + \frac{1}{v_n} \tag{12.6}$$

Alternatively, equation (12.6) may be written in terms of the reciprocal of specific volume, i.e. density

$$\rho = \rho_a + \rho_b + \ldots + \rho_n \tag{12.7}$$

The Gibbs–Dalton law may be expressed as follows—

Internal energy

$$\begin{aligned} U &= U_a + U_b + \ldots + U_n \\ mu &= m_a u_a + m_b u_b + \ldots + m_n u_n \end{aligned} \tag{12.8}$$

Enthalpy

$$\begin{aligned} H &= H_a + H_b + \ldots + H_n \\ h &= m_a h_a + m_b h_b + \ldots + m_n h_n \end{aligned} \tag{12.9}$$

Entropy

$$\begin{aligned} S &= S_a + S_b + \ldots + S_n \\ ms &= m_a s_a + m_b s_b + \ldots + m_n s_n \end{aligned} \tag{12.10}$$

The application of the above equations derived directly from the Dalton and Gibbs–Dalton laws will be investigated under two main headings—

(*a*) Mixtures of perfect or semi-perfect gases (Chapter 10)
(*b*) Mixtures of perfect or semi-perfect gases and vapours (Chapter 8)

12.3 Mixtures of Perfect or Semi-perfect Gases

12.3.1 Gas constant of a mixture

Each constituent i of a mixture of perfect or semi-perfect gases obeys the state equation for unit mass

$$p_i v_i = R_i T_i \tag{12.11a}$$

or for mass m_i

$$p_i V_i = m_i R_i T_i \tag{12.11b}$$

where, from Dalton's law,

$$V_i = V \tag{i}$$

$$T_i = T \tag{ii}$$

Substituting from equations (12.11b) and auxiliary equations (i) and (ii) in equation (12.1)

$$p = \frac{m_a R_a T}{V} + \frac{m_b R_b T}{V} + \ldots + \frac{m_n R_n T}{V}$$

i.e.

$$p = \sum_{i=1}^{n} (m_i R_i) \frac{T}{V} \qquad \text{(iii)}$$

Equation (iii) connects the three properties, pressure p, temperature T, volume V, and only these are variables, whilst the term

$$\sum_{i=1}^{n} (m_i R_i)$$

is clearly a constant depending on (*a*) the gravimetric composition of the mixture, (*b*) the gas constants of the individual constituents. Comparing equation (iii) with the general state equation for perfect or semi-perfect gases

$$p = (mR) \frac{T}{V} \qquad \text{(iv)}$$

it is evident that these two equations are of the same form, i.e. the mixture, like the individual constituents, behaves as a perfect or semi-perfect gas.

Hence, comparing equations (iii) and (iv), and equating the two constant terms

$$\sum_{i=1}^{n} (m_i R_i) = mR \qquad \text{(v)}$$

m, the mass of mixture, is clearly given by

$$m = \sum_{i=1}^{n} (m_i) \qquad \text{(vi)}$$

so that the characteristic gas constant of the mixture, R, becomes

$$R = \sum_{i=1}^{n} (m_i R_i) \Big/ \sum_{i=1}^{n} (m_i) \qquad (12.12)$$

which may now be substituted in equation (iv), the state equation for the mixture, i.e.

$$pV = mRT \qquad (12.13)$$

Since, according to equation (12.13) a mixture of perfect or semi-perfect gases behaves in a manner analogous to that of the individual constituents, see equation (12.11), it may therefore itself be regarded as a perfect or semi-perfect gas.

Equation (12.12) shows that the gas constant R of the mixture is the MASS OR WEIGHTED MEAN of the individual gas constants.

12.3.2 Equivalent molecular weight

The equivalent molecular weight of the mixture may be inferred from equation (12.13), i.e. the characteristic state equation of the mixture, in conjunction with equation (3.7), i.e. the relationship between the universal gas constant G and the particular gas constant R_i for perfect or semi-perfect gases in terms of the molecular weight M_i, namely

$$M_i = \frac{G}{R_i} \qquad (12.14)$$

Since the perfect (or semi-perfect) gas behaviour of the mixture as a whole has already been established by analogy between equations (12.11) for individual constituents and (12.13) for the mixture, it follows that a similar analogy will hold between equation (12.14) and a corresponding equation for the mixture, i.e.

$$M = \frac{G}{R} \qquad (12.15)$$

M is defined as the equivalent molecular weight of the mixture whilst the mixture gas constant R is given by equation (12.12), so that

$$M = G \sum_{i=1}^{n}(m_i) \Big/ \sum_{i=1}^{n}(m_i R_i) \qquad (12.15a)$$

Equation (12.15a) is the most convenient form for the evaluation of the molecular weight of the mixture given its gravimetric composition. The significance of equation (12.15a) becomes at once evident by making the substitutions

$$R = \sum_{i=1}^{n}(m_i R_i) \Big/ \sum_{i=1}^{n}(m_i) \quad \text{from equation (12.12)}$$

and $R_i = \dfrac{G}{M_i}$ from equation (12.14)

Thus

$$R = \sum_{i=1}^{n}\left(m_i \frac{G}{M_i}\right) \Big/ \sum_{i=1}^{n}(m_i) = G \sum_{i=1}^{n}\left(\frac{m_i}{M_i}\right) \Big/ \sum_{i=1}^{n}(m_i)$$

where $m_i/M_i = n_i$, the number of kilogramme-molecules of a particular constituent.

Hence

$$R = G\sum_{i=1}^{n} n_i \Big/ \sum_{i=1}^{n} m_i = \frac{nG}{m}$$

whence, substituting finally in equation (12.15)

$$M = \frac{mG}{nG} = \frac{m}{n}$$

or
$$nM = m \tag{12.16}$$

Equation (12.16) expresses the fact that the mass of mixture is equal to the product of the total number of kg-mols and the equivalent molecular weight. This behaviour is evidently analogous to that for a single constituent for which

$$n_i M_i = m_i \tag{12.17}$$

Having determined the gas constant and molecular weight for the mixture, it remains to establish expressions for the specific heats at constant volume and constant pressure.

12.3.3 Specific heat at constant volume

Differentiating equation (12.8) and substituting for

$$(c_v)_i = \frac{du_i}{dT}$$

from equation (10.13b), i.e. the definition of the specific heat at constant volume for a single constituent

$$m\frac{du}{dT} = m_a\frac{du_a}{dT} + m_b\frac{du_b}{dT} + \ldots + m_n\frac{du_n}{dT}$$

or
$$mc_v = m_a(c_v)_a + m_b(c_v)_b + \ldots + m_n(c_v)_n$$

i.e.
$$c_v = \sum_{i=1}^{n} [m_i(c_v)_i] \Big/ \sum_{i=1}^{n} (m_i) \tag{12.18}$$

Hence the specific heat at constant volume of the mixture, like the gas constant given by equation (12.12), is the mass mean of the individual specific heat at constant volume.

12.3.4 Specific heat at constant pressure

Adopting the same procedure as for the specific heat at constant volume, i.e. differentiating equation (12.9) and substituting for

$$(c_p)_i = \frac{dh_i}{dT}$$

from equation (10.14b), the equivalent expression for the specific heat at constant pressure of the mixture becomes

$$c_p = \sum_{i=1}^{n} [m_i(c_p)_i] \Big/ \sum_{i=1}^{n} (m_i) \qquad (12.19)$$

so that, again, the specific heat at constant pressure of the mixture is the mass mean of the individual specific heats.

12.3.5 Volumetric and gravimetric analyses of mixtures

In many practical cases, particularly in the analysis of products of combustion, we are required to infer the gravimetric composition of a gas mixture from an experimentally determined volumetric composition. The latter is a more convenient method of assessing the proportion of the consitutuents since, given suitable methods of separating the constituents, their individual volumes under readily measurable conditions of pressure and temperature may be determined directly. The most important of such experimental methods, known as the ORSAT ANALYSIS, will be discussed in a later section.

Consider a mixture of n distinct gases and assume that, by suitable methods of separating the constituents, each occupies a "compartment" of its own in a suitably constructed container.

The volume of the container as a whole is V and the individual constituents occupy volumes V_a, V_b . . . V_n. It will be assumed further that the individual constituents are all at the same pressure p and temperature T. The volumetric analysis of the mixture is then given by the partial volumes V_a, V_b . . . V_n, resulting in the percentage volumetric analysis

$$\frac{V_a}{V} \times 100, \quad \frac{V_b}{V} \times 100, \quad \ldots, \quad \frac{V_n}{V} \times 100 \qquad (12.20)$$

Hence, applying the equation of state (12.11b) to each constituent, and the mixture equation of state (12.13) for the mixture as a whole, to the equation expressing the total volume of the container

$$V = V_a + V_b + \ldots + V_n \qquad \text{(i)}$$

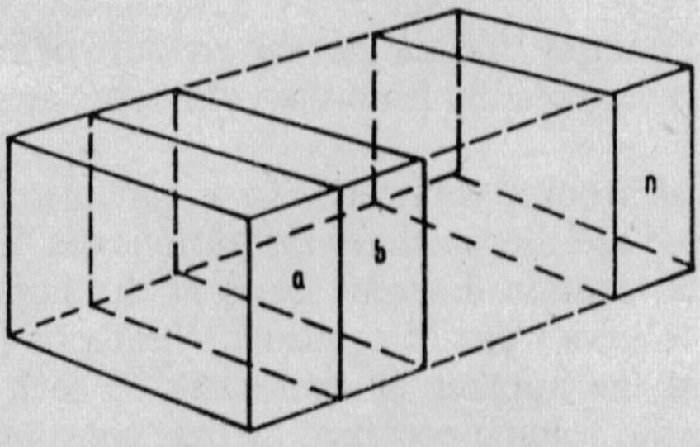

Fig. 12.1 Illustrating volumetric mixture composition

(see *Fig.* 12.1) we obtain

$$\frac{mRT}{p} = \frac{m_aR_aT}{p} + \frac{m_bR_bT}{p} + \ldots + \frac{m_nR_nT}{p} \tag{ii}$$

Combining equations (i) and (ii), dividing each term of the r.h.s. in turn by the term on the l.h.s.

$$\frac{V_a}{V} = \frac{m_aR_a}{mR}, \quad \frac{V_b}{V} = \frac{m_bR_b}{mR}, \quad \ldots, \quad \frac{V_n}{V} = \frac{m_nR_n}{mR} \tag{12.21}$$

From the partial pressure equation (12.1), again applying equations (12.11b) and (12.13), but now to the fully mixed gases instead of to the individual constituents, each in its own "compartment,"

$$p = p_a + p_b + \ldots + p_n \tag{iii}$$

and after substitution from equations (12.11b) and (12.13)

$$\frac{mRT}{V} = \frac{m_aR_aT}{V} + \frac{m_bR_bT}{V} + \ldots + \frac{m_nR_nT}{V} \tag{iv}$$

Combination of equations (iii) and (iv) leads to a relationship between partial pressure and gravimetric composition

$$\frac{p_a}{p} = \frac{m_aR_a}{mR}, \quad \frac{p_b}{p} = \frac{m_bR_b}{mR}, \quad \ldots, \quad \frac{p_n}{p} = \frac{m_nR_n}{mR} \tag{12.22}$$

Combination of equations (12.21) and (12.22) leads to the following even simpler relationship between partial pressure and volumetric composition

$$\frac{p_a}{p} = \frac{V_a}{V}, \quad \frac{p_b}{p} = \frac{V_b}{V}, \quad \ldots, \quad \frac{p_n}{p} = \frac{V_n}{V} \tag{12.23a}$$

or

$$\frac{p_a}{p_b} = \frac{V_a}{V_b}, \quad \frac{p_b}{p_c} = \frac{V_b}{V_c}, \quad \text{etc.} \tag{12.23b}$$

Equations (12.23) imply that the partial pressure of each constituent may be obtained very readily from the volumetric analysis expressed by (12.20).

The conversion from a volumetric to a gravimetric analysis may now be completed as follows. Since the various gas "compartments" in *Fig*. 12.1 each contain different gases at the same pressure and temperature, it follows from Avogadro's hypothesis (see Chapter 9, section 9.4) that the number of molecules of each constituent is proportional to the volume occupied by that constituent, and hence to the volumetric analysis expressed by equation (12.20). This, in turn, is proportional to the partial pressure ratio (from equation (12.23b)), i.e.

$$\frac{n_i}{\sum_{i=1}^{n} n_i} = \frac{v_i}{V} = \frac{p_i}{p} \tag{12.24a}$$

Hence the mass of each constituent present, which is proportional to the product $n_i M_i'$ or $N_i M_i$ (where M_i' is the absolute mass of each molecule and M_i is the kilogramme molecular weight) may also be obtained from the volumetric analysis of the mixture. Hence we may write

$$m_a = M_a\left(\frac{V_a}{V}\right) \times K, \quad m_b = M_b\left(\frac{V_b}{V}\right) \times K, \quad \ldots,$$

$$m_n = M_n\left(\frac{V_n}{V}\right) \times K \tag{v}$$

and $$m = \sum_{i=1}^{n}(m_i) = \sum_{i=1}^{n}\left[M_i\left(\frac{V_i}{V}\right)K\right] \tag{vi}$$

where K is a constant.

Combining equations (v) and (vi) the percentage gravimetric analysis corresponding to the volumetric analysis of equation (12.20) becomes

$$\frac{m_a}{m} \times 100 = M_a\left(\frac{V_a}{V}\right) \times 100 \Big/ \sum_{i=1}^{n}\left(M_i \frac{V_i}{V}\right)$$

$$\frac{m_b}{m} \times 100 = M_b\left(\frac{V_b}{V}\right) \times 100 \Big/ \sum_{i=1}^{n}\left(M_i \frac{V_i}{V}\right)$$

$$\frac{m_n}{m} \times 100 = M_n\left(\frac{V_n}{V}\right) \times 100 \Big/ \sum_{i=1}^{n}\left(M_i \frac{V_i}{V}\right) \tag{12.24b}$$

the constant K in equations (v) and (vi) vanishing after division.

Equation (12.24b) suggests a very convenient tabular method for transforming volumetric into gravimetric analyses. This is best illustrated in tabular form (Table 12.1).

Table 12.1

(1) *Constituent*	(2) *Percentage vol.*	(3) *Percentage vol.* × M_i	(4) *Percentage by mass*
a	$\frac{V_a}{V} \times 100$	$\left(\frac{V_a}{V} \times 100\right) M_a$	$100\left(\frac{V_a}{V} \times 100\right) M_a \Big/ \left(\sum_{i=1}^{n} \text{col. } 3\right)$
b	$\frac{V_b}{V} \times 100$	$\left(\frac{V_b}{V} \times 100\right) M_b$	$100\left(\frac{V_b}{V} \times 100\right) M_b \Big/ \left(\sum_{i=1}^{n} \text{col. } 3\right)$
n	$\frac{V_c}{V} \times 100$	$\left(\frac{V_n}{V} \times 100\right) M_n$	$100\left(\frac{V_n}{V} \times 100\right) M_n \Big/ \left(\sum_{i=1}^{n} \text{col. } 3\right)$
	$\Sigma = 100$	$\Sigma\left(\frac{V_i}{V} \times 100\right) M_i$	$\Sigma = 100\%$

EXAMPLE 12.1

The percentage volumetric composition of air is approximately as follows: nitrogen (N_2) 78·09, oxygen (O_2) 20·95, argon (A) 0·96. Given that the molecular weights of these three constituents are respectively 28, 32 and 40, find the percentage gravimetric composition, and the equivalent molecular weight.

Table 12.2

Constituent	*Percentage volume*	*Percentage volume* × M_i	*Percentage by mass (gravimetric)*
Nitrogen (N_2)	78·09	78·09 × 28 = 2,186·52	75·75
Oxygen (O_2)	20·95	20·95 × 32 = 670·40	22·92
Argon (A)	0·96	0·96 × 40 = 38·40	1·33
	100·00	2,895·32	100·00

Whilst the expressions for the mixture gas constant R, and the two specific heats c_v and c_p in terms of the gravimetric composition (equations (12.12),

(12.18) and (12.19)) are of fairly simple form, each of these constants being the weighted mean of the respective individual constants, the expression for the equivalent molecular weight (equation (12.15a)) may be simplified considerably by expressing it in terms of the volumetric composition.

Thus, substituting for

$$\sum_{i=1}^{n}(m_i) = \sum_{i=1}^{n}\left(\frac{pV_i}{R_iT}\right) = \frac{p}{T}\sum_{i=1}^{n}\left(\frac{V_i}{R_i}\right) \qquad \text{(i)}$$

and

$$\sum_{i=1}^{n}(m_iR_i) = \sum_{i=1}^{n}\left(\frac{pV_i}{T}\right) = \frac{p}{T}\sum_{i=1}^{n}(V_i) \qquad \text{(ii)}$$

in equation (12.15a),

$$M = G\sum_{i=1}^{n}\left(\frac{V_i}{R_i}\right)\Bigg/\sum_{i=1}^{n}(V_i) \qquad \text{(iii)}$$

But from equation (12.14)

$$G/R_i = M_i \qquad \text{(iv)}$$

where M_i is the molecular weight of constituent (i), whence substituting from (iv) in (iii)

$$M = \sum_{i=1}^{n}(M_iV_i)\Bigg/\sum_{i=1}^{n}(V_i) = \sum_{i=1}^{n}\left(M_i\frac{V_i}{V}\right)\Bigg/\sum_{i=1}^{n}\left(\frac{V_i}{V}\right) = \sum_{i=1}^{n}\left(M_i\frac{V_i}{V}\right)$$

since $\sum_{i=1}^{n}\left(\frac{V_i}{V}\right) = 1.$ (12.15b)

Hence the equivalent molecular weight of the mixture is equal to the weighted mean of the individual molecular weights, in terms of the volumetric, not of the gravimetric composition. Inspection of Table 12.1, and comparison with equation (12.25) indicates that the summated third column of that table divided by a hundred gives the equivalent molecular weight of the mixture.

Thus Table 12.2 gives the equivalent molecular weight of air as 28·9532.

12.3.6 Adiabatic mixing of gases

Consider again the rigid container of *Fig.* 12.1 and let it be surrounded by insulated walls. Let the partitions between the constituents be removed, so that the resultant process is one of ADIABATIC MIXING.

Before the removal of the partitions each constituent is at pressure p and temperature T. Since the process is adiabatic, and no useful work can be done as the external walls of the container are stationary and the internal partitions do not move in the direction of the pressure exerted on them, it follows from the first law of thermodynamics, or more specifically from the non-flow energy equation (N.F.E.E. equation (4.4b)) that the internal energy of the gases before and after mixing must be the same. Since for each of the gases, being perfect or semi-perfect, the internal energy is a function of temperature only (see equation (10.2)), each gas being at the common temperature T before mixing, and since after mixing the internal energy of the mixture is equal to the sum of the component internal energies in the mixed state (equation (12.8)) and the components are once again of necessity at a common temperature T', it follows that the final temperature T' must be equal to the initial temperature T, or

$$T = T' \tag{12.25}$$

The resultant pressure p' may be evaluated as follows. Before mixing

$$V = V_a + V_b + \ldots + V_n \tag{i}$$

$$p = p_a = p_b = \ldots = p_n \tag{ii}$$

After mixing, applying the partial pressure law, equation (12.1)

$$p' = p_a + p_b + \ldots + p_n \tag{iii}$$

$$V = V_a = V_b = \ldots = V_n \tag{iv}$$

But since the final temperature T' equals the initial temperature T, applying the state equation (12.11b) to each constituent before and after mixing

$$V_a = \left(\frac{m_a RT}{p}\right), \quad p_a = \left(\frac{m_a RT}{V}\right)$$

or

$$pV_a = p_a V, \quad pV_b = p_b V, \text{ etc.} \tag{v}$$

Substituting from equation (v) in equation (iii)

$$p' = \frac{p}{V}(V_a + V_b + \ldots + V_n) = p\frac{V}{V}$$

or

$$p' = p \tag{12.26}$$

i.e. the final mixture pressure is equal to the pressure of each of the individual constituents before mixing.

An interesting result may be deduced mathematically for the case of perfect as opposed to semi-perfect gases, although in fact the result thus obtained applies equally to perfect and semi-perfect gases.

It is clear from previous considerations (see Chapter 5, section 5.5) that the process of mixing by successive or simultaneous removal of the partitions of the gas container is an irreversible one. Hence, according to the 7th corollary of the second law (see Chapter 7, section 7.6) it must be accompanied by an increase of entropy.

Applying the perfect gas entropy equation in that form which connects entropy with volume and temperature, to any one of the constituents (equation (10.31))

$$\Delta S_i = m_i R_i \ln\left(\frac{V}{V_i}\right) + m_i (c_v)_i \ln\left(\frac{T'}{T}\right) = m_i R_i \ln\left(\frac{V}{V_i}\right) \quad (12.27)$$

since $T = T'$, equation (12.25).

But in each case $(V/V_i) > 1$, since the volume occupied initially is less than that of the container as a whole, so that the entropy change indicated by equation (12.27) is always positive. Hence for the mixing process as a whole

$$\sum_{i=1}^{n} \Delta S_i = \sum_{i=1}^{n} m_i R_i \ln\left(\frac{V}{V_i}\right) > 0 \quad (12.28)$$

Equation (12.28) thus indicates that the mixing process is accompanied by an increase of entropy, as predicted by the second law.

EXAMPLE 12.2

Adiabatic mixing with gain of entropy

Two containers, each 0·7 m³ in volume, are isolated from each other by a partition. One contains oxygen, the other nitrogen, each at 7 bar and 90°C. The partition is removed and mixing occurs. Calculate the end state of the mixture, as well as the entropy change for adiabatic mixing.

SOLUTION

From equation (12.25),

$$T = T', \text{ i.e. mixture temperature } T' = 90°\text{C} = 363 \text{ K}$$

From equation (12.26),

$$p = p', \text{ i.e. mixture pressure } p' = 7 \text{ bar}$$

Partial pressure of oxygen and nitrogen, from equation (12.23b),

$$\frac{p_{O_2}}{p_{N_2}} = \frac{V_a}{V_b}$$

and since $V_a = V_b = 0{\cdot}7 \text{ m}^3$

$$p_{O_2} = p_{N_2}$$

whilst, from equation (12.1),

$$p_{O_2} + p_{N_2} = p$$
$$p_{O_2} = p_{N_2} = \tfrac{1}{2}p' = 3{\cdot}5 \text{ bar}$$

The entropy change may be calculated from equation (12.28). Since

$$m_{O_2} = \frac{p_{O_2}V}{R_{O_2}T'} = \frac{3{\cdot}5 \times 10^2 \times 1{\cdot}4}{(8{\cdot}3143/32)363} = 5{\cdot}20 \text{ kg}$$

and

$$m_{N_2} = \frac{p_{N_2}V}{R_{N_2}T'} = \frac{3{\cdot}5 \times 10^2 \times 1{\cdot}4}{(8{\cdot}3143/28)363} = 4{\cdot}55 \text{ kg}$$

$$\Sigma\Delta S = m_{O_2}R_{O_2} \ln\left(\frac{p'}{p_{O_2}}\right) + M_{N_2}R_{N_2} \ln\left(\frac{p'}{p_{N_2}}\right)$$
$$= (5{\cdot}20 \times 0{\cdot}2602 + 4{\cdot}55 \times 0{\cdot}2972) \ln(7{\cdot}0/3{\cdot}5)$$
$$= 1{\cdot}87 \text{ kJ/K}$$

Hence mixing leads to an entropy gain as predicted.

Consider now the special case of a "mixture" of two identical gases, i.e. a process in which a partition initially dividing two portions *a* and *b* of the same gas at the same pressure *p* and temperature *T*, is suddenly removed.

It is clear from equation (12.27) that since, in the present instance, no increase in specific volume occurs, the pressure and temperature of the gas remaining unchanged, there is no increase of entropy. Yet the process would appear to be irreversible, as before.

To explain this apparent paradox, known as Gibbs' paradox, it is necessary to refer to the basic definition of irreversibility in Chapter 5. The present process, unlike the previous one, does not involve any change of state and hence does not lend itself, under any circumstances, to the extraction of useful work. In the case of mixing of different gases, however, each gas undergoes a significant, i.e. measurable, change of state which, if brought about reversibly by special means, lends itself to the production of useful work.

The production of useful work as the result of an adiabatic mixing process discussed above is conditional on the use of an artifice, namely the semi-permeable membrane which, whilst allowing the passage of one gas, is impermeable to the other gas of the mixture.

Consider the adiabatic container (*Fig*. 12.2) with a single partition only in the form of a semi-permeable membrane, dividing a volume of gas *a* from an equal volume of gas *b*, both at the same temperature. This membrane allows the passage of gas from left to right, but not vice versa. In addition there is a pair of pistons rigidly connected, of which piston I is completely impermeable to either gas, whilst piston II is permeable to gas *b*. As the piston assembly is moved to the right, the four spaces formed will contain respectively, in order from left to right

space 1: vacuum,
space 2: gas *a* only,
space 3: gas mixture *a* and *b* with increasing proportion of *a*,
space 4: gas *b* only.

As the piston assembly approaches the limit of its travel, space 1 will occupy the whole of the space originally occupied by gas *a*, while space *b* will now contain a mixture of the gases *a* and *b*.

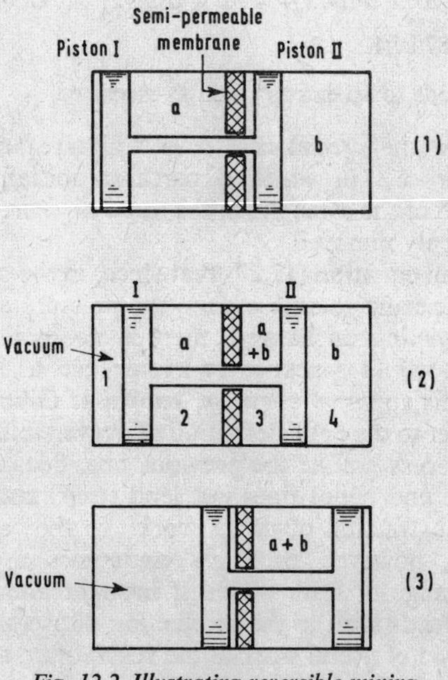

Fig. 12.2 Illustrating reversible mixing

The piston motion is infinitely slow and the temperature during the mixing process will, since both gases are perfect, remain constant. Hence, from the Gibbs–Dalton law, the volume and temperature of the separate gases, initially equal, will also be the volume and temperature of the final mixture. If the initial pressures of the unmixed gases are p_a and p_b, these will therefore also be their partial pressures in the mixture.

At any intermediate stage of the mixing process the net force on piston I is given by

$$A(p_a - 0) = A\,p_a \quad \text{to the left}$$

whilst that on piston II is given by

$$A[(p_a + p_b) - p_b] = A\,p_a \quad \text{to the right}$$

Hence the piston assembly as a whole is balanced, and no net work will be done by or on the piston assembly. Furthermore, the process will be fully reversible, since an infinitesimal external force will reverse the process.

Nevertheless, the net result of the process described above is to create a situation in which, by making use of the vacuum created, external work can be done on the surroundings, thus demonstrating that the unmixed gases contain available energy which can be converted into useful work (see also Chapter 14), and that the normal adiabatic mixing process without work is therefore, by definition, irreversible.

12.4 Mixtures of Perfect or Semi-perfect Gases and Vapours

As already stated in the introduction to the present chapter, the most important case of mixtures of gases and vapours is that of the air–water vapour mixture, with the water vapour either in the "unsaturated" or "saturated" form. The former may be regarded as a single-phase mixture in which the water vapour is at sufficiently low partial pressure to be treated, without serious loss of accuracy, as a perfect gas (see section 3.5 and Chapter 9)*, whilst the latter is a two-phase mixture (gas, water vapour and liquid water) in which the water vapour cannot generally be treated as a perfect gas. The mixture of air and unsaturated water vapour obeys the Gibbs–Dalton law, whilst the mixture of air and saturated vapour mixture has to be treated by special methods. It is understood, of course, that the methods outlined here apply equally to mixtures of other gases and vapours.

* It will be recalled that superheated vapours approach perfect gas behaviour when

(*a*) the temperature is well above the critical temperature T_c
(*b*) the absolute pressure is very low.

It is the latter condition which is fulfilled in the present case.

12.4.1 General discussion and terminology

The study of the behaviour of mixtures of air and water vapour is known as HYGROMETRY or PSYCHOMETRY. Several special terms are in common use in this subject, and these will be discussed in conjunction with the various physical and thermodynamic characteristics of air–water vapour mixtures in the present section.

The general behaviour of mixtures of air and water vapour is best understood by referring to the temperature–entropy diagram for water (*Fig*. 12.3).

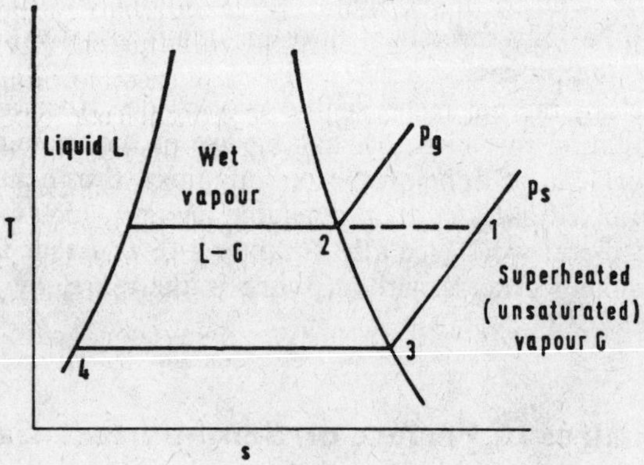

Fig. 12.3 Phase diagram for H_2O *in air–water mixtures*

For any mixture, whether saturated or unsaturated, the partial pressures of air and water vapour, respectively, will be denoted by p_a and p_s. These partial pressures will, under either set of conditions, provide a measure of the gravimetric composition of the mixture.

Consider first an unsaturated mixture of air and water vapour at atmospheric pressure, i.e. one in which liquid water is absent, and let the partial pressure of the water vapour be p_s. Since liquid water is absent, the vapour exists in superheated form, and due to the much smaller mass fraction of water vapour compared with that of air, the partial pressure of the vapour, p_s, will also be much smaller than that of the air, p_a, being normally a small fraction of one bar. Under such conditions, typified by point 1 (temperature T_1) in *Fig*. 12.3, the vapour may be treated as a perfect gas. Hence, the Gibbs–Dalton law may be applied directly, so that the gravimetric proportions of water vapour and air are given by equation (12.22)

$$\omega = \frac{m_s}{m_a} = \frac{p_s}{p_a}\frac{R_a}{R_s} = \frac{v_a}{v_s} \qquad (12.29)$$

where R_s is the characteristic gas constant for water vapour, i.e.

$$R_s = \frac{G}{M_s} = \frac{8{\cdot}3143}{18} = 0{\cdot}4624 \text{ kJ/kg-K}$$

and the symbol ω is used to denote the mass fraction of water vapour in relation to dry air, known as the MOISTURE CONTENT or SPECIFIC HUMIDITY.

Substituting for $R_a = 0{\cdot}2870$ kJ/kg-K and $p_a = (p - p_s)$ (from Dalton's law), where p is the barometric pressure, in equation (12.29), the following simplified expression is obtained,

$$\omega = 0{\cdot}622 \frac{p_s}{p - p_s} \qquad (12.30)$$

Still considering the mixture with water vapour at state point 1, it is clear that since the mixture is unsaturated and the water vapour superheated, the MOISTURE CONTENT could be increased at the given temperature T_1, by progressive injection of water until the mixture becomes saturated, i.e. incapable of holding more water vapour at that temperature. This saturated condition of the vapour is given by point 2 in *Fig*. 12.3. The partial pressure of the vapour is now p_g, the saturation pressure at the temperature $T_2 = T_1$, and this exceeds the original partial pressure p_s at point 1.

From equation (12.30), and assuming that the Gibbs–Dalton law on which that equation is based still holds at point 2, i.e. under saturated conditions, the ratio of the moisture content under unsaturated conditions at point 1 to that under saturated conditions at point 2 becomes

$$\phi = \frac{(m_s)_1/m_a}{(m_s)_2/m_a} = \frac{(p_s)_1[p - (p_s)_2]}{[p - (p_s)_1](p_s)_2} \simeq \frac{(p_s)_1}{(p_s)_2} = \frac{(p_s)_1}{(p_g)_2} \qquad (12.31a)$$

where $(p_g)_2$ refers to the saturation pressure at point 2. The above approximation is very nearly correct since, under all conditions

$$p_s \ll p$$

In general, we may write equation (12.31a) in the form

$$\phi = \frac{m_s}{(m_s)_g} = \frac{p_s}{p_g} \qquad (12.31b)$$

The symbol ϕ in equation (12.31b) denotes the RELATIVE HUMIDITY, i.e. the ratio of the actual moisture content, ω_1, to that under saturated conditions, ω_2. Equation (12.31) indicates that the relative humidity ϕ may be regarded either as the ratio of actual to saturated mass content of water vapour, or alternatively as the ratio of actual to saturated partial pressure of water vapour.

Considering now the alternative process path 1–3 (*Fig.* 12.3), i.e. cooling of the unsaturated mixture at point 1 at constant mixture pressure and without change of composition of the mixture, such cooling is possible down to the limiting condition 3, known as the DEW POINT, i.e. that point at which the mixture becomes saturated with moisture and any further cooling from which would produce partial condensation of water vapour along the path 3–4. It is evident from inspection of equation (12.30) that, during the process 1–3, not only the composition, but also the partial pressure of the vapour, p_s, remains constant since reduction in temperature leads to a corresponding reduction in saturation pressure. The dew point is in fact that point at which the partial pressure p_s becomes equal to the saturated vapour pressure $(p_g)_d$, the second suffix d referring to the dew point.

Substituting for $(p_g)_d = p_s$ in equation (12.31b), we obtain the following alternative expression for the relative humidity ϕ

$$\phi = \frac{m_s}{(m_s)_g} = \frac{(p_g)_d}{p_g} \tag{12.31c}$$

The importance of the dew point lies in the fact that it provides a simple experimental basis for the determination of the relative humidity. The dew point may readily be observed, for example, by cooling a metal plate and observing the temperature at which dew begins to form.

The relative humidity may also be determined by measuring the so-called WET-AND-DRY-BULB TEMPERATURES. If unsaturated air is caused to flow over a wetted thermometer bulb, partial evaporation of the water on the bulb will occur, the air downstream of the thermometer being assumed to be saturated with water vapour. Due to the abstraction of the latent heat of evaporation of the water thus absorbed into the airstream from the surroundings, an equilibrium temperature lower than that of the air is recorded. Clearly the resultant difference, known as the wet-bulb depression, is greatest when the air is initially completely dry, i.e. capable of absorbing a maximum amount of moisture, and zero when the air is initially saturated. The actual temperature of the airstream may be recorded with a mercury-in-glass thermometer on which the bulb is kept dry, thus eliminating the possibility of evaporation from the bulb and the resultant evaporative cooling. The wet-and-dry-bulb thermometers are usually incorporated in a simple instrument, the wet-and-dry-bulb hygrometer (*Fig.* 12.4).

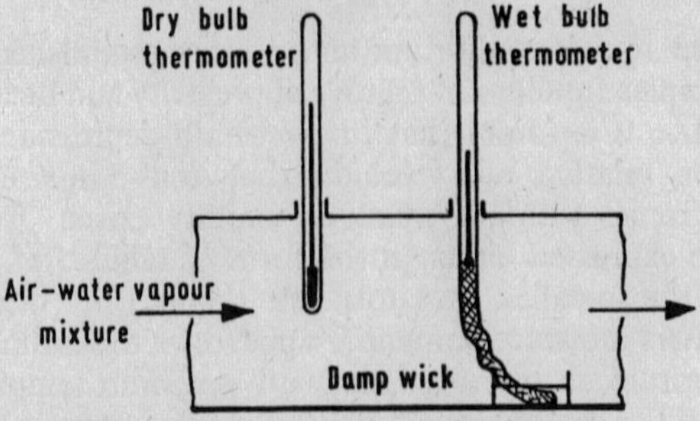

Fig. 12.4 Wet-and-dry-bulb hygrometer

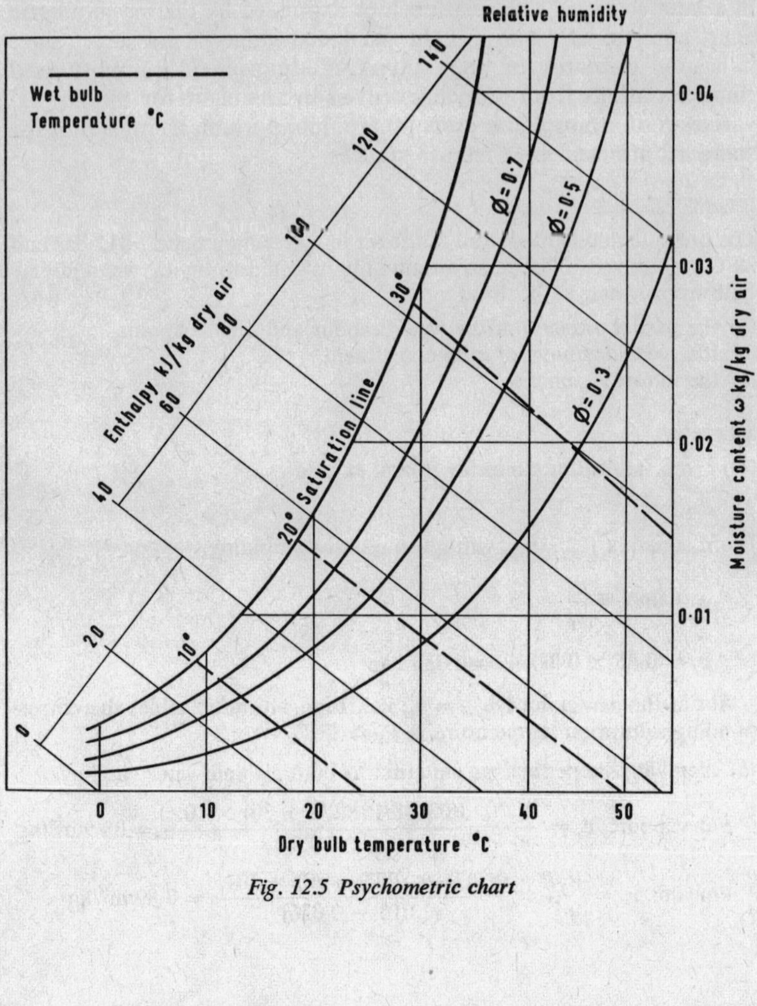

Fig. 12.5 Psychometric chart

Clearly the actual equilibrium temperature established at the wet bulb is a complex function of relative air velocity and local geometry, but in practice it is found that the wet-bulb depression is sensibly constant for relative air velocities between 3 m/s and 40 m/s. Within this range a unique relative humidity exists. This relationship may be expressed either in the form of tables (ref. 12.1) or in the form of the so-called PSYCHOMETRIC CHART (*Fig*. 12.5).

On this chart moisture content ω appears as the ordinate and dry bulb temperature as the abscissa, with wet bulb temperature and relative humidity ϕ as parameters. On more comprehensive charts lines of constant enthalpy h per pound of *dry* air are added. The computation of enthalpy for air–water vapour mixtures is discussed in a later section. The relationships expressed by the psychometric chart apply strictly only for the particular value of pressure. Since it is used primarily in the analysis of atmospheric humidity, and since deviations from the values given by the chart for the normal variation of atmospheric pressure are unimportant, a single plot for standard atmospheric pressure suffices.

EXAMPLE 12.3

The pressure and temperature of the air in a greenhouse are 1·015 bar and 30°C respectively. The relative humidity is found to be, by wet-and-dry-bulb hygrometer, 85%, Find

(*a*) the partial pressure of the water vapour and the dew point,
(*b*) the specific volume of each constituent,
(*c*) the moisture content.

SOLUTION

(*a*) From saturation tables for steam, at 30°C,

$$p_g = 0{\cdot}04242 \text{ bar}$$

From equation (12.31b), with given relative humidity,

$$\phi = 0{\cdot}85 = \frac{p_s}{p_g}$$

$$p_s = 0{\cdot}85 \times 0{\cdot}04242 = 0{\cdot}036 \text{ bar}$$

But at the dew point $(p_g)_d = p_s$, i.e. from saturation tables the corresponding saturation temperature is $T_d \simeq 27°\text{C}$.

(*b*) Applying the perfect gas equation to both air and water vapour,

$$\text{For vapour: } v_s = \frac{R_s T}{p_s} = \frac{(8{\cdot}3143/18)(273 + 30) \times 10^{-2}}{0{\cdot}036} = 38{\cdot}9 \text{ m}^3/\text{kg}$$

$$\text{For air: } v_a = \frac{R_a T}{p_a} = \frac{0{\cdot}2871 \times (273 + 30) \times 10^{-2}}{(1{\cdot}015 - 0{\cdot}036)} = 0{\cdot}89 \text{ m}^3/\text{kg}$$

(*c*) From equation (12.29),

$$\text{Moisture content } \omega = \frac{v_a}{v_s} = \frac{0{\cdot}89}{38{\cdot}9} = 0{\cdot}023 \text{ kg}_s/\text{kg}_a$$

Alternatively, from equation (12.30)

$$\omega = 0{\cdot}622 \frac{p_s}{p - p_s} = 0{\cdot}622 \times \frac{0{\cdot}036}{0{\cdot}979}$$

$$= 0{\cdot}023 \text{ kg}_s/\text{kg}_a$$

12.4.2 Unsaturated mixtures of air and water vapour

Unsaturated mixtures of air and water vapour may be regarded as single-phase mixtures, in which liquid water is absent, and in which the water vapour exists in superheated form at sufficiently low pressure to be treated, for the purpose of Dalton's law, as a perfect gas. However, when evaluating internal energy, enthalpy and entropy of unsaturated air–water vapour mixtures, it is advisable to use tables or charts for superheated steam, since although the perfect gas relationships can be applied with reasonable accuracy to the evaluation of differences between these extensive properties within the superheated range, they cannot be used to find their values relative to the standard datum, i.e. $p = 0{\cdot}006112$ bar, $T = 293{\cdot}16$ K.

Referring to the equations of section 12.2 which apply quite generally to mixtures of gases as well as mixtures of gases and condensable vapours, it is clear that mixtures of air and superheated water vapour obey these equations without restriction.

However, the equations of section 12.3 for mixtures of perfect or semi-perfect gases have to be modified in certain instances. Equation (12.12) for the characteristic gas constant, and (12.15a) for the equivalent molecular weight of the mixture may be retained, i.e. using the suffix notation for air–water vapour mixtures—

$$R = \frac{m_a R_a + m_s R_s}{m_a + m_s} = \frac{0{\cdot}2871 m_a + 0{\cdot}4624 m_s}{m_a + m_s} \tag{12.32}$$

and

$$M = G\left(\frac{m_a + m_s}{m_a R_a + m_s R_s}\right) = G\left(\frac{m_a + m_s}{0{\cdot}2871 m_a + 0{\cdot}4624 m_s}\right) \tag{12.33}$$

Likewise the partial pressure equation (12.22) may be retained in the form

$$\frac{p_s}{p_a} = \frac{m_s R_s}{m_a R_a} = \frac{0{\cdot}4624 m_s}{0{\cdot}2871 m_a} = \frac{1}{0{\cdot}622}\,\omega \tag{12.34}$$

(see also equation (12.30)) as may also equation (12.23b) for partial volumes

$$\frac{p_s}{p_a} = \frac{V_s}{V_a} \tag{12.35}$$

However, equations (12.18) for the specific heat at constant volume, (12.19) for the specific heat at constant pressure and (12.27) for entropy cannot be used, except for the determination of differences in internal energy, enthalpy and entropy within the unsaturated mixture (superheated vapour) range, as already stated.

These procedures are best illustrated by means of an example.

EXAMPLE 12.4

In part of an air conditioning plant 20 m³/min of air at 32°C, 60% relative humidity, are cooled to 24°C, the atmospheric pressure being 1 bar.

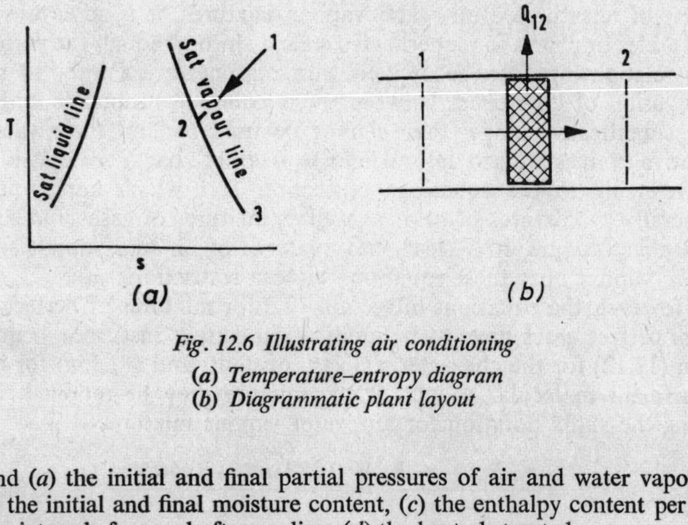

Fig. 12.6 Illustrating air conditioning
(a) Temperature–entropy diagram
(b) Diagrammatic plant layout

Find (*a*) the initial and final partial pressures of air and water vapour, (*b*) the initial and final moisture content, (*c*) the enthalpy content per kg of mixture before and after cooling, (*d*) the heat abstracted.

SOLUTION

Partial pressures. From equation (12.31b)

At point 1

Relative humidity $\phi = 0{\cdot}6$
Saturation pressure at 32°C (from tables) $(p_g)_1 = 0{\cdot}04754$ bar.
Therefore, partial pressure $(p_s)_1 = 0{\cdot}6 \times 0{\cdot}04754 = 0{\cdot}0286$ bar.

Dew point

Partial pressure $(p_g)_d = (p_s)_1 = 0{\cdot}0286$ bar.
Therefore $(T_g)_d = 23{\cdot}3$°C, by interpolation from tables.

At point 2

$T_2 = 24°C > (T_g)_d$ i.e. mixture is still unsaturated.

(*a*) *Partial pressures*

$$(p_s)_2 = (p_s)_1 = 0{\cdot}0286 \text{ bar}$$

But equality of partial water vapour pressures at points 1 and 2 implies equality of the partial pressure of air. Therefore

$$(p_a)_2 = (p_a)_1 = p - (p_s)_1$$
$$= 1 - 0{\cdot}0286 = 0{\cdot}9714 \text{ bar} \quad \textit{Ans.}$$

(*b*) *Moisture content*

From equation (12.30), constancy of partial pressure implies constancy of moisture content, i.e.

$$\omega_1 = \omega_2 = 0{\cdot}622 \frac{p_s}{p - p_s} = 0{\cdot}622 \frac{0{\cdot}0286}{1 - 0{\cdot}0286}$$
$$= 0{\cdot}0183 \text{ kg } H_2O/\text{kg air} \quad \textit{Ans.}$$

(*c*) *Enthalpy content*, relative to $p_0 = 0{\cdot}006112$ bar, $T_0 = 273{\cdot}16$ K

At point 1

$$H_1 = m_a(h_a)_1 + (m_s)_1(h_s)_1 = m_a(c_p)_a(T_1 - 0{\cdot}01) + (m_s)_1(h_s)_1 \qquad \text{(i)}$$
$$= (1 \times 1{\cdot}005)(32 - 0{\cdot}01) + 0{\cdot}0183 \times 2{\cdot}555$$
$$= 32{\cdot}19 + 46{\cdot}18$$
$$= 78{\cdot}99 \text{ kJ/kg of dry air}$$

where $(h_s)_1 = 2{\cdot}555$ kJ/kg by interpolation from superheat tables.

At point 2

$$H_2 = m_a(h_a)_2 + (m_s)_2(h_s)_2 = m_a(c_p)_a(T_2 - 0{\cdot}01) + (m_s)_2(h_s)_2 \qquad \text{(ii)}$$
$$= (1 \times 1{\cdot}005)(24 - 0{\cdot}01) + 0{\cdot}0183 \times 2540$$
$$= 24{\cdot}15 + 46{\cdot}50$$
$$= 70{\cdot}65 \text{ kJ/kg of dry air}$$

where $(h_s)_2 = 2540$ kJ/kg by interpolation from superheat tables.

N.B. Use of the perfect gas equation (10.28) for the evaluation of the enthalpy of the superheated steam would have led to the following result

$$(c_p)_s = \frac{\gamma}{\gamma - 1} R$$

where $\gamma = 1{\cdot}333$ (since H_2O is triatomic, equation (9.5c)) and

$$R = \frac{G}{M_s} = \frac{8{\cdot}3143}{18} = 0{\cdot}4624 \text{ kJ/kg-K}$$

so that, substituting in (i) above,

$$\begin{aligned} H_1 &= m_a(h_a)_1 + (m_s)_1(h_s)_1 = [m_a(c_p)_a + (m_s)_1(c_p)_s](T_1 - 0{\cdot}01) \\ &= (1 \times 1{\cdot}005 + 0{\cdot}0183 \times 1{\cdot}849)(32 - 0{\cdot}01) \\ &= 32{\cdot}19 + 1{\cdot}088 \\ &= 33{\cdot}27 \text{ kJ/kg of dry air} \end{aligned}$$

(cf. 78·97 kJ/kg of dry air by previous evaluation) and substituting in (ii) above

$$\begin{aligned} H_2 &= m_a(h_a)_2 + (m_s)_2(h_s)_2 = [m_a(c_p)_a + (m_s)_2(c_p)_s](T_2 - 0{\cdot}01) \\ &= (1 \times 1{\cdot}005 + 0{\cdot}0183 \times 1{\cdot}849)(24 - 0{\cdot}01) \\ &= 24{\cdot}15 + 0{\cdot}80 \\ &= 24{\cdot}95 \text{ kJ/kg of dry air} \end{aligned}$$

(cf. 70·65 kJ/kg of dry air by previous calculation).

It is evident from the above that the use of the perfect gas equations for water vapour introduces severe errors into the enthalpy calculations, due to neglect of the phase change between 0·01°C on the one hand, and T_1 and T_2 on the other. However, if enthalpy *differences* only are required, the accuracy is adequate, i.e.

$$\Delta H_s = -(m_s)_1(h_s)_1 + (m_s)_2(h_s)_2 = -46{\cdot}78 + 46{\cdot}50 = -0{\cdot}28 \text{ kJ/kg}$$

of air by original method or

$$-1{\cdot}08 + 0{\cdot}80 = -0{\cdot}28 \text{ kJ/kg}$$

of air by perfect gas method.

In practice, the most satisfactory method for purposes of numerical evaluation, and to avoid the difficulty of interpolation, is to use property tables for saturated steam in order to evaluate the saturation enthalpy at the dew point (in the present case 23·3°C) and to add to this value the superheat using the perfect gas equation, or, in symbols

$$h_s = (h_g)_d + (c_p)_s[T - (T_g)_d]$$

$$\begin{aligned} (h_s)_1 &= 2543{\cdot}54 + 1{\cdot}849(32 - 24) \\ &= 2558{\cdot}30 \text{ kJ/kg of } H_2O \end{aligned} \qquad \text{(iii)}$$

$$\begin{aligned} (h_s)_2 &= 2543{\cdot}54 + 1{\cdot}849(24 - 23{\cdot}3) \\ &= 2544{\cdot}83 \text{ kJ/kg of } H_2O \end{aligned} \qquad \text{(iv)}$$

$$\Delta H_s = -0{\cdot}0183(2558{\cdot}30 - 2544{\cdot}83) = -0{\cdot}25 \text{ kJ/kg} \quad \textit{Ans.}$$

(*d*) *Heat abstracted*

$$\begin{aligned} \Delta q &= H_2 - H_1 = \Delta H_a + \Delta H_s \qquad \text{(v)} \\ &= (24{\cdot}15 - 32{\cdot}19) - 0{\cdot}25 \\ &= -8{\cdot}29 \text{ kJ/kg of air} \end{aligned}$$

But since the plant handles 20 m^3/min of air–water vapour mixture, and

$$m_a = \frac{p_a V_a}{R_a T}$$

where $V = V_a = 20$ m^3/min (equation (12.2)), $p_a = 0{\cdot}9714$ bar, $T = 32$°C $= 305{\cdot}15$ K then

$$\begin{aligned} m_a &= \frac{0{\cdot}9714 \times 20 \times 10^2}{0{\cdot}2871 \times 305{\cdot}15} \\ &= 222 \text{ kg/min} \\ \Delta Q &= m_a \Delta q \quad \text{equation (v), above} \\ &= -222 \times 8{\cdot}29 \\ &= -1840 \text{ kJ/min} \quad \textit{Ans.} \end{aligned}$$

12.4.3 Saturated mixtures of air and water vapour (air and saturated water vapour)

Saturated mixtures, as opposed to unsaturated mixtures, are characterized by the presence of liquid water. Hence the partial pressure of the water vapour in such mixtures equals the saturation pressure *at the temperature of the mixture.*

Although the Dalton and Gibbs–Dalton relationships still apply in the sense that partial pressures, internal energies, enthalpies and entropies are additive, as they were in the case of mixtures of gases, and of mixtures of gases and unsaturated vapours, all values appertaining to steam should now be taken from tables for saturated steam, with the exception of the evaluation of moisture content which is based for convenience, and with little error, on equation (12.29) or the related equation (12.30).

The methods are best illustrated by a simple example based on the dew point conditions of the unsaturated air–water vapour mixtures of example 12.4.

EXAMPLE 12.5

Evaluate (*a*) the partial pressures of air and water vapour, (*b*) the moisture content, (*c*) the enthalpy content of a mixture of air and water vapour at its dew point temperature of 20°C when the atmospheric pressure is 1·013 bar.

SOLUTION

(*a*) *Partial pressure*

Partial pressure of water vapour

$$p_s = (p_g)_d = 0{\cdot}02337 \text{ bar}$$

(from tables for saturated steam, see also example 12.3). Therefore

$$p_a = p - p_s = 1{\cdot}0130 - 0{\cdot}02337 = 0{\cdot}98963 \text{ bar} \quad \textit{Ans.}$$

(*b*) *Moisture content*

From equation (12.30)

$$\omega = 0{\cdot}622 \frac{p_s}{p - p_s} = 0{\cdot}622 \frac{0{\cdot}02337}{0{\cdot}98963}$$

$$= 0{\cdot}0147 \text{ kg } H_2O\text{/kg air} \quad \textit{Ans.}$$

(*c*) *Enthalpy content*

$$H = m_a h_a + m_s h_s$$
$$= m_a(c_p)_a[(T_g)_d - 492)] + m_s(h_g)_d$$
$$= 1{\cdot}005[20 - 0{\cdot}01] + 0{\cdot}0147 \times 2537{\cdot}6$$
$$= 20{\cdot}00 + 37{\cdot}23$$
$$= 57{\cdot}23 \text{ kJ/kg of dry air} \quad \textit{Ans.}$$

where $(h_g)_d = 2537{\cdot}6$ kJ/kg H_2O by interpolation from saturated steam tables.

Further aspects of the behaviour of saturated mixtures of air and water vapour are illustrated by the following non-flow example.

EXAMPLE 12.6

A cylinder sealed by a sliding piston contains 0·136 kg of wet steam having a dryness fraction of 0·2, and 0·181 kg of air, the temperatures of the mixture being 80°C. Find the volume of the cylinder, and the pressure of the mixture, as well as the temperature to which the mixture must be raised to evaporate all the water, assuming that mixture pressure remains constant, while the volume increases as a result of motion of the piston.

SOLUTION

Initial conditions

Mass of saturated steam $m_s = 0{\cdot}2 \times 0{\cdot}136 = 0{\cdot}0272$ kg

Mass of saturated water $m_w = 0{\cdot}136 - 0{\cdot}0272 = 0{\cdot}1088$ kg

Partial pressure of steam

$$(p_s)_1 = (p_g)_1 = 0{\cdot}4736 \text{ bar}$$

(from tables for saturated steam).

Specific volume of steam (from tables for saturated steam)

$$v_s = 3{\cdot}408 \text{ m}^3\text{/kg}$$

Therefore volume occupied by water vapour

$$V_s = m_s v_s = 0{\cdot}0272 \times 3{\cdot}408 = 0{\cdot}0925 \text{ m}^3$$

Specific volume of saturated water

$$v_w = 1{\cdot}029 \times 10^{-3} \text{ m}^3\text{/kg}$$

Therefore volume occupied by water

$$V_w = m_w v_w = 0{\cdot}1088 \times 1{\cdot}029 \times 10^{-3} = 1{\cdot}118 \times 10^{-4}\ \text{m}^3$$

i.e. Volume of cylinder $(V_{cyl})_1 = V_w + V_s = 0{\cdot}09262\ \text{m}^3$
Partial pressure of air

$$(p_a)_1 = \frac{m_a R_a T_1}{V_s} = \frac{0{\cdot}181 \times 0{\cdot}2871 \times 353 \times 10^{-2}}{0{\cdot}0925} = 1{\cdot}98\ \text{bar}$$

i.e. Pressure of mixture $= [(p_s) + (p_a)]_1 = p_1 = 2{\cdot}4536$ bar

Final condition

Let cylinder volume $= (V_{cyl})_2$
Let cylinder temperature $= T_2$
Partial pressure of air

$$(p_a)_2 = \frac{m_a R_a T_2}{(V_{cyl})_2} = \frac{0{\cdot}181 \times 0{\cdot}2871 T_2 \times 10^{-2}}{(V_{cyl})_2}\ \text{bar} \qquad \text{(i)}$$

But since the total mixture pressure remains constant,

$$p_2 = p_1 = (p_a)_2 + (p_s)_2 = 2{\cdot}4536\ \text{bar} \qquad \text{(ii)}$$

Finally, since the moisture content of the mixture in its final state is five times that in the initial state, applying equation (12.30)

$$\frac{(p_s)_2}{p_2 - (p_s)_2} = 5\left[\frac{(p_s)_1}{p_1 - (p_s)_1}\right]$$

i.e. $$\frac{(p_s)_2}{2{\cdot}4536 - (p_s)_2} = 5 \times \frac{0{\cdot}4736}{1{\cdot}980} = 1{\cdot}195$$

whence $(p_s)_2 = 1{\cdot}335$ bar (iii)

i.e. $T_2 = 107{\cdot}9°\text{C} = 380{\cdot}9$ K (iv)

(from tables for saturated steam)
From equation (ii)

$$(p_a)_2 = p_2 - (p_s)_2 = 2{\cdot}4536 - 1{\cdot}335 = 1{\cdot}1186\ \text{bar} \quad \textit{Ans.} \qquad \text{(v)}$$

From equations (iv), (v) and (ii)

$$(V_{cyl})_2 = \frac{0{\cdot}181 \times 0{\cdot}2871 \times 380{\cdot}9 \times 10^{-2}}{1{\cdot}1186} = 1{\cdot}78\ \text{m}^3 \quad \textit{Ans.}$$

Hence the mixture temperature has to be raised to 107·9°C when the cylinder volume becomes 1·78 m³ (neglecting volume occupied by water).

12.4.4 Applications of relationships for saturated and unsaturated mixtures of air and water vapour

(*a*) THE ORSAT APPARATUS (*Fig.* 12.7)

The Orsat apparatus is a device commonly used in the laboratory to determine the *volumetric* composition of products of combustion, for example, in boilers or internal combustion engines, the results being used to assess the quality of the combustion process. (For details of such analyses, see Chapter 13, section 13.2.)

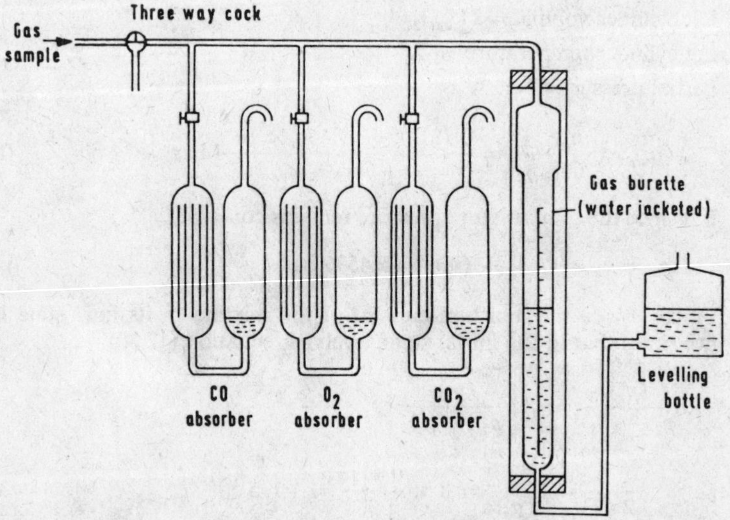

Fig. 12.7 Orsat apparatus

It consists essentially of three absorption vessels filled with different reagents—

(*a*) potassium hydroxide (KOH) solution for the absorption of carbon dioxide (CO_2),
(*b*) pyrogallic acid in potassium hydroxide for the absorption of oxygen (O_2),
(*c*) cuprous chloride solution for the absorption of carbon monoxide (CO),

which remove the above three constituents from the products of combustion in the order stated. The products are trapped in a graduated cylinder at atmospheric pressure. After each absorption the pressure in the graduated cylinder is restored to its original value by means of the "levelling bottle" shown. Hence, by recording the volume of gas remaining in the graduated cylinder after successive absorptions, the volumetric fraction of each constituent, in relation to the original total volume of products, may be determined. This, of course, constitutes the volumetric analysis which, by the method of section 12.3.5, may readily be converted into a gravimetric analysis, if required.

A complication arises from the fact that, since the mixture is permanently in contact with water, it is clearly saturated with water vapour. However, it will be shown that the results may be used as if the water vapour were completely absent, i.e. as if we were dealing with *dry* products of combustion, the resultant analysis being the *dry* analysis.

Its evaluation is based on the tabular procedure of section 12.3.5. Hence, if a, b, c denote the percentages of CO_2, O_2 and CO, the balance, i.e. $d = 100 - (a + b + c)$, is assumed to be nitrogen. In tabular form, see Table 12.3.

Table 12.3

(1) *Constituent*	(2) *Percentage of volume*	(3) *Percentage by volume × M*	(4) *Percentage by mass*
CO_2	a	$a \times 44$	$100a \times 44/\Sigma(3)$
O_2	b	$b \times 32$	$100b \times 32/\Sigma(3)$
CO	c	$c \times 28$	$100c \times 28/\Sigma(3)$
N_2	d	$(d) \times 28$	$100d \times 28/\Sigma(3)$
	100·00	$\Sigma(3)$	$\Sigma(4) = 100{\cdot}00$

However, in the presence of water vapour the above analysis must be modified as follows:

Since the temperature of the trapped gas is constant, the partial pressure of the water vapour remains constant at its saturation value, i.e. $p_s = p_g$. Applying equation (12.23a), and since the mixture pressure p remains constant, from Avogadro's hypothesis

$$\frac{p_s}{p} = \frac{V_s}{V} = \frac{n_s}{n} \tag{12.36}$$

where V = total mixture volume at any stage of the absorption process, V_s = partial volume of water vapour, n_s = number of kg-mols of water vapour, n = total number of kg-mols.

In fact, it is more convenient to work in terms of mol numbers, so that converting Table 12.3 and including the water vapour, Table 12.4 is obtained.

Table 12.4

(1) Constituent	(2) Percentage by volume	(3) No. of kg-mol	(4) No. of kg-mol × M	(5) Percentage by mass
CO_2	a'	$n_{a'}$	$n_{a'} \times 44$	$100\,a' \times 44/\Sigma(4)$
O_2	b'	$n_{b'}$	$n_{b'} \times 32$	$100\,b' \times 32/\Sigma(4)$
CO	c'	$n_{c'}$	$n_{c'} \times 28$	$100\,c' \times 28/\Sigma(4)$
N_2	d'	$n_{d'}$	$n_{d'} \times 28$	$100\,d' \times 28/\Sigma(4)$
H_2O	s	n_s	$n_s \times 18$	$100\,s \times 18/\Sigma(4)$
	$\Sigma(2) = 100$	$\Sigma(3)$	$\Sigma(4)$	$\Sigma(5) = 100$

Applying equation (12.36), we have the following situation—

(i) Before first absorption: $\dfrac{p_s}{p} = \dfrac{n_s}{n} = \dfrac{s}{100}$

(ii) After first absorption: $\dfrac{p_s}{p} = \dfrac{n_s - (n_s)_1}{n - [n_{a'} + (n_s)_1]} = \dfrac{n_s}{n}$

where $(n_s)_1$ = number of kg-mols of water vapour condensed during first absorption,

or rewriting in terms of percentage volume

$$\frac{s - s_1}{100 - (a' + s_1)} = \frac{s}{100}$$

whence

$$s_1 = \frac{a's}{100 - s} \qquad \text{(i)}$$

Hence the percentage contraction of the trapped volume of gas after the first absorption becomes

$$a' + s_1 = a'\left(1 + \frac{s}{100 - s}\right) = \frac{100a'}{100 - s} \qquad \text{(ii)}$$

But this is the percentage by volume of CO_2 in the *dry* sample, i.e. a in Table 12.3.

(iii) After second absorption

$$\frac{p_s}{p} = \frac{n_s - (n_s)_1 - (n_s)_2}{n - [(n_{a'} + n_{b'} + (n_s)_1 + (n_s)_2]} = \frac{n_s}{n}$$

or in terms of percentage volumes

$$\frac{s - s_1 - s_2}{100 - (a' + b' + s_1 + s_2)} = \frac{s}{100}$$

giving, after substitution for *s*, from (i)

$$s_2 = \frac{b's}{100 - s} \tag{iii}$$

so that the percentage contraction after the second absorption becomes

$$b' + s_2 = b'\left(1 + \frac{s}{100 - s}\right) = \frac{100b'}{100 - s} \tag{iv}$$

But this is the percentage by volume of O_2 in the dry sample, i.e. *b* in Table 12.3.

The argument may be repeated for the third absorption. The important conclusion to be drawn from the above discussion is that although the actual samples collected and absorbed are all saturated with water vapour, the volumetric analysis obtained by this means is the *dry* analysis.

(*b*) Cooling towers (*Fig.* 12.8 and 12.9)

Cooling towers are used in steam power stations on sites where the amount of cooling water available for condensing the steam leaving

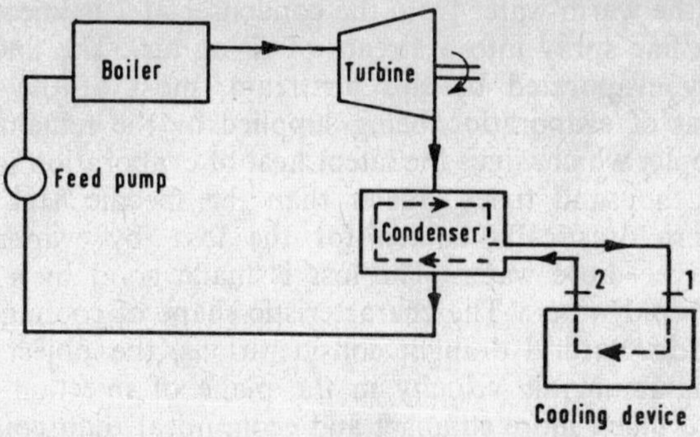

Fig. 12.8 Steam plant with cooling tower

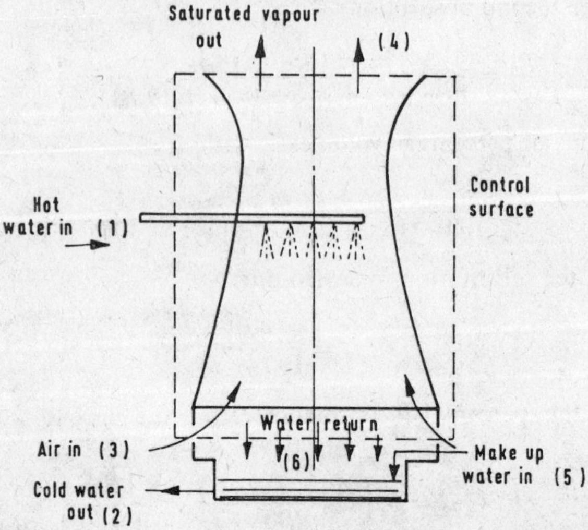

Fig. 12.9 Cooling tower

the turbine is limited. The purpose of the condenser and its associated cooling tower is best understood with reference to section 4.8. The efficiency and output of a steam plant are both greatly increased if the steam leaving the turbine is condensed before being returned to the boiler as feed water. The large quantity of heat to be abstracted from the steam in the condenser is absorbed by a cooling water flow of the order of fifteen to twenty times that of the steam flow through the plant. Since such enormous quantities of cooling water cannot, usually, be supplied, recirculation of the greater part of the water employed for this purpose is essential. In most cases the necessary cooling device 1–2 takes the form of an evaporative cooling tower, in which the warm water from the condenser at 1 is injected in the form of a fine spray into a stream of rising air. The fine droplets are partly evaporated by this airstream, most of the necessary latent heat of evaporation being supplied by the remaining water in the droplet which, since the latent heat of evaporation is approximately a thousand times greater than the specific heat of water, is therefore drastically cooled for the loss (by evaporation) of comparatively little water. The loss is made good by a make up supply of cold water. The characteristic shape of cooling towers is derived from natural draught considerations, the object being to induce maximum air velocity in the plane of injection. Cooling towers are much more compact and economical than conventional tube type heat exchangers and, in addition, the theoretical minimum temperature of the water is the wet-bulb, and not the dry-bulb temperature.

With reference to *Fig.* 12.9, cooling water at maximum temperature enters the spraying unit at 1 and is met by a rising air stream which enters the base of the cooling tower under ambient conditions at 3. While the air is rising it is progressively heated by the falling water and, because of the rise in temperature, its theoretical maximum moisture content also rises. Hence continuous evaporation from the surface of the droplet accompanies the downward journey of the latter and the upward journey of the air. It is usually assumed that the air leaving the tower at 4 is saturated with water vapour. The make up supply enters at 5. The method of analysis is best illustrated by an example.

EXAMPLE 12.7

Air enters the base of a natural draught cooling tower at the rate of 1 700 m^3/min, at a pressure and temperature of 1 bar and 16°C, and with a relative humidity of 60 per cent. It leaves saturated at 30°C, while the temperature of the entering and leaving cooling water are respectively 50°C and 18°C. Find the mass flow of water entering the tower and the percentage loss by evaporation.

SOLUTION (The notation of *Fig.* 12.9 is used)

At section 3

$$(p_s)_3 = \phi_3(p_g)_3 = 0{\cdot}6 \times 0{\cdot}01817 = 0{\cdot}01089 \text{ bar}$$

$$(m_s)_3 = \frac{(p_s)_3 V}{R_s \times T_s} = \frac{0{\cdot}01089 \times 1700}{(8{\cdot}3143/12) \times 289 \times 10^{-2}} = 13{\cdot}80 \text{ kg/min}$$

$$(m_a)_3 = \frac{[(p - (p_s)_3]V}{R_a T_3} = \frac{(1 - 0{\cdot}01089) \times 1700}{0{\cdot}2871 \times 289 \times 10^{-2}} = 202 \text{ kg/min}$$

$$= (m_a)_4$$

At section 4, using equation (12.30)

$$(p_s)_4 = (p_g)_4 = 0{\cdot}04242 \text{ bar}$$

$$\omega_4 = 0{\cdot}622 \frac{(p_s)_4}{p - (p_s)_4} = 0{\cdot}622 \times \frac{0{\cdot}04242}{1 - 0{\cdot}04242} = 0{\cdot}0275 \text{ kg}_s/\text{kg}_a$$

i.e. $(m_s)_4 = (m_a)_4\omega_4 = (m_a)_3\omega_4 = 0{\cdot}0275 \times 202 = 55{\cdot}50$ kg/min

Therefore water lost by evaporation

$$\Delta m_w = (m_s)_4 - (m_s)_3 = 55{\cdot}50 - 13{\cdot}80 = 41{\cdot}70 \text{ kg/min}$$

i.e. $(m_w)_6 = (m_w)_1 - 41{\cdot}70$ kg/min

This example, unlike those discussed previously, involves the application of the first law in the form of the S.F.E.E. to the two-phase system of the cooling tower. Regarding the system as an adiabatic one without work transfer the S.F.E.E. may be written as

(enthalpy flow out) – (enthalpy flow in) = 0

i.e.

$$[\{(m_a)_4(h_a)_4 + (m_s)_4(h_s)_4\} + (m_w)_6(h_w)_6)] - [(m_w)_1(h_w)_1 + \{(m_a)_3(h_a)_3 + (m_s)_3(h_s)_3\}] = 0 \quad \text{(i)}$$

Note that the control surface in *Fig.* 12.9 excludes the make up supply 5 and the flow return which could be included in a separate energy balance.

The following enthalpy values may now be substituted in (i)—

$(h_w)_1 = (h_f)_1 = 209{\cdot}3$ kJ/kg (from saturated steam tables at 50°C)

$(h_w)_6 = (h_f)_6 = 75{\cdot}5$ kJ/kg (from saturated steam tables at 18°C)

$$(h_s)_3 = (h_g)_{d_3} + (c_p)_s[T_3 - (T_d)_3]$$

where suffix d_3 refers to saturation, i.e. dew point conditions at the partial pressure $(p_s)_3 = 0{\cdot}01089$ bar (see above), i.e.

$(h_s)_3 = 2516{\cdot}8 + 1{\cdot}85(16 - 8{\cdot}7)$

$= 2530{\cdot}3$ kJ/kg (from tables for saturated steam)

$(h_s)_4 = (h_g)_4 = 2555{\cdot}7$ kJ/kg (from tables for saturated steam)

$$(m_a)_4(h_a)_4 - (m_a)_3(h_a)_3 = (m_a)_3(c_p)_a(T_4 - T_3) = 202 \times 1{\cdot}004(30 - 16)$$

Substituting these values together with the appropriate mass flows in equation (i) above

$$[(202 \times 1{\cdot}004 \times 14) + 55{\cdot}50 \times 2555{\cdot}7 + ((m_w)_1 - 41{\cdot}70)75{\cdot}5] - [(m_w)_1 \times 209{\cdot}3 + 13{\cdot}80 \times 2530{\cdot}3] = 0$$

whence $(m_w)_1 = 799$ kg/min *Ans.*

$$\text{Percentage loss by evaporation} = \frac{\Delta m_w}{(m_w)_1} \times 100 = \frac{41{\cdot}7}{799} \times 100 = 5{\cdot}23\% \quad \textit{Ans.}$$

(*c*) Air-conditioning systems

Temperature and humidity control in modern office buildings constitutes a most important application of the methods of analysis for mixtures of air and water vapour. Depending on geographical location such systems have to provide cooling and dehumidification in summer and heating and moisture addition in winter, or one or other of these two functions.

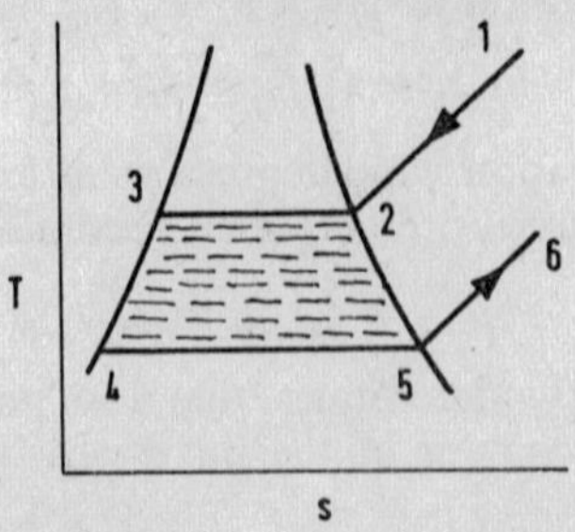

Fig. 12.10 Phase diagram for air-conditioning system

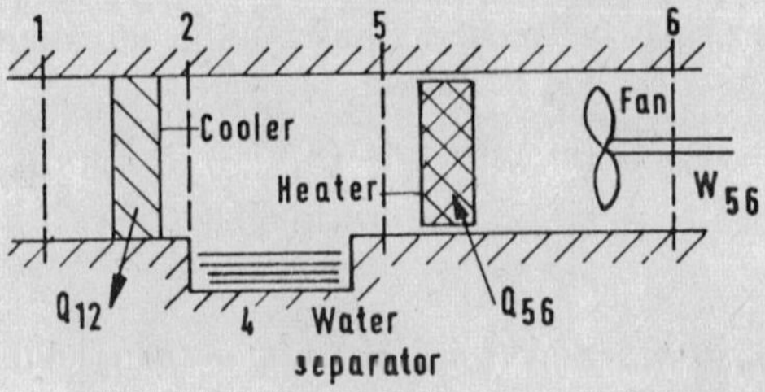

Fig. 12.11 Air-conditioning plant

(i) *Cooling and moisture abstraction.* The air is initially at temperature T_1, the state of the moisture in the air being given by point 1. Since 1 lies generally to the right of the dew point 2, the vapour at 1 is superheated, and the mixture unsaturated. Since it is required to remove moisture, the mixture has to be cooled at least to the dew point temperature T_2, and subsequently below, to temperature $T_4 = T_5$, so that the difference between the moisture content of the saturated mixture at the original dew point 2 and that at the new dew point 5 appears as condensate at the lower temperature $T_4 = T_5$, the condensation paths for successive moisture aggregates being indicated by the dotted lines parallel to the original condensation path 2–3–4. If the required moisture content is low, the dew point temperature T_5 may be so low that subsequent heating of the mixture is required, the final required room temperature being T_6 and this moisture content being ω_6 or relative humidity ϕ_6.

The plant required for this process is shown diagrammatically in *Fig.* 12.11, and consists of a cooler, condensate collector, heater and extraction fan, all enclosed in suitable ducting. The states 2 and 3 (*Fig.* 12.10) occur between sections 1 and 5, in *Fig.* 12.11, but cannot be conveniently shown.

Let the final condition be given by T_6, ϕ_6.

Therefore $$(p_s)_6 = (p_s)_5 = \phi_6(p_g)_6 \qquad \text{(from 12.31b)} \quad \text{(i)}$$

where $(p_g)_6$ = saturation vapour pressure at temperature T_6.

The dew point partial pressure then becomes

$$(p_d)_5 = (p_g)_5 = \phi_6(p_g)_6 \qquad \text{(ii)}$$

and this immediately determines the dew point temperature T_5, while the enthalpy content of the mixture at the final condition 6 becomes

$$(m_a)_6(h_a)_6 + (m_s)_6(h_s)_6$$
$$= (m_a)_6(h_a)_6 + (m_s)_6[(h_g)_5 + (c_p)_s(T_6 - T_5)] \qquad \text{(iii)}$$

Applying the S.F.E.E. between sections 5 and 6 and assuming that the extraction fan work W_{56} is known

$$Q_{56} - W_{56} = (m_a)_6[(h_a)_6 - (h_a)_5] + (m_s)_6[(h_s)_6 - (h_s)_5]$$
$$= [m_a(c_p)_a + (m_s)_6(c_p)_s](T_6 - T_5) \qquad \text{(iv)}$$

where

$$(m_a)_5 = (m_a)_6 = m_a \text{ and } (m_s)_5 = (m_s)_6 = (m_a)_6\left[0{\cdot}622\frac{(p_s)_6}{p - (p_s)_6}\right] \qquad \text{(v)}$$

Equation (iv) may be solved for the required quantity of heat to be transferred, Q_{56}.

Applying the S.F.E.E. between sections 1 and 5 and including the condensate at 4,

$$Q_{15} = = m_a[(h_a)_5 - (h_a)_1] + (m_s)_5(h_s)_5 - (m_s)_1(h_s)_1 + (m_w)_4(h_w)_4 \qquad \text{(vi)}$$

The various quantities in equation (vi) may readily be evaluated given the initial conditions T_1, ϕ_1.

Therefore $$(p_s)_1 = (p_s)_2 = \phi_1(p_g)_1 \qquad \text{(vii)}$$

and this immediately determines the dew point temperature T_2, where $(p_g)_1$ = saturation pressure at temperature T_1, giving

$$(m_s)_1 = 0{\cdot}622\left[\frac{(p_s)_1}{p - (p_s)_1}\right](m_a)_1 \qquad \text{(viii)}$$

where $$(m_a)_1 = (m_a)_5 = (m_a)_6 = m_a$$

But $$(m_w)_4 = (m_s)_1 - (m_s)_5 \qquad \text{(ix)}$$

where $(m_s)_5$ is given by equation (v).

The enthalpy content of the mixture at the initial condition 1 therefore becomes

$$m_a(h_a)_1 + (m_s)_1(h_s)_1 = m_a(h_a)_1 + (m_s)_1[(h_g)_2 + (c_p)_s(T_2 - T_1)] \quad \text{(x)}$$

while that at section 5 may be written

$$(m_a)_5(h_a)_5 + (m_s)_5(h_s)_5 = m_a(h_a)_5 + (m_s)_5(h_g)_5 \quad \text{(xi)}$$

and that of the water separated out at 4 becomes

$$(m_a)_4(h_f)_5 = [(m_s)_1 - (m_s)_5](h_f)_5 \quad \text{(xii)}$$

Substituting from (viii), (ix), (x), (xi) and (xii) in the S.F.E.E., equation (vi), the latter simplifies to the form

$$Q_{15} = m_a(c_p)_5(T_5 - T_1) + (m_s)_5(h_g)_5 - (m_s)_1[(h_g)_2 + (c_p)_2(T_2 - T_1)] + [(m_s)_1 - (m_s)_5](h_f)_5 \quad \text{(xiii)}$$

Equation (xiii) may be used to evaluate the unknown negative heat transfer Q_{15}, so that the problem may be regarded as completely solved.

(ii) *Heating and moisture addition* (*Figs.* 12.12 and 12.13). This process can be performed in a variety of ways. One of the simplest is illustrated in *Figs.* 12.12 and 12.13. The air is initially at temperature T_1, the state of the moisture in the air being given by point 1. As in the previous case, the vapour is superheated and the mixture unsaturated. Heating of the mixture results in a rise in temperature to T_2 which is not necessarily equal to the finally required temperature T_4. The partial pressure of the water vapour remains constant at $(p_s)_1$, i.e.

$$(p_s)_2 = (p_s)_1 = \phi_i(p_g)_1 = (p_d)_0 \quad \text{(i)}$$

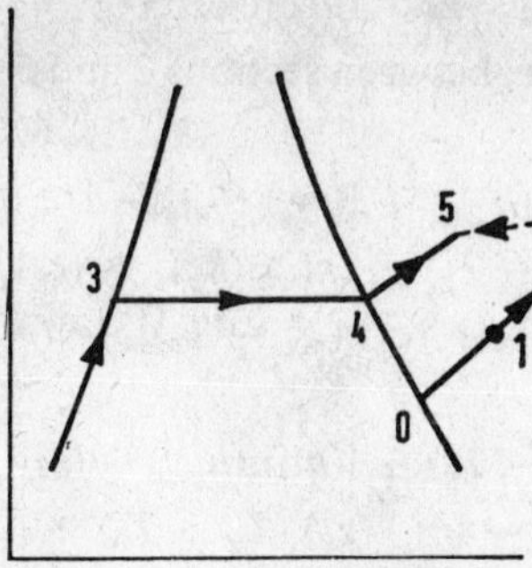

Fig. 12.12 Phase diagram for heating and moisture addition

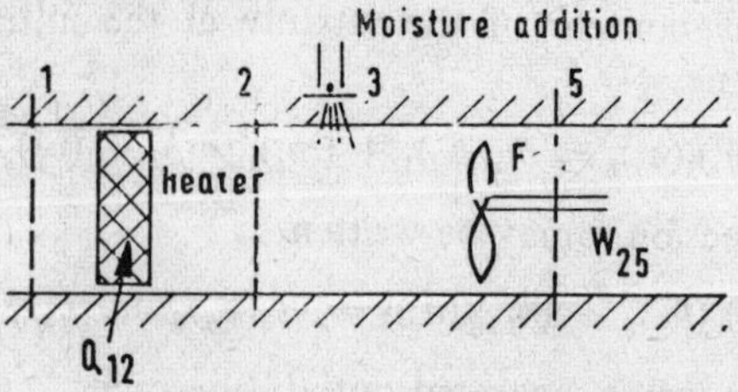

Fig. 12.13 Heating and moisture addition plant

where $(p_g)_1$ is the saturation vapour pressure at temperature T_1 and d_0 refers to the dew point 0, since there is no change in moisture content, i.e.

$$\omega_1 = \omega_2 = 0{\cdot}622 \frac{(p_s)_1}{p - (p_s)_1} \tag{ii}$$

Thereafter water at state 3 is injected in the form of a spray, to obtain the final specified conditions T_5 and ϕ_5 since

$$(p_s)_5 = \phi_5(p_g)_5 = (p_d)_4 \tag{iii}$$

where $(p_g)_5$ is the saturation vapour pressure at temperature T_5 and $(p_d)_4$ refers to the dew point 4 and

$$\omega_5 = 0{\cdot}622 \frac{(p_s)_5}{p - (p_s)_5} \tag{iv}$$

The quantity of water to be injected is obtained from (ii) and (iv) as

$$(m_w)_3 = (m_s)_5 - (m_s)_1 = (\omega_5 - \omega_1)m_a \tag{v}$$

Applying the S.F.E.E. between sections 1 and 2,

$$\begin{aligned} Q_{12} &= m_a[(h_a)_2 - (h_a)_1] + (m_s)_1[(h_s)_2 - (h_s)_1] \\ &= [m_a(c_p)_a + (m_s)_1(c_p)_s]\,(T_2 - T_1) \end{aligned} \tag{vi}$$

and applying the S.F.E.E. between sections 2 and 5 allowing for the fan work W_{25}

$$\begin{aligned} -W_{25} &= m_a[(h_a)_5 - (h_a)_2] + [(m_s)_5(h_s)_5 - (m_s)_2(h_s)_2] + (m_w)_3(h_w)_3 \\ &= m_a(c_p)_a(T_5 - T_2) + (m_s)_5[(h_g)_4 + (c_p)_s(T_5 - T_4)] \\ &\quad - (m_s)_1[(h_g)_0 + (c_p)_s(T_2 - T_0)] + [(m_s)_5 - (m_s)_1](h_f)_3 \end{aligned}$$

i.e.

$$\begin{aligned} -W_{25} &= m_a(c_p)_a(T_5 - T_2) + [(m_s)_5(h_g)_4 - (m_s)_1(h_g)_0] \\ &+ (m_s)_5(c_p)_s(T_5 - T_4) - (m_s)_1(c_p)_s(T_2 - T_0) + [(m_s)_5 - (m_s)_1](h_f)_3 \end{aligned} \tag{vii}$$

Assuming that the fan work W_{25} is known, equation (viii) may be used to evaluate the unknown intermediate temperature T_2 so that, substituting in equation (vi) it is possible to solve for the unknown heat transfer Q_{12}. The quantity of water to be injected is obtained from equation (v), as already stated.

(iii) *Comprehensive air-conditioning plant.* Clearly, by combining the systems indicated diagrammatically in *Figs.* 12.11 and 12.13 a comprehensive air-conditioning system would be obtained. With the exception of a water-injection device, the system shown in *Fig.* 12.11 contains all the required components.

REFERENCE

12.1 Institution of Heating and Ventilating Engineers, *Tables of Hygrometric Data for Air* (1954).

EXERCISES

12.1 State the law of Dalton relating to the pressures exerted by the components in a mixture of gases. Demonstrate that this law is consistent with the results of the kinetic theory.

A Marcet boiler, having a volume of 0·1 m^3, is filled with 0·227 kg of water, the ambient temperature and pressure being 289 K and 1·02 bar respectively. Calculate the boiler pressure when the temperature is 400 K if the air bleed cock is closed prior to heating, thereby retaining the air in the free space above the water surface. The change in volume of the stored water during evaporation and the presence of water vapour in the free space prior to heating may be neglected.

(*Ans:* 3·88 bar)

12.2 The dry flue gas analysis, by volume, obtained during the acceptance trial of a coal-fired boiler was as follows: CO_2, 8·3%; CO, 1·4%; N_2, 80·3%; O_2, 10%. Assuming that the combustible part of the fuel consists of hydrogen and carbon only and that the oxygen and ash content, by weight, were 8% and 6·5% respectively, estimate the proportion of carbon in the fuel.

Calculate the higher calorific value of the fuel if the calorific value of carbon is 33870 kJ/kg and the higher calorific value of hydrogen is 144000 kJ/kg.

(*Ans:* 0·705; 45500 kJ/kg)

12.3 Argon, required for a low-pressure discharge tube gas filling system, is stored in a flask having a volume of 0·06 m^3. The total volume of the tubes and valves in the apparatus is 0·003 m^3 and the flask is isolated from the apparatus by a cock in the neck of the flask. The isolating cock is accidentally opened before the filling system is completely evacuated so that the air in the residual volume of the apparatus becomes mixed with, and contaminates, the argon in the flask. If it is known that the pressure in the flask was 0·5 bar and the pressure in the apparatus was 0·01 bar just prior to the opening of the cock, calculate the degree of contamination, expressing the level as a percentage by volume of air in argon.

Comment on the reversibility of the process and evaluate the change of entropy of the argon. The gas constant R for argon is assumed to be 0·208 kJ/kg-K.

(*Ans:* 0·0999%; 0·00487 kJ/K)

12.4 A gaseous mixture, details of which are given in the table below, is to be compressed from 1 bar and 15°C to 4 bar in a turbo machine. Calculate the required horsepower if the efficiency of the compression, which is ideally isentropic, is 72% and the drive efficiency is 90%: the mass flow is 9 kg/s. Find also the volumetric analysis at the final pressure.

Constituent	c_p kJ/kg-K	c_v kJ/kg-K	*Partial pressure in mixture at 1 bar*
Hydrogen	14·36	10·21	0·3333
Oxygen	0·917	0·658	0·0667
Carbon Monoxide	1·042	0·745	0·6000
			1·0000

(*Ans:* H_2, 3·46%; O_2, 10·94%; CO_2, 85·60%, 3920 hp)

12.5 A mixture of gases has the following volumetric analysis:

Constituent	*Molecular weight*	*Volumetric analysis %*
CO_2	44	12·3
O_2	32	6
N_2	28	81·7

One kilogramme of the mixture exists in an insulated cylinder where the pressure is 9·65 bar and the temperature is 27°C. Calculate the volume of the cylinder and the partial pressures exerted by the constituents. The mean specific heats of the constituent gases are as follows:

	c_p kJ/kg-K	c_v kJ/kg-K
CO_2	0·850	0·660
O_2	0·917	0·657
N_2	1·039	0·740

The cylinder is connected through a pipe and cock of negligible volume to an identical cylinder which is evacuated. The cock is opened; calculate the final equilibrium pressure and temperature of the mixture.

(*Ans:* p_{CO_2} = 1·19 bar, p_{O_2} = 0·58 bar, p_{N_2} = 7·89 bar
p_2 = 4·83 bar, T_2 = 27°C)

12.6 The volumetric analysis for a gas mixture consists of N_2 70%, CO_2 20%, CO 10%. If the pressure and temperature of the mixture are 1·38 bar and 38°C respectively, determine (*a*) the partial pressures, (*b*) the gravimetric analysis, (*c*) the gas constant for the mixture.

(*Ans:* (*a*) 0·965, 0·276, 0·138 bar; (*b*) 62·8%, 28·2%, 9%; (*c*) 0·2641 kJ/kg-K)

12.7 The gravimetric analysis of a gas mixture is H_2 10%, N_2 30%, CO 40%, CO_2 20%. If the pressure and temperature for the mixture are 0·69 bar and 21°C respectively, determine for the mixture

(*a*) the partial pressures
(*b*) internal energy
(*c*) c_p and c_v
(*d*) enthalpy
(*e*) molecular weight
(*f*) gas constant.

(*Ans:* (*a*) 0·433, 0·093, 0·124, 0·04 bar, (*b*) 457·5 kJ/kg, (*c*) $c_p = 2{\cdot}23$ kJ/kg-K, $c_v = 1{\cdot}55$ kJ/kg-K, (*d*) 679·8 kJ/kg, (*e*) 12·6, (*f*) 0·6605 kJ/kg-K)

12.8 Air enters the base of a natural draught cooling tower at the rate of 570 m³/min with a temperature and pressure of 15·5°C and 1·02 bar respectively, and a relative humidity of 40%. If air leaves the tower at 36°C in a saturated condition, and water enters at the rate of 453 kg/min, find the percentage loss of water by evaporation.
(*Ans:* 5·07%)

12.9 A closed vessel contains 0·180 kg wet steam of dryness fraction 0·1 and 0·227 kg air. The temperature of the mixture is 250·3°F. Determine the volume and total pressure of the vessel. Take the specific volume of water at 121°C as 0·001062 m³/kg.

To what temperature must the mixture be raised for all the water to be evaporated and what then is the corresponding pressure of the mixture?
(*Ans:* 0·0158 m³, 18·6 bar, 220°C, 43·2 bar)

12.10 Sketch and describe the Orsat apparatus for analysing the composition of exhaust gases.

Define the term mean (or apparent) molecular weight of a mixture of gases. Show that for a mixture of perfect gases, the gas constant R of the mixture is given by:

$$R = \frac{G\Sigma n_i}{\Sigma n_i M_i}$$

where n_i and M_i respectively are the number of moles and the molecular weight of a constituent in the mixture.

The volumetric composition of a mixture of gases is found to be: 80% N_2, 12% CO_2, and 8% O_2. Determine (*a*) the composition on a gravimetric basis, (*b*) the mean molecular weight of the mixture, (*c*) the gas constant R of the mixture.

(*Ans:* (*a*) 74·1, 17·5, 8·4%; (*b*) 30·24; (*c*) 0·2741 kJ/kg-K)

12.11 Define the following terms:

(i) relative humidity,
(ii) moisture content.

Show that the moisture content ω of atmospheric air is given by the expression:

$$\omega = 0{\cdot}622\frac{p_s}{p_a}$$

where p_s = pressure of water vapour and p_a = pressure of dry air.

A natural draught cooling tower takes in 1300 m³/min of air at a pressure and temperature 1·013 bar and 14·5°C. If the air enters with a relative humidity of 60% and leaves just saturated at 34°C, calculate:

(*a*) mass of dry air passing through the tower per minute,
(*b*) mass of water evaporated per minute.

(*Ans:* (*a*) 1580 kg/min; (*b*) 41·5 kg/min)

12.12 An air conditioning plant is required to deliver 17 m³/min of air of 21°C with a relative humidity of 61%, the atmosphere being saturated with moisture and at a temperature and pressure of 38°C and 1·013 bar respectively. The plant consists of a cooler, a water extraction unit, and a heater, followed by a suction fan. If the cooler reduces the temperature of the mixture to 13°C, find the rate of moisture extraction, and show that the given temperature at the cooler is consistent with the required conditions beyond the heater. Pressure changes in the system may be neglected and the moisture in unsaturated air may be treated as a perfect gas.

$c_{p\ steam}$ = 1·861 kJ/kg-K, and R_{steam} = 0·461 kJ/kg-K.

(*Ans:* 0·951 kg/min)

12.13 State the energy equation for a steady flow system, through which more than one fluid passes.

Part of an air conditioning system consists of a duct in which are mounted a fan, a cooler and a water extraction unit. 17 m³/min of air enter at the fan end of the duct at 1·013 bar and 32°C. The air is saturated with moisture and contains 0·031 kg of water vapour per kg of air. The temperature downstream of the cooler is 13°C. As a result of this cooling, 69% of the moisture initially present separates out as water in the extracting unit.

Treating the air as a perfect gas, but using tabulated properties for steam and water, find the rate of heat extraction if the fan absorbs 1 hp. Pressure change may be neglected.

12.14 A vessel A of volume 0·285 m^3 contains 0·91 kg of nitrogen at 65°C. Vessel B of volume 0·427 m^3 contains 0·68 kg of oxygen at 38°C. The two vessels are suddenly put into communication with each other. Assuming perfect gas behaviour and absence of chemical action, and that the process as a whole is adiabatic, find

(*a*) the initial pressure of both nitrogen and oxygen,
(*b*) the final pressure and temperature of the mixture,
(*c*) the partial pressures of the two constituents.

Evaluate the entropy change during the mixing process and comment on the nature of irreversibility.

Universal gas constant $G = 8{\cdot}3143$ kJ/kg-mol-K
Molecular weight of oxygen: 32
Molecular weight of nitrogen: 28
Specific heat ratio for both gases: $\gamma = 1{\cdot}4$.

(*Ans:* (*a*) 32·3 and 19·5 bar; (*b*) 2·08 bar, 53·5°C;
(*c*) 1·27 and 0·82 bar)

12.15 The volumetric analysis of the dry products of combustion of an oil engine is CO_2 12·48%, N_2 80·44%, O_2 7·08%.

Assuming that each of the constituents behaves as a perfect gas, find the partial pressures of the constituents in the mixture, when the total pressure is 1·013 bar.

Find also the gas constant for the mixture.

Atomic weights: C = 12; O = 16; N = 14.

(*Ans:* 0·126, 0·815, 0·072 bar; 0·2741 kJ/kg-K)

12.16 Derive the relationship between the partial pressures, partial volumes, and mole fractions of the constituents of a mixture of perfect gases.

In a sample of gas, the moisture constitutes a certain mole fraction x, equal to that required for saturation of the total. Show that absorption of volume fractions a and b of constituents A and B gives the correct analysis of the dry sample, in spite of the presence of moisture.

The temperature and pressure during the absorption process may be assumed constant and the water vapour may be treated as a perfect gas.

12.17 Deduce the relationship between the volumetric and gravimetric analyses of gas mixtures and discuss briefly the practical suitability of these two methods.

The volumetric analysis of a mixture of methane and air is found to be methane CH_4 41%, O_2 12·4%, N_2 46·6%.

Find the mole fractions of the various constituents, the gravimetric analysis, the apparent molecular weight of the mixture, and the fractional partial pressures of the constituents in the original mixture.

Atomic weights: C = 12, H = 1, O = 16, N = 14.
Universal gas constant: $G = 1.545$ kJ/kg-mol-K.

(*Ans:* mole fraction = vol fraction; percentage mass: 27·80, 16·75, 55·45; partial pressures = vol fraction)

12.18 State Dalton's law of partial pressures and discuss its bearing on hygrometry. Explain briefly the most important terms used.

Air enters the base of a natural draught cooling tower at the rate of 570 m^3/min, with a temperature and pressure of 15·5°C and 1·013 bar respectively, and a relative humidity of 40%. If air leaves the tower at 35·5°C in a saturated condition, and water enters at the rate of 435 kg/min, find the percentage loss of water by evaporation.
(*Ans:* 5·067%)

12.19 Show qualitatively how Dalton's law of partial pressures may be referred from the kinetic theory of gases.

Hence derive expressions for the partial pressures exerted by the constituents of a mixture of gases

(*a*) in terms of gravimetric analysis,
(*b*) in terms of the volumetric analysis.

A closed vessel of 0·14 m^3 capacity contains dry air at 0·345 bar and 27°C. Water at the same temperature is allowed to drip into the vessel until the air is saturated. Find the mass of water added and the resultant pressure in the vessel, assuming that the temperature remains constant at 27°C.
(*Ans:* 0·0036 kg, 0·3805 bar)

12.20 State Dalton's law of partial pressures and discuss its bearing on hygrometry, giving a brief explanation of the most important terms used.

The pressure and temperature in a room are 1 bar and 26·5°C respectively. If the relative humidity is 40% find:

(*a*) the partial pressures of water vapour and air,
(*b*) the dew point,
(*c*) the specific volume of each constituent.

Universal gas constant $G = 8{\cdot}3143$ kJ/kg-mol-K.
Molecular weight of water vapour = 18.
Molecular weight of air = 29.
(*Ans:* (*a*) 0·0138, 0·9862 bar; (*b*) 11·7°C; (*c*) 100, 0·875 m^3/kg)

Chapter 13

Processes involving Chemical Change: Combustion

In Chapter 12 we found that the methods of analysis for pure substances, i.e. substances having a single chemical structure, established in the earlier parts of this book, could be extended, with the aid of the Dalton and Gibbs–Dalton laws, to mixtures of two or more pure substances. In the present chapter we shall extend these investigations further, to processes involving not only several chemical substances, but also chemical change. Not surprisingly, we shall find that processes are still governed by the first and second laws of thermodynamics, in addition to which we shall invoke the principle of the conservation of mass.

The practical importance of such processes is immense since, in the form of combustion processes, usually involving air and a hydrocarbon fuel, they form the basis of by far the greatest part of the world's power production. We have seen in Chapter 6, section 6.2, that heat engines, i.e. devices producing work (or power, which by definition is the rate at which work is being done), do so by receiving heat from a high temperature source and rejecting a proportion of this heat to a low temperature sink. While examples of the latter are found in abundance in nature, e.g. the sea, rivers, lakes, or the atmosphere, high temperature heat sources are rare. We therefore have to create such sources, or their equivalent, by artificial means, the most common being combustion processes in which chemical energy is converted into internal energy (in the sense used throughout this book), i.e. energy in virtue of temperature.

In the steam power cycle (see section 4.8; *Fig.* 13.1) the heat released by combustion of the fuel in the boiler furnace creates a high temperature reservoir in the form of the combustion gases. Heat is absorbed from these gases by the working medium (water) in the boiler drum, superheater, etc., part of the heat thus absorbed being converted into work as a result of expansion of the steam in a turbine or reciprocating engine, and the balance being rejected in the condenser through the medium of condenser cooling water to the cold sink, i.e. the atmosphere (see cooling towers, section 12.4.4, *Fig*. 12.8) or alternatively sea or river water. The working medium clearly operates in a cycle.

Alternatively, in the internal combustion engine, such as the "open cycle" gas turbine (*Fig*. 13.2), we dispense with the external high temperature heat source and instead use the products of combustion both as an equivalent heat source and as a working fluid.

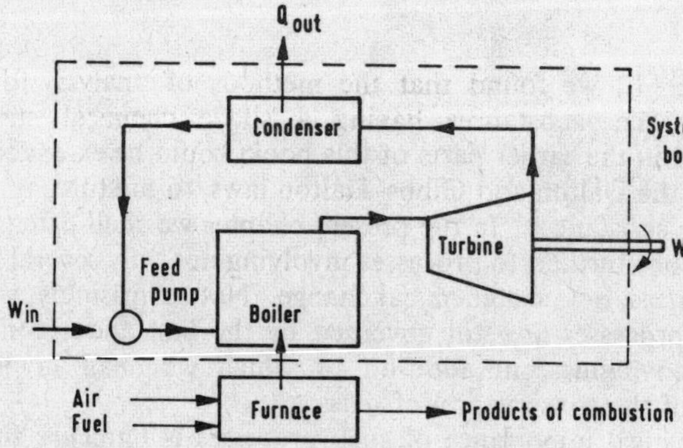

Fig. 13.1 Steam power cycle

The working fluid clearly does not operate in a cycle. Since such a cycle operates without heat absorption from, and rejection to, external heat reservoirs, and instead involves mass transfer in the form of air and fuel flow to the system and flow of products of combustion from the system defined by the boundaries in *Fig*. 13.2 and since, as already stated, the working medium does not operate in a cycle, the unit does not constitute a heat engine according to the definition of section 6.2. That it is, nevertheless, essentially a heat engine will readily be seen from the equivalent scheme, *Fig*. 13.3, in which the working fluid, now pure air, is continuously circulating, while heat is transferred from the external combustion chamber to the working fluid through the heater and from the working fluid to the heat sink, i.e. the atmosphere, through the cooler.

Our large scale power requirements are met primarily by steam power plant, while smaller units, particularly for vehicles and mobile equipment generally, operate on the internal combustion principle.

A quantitative treatment of combustion processes for the evaluation of such end effects as maximum cycle temperature, maximum

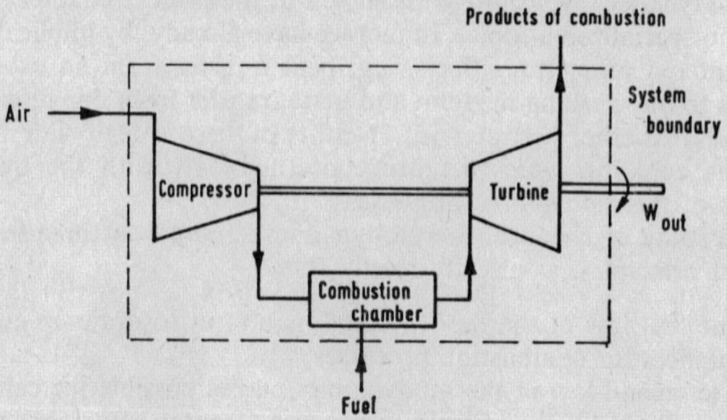

Fig. 13.2 Gas turbine open cycle

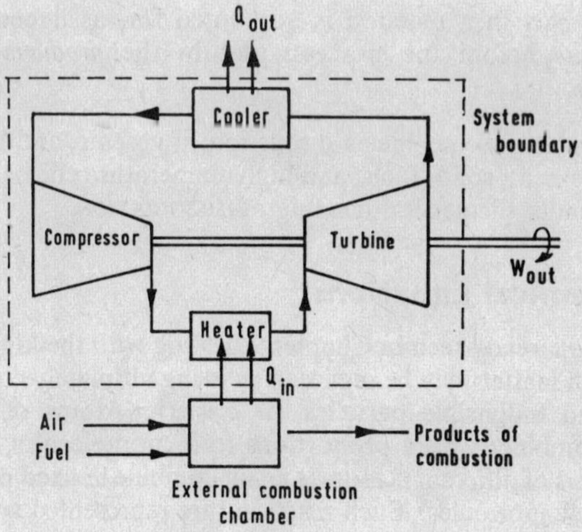

Fig. 13.3 Closed gas turbine cycle equivalent to simple open cycle in Fig. 13.2

available energy, etc., is clearly most important. It is the primary object of the present chapter to present such a working knowledge of chemical thermodynamics.

In addition to the end effects of a chemical reaction, such as those just described, it is often required to calculate the rate at which such chemical reactions proceed, since this determines the size of equipment required for a given application. Such RATE EFFECTS are dealt with by REACTION KINETICS and this, unlike chemical thermodynamics, will not be discussed in the present chapter since it is too specialized a topic. In fact we have already, by implication, encountered other rate effects, e.g. heat transfer from an external source to the working medium and mass transfer from one reservoir of fluid to another (Chapter 12). Neither of these is dealt with in the present book, in which we are concerned only with the overall effects of thermodynamic processes.

The study of chemical thermodynamics is based on three fundamental principles, as already briefly stated—

(*a*) The first law of thermodynamics enables us to draw up energy balances for combustion processes.
(*b*) The second law of thermodynamics makes possible the calculation of *maximum available energy* (see Chapter 14), since we can lay down the conditions for minimum irreversibility.
(*c*) The principle of conservation of mass enables us to formulate chemical equations, in which each of the chemical elements taking part in a reaction is accounted for, as it occurs in the *reactants* before the process, and in the *products* after the process.

The chapter also includes a discussion of gaseous and liquid (and to a lesser extent, solid) fuels, and high temperature effects in gaseous mixtures under the general heading of DISSOCIATION.

13.1 Chemical Equations*

As we have already seen in Chapter 9 dealing with the kinetic theory of gases, all matter may be regarded as being ultimately composed of minute and indivisible particles, or *atoms*†. Atoms of the same element combine in fixed proportions to form molecules, and atoms or molecules of different ELEMENTS again combine in fixed proportions to form new molecules. Such reactions are represented symbolically by CHEMICAL FORMULAE expressing the number and composition of atoms or molecules of the reactants (usually on the l.h.s. of the chemical equation) and of the products (usually on the r.h.s. of the

* It is assumed that the student has an elementary knowledge of chemistry and that, in particular, he is familiar with the standard notation for the chemical elements.

† As we all know the atom is not, in fact, indivisible. In fact as long ago as 1906 Rutherford "split" the helium atom by bombarding it with α particles, while power production in nuclear reactors depends on the bombardment of uranium 235 atoms with neutrons. However, since we are not here concerned with energy release as a result of changes in *atomic structure*, but merely with the "conventional" and in practice still overwhelmingly more important case of energy release by chemical processes, i.e. by changes in *molecular* structure, the above statement is adequate for our purposes.

equation). In combustion equations these reactants consist of the FUEL on the one hand, and the OXIDANT on the other.

The chief constituents of most of the common fuels are carbon (C), hydrogen (H_2), and to a much smaller extent sulphur (S). Combustion involves an oxidation reaction, in which the necessary oxygen (O_2) is usually provided by air, the latter being a mixture of oxygen and nitrogen (N_2) and a very small proportion of the rare gases argon (A) and xenon (Xe) (see Chapter 12, example 12.1). Since only the oxygen (O_2) in the air takes an active part in the chemical reaction, the passive constituents may either be omitted altogether from the chemical equations, or alternatively they appear unchanged on both sides. In order to be able to formulate the chemical equation for a particular reaction, we must know the composition by mass (gravimetric) or by volume (volumetric) of the fuel, as well as the corresponding composition of the oxidant.

In order to proceed to the full combustion equation for a given fuel and oxidant, it is best to formulate the combustion equations for the individual constituents.

Thus, for the individual constituents carbon (C), hydrogen (H_2) and sulphur (S), the combustion equations when pure oxygen (O_2) is used as the oxidant may be written as follows—

$$\textit{Carbon: } C + O_2 = CO_2 \tag{13.1}$$

i.e. 1 atom of carbon + 1 molecule of oxygen form 1 molecule of carbon dioxide.

$$\textit{Hydrogen: } 2H_2 + O_2 = 2H_2O \tag{13.2}$$

i.e. 2 molecules of hydrogen + 1 molecule of oxygen form 2 molecules of water.

$$\textit{Sulphur: } S + O_2 = SO_2 \tag{13.3}$$

i.e. 1 atom of sulphur + 1 molecule of oxygen form 1 molecule of sulphur dioxide.

Each of the above equations clearly complies with the principle of conservation of mass, but since atoms or molecules represent inconveniently small masses, we invariably associate chemical equations with macroscopic rather than microscopic quantities, by substituting "kilogramme-molecules" or "kg-mol" for individual molecules. (One kg-mol was defined in Chapter 3, equation (3.6) as the molecular weight of a substance in kilogrammes.)

Hence given the atomic weight: C = 12, O = 16, H = 1, S = 32, we now interpret equations (13.1) to (13.3) as follows—

$$1 \text{ kg-mol of C} + 1 \text{ kg-mol of } O_2 = 1 \text{ kg-mol of } CO_2 \tag{13.1a}$$

i.e.
$$12 \text{ kg of C} + 32 \text{ kg of } O_2 = 44 \text{ kg of } CO_2 \tag{13.1b}$$

or $$1 \text{ kg of C} + \frac{32}{12} \text{ kg of } O_2 = \frac{44}{12} \text{ kg of } CO_2 \qquad (13.1c)$$

$$2 \text{ kg-mol of } H_2 + 1 \text{ kg-mol of } O_2 = 2 \text{ kg-mol of } H_2O \qquad (13.2a)$$

i.e. $$4 \text{ kg of } H_2 + 32 \text{ kg of } O_2 = 36 \text{ kg of } H_2O \qquad (13.2b)$$

or $$1 \text{ kg of } H_2 + 8 \text{ kg of } O_2 = 9 \text{ kg of } H_2O \qquad (13.2c)$$

$$1 \text{ kg-mol of S} + 1 \text{ kg-mol of } O_2 = 1 \text{ kg-mol of } SO_2 \qquad (13.3a)$$

i.e. $$32 \text{ kg of S} + 32 \text{ kg of } O_2 = 64 \text{ kg of } SO_2 \qquad (13.3b)$$

or $$1 \text{ kg of S} + 1 \text{ kg of } O_2 = 2 \text{ kg of } SO_2 \qquad (13.3c)$$

Hence, given the composition by mass (or gravimetric composition) of the fuel to be burnt, e.g. $x\%$ of carbon, $y\%$ of hydrogen, $z\%$ of sulphur, $p\%$ of non-combustible constituents, we may combine equations (13.1c) to (13.3c) and apply them to 1 kg of fuel burning under chemically correct or stoichiometric conditions, as follows—

$$\underbrace{\underset{C}{\frac{x}{100}} + \underset{O_2}{\left(\frac{x}{100} \times \frac{32}{12}\right)}}_{\text{combustion of carbon}} + \underbrace{\underset{H_2}{\frac{y}{100}} + \underset{O_2}{\frac{8y}{100}}}_{\text{combustion of hydrogen}} + \underbrace{\underset{S}{\frac{z}{100}} + \underset{O_2}{\frac{z}{100}}}_{\text{combustion of sulphur}} + \underbrace{\frac{p}{100}}_{\text{non-combustible constituents}} \text{ kg}$$

$$= \underset{CO_2}{\left(\frac{x}{100} \times \frac{44}{12}\right)} + \underset{H_2O}{\left(\frac{y}{100} \times 9\right)} + \underset{SO_2}{\left(\frac{z}{100} \times 2\right)} + \underset{\text{non-combustible constituents}}{\frac{p}{100}} \text{ kg} \qquad (13.4)$$

or

$$\underbrace{\left[\frac{x}{100} + \frac{y}{100} + \frac{z}{100} + \frac{p}{100}\right]}_{\text{1 kg of fuel}} + \underbrace{\left[\left(\frac{x}{100} \times \frac{32}{12}\right) + \left(\frac{y}{100} \times 8\right) + \frac{z}{100}\right]}_{\text{kg of oxygen required for combustion}} + \frac{p}{100}$$

$$= \underbrace{\underset{CO}{\left(\frac{x}{100} \times \frac{44}{12}\right)} + \underset{H_2O}{\left(\frac{y}{100} \times 9\right)} + \underset{SO_2}{\left(\frac{z}{100} \times 2\right)}}_{\text{kg of products of combustion}} + \underbrace{\frac{p}{100}}_{\text{kg of non-combustible constituents}} \qquad (13.4a)$$

Equation (13.4a) refers to the most general case of a fuel in which the constituents occur in random proportions x, y, z, p as indicated. On the other hand many liquid and gaseous fuels may be represented by chemical formulae.

Hydrocarbon fuels in particular have the general chemical formula C_mH_n so that applying equations (13.1) and (13.2), as well as equations (13.1b) and (13.2b), the combustion equation for such fuels may be written

$$C_mH_n + (m + \tfrac{1}{4}n)O_2 = m\,CO_2 + \tfrac{1}{2}n\,H_2O \qquad (13.5)$$

or

$$1 \text{ kg-mol of fuel} + (m + \tfrac{1}{4}n) \text{ kg-mol of } O_2 = m \text{ kg-mol of } CO_2 + \tfrac{1}{2}n \text{ kg-mol of } H_2O \qquad (13.5a)$$

Written as a mass balance, and using equations (13.1b) and (13.2b) we have $12m$ kg of C + n kg of H_2 + $32(m + \frac{1}{4}n)$ kg of O_2 = $44m$ kg of CO_2 + $9n$ kg of H_2O, or

$$(12m + n) \text{ kg of fuel} + 32(m + \tfrac{1}{4}n) \text{ kg of } O_2 = 44m \text{ kg of } CO_2 + 9n \text{ kg of } H_2O \qquad (13.5b)$$

Alternatively, dividing both sides of equation (13.5b) by $(12m + n)$

$$1 \text{ kg of fuel} + \frac{32(m + \frac{1}{4}n)}{12m + n} \text{ kg of } O_2 = \frac{44m}{12m + n} \text{ kg of } CO_2 + \frac{9n}{12m + n} \text{ kg of } H_2O \qquad (13.5c)$$

EXAMPLE 13.1

State the combustion equation and the corresponding mass balance for benzene (C_6H_6) burning in oxygen in chemically correct or STOICHIOMETRIC proportions.

SOLUTION

Combustion equation, equation (13.5)

$$C_6H_6 + (6 + \tfrac{6}{4})O_2 = 6\,CO_2 + \tfrac{6}{2}\,H_2O$$

or

$$C_6H_6 + 7\tfrac{1}{2}\,O_2 = 6\,CO_2 + 3\,H_2O \qquad \text{(i)}$$

Mass balance, equation (13.5b)

$$[(12 \times 6) + 6] \text{ kg of fuel} + (7\tfrac{1}{2} \times 32) \text{ kg of } O_2 = (6 \times 44) \text{ kg of } CO_2 + (3 \times 18) \text{ kg of } H_2O$$

i.e. 78 kg of fuel + 240 kg of O_2 = 264 kg of CO_2 + 54 kg of H_2O (ii)

or 1 kg of fuel + $\frac{240}{78}$ kg of O_2 = $\frac{264}{78}$ kg of CO_2 + $\frac{54}{78}$ kg of H_2O (iii)

Equation (ii) may be expressed as

$$318 \text{ kg of reactants} = 318 \text{ kg of products} \qquad \text{(iv)}$$

i.e. mass balance is clearly satisfied.

In practice, at least two complications arise—

(1) Fuels are rarely burnt in pure oxygen*, air being by far the most common oxidant.

Very approximately (see example 12.1, Chapter 12) the VOLUMETRIC COMPOSITION of air is 21% O_2 and 79% N_2 while the corresponding GRAVIMETRIC COMPOSITION is 23·3% O_2 and 76·7% N_2. Hence for every kg of O_2 involved in a chemical reaction, 76·7/23·3 = 3·29 kg of N_2 are carried over from the reactant to the product side of the combustion equation as a passive constituent.† The resultant modification of equations (13.4a) and (13.5) is immediately obvious.

The first equation (for fuels of "random" composition) becomes

$$1 \text{ kg of fuel} + \left[\left(\frac{x}{100} \times \frac{32}{12}\right) + \left(\frac{y}{100} \times 8\right) + \frac{z}{100}\right] \text{kg of } O_2$$

$$+ \left[\left(\frac{x}{100} \times \frac{32}{12}\right) + \left(\frac{y}{100} \times 8\right) + \frac{z}{100}\right] \times 3{\cdot}29 \text{ kg of } N_2$$

$$= \left(\frac{x}{100} \times \frac{44}{12}\right) \text{kg } CO_2 + \left(\frac{y}{100} \times 9\right) \text{kg } H_2O + \left(\frac{z}{100} \times 2\right) \text{kg } SO_2$$

$$+ \left[\left(\frac{x}{100} \times \frac{32}{12}\right) + \left(\frac{y}{100} \times 8\right) + \frac{z}{100}\right] \times 3{\cdot}29 \text{ kg } N_2$$

$$+ \frac{p}{100} \text{ kg non-combustible constituents} \qquad (13.6)$$

while equations (13.5) and (13.5c) for hydrocarbon fuels becomes per kg-mol of fuel:

$$C_mH_n + (m + \tfrac{1}{4}n)\, O_2 + (m + \tfrac{1}{4}n) \frac{79}{21} N_2$$

$$= m\, CO_2 + \tfrac{1}{2}n\, H_2O + (m + \tfrac{1}{4}n) \frac{79}{21} N_2 \qquad (13.7a)$$

* A well known example of the use of pure oxygen is the rocket motor, i.e. the rocket carries its own atmosphere into space, where the air is so rarified that it cannot be used to sustain the combustion reaction.

† Nitrogen is in fact not a completely passive constituent, but may under certain conditions form nitrous oxide according to the equation—

$$N_2 + O_2 = 2NO$$

per kg of fuel:

$$1 \text{ kg of fuel} + 32\,\frac{m + \frac{1}{4}n}{12m + n}\text{ kg } O_2 + 32\left(\frac{m + \frac{1}{4}n}{12m + n}\right) \times 3{\cdot}29 \text{ kg } N_2$$

$$= \left(\frac{44m}{12m + n}\right)\text{ kg } CO_2 + \left(\frac{9n}{12m + n}\right)\text{ kg } H_2O$$

$$+ 32\left(\frac{m + \frac{1}{4}n}{12m + n}\right) \times 3{\cdot}29 \text{ kg } N_2 \quad (13.7\text{b})$$

(2) Fuels are rarely burnt in chemically correct or STOICHIOMETRIC proportions.

Instead the reacting mixture is either WEAK, denoting an excess of oxidant over that required under stoichiometric conditions, or alternatively RICH, denoting a deficiency of oxidant.

When formulating the modified equations, it is convenient to express the departure from stoichiometric conditions as a percentage of excess oxidant where e may be positive (weak mixture) or negative (rich mixture).* The resultant forms of the combustion equations (13.6) and (13.7), which already include the modification for nitrogen, are respectively—

For fuels of "random" composition (equation (13.6))

$$1 \text{ kg of fuel} + \left[\left(\frac{x}{100} \times \frac{32}{12}\right) + \left(\frac{y}{100} \times 8\right) + \frac{z}{100}\right]\left(1 + \frac{e}{100}\right)\text{ kg } O_2$$

$$+ \left[\left(\frac{x}{100} \times \frac{32}{12}\right) + \left(\frac{y}{100} \times 8\right) + \frac{z}{100}\right]\left(1 + \frac{e}{100}\right) \times 3{\cdot}29 \text{ kg } N_2$$

$$= a \text{ kg } CO_2 + b \text{ kg } CO + \left(\frac{y}{100} \times 9\right)\text{ kg } H_2O + \left(\frac{2z}{100}\right)\text{ kg } SO_2$$

$$+ c \text{ kg } O_2 + \left[\left(\frac{x}{100} \times \frac{32}{12}\right) + \left(\frac{y}{100} \times 8\right) + \frac{z}{100}\right]\left(1 + \frac{e}{100}\right) \times 3{\cdot}29 \text{ kg } N_2$$

$$+ \frac{p}{100}\text{ kg non-combustible constituents} \quad (13.8)$$

For hydrocarbon fuels (equations (13.7a) and (b))
per kg-mol of fuel:

$$C_mH_n + \left(m + \tfrac{1}{4}n\right)\left(1 + \frac{e}{100}\right) O_2 + \left(m + \tfrac{1}{4}n\right)\left(1 + \frac{e}{100}\right)\frac{79}{21} N_2$$

$$= a\,CO_2 + b\,CO + \tfrac{1}{2}n\,H_2O + c\,O_2 + \left(m + \tfrac{1}{4}n\right)\left(1 + \frac{e}{100}\right) \times \frac{79}{21} N_2$$

$$(13.9\text{a})$$

* Petrol engines often operate with slightly rich mixtures, compression ignition or diesel engines invariably with weak mixtures.

per kg of fuel:

$$1 \text{ kg of fuel} + 32\left(\frac{m + \frac{1}{4}n}{12m + n}\right)\left(1 + \frac{e}{100}\right) \text{kg } O_2$$

$$+ 32\left(\frac{m + \frac{1}{4}n}{12m + n}\right)\left(1 + \frac{e}{100}\right) \times 3{\cdot}29 \text{ kg } N_2$$

$$= a' \text{ kg } CO_2 + b' \text{ kg } CO + \frac{9n}{12m + n} \text{kg } H_2O + c' \text{ kg } O_2$$

$$+ \frac{32(m + \frac{1}{4}n)}{12m + n}\left(1 + \frac{e}{100}\right) \times 3{\cdot}29 \text{ kg } N_2 \qquad (13.9b)$$

Equations (13.8) and (13.9) allow for the possibility of incomplete combustion.*

In experimental combustion analysis we proceed in reverse order.

This statement implies that a VOLUMETRIC PRODUCT ANALYSIS is used to evaluate a GRAVIMETRIC REACTANT ANALYSIS (in the case of liquid or solid fuels) or a VOLUMETRIC REACTANT ANALYSIS (in the case of gaseous fuels).

The object of such an analysis is to assess the conditions under which combustion takes place, i.e. whether with a rich, chemically correct, or weak mixture. Quite apart from this, such a product analysis immediately indicates incomplete (and hence thermodynamically wasteful) combustion, primarily by the presence of carbon monoxide (CO).

The relevant procedures are outlined in section 13.2.

EXAMPLE 13.2

Note: This example is based on the same fuel, viz. benzene C_6H_6, as example 13.1. The following complications are introduced—

(*a*) air is used as the oxidant instead of pure oxygen,
(*b*) the fuel is burnt in excess air.

Benzene is burnt in 20 per cent excess air compared with the chemically correct (or stoichiometric) amount. Find (*a*) the air fuel ratio (by mass) for stoichiometric combustion, (*b*) the percentage gravimetric composition of the wet products when 20 per cent excess air is used, also the percentage volumetric composition of the dry products.

* Incomplete combustion is indicated primarily by the presence of carbon monoxide (CO). Because of the very great affinity of hydrogen for oxygen, free hydrogen (H_2) is most unlikely to be found in the products of combustion.

SOLUTION

(a) Air–fuel ratio for stoichiometric combustion

From equation (13.7b), on basis of 1 kg of fuel ($m = 6$, $n = 6$)

$$1\text{ kg of fuel} + 32\left[\frac{6 + 6/4}{(12 \times 6) + 6}\right]\text{kg } O_2 + 32\left[\frac{6 + 6/4}{(12 \times 6) + 6}\right]3{\cdot}29\text{ kg } N_2$$

$$= \frac{44 \times 6}{(12 \times 6) + 6}\text{kg } CO_2 + \frac{9 \times 6}{(12 \times 6) + 6}\text{kg } H_2O + \left[32\,\frac{6 + 6/4}{(12 \times 6) + 6}\right] \times 3{\cdot}29\text{ kg } N_2 \qquad \text{(i)}$$

i.e. $1\text{ kg of fuel requires } 32\left[\frac{6 + 6/4}{(12 \times 6) + 6}\right](1 + 3{\cdot}29)$

$$= 13{\cdot}2\text{ kg of air } [3{\cdot}08\text{ kg } O_2 + 10{\cdot}12\text{ kg } N_2]$$

or $$\text{air–fuel ratio } A/F = \frac{13{\cdot}2}{1} = 13{\cdot}2$$

(b) Product analysis

From equation (13.9b), on basis of 1 kg of fuel, and assuming complete combustion, i.e. absence of carbon monoxide ($b = 0$)

$$1\text{ kg of fuel} + (3{\cdot}08 \times 1{\cdot}2)\text{ kg } O_2 + (10{\cdot}12 \times 1{\cdot}2)\text{ kg } N_2$$

$$= \frac{44 \times 6}{(12 \times 6 + 6)}\text{kg } CO_2 + \frac{9 \times 6}{(12 \times 6) + 6}\text{kg } H_2O + 3{\cdot}08 \times 0{\cdot}2\text{ kg } O_2 + 10{\cdot}12 \times 1{\cdot}2\text{ kg } N_2$$

$$= 3{\cdot}39\text{ kg } CO_2 + 0{\cdot}69\text{ kg } H_2O + 0{\cdot}62\text{ kg } O_2 + 12{\cdot}15\text{ kg } N_2 \qquad \text{(ii)}$$

or 16·84 kg reactants = 16·85 kg products. Equation (ii) thus clearly satisfies the condition of mass balance.

The r.h.s. of equation (ii) immediately provides the required gravimetric composition of the *wet* products.

In tabular form

Constituent	*Mass* m_i	*Percentage mass*
CO_2	3·39	20·10
H_2O	0·69	4.09
O_2	0·62	3·68
N_2	12·15	72·13
	16·85	100·00

The corresponding volumetric composition of the dry products is now obtained by dividing the masses of the dry constituents by their respective molecular weights in order to obtain the kg-mol numbers and hence the volumetric (or kg-mol) proportions

Constitutent	*Mass* m_i	M	$m_i/M = n_i$	*Percentage volume*
CO_2	3·39	44	0·0772	14·51
O_2	0·62	32	0·0194	3·65
N_2	12·15	28	0·4341	81·84
			0·5307	100·00

13.2 Experimental Combustion Analysis

The most common experimental method for the determination of the volumetric composition of products of combustion, Orsat analysis, has already been described in section 12.4.4.

The resultant analysis is a so-called "dry" analysis, i.e. it gives the volumetric proportions of carbon dioxide (CO_2), carbon monoxide (CO), oxygen (O_2) and nitrogen (N_2) of the mixture of products of combustion from which all moisture may be assumed to have been removed.* The tabular method for converting the dry volumetric product analysis to the equivalent gravimetric analysis has also been given in section 12.3.5, Tables 12.1 and 12.2 in the same section.

The next step is to strike a mass balance between products and reactants, using equations (13.8) or (13.9), in the case of liquid or solid fuels, or a volumetric or kg-mol balance using the reaction equation itself in the case of gaseous fuels, in order to establish the percentage e of excess air. (It will be assumed from now on that the fuel is always burnt in air, rather than in pure oxygen.)

The methods used will be illustrated by means of two examples, the first dealing with a gaseous fuel having a "random" volumetric composition, burning with excess air, and the second with a liquid fuel having a definite chemical composition, burning with a deficiency of air.

EXAMPLE 13.3

A gaseous fuel has the following volumetric composition—

$$CO\ 12{\cdot}50\%,\ H_2\ 44{\cdot}00\%,\ CH_4\ 27{\cdot}50\%,\ O_2\ 1{\cdot}40\%,\ N_2\ 14{\cdot}60\%$$

* It was pointed out in section 12.4.4 that, in fact, the mixture being analysed is saturated with water vapour throughout. Nevertheless, the numerical proportions of the constituent are those in the dry mixture.

The dry volumetric analysis of the products is as follows—

$$CO_2\ 9{\cdot}39\%,\ N_2\ 86{\cdot}76\%,\ O_2\ 3{\cdot}85\%$$

Find (*a*) the percentage of excess air, (*b*) the percentage gravimetric composition of the *wet* products of combustion.

SOLUTION

Since we are dealing with a gaseous fuel, it is unnecessary to convert the volumetric product analysis into a gravimetric analysis. Furthermore, it was shown in section 12.3.5 that the volume fractions of the constituents from volumetric analysis are numerically equal to the kg-mol fractions.

Hence, using the combustion equations (13.1) and (13.2) for carbon and hydrogen, it is best to write down first the stoichiometric combustion equation for the separate constituents using pure oxygen only and subsequently to combine them in order to evaluate the air requirements.

For carbon monoxide (CO)

$$2\,CO + O_2 = 2\,CO_2$$

i.e. $$0{\cdot}125 \text{ kg-mol } CO + \frac{0{\cdot}125}{2} \text{ kg-mol } O_2 = 0125 \text{ kg-mol } CO_2 \qquad \text{(i)}$$

For hydrogen (H_2)

$$2\,H_2 + O_2 = 2\,H_2O$$

i.e. $$0{\cdot}44 \text{ kg-mol } H_2 + 0{\cdot}22 \text{ kg-mol } O_2 = 0{\cdot}44 \text{ kg-mol } H_2O \qquad \text{(ii)}$$

For methane (CH_4)

$$CH_4 + 2\,O_2 = CO_2 + 2\,H_2O$$

i.e. $$0{\cdot}275 \text{ kg-mol } CH_4 + 0{\cdot}550 \text{ kg-mol } O_2$$
$$= 0{\cdot}275 \text{ kg-mol } CO_2 + 0{\cdot}550 \text{ kg-mol } H_2O \qquad \text{(iii)}$$

The oxygen in the fuel may be deduced from the required kg-mol number given by the separate equations (i), (ii) and (iii), above, i.e. number of kg-mol of O_2 required $= \frac{1}{2}(0{\cdot}125) + 0{\cdot}220 + 0{\cdot}550 - 0{\cdot}014 = 0{\cdot}819$ kg-mol. Hence the complete combustion equation, allowing for the presence of nitrogen, as well as excess air becomes

$$0{\cdot}125\,CO + 0{\cdot}44\,H_2 + 0{\cdot}275\,CH_4 + 0{\cdot}014\,O_2 + 0{\cdot}146\,N_2$$

$$+ 0{\cdot}819\left(1 + \frac{e}{100}\right) O_2 + 0{\cdot}819\left(1 + \frac{e}{100}\right)\frac{79}{21} N_2$$

$$= 0{\cdot}400\,CO_2 + 0{\cdot}990\,H_2O + 0{\cdot}819\frac{e}{100}\,O_2 + 0{\cdot}819\left(1 + \frac{e}{100}\right)\frac{79}{21} N_2 \qquad \text{(iv)}$$

(*a*) The percentage excess air may now be estimated by comparing the composition of the dry products on the r.h.s. of equation (iv), i.e. excluding the 0·990 kg-mol of H_2O, with the given volumetric analysis.

Constituent	*Number of kg-mol*	*Percentage volumetric composition*
CO_2	$0{\cdot}400$	9·39
O_2	$0{\cdot}819\,\dfrac{e}{100}$	3·85
N_2	$3{\cdot}08\left(1+\dfrac{e}{100}\right)$	86·76
	$3{\cdot}48 + 3{\cdot}90\,\dfrac{e}{100}$	100·00

Comparing column 2 with column 3

$$\text{For } O_2 \qquad \left(\frac{0{\cdot}819\,e/100}{3{\cdot}48 + 3{\cdot}90\,e/100}\right)100 = 3{\cdot}85$$

whence $$66{\cdot}9e = 1340$$

i.e. $$e = 20{\cdot}02\% \qquad \text{(v)}$$

$$\text{For } N_2 \qquad \frac{3{\cdot}08(1 + e/100)}{3{\cdot}48 + 3{\cdot}90\,e/100} = 86{\cdot}76$$

whence $$0{\cdot}30e = 6$$

i.e. $$e = 20{\cdot}00\% \qquad \text{(vi)}$$

The two results (v) and (vi) have been obtained independently and by their close agreement prove that the given dry volumetric analysis is very accurate. In principle, however, one of the two given figures, i.e. either percentage O_2 or percentage N_2, is redundant. Nevertheless, the evaluation of e from both figures is useful as a check on the given data.

An arithmetically simpler method, applicable only when combustion is complete, as is the case in the present example, consists in comparing corresponding entries in columns 2 and 3 of the above Table for O_2 and N_2, i.e.

$$\frac{0{\cdot}819\,e/100}{3{\cdot}08(1 + e/100)} = \frac{3{\cdot}85}{86{\cdot}76}$$

or $$e = \frac{11{\cdot}88}{0{\cdot}590} = 20{\cdot}1$$

(agreeing substantially with the earlier result).

(*b*) The percentage gravimetric composition of the wet products may now be evaluated by a simple tabular method, substituting the value $e = 20$ in the r.h.s. of equation (iv) (see also Tables 12.1 and 12.2, section 12.3.5).

The last column gives the required percentage gravimetric composition.

Constituent	*Mol number* *n*	*Percentage kg-mol*	*M*	*Percentage kg-mol* × M_i	*Percentage by mass*
CO_2	0·400	7·65	44	7·65 × 44 = 337	12·30
O_2	0·164	3·13	32	3·13 × 32 = 100	3·65
N_2	3·695	70·32	28	70·32 × 28 = 1965	71·65
H_2O	0·990	18·90	18	18·9 × 18 = 340	12·40
	Σ = 5·249	Σ = 100·00		Σ = 2,742	Σ = 100·00

EXAMPLE 13.4

Note: This example is again based on the same fuel, viz. benzene C_6H_6 as the earlier examples 13.1 and 13.2. The following complications are introduced—

(*a*) We proceed from a given product analysis to a reactant analysis instead of vice versa,
(*b*) Combustion is incomplete, i.e. carbon monoxide (CO) is present in the products.

The volumetric analysis of the *dry* products when benzene is burnt in air is as follows—

$$CO_2\ 10{\cdot}60\%,\ CO\ 10{\cdot}45\%,\ N_2\ 78{\cdot}95\%$$

Find (*a*) the percentage excess air used (positive or negative) in relation to the stoichiometric quantity, (*b*) the gravimetric analysis of the wet products.

SOLUTION

The percentage excess air may be evaluated either from a *carbon balance* or an *"oxygen–hydrogen" balance*, using equation (13.9a), on the basis of 1 kg-mol of fuel.

Carbon balance

From equation (13.9a),

1 kg-mol of fuel (C_6H_6) is associated with *a* kg-mol of CO_2 and *b* kg-mol of CO. Now $m = n = 6$ and from volumetric analysis (i.e. mol analysis) of *dry* products,

$$\frac{a}{b} = \frac{10{\cdot}60}{10{\cdot}45} \quad \text{or} \quad b = 0{\cdot}985a \qquad \text{(i)}$$

Now 1 kg-mol of fuel is associated with $12n = 72$ kg C (ii)
while *a* kg-mol CO_2 are associated with $12a$ kg C (iii)
and *b* kg-mol CO are associated with $12b = 11{\cdot}81a$ kg C (iv)

Since, clearly, the mass of carbon associated with products and reactants must be the same, then

$$72 = 12a + 11{\cdot}81a \quad \text{i.e.} \quad a = 3{\cdot}02 \tag{v}$$

$$b = 2{\cdot}98 \tag{vi}$$

But from volumetric product analysis 10·60 kg-mol CO_2 are associated with 78·95 kg-mol N_2, whence

3·02 kg-mol CO_2 are associated with 3·02 × (78·95/10·6) = 22·4 kg-mol N_2

or 1 kg-mol of fuel is associated with 22·4 kg-mol N_2 (vii)

But from equation (13.9a)

$$22{\cdot}4 = \left(m + \tfrac{1}{4}n\right)\left(1 + \frac{e}{100}\right)\frac{79}{21}$$

$$= 7{\cdot}5\left(1 + \frac{e}{100}\right)\frac{79}{21}$$

whence

$$\frac{e}{100} = 0{\cdot}795 - 1 = -\,0{\cdot}205$$

or

$$e = -\,20{\cdot}5 \tag{viii}$$

Hence* the fuel–air mixture is 20·5 per cent *rich*.

Oxygen–hydrogen balance

From equation (13.9a),

1 kg-mol of fuel (C_mH_n) is associated with a kg-mol of CO_2, b kg-mol of CO, and $\left(m + \tfrac{1}{4}n\right)\frac{79}{21}\left(1 + \frac{e}{100}\right)$ kg-mol of N_2 in the dry product.

Now

a kg-mol CO_2 are associated with $32a$ kg O_2 (i)

b kg-mol CO are associated with $16b$ kg O_2 (ii)

$$\left(m + \tfrac{1}{4}n\right)\left(1 + \frac{e}{100}\right)\frac{79}{21}\ \text{kg-mol } N_2$$

are associated with

$$32\left(m + \tfrac{1}{4}n\right)\left(1 + \frac{e}{100}\right)\ \text{kg } O_2 \text{ in air} \tag{iii}$$

Hence the difference between (iii) and the sum of (i) and (ii) is the O_2 which has combined with H_2 and therefore does not appear in the *dry* products.

* The negative value of e indicates a deficiency of air, i.e. a *rich* mixture.

But from (13.9a),

1 kg-mol of fuel is associated with $\frac{1}{2}n$ kg-mol or $9n$ kg H_2O and $9n$ kg H_2O are associated with $(9n)\dfrac{16}{18} = 8n$ kg O_2 whence

$$32\left(m + \tfrac{1}{4}n\right)\left(1 + \frac{e}{100}\right) - (32a + 16b) = 8n \qquad \text{(iv)}$$

But from the volumetric product analysis

$$\frac{a}{b} = \frac{10{\cdot}60}{10{\cdot}45} \quad \text{or} \quad b = 0{\cdot}985a \qquad \text{(v)}$$

and a kg-mol CO_2 are associated with

$$a \times \frac{78{\cdot}95}{10{\cdot}60} = \left(m + \tfrac{1}{4}n\right)\left(1 + \frac{e}{100}\right)\frac{79}{21} \text{ kg-mol } N_2 \qquad \text{(vi)}$$

Hence, substituting from (iii), (v) and (vi) in (iv)

$$\frac{21}{79} \times 32\left(a \times \frac{78{\cdot}95}{10{\cdot}60}\right) - (32a + 15{\cdot}75a) = 8 \times 6 = 48$$

or
$$(63{\cdot}2 - 32 - 15{\cdot}75)\,a = 48$$

or
$$a = \frac{48}{15{\cdot}45} = 3{\cdot}10 \qquad \text{(vii)}$$

$$b = 2{\cdot}90 \qquad \text{(viii)}$$

Substituting from (vii) and (viii) in (iv)

$$32\,(7{\cdot}5)\left(1 + \frac{e}{100}\right) - (32 \times 3{\cdot}10 + 16 \times 2{\cdot}90) = 48$$

whence
$$\frac{e}{100} = 0{\cdot}805 - 1 = -\,0{\cdot}195$$

or
$$e = -\,19{\cdot}5 \qquad \text{(ix)}$$

Hence the fuel–air mixture is 19·5 per cent rich.

The results of the two analyses are thus in substantial agreement with each other, proving that the given volumetric composition of the dry products is substantially correct.

In general, the carbon balance is to be preferred to the hydrogen–oxygen balance in being the simpler method. Nevertheless it is useful to employ the second method as a check on the result of the first as well as on the accuracy of the given product analysis.

Note: Had a fuel of "random" composition been given instead of a hydrocarbon fuel of given chemical composition C_mH_n, equation (13.9b) would have been used in place of equation (13.9a), the volumetric composition of the dry products being first converted into a gravimetric composition. It is left to the student to carry out this exercise for a fuel containing 72/78 = 0·923 kg of carbon and 6/78 = 0·077 kg of hydrogen per kg of fuel, this being identical with the gravimetric composition of benzene.

However, a very much quicker solution, in this case, is obtained by striking a *carbon–nitrogen balance* for both reactants and products, the mass ratio of these two elements clearly being the same before and after the reactions.

From the given volumetric product analysis

$$\left(\frac{m_C}{m_N}\right)_{products} = \frac{12(10{\cdot}6 + 10{\cdot}45)}{28 \times 78{\cdot}95} = 0{\cdot}1145 \qquad \text{(i)}$$

while, with an air–fuel ratio A/F, with the given gravimetric composition of the fuel

$$\frac{m_C}{m_N} = \frac{72/78}{(A/F)(76{\cdot}7/100)} = \frac{1{\cdot}205}{A/F} \qquad \text{(ii)}$$

Equating (i) and (ii)

$$A/F = 10{\cdot}53 \qquad \text{(iii)}$$

In order to determine the percentage excess air we have to refer back to the stoichiometric combustion equation (13.7b), and example 13.2, giving

$$\left(\frac{A}{F}\right)_{stoi.} = 13{\cdot}2$$

whence

$$\left(1 + \frac{e}{100}\right)\left(\frac{A}{F}\right)_{stoi.} = \frac{A}{F}$$

or

$$1 + \frac{e}{100} = \frac{10{\cdot}53}{13{\cdot}2} = 0{\cdot}80$$

or

$$e = -20$$

The solution once again agrees almost exactly with those previously derived.

13.3 First Law of Thermodynamics Applied to Combustion Processes

13.3.1 Closed and open system equations

In Chapter 4 we studied in detail the application of the first law of thermodynamics to closed and open systems, the working fluid being a pure substance of fixed chemical composition. These analyses resulted in the non-flow energy equation (N.F.E.E.) for closed systems (equation (4.4b)) and the steady flow energy equation (S.F.E.E.) for open systems (equation (4.13)).

Similar procedures may be applied to reacting systems, provided that allowance is made for the heat released by the combustion process. In order to establish suitable equations for closed system and open system combustion processes (e.g. in a motor car engine cylinder on the one hand and in a gas turbine combustion chamber on the other) it is convenient to reduce all such processes to one of two basic processes, namely

(1) The *constant temperature–constant volume process* for closed system combustion (isothermal–isochoric process).
(2) The *constant temperature–constant pressure process* for steady flow open system combustion (isothermal–isobaric process).

In each case the heat released by the chemical reaction is removed from the system by suitable means. These processes are approached very closely by the isothermal bomb calorimeter for the former case and the Junkers gas calorimeter for the latter (see section 13.4).

Thus for a closed system (*Fig.* 13.4) applying the N.F.E.E. to the isothermal–isochoric process and denoting the states before and

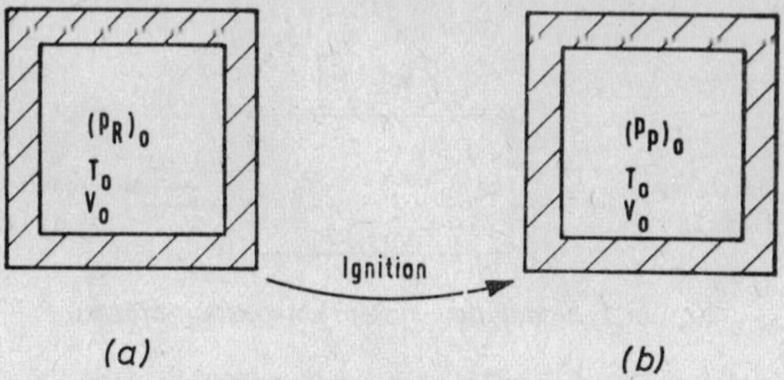

Fig. 13.4 Closed system combustion process

after the reaction by suffixes R_0 and P_0 respectively (where R refers to reactants and P to products), with fixed temperature T_0 and volume V_0, we have

$$Q_{R_0-P_0} - W_{R_0-P_0} = (U_P)_0 - (U_R')_0 \qquad (13.10)^*$$

* The internal energy and enthalpy of the reactants are denoted respectively by the symbols U' and H' as opposed to U and H to emphasize the fundamental difference between these quantities.

In the case of the products, U and H refer respectively to *sensible* internal energy and enthalpy, i.e. energy of the working fluid by virtue of temperature only. In the case of the reactants, however, U' and H' include in addition the chemical energy about to be released as heat in the ensuing chemical reaction.

If, as is often the case, combustion is incomplete, the products will contain some chemical energy, and we should therefore employ the symbols U' and H' for the products also. In the following, it will be assumed that combustion is complete unless otherwise stated.

where $W_{R_0-P_0} = 0$ (no external work)
and $Q_{R_0-P_0}$ = heat transferred *to* system during process.

This latter quantity, which is clearly *negative* according to the agreed sign convention for heat and work (see section 2.13), is also referred to as

the internal energy of combustion
or the constant volume heat of combustion } at temperature T_0

and denoted by the symbol ΔU_0, i.e.

$$Q_{R_0-P_0} = (U_P)_0 - (U'_R)_0 = \Delta U_0 \qquad (13.11)^*$$

Similarly, for the steady flow system applying the S.F.E.E. (4.13) to the isothermal–isobaric process (*Fig.* 13.5)

$$Q'_{R_0-P_0} - W'_{R_0-P_0} = (H_P)_0 - (H'_R)_0 \qquad (13.12)$$

where $W'_{R_0-P_0} = 0$
and $Q'_{R_0-P_0}$ = heat transferred *to* system during process.

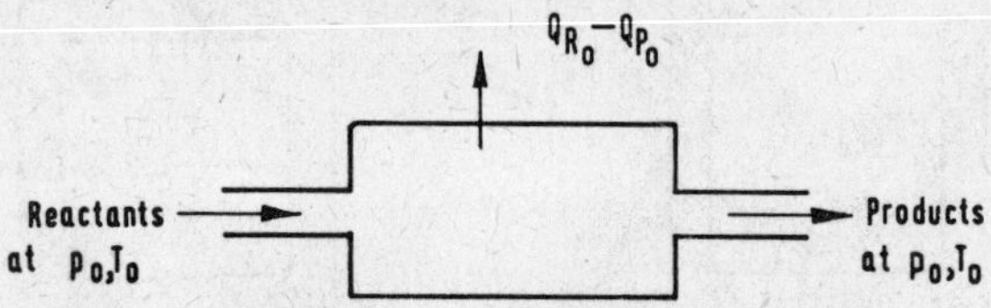

Fig. 13.5 Steady flow system combustion process

This latter quantity, which is again *negative*, is also referred to as

the enthalpy of combustion
or the constant pressure heat of combustion } at temperature T_0

and denoted by the symbol ΔH_0, i.e.

$$Q'_{R_0-P_0} = (H_P)_0 - (H'_R)_0 = \Delta H_0 \qquad (13.13)^*$$

The great advantage of this procedure is that most combustion processes may be reduced to one or other of the above standard processes by changing the state of the reactants from any given initial state, T_1, V_1 (for the closed system) or T_1, p_1 (for the open system)

* The internal energy of combustion, ΔU_0, and enthalpy of combustion, ΔH_0, are essentially determined by temperature only. Hence it is sufficient to specify a standard reaction temperature T_0, usually 25°C.

to the state T_0, V_0 (closed system) or T_0, p_0 (open system) without chemical reaction, while similarly any given final state of the products, i.e. T_2, V_2 for closed systems or T_2, p_2 for open systems, may again be attained by changing the state of the products without chemical change from the reference state to one or other of the above final states, applying the original forms of the non-flow and steady flow energy equations.

Thus, for a process involving successive changes as follows—

(*a*) Reactants from initial state T_1, V_1 (T_1, p_1) to reference state T_0, V_0 (T_0, p_0) without chemical change;
(*b*) Reactants at reference state T_0, V_0 (T_0, p_0) to products at reference state T_0, V_0 (T_0, p_0) with chemical change;
(*c*) Products at reference state T_0, V_0 (T_0, p_0) to products at final state T_2, V_2 (T_2, p_2) without chemical change;

we may write:

For *closed system combustion* (equations (13.10) and (13.11))

$$\begin{aligned}(U_P)_2 - (U'_R)_1 &= [(U_P)_2 - (U_P)_0] + [(U_P)_0 - (U'_R)_0] + [(U'_R)_0 - (U'_R)_1] \\ &= [(U_P)_2 - (U_P)_0] + \Delta U_0 + [(U'_R)_0 - (U'_R)_1] \qquad (13.14)\end{aligned}$$

where, assuming reactants and products to be perfect or semi-perfect gases as defined respectively by equations (10.1) [state equation $pv = f_1(T) = RT$] and (10.2) [internal energy $u = f_2(T) = c_vT$],

$$(U_P)_2 - (U_P)_0 = \sum_P m_i(c_v)_i(T_2 - T_0) \qquad \text{(i)}$$

$$(U'_R)_0 - (U'_R)_1 = \sum_R m_i(c_v)_i(T_0 - T_1) \qquad \text{(ii)}$$

where m_i and $(c_v)_i$ refer respectively to the mass and mean specific heat at constant volume, between appropriate temperature limits, of constituents.

Similarly, for *open system combustion*

$$\begin{aligned}(H_P)_2 - (H'_R)_1 &= [(H_P)_2 - (H_P)_0] + [(H_P)_0 - (H'_R)_0] + [(H'_R)_0 - (H'_R)_1] \\ &= [(H_P)_2 - (H_P)_0] + \Delta H_0 + [(H'_R)_0 - (H'_R)_1] \qquad (13.15)\end{aligned}$$

where
$$(H_P)_2 - (H_P)_0 = \sum_P m_i(c_p)_i(T_2 - T_0) \qquad \text{(iii)}$$

$$(H'_R)_0 - (H'_R)_1 = \sum_R m_i(c_p)_i(T_0 - T_1) \qquad \text{(iv)}$$

Experiment has shown that, at normal pressures, ΔU_0 and ΔH_0 are functions of temperature only, although it is important to specify the phase of any individual reactant or product which may be in doubt. Likewise the internal energy and enthalpy of combustion for a process in which reactants at state 1 are converted into products at state 2 are functions of temperature only, bearing in mind the fact that the reactant and product internal energy and enthalpy difference are themselves only temperature-dependent (auxiliary equations (i), (ii), (iii) and (iv)). The latter may be derived from suitable tabulations (Keenan & Kaye—ref. 13.1, Geyer & Bruges—ref. 13.2) in which u and h values (in kJ/kg-mol) are given as functions of temperature for the common gases, or alternatively from rather less accurate charts giving mean volumetric heats (in kJ/kg-mol) between a fixed datum temperature and the required product or reactant temperature, again for a wide range of common gases (see *Fig*. 10.4 volumetric heat chart).

EXAMPLE 13.5

Find the enthalpy and internal energy of combustion of a stoichiometric mixture of coal gas and air, (1) when reactants and products are at $T_0 = 25°C$, (2) when reactants and products are $T_1 = T_2 = 540°C$.

The volumetric composition of coal gas and specific heat at constant pressure and enthalpies of combustion of the constituents are given.

Volumetric composition

	Per cent		*Per cent*
H_2	48·0	CO	19·0
CH_4	21·0	CO_2	4·5
C_6H_6	1·5	N_2	6·0

Specific heats (kJ/kg-K)

	at 25°C	*over range* 25°C *to* 540°C
H_2	14·580	15·600
CH_4	3·212	4·132
C_6H_6	1·809	2·830
CO	1·039	1·140
CO_2	0·761	1·050
N_2	1·039	1·069
H_2O	1·880	1·989
O_2	0·910	0·988

Enthalpy of combustion ΔH_0 at 25°C

	M	kJ/kg	
H_2	2	−120900	(H_2O in products as vapour)
C_6H_6	78	−44550	(H_2O in products as vapour)
CO	28	−10100	
CH_4	16	−50200	(H_2O in products as vapour)

SOLUTION

Combustion equation

$$[0{\cdot}48\ H_2 + 0{\cdot}21\ CH_4 + 0{\cdot}015\ C_6H_6 + 0{\cdot}19\ CO + 0{\cdot}045\ CO_2 + 0{\cdot}06\ N_2]_{1\ kg\ mol\ of\ gas} + [0{\cdot}8675\ O_2 + 3{\cdot}262\ N_2]_{air}$$

$$= \underset{\text{from gas}}{[0{\cdot}045\ CO_2 + 0{\cdot}06\ N_2]} + \underset{\text{products of combustion}}{[0{\cdot}945\ H_2O + 0{\cdot}49\ CO_2 + 3{\cdot}262\ N_2]}$$

$$= 0{\cdot}945\ H_2O + 0{\cdot}535\ CO_2 + 3{\cdot}322\ N_2 \qquad \text{(i)}$$

or $\qquad 5{\cdot}120$ kg-mol reactant → $4{\cdot}800$ kg-mol products $\qquad$ (ii)

(*a*) *Enthalpy of combustion per kg-mol of gas at* 25°C

$$\Delta H_0 = -[\underset{H_2}{0{\cdot}48 \times 2 \times 120900} + \underset{CH_4}{0{\cdot}21 \times 16 \times 50200} + \underset{C_6H_6}{0{\cdot}015 \times 78 \times 44550} + \underset{CO}{0{\cdot}19 \times 28 \times 10100}]$$

$$= -390500 \text{ kJ/kg-mol} \quad \textit{Ans.}$$

(*b*) *Internal energy of combustion per kg-mol of gas at* 25°C

From equations (13.11) and (13.13), and perfect gas relationship between internal energy and enthalpy, equations (10.16)

$$(H_P)_0 = (U_P)_0 + n_P G(T_0 - T_d)$$
$$(H'_R)_0 = (U'_R)_0 + n_R G(T_0 - T_d)$$

where datum temperature $T_d \neq$ standard combustion temperature T_0, unless specifically stated, i.e.

$$(H_P)_0 - (H'_R)_0 = (U_P)_0 - (U'_R)_0 + (n_P - n_R)G(T_0 - T_d)$$

or

$$\Delta H_0 = \Delta U_0 + (n_P - n_R)G(T_0 - T_d) \qquad \text{(iv)}$$

where $n_P = 4{\cdot}800$ and $n_R = 5{\cdot}120$ (see equation (ii))

Therefore, substituting from (iii) in (iv) with $G = 8{\cdot}3143$ kJ/kg-mol-K and $T_0 = 298\ K = T_d$

$$\Delta U_0 = -390500 + 0 = -390500 \text{ kJ/kg-mol} \quad \textit{Ans.}$$

i.e.

$$\Delta U_0 = \Delta H_0 \qquad \text{(v)}$$

(*c*) *Enthalpy of combustion per kg-mol of gas at* 540°C

From equation (13.15)

$$(H_P)_2 - (H'_R)_1 = \Delta H_0 + \sum_P m_i(c_p)_i(T_2 - T_0) - \sum_R m_i(c_p)_i(T_2 - T_0)$$

$$= -390500$$

$$+ 515[(\underset{H_2O}{0{\cdot}945 \times 18 \times 1{\cdot}989}) + (\underset{CO_2}{0{\cdot}535 \times 44 \times 1{\cdot}050})$$

$$+ (\underset{N_2}{3{\cdot}322 \times 28 \times 1{\cdot}069})]$$

$$- 515[(\underset{H_2}{0{\cdot}48 \times 2 \times 15{\cdot}60}) + (\underset{CH_4}{0{\cdot}21 \times 16 \times 4{\cdot}132}) + (\underset{C_6H_6}{0{\cdot}015 \times 78 \times 2{\cdot}830})$$

$$+ (\underset{CO}{0{\cdot}19 \times 28 \times 1{\cdot}140}) + (\underset{CO_2}{0{\cdot}045 \times 44 \times 1{\cdot}050}) \times (\underset{N_2}{3{\cdot}322 \times 28 \times 1{\cdot}069})$$

$$+ (\underset{O_2}{0{\cdot}8675 \times 32 \times 0{\cdot}988})$$

$$= -390500 + 78990 - 86050$$

$$= -397560 \text{ kJ/kg-mol} \quad \textit{Ans.}$$

(*d*) *Internal energy of combustion per kg-mol of gas at* 540°C

From equation (iv) above,

$$(U_P)_2 - (U'_R)_1 = (H_P)_2 - (H'_R)_1 - [n_P G(T_2 - T_d) - n_R G(T_1 - T_d)]$$

where $T_2 = T_1 = 540°\text{C}$, $T_d = 25°\text{C}$

i.e.

$$(U_P)_2 - (U'_R)_1 = (H_P)_2 - (H'_R)_1 - (4{\cdot}8 - 5{\cdot}12)515 \times 8{\cdot}3143$$

$$= (H_P)_2 - (H'_R)_1 + 1371$$

$$= -396189 \text{ kJ/kg-mol} \quad \textit{Ans.}$$

The above example underlines the fact that differences between ΔH_0 and ΔU_0 are not large. Furthermore, neither ΔH_0 nor ΔU_0 are affected by non-reacting constituents of the reactants or products, e.g. nitrogen or excess oxygen. This is important, since large amounts of excess air, desirable for complete combustion and hence high efficiency in diesel engines or as a means of limiting maximum temperatures in gas turbines, can be used without adverse effect.

13.3.2 Energy–Temperature diagrams

Since the internal energy and enthalpy changes in the closed and open system combustion reactions described in the previous sub-section are substantially functions of temperature only, it is possible to represent these processes very conveniently in *U–T* diagrams for closed system reactions and *H–T* diagrams for open system reactions. Such diagrams can be arranged to include the effect of phase changes of certain constituents of the reactants of products, e.g. liquid benzene to benzene vapour before the reaction which raise the energy level of the reactants, or water vapour to liquid water after the reaction which lower the energy level of the products.

In addition, the effects of dissociation of the reactants and products at elevated temperatures may be included (see section 13.6). *Fig.* 13.6*a* and 13.6*b* are typical though simplified internal energy–temperature and enthalpy–temperature diagrams. The U–T (or H–T) curve for the products is based on the use of tables or charts in conjunction with the appropriate semi-perfect gas relationship—

$$(U_P)_2 - (U_P)_d = \sum_P m_i(c_v)_i(T_2 - T_d) \tag{13.16}$$

and

$$(H_P)_2 - (H_P)_d = \sum_P m_i(c_p)_i(T_2 - T_d) \tag{13.17}$$

where suffixes 2 and d refer respectively to the actual product and arbitrary datum temperature, i.e. T_2 and T_d.

The corresponding reactant curves are based on similar information and expressions, but are displaced upwards from the product curve by amounts ΔU_0 and ΔH_0, respectively, both at the reference temperature T_0. This upward displacement indicates the fact that, at the reference temperature T_0, the energy level of the reactants exceeds that of the products, by virtue of their chemical energy, the difference being released to the surroundings as heat (and part of it being available for conversion into useful work).

With the aid of these "energy" diagrams we may readily evaluate the energy released in any specified closed or open system combustion process with initial reactant temperature T_1 and final product temperature T_2 by graphical determination of the quantities ΔU_{12} for closed system processes from *Fig.* 13.6*a* or ΔH_{12} for open system processes from *Fig.* 13.6*b*.

The two "ideal" processes discussed with reference to *Figs.* 13.4 and 13.5, namely the isothermal–isochoric (constant temperature–constant volume) closed system (bomb calorimeter) process and the isothermal–isobaric (constant temperature–constant pressure) open system (gas calorimeter) process are represented respectively by the vertical lines R_0—P_0 in *Figs.* 13.6*a* and *b*, indicating that the heat released is respectively ΔU_0 and ΔH_0 in the two cases.

If, on the other hand, the combustion process is carried out adiabatically, i.e. without heat release to the surroundings, we can see intuitively that the temperature in both cases will rise. Furthermore, if the process takes place without work transfer to the surroundings

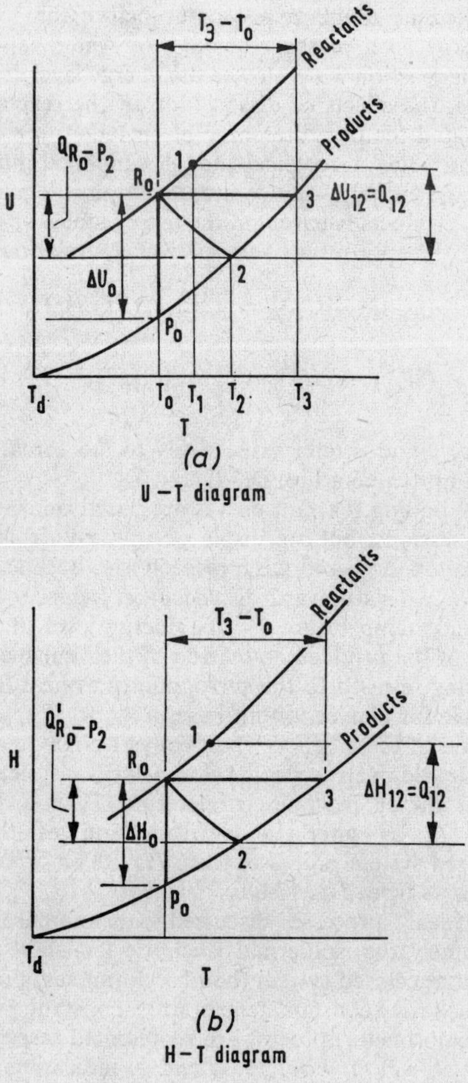

Fig. 13.6 U–T and H–T diagrams

application of the modified non-flow and steady flow energy equations (13.10) and (13.11) between the initial state R_0 and the final state 3 gives for closed systems

$$Q_{R_0-3} - W_{R_0-3} = U_3 - (U'_R)_0 = 0 \qquad (13.10a)$$

and for open systems

$$Q'_{R_0-3} - W'_{R_0-3} = H_3 - (H'_R)_0 = 0 \qquad (13.11a)$$

i.e. the adiabatic processes are both represented by horizontal lines in *Figs.* 13.6*a* and *b*, indicating the maximum possible temperature rise of magnitude $(T_3 - T_0)$ which may be read directly from the diagram. The isothermal process $(R_0 - P_0)$ and the adiabatic process $(R_0 - 3)$ may be regarded as limiting cases of the combustion process *without work transfer*. Still considering workless processes, an infinite number of intermediate processes can be visualized, namely $(R_0 - 2)$, *Fig.* 13.6, in which both heat transfer and temperature rise occur. Thus, if the heat loss Q_{R_0-2} from the combustion chamber is known then by displacing the end state 2 vertically downwards from the initial state R_0 by the interval Q_{R_0-2} the end temperature T_2 immediately follows. This is clearly less than the maximum temperature T_3 for the adiabatic process. For the general case of combustion *without work transfer*, as represented* by the path $(1-R_0-P_0-2)$, the energy diagrams 13.6*a* and *b* again furnish the required heat transfer $Q_{12}(= \Delta U_{12}$ or $\Delta H_{12})$ and temperature rise $(T_2 - T_1)$ in simple and elegant fashion, as indicated.

13.3.3 Departures from simplified combustion conditions

In practice the following complications arise—

(1) CERTAIN CONSTITUENTS OF THE REACTANTS AND PRODUCTS MAY EXIST IN ONE OF TWO PHASES, I.E. GASEOUS OR LIQUID

This complication has already been discussed briefly. In the case of the spark ignition or petrol engine, the fuel is initially in liquid form, but is transformed into vapour in the carburettor, thereby absorbing its latent heat of evaporation from the surroundings. The reactant energy curve in *Figs.* 13.6*a* and *b* has been tacitly assumed to apply to reactants in gaseous form. Allowance for this phase change may

* The combustion actually proceeds between initial state 1 and end state 2 in an unspecified manner. The substitution of path $(1-R_0-P_0-2)$ for this actual path is an artifice designed to facilitate the evaluation of the energy change ΔU_{12} or ΔH_{13} but is justified in view of the fact that we are dealing with *property* changes.

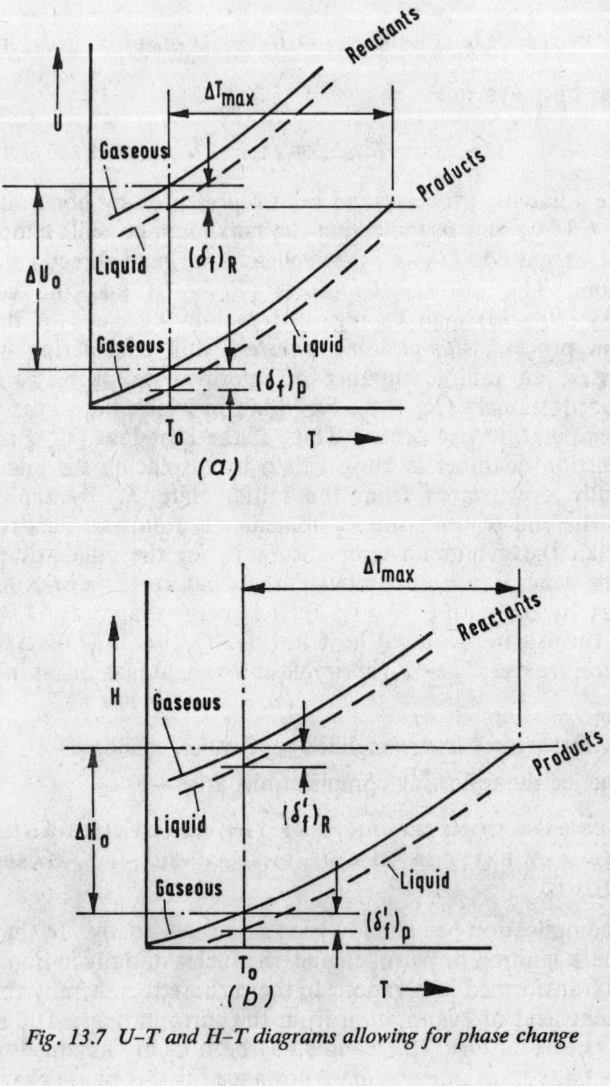

Fig. 13.7 U–T and H–T diagrams allowing for phase change

be made by displacing the reactant curve downwards as shown in *Fig.* 13.7, for the closed system process

$$(\delta_f)_R = \sum_R (m_f)_i (u_{fg})_i$$

and for the open system process

$$(\delta'_f)_R = \sum_R (m_f)_i (h_{fg})_i$$

where f_R denotes liquid (fluid) constituents of the fuel,
u_{fg} denotes internal energy of evaporation,
h_{fg} denotes enthalpy of evaporation.

Similarly, the product energy curves imply products in gaseous form, whereas in practice certain constituents, chiefly water, may exist either in gaseous or liquid form. Allowance for condensation of water vapour may be made by displacing the product energy curve downwards by

$$(\delta_f)_P = m_{H_2O}(u_{fg})_{H_2O} \quad \text{(i)}$$

for the closed system process

or $$(\delta'_f)_P = m_{H_2O}(h_{fg})_{H_2O} \quad \text{(ii)}$$

for the open system process (*Fig.* 13.7*b*).

These corrections are indicated qualitatively in *Figs.* 13.7*a* and *b*.

It is clear that the heat release during combustion may be considerably diminished or augmented by the phase changes indicated above. The most favourable situation arises when the condensible constituents of the reactants are in gaseous and those of the products in liquid form, while the reverse situation (as actually encountered in the cylinder of the petrol engine) will lead to the least favourable situation, i.e. the smallest quantity of heat released. The same arguments apply to the temperature rise ΔT_{max} for the adiabatic combustion process.

In practice such changes are allowed for in a somewhat approximate manner by specifying a higher or lower calorific value of the fuel (H.C.V. or L.C.V.), see section 13.4.

EXAMPLE 13.6

In the case of the stoichiometric combustion of coal gas given in example 13.5, find the enthalpy and internal energy of combustion when reactants and products are at 25°C, but when the water vapour in the products exists in liquid instead of vapour form.

SOLUTION

From *Fig*. 13.7 it is evident that condensation of the water vapour in the products increases the enthalpy of combustion by $m_{H_2O}(h_{fg})_{H_2O}$ and similarly the internal energy of combustion by $m_{H_2O}(u_{fg})_{H_2O}$.

(*a*) Enthalpy of combustion

$$\Delta H'_0 = -(390\,500 + 0{\cdot}945 \times 18 \times 2\,441{\cdot}8)$$
$$= -432\,150 \text{ kJ/kg-mol} \quad \textit{Ans.}$$

(*b*) Internal energy of combustion

$$\Delta U'_0 = -390\,500 + 0{\cdot}945 \times 18 \times 2\,304{\cdot}3)$$
$$= -429\,690 \text{ kJ/kg-mol} \quad \textit{Ans.}$$

(2) WORK TRANSFER OR OTHER ENERGY TRANSFORMATIONS MAY OCCUR DURING COMBUSTION

In the case of closed system processes, any departure from the assumed constant volume (isochoric) condition will result in work being done on the surroundings, e.g. in the reciprocating internal combustion engine in which the cylinder volume and pressure change very considerably during the combustion process. Hence the modified non-flow energy equation (13.10) combined with equation (13.14) when applied to a process in which reactants at state 1 change to products at state 2 becomes

$$Q_{12} - W_{12} = \Delta U_{12}$$

or $$Q_{12} - \int_1^2 p\,dV = [(U_P)_2 - (U_P)_0] + \Delta U_0 + [(U'_R)_0 - (U'_R)_1] \tag{13.18}$$

In the case of open systems, e.g. in the gas turbine combustion chamber while work transfer during combustion is usually absent, there may be considerable changes in kinetic energy between the entry and exit sections (see *Fig*. 13.5). Such changes result in the following form of the modified steady flow energy equation (13.12), combined with equation (13.15)

$$Q_{12} - \tfrac{1}{2}\Delta c^2 = \Delta H_{12}$$

or $$Q_{12} - \tfrac{1}{2}(c_2^2 - c_1^2) = [(H_P)_2 - (H_P)_0] + \Delta H_0 + [(H'_R)_0 - (H'_R)_1] \tag{13.19}$$

Clearly, in order to achieve the maximum heat release (or temperature rise) in a gas turbine or other steady flow combustion chamber, kinetic energy changes must be kept to a minimum.

(3) AT ELEVATED TEMPERATURES, THE COMPOSITION OF THE PRODUCTS OF COMBUSTION CHANGES DUE TO DISSOCIATION

The energy–temperature diagrams (*Figs.* 13.6 and 13.7) apply only to combustion in accordance with the chemical equations of section 13.1, and should be regarded as being most suitable for the evaluation of processes involving stoichiometric or weak mixtures. Under such conditions, the chemical equations of section 13.1 indicate the presence of CO_2 and H_2O (and to a lesser extent SO_2) in the product gases as the end product of the oxidation reactions for C, H_2 and S, as described by equations (13.1), (13.2) and (13.3). However, at high product temperature these reactions do not proceed fully in the direction of the above end products. Instead, CO_2 is partly *dissociated* according to the reaction

$$CO_2 \rightarrow CO + \tfrac{1}{2}O_2 \qquad (13.20)$$

if sufficient energy is supplied to the CO_2 molecule.
Similarly, water vapour partly dissociates according to the reaction

$$H_2O \rightarrow H_2 + \tfrac{1}{2}O_2 \qquad (13.21)$$

a temperature-dependent state of equilibrium being reached in accordance with principles to be discussed in section 13.6.

The effect of dissociation is thus identical with incomplete combustion, so that the actual heat release or adiabatic temperature rise will be less than their equivalent values for chemically complete combustion. Furthermore the product energy curves in *Figs.* 13.6 and 13.7 will be affected by the resultant change in composition of the product gases, i.e. by the different value of the internal energy or enthalpy as given respectively by equations (13.16) and (13.17).

13.4 Calorific Value of Fuels

In section 13.3.1 the internal energy and enthalpy of combustion, i.e. ΔU_0 and ΔH_0, were strictly defined in terms of the initial state of the reactants and the final state of the products. It was also shown that these quantities were essentially dependent on temperature only, and that their numerical values did not differ greatly.

In practice these two quantities, both of which were seen to be *negative* (representing heat abstracted from the reacting system), are replaced by the *calorific value*, an equivalent positive quantity which, however, is less stringently defined in view of the various approximations stated above.

The calorific value (C.V.) is the number of heat units released when unit mass of the fuel is burnt completely in a calorimeter under specified conditions.

The calorific value of liquid and solid fuel is usually determined with the aid of a bomb calorimeter (*Fig.* 13.8), i.e. a gas-tight steel

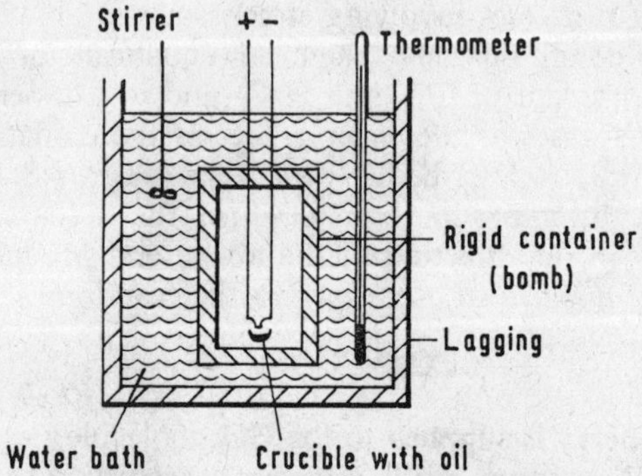

Fig. 13.8 Bomb calorimeter

vessel immersed in a conventional water calorimeter. The reaction corresponds very nearly to the constant temperature, constant volume reaction forming the basis for the definition of the internal energy of combustion (equation (13.11)), i.e.

$$(\text{C.V.})_{bomb} \simeq -\Delta U_0 \qquad (13.22)$$

The approximation is introduced by the fact that the temperature of the calorimeter and the content actually rises by a small amount (5–6°C) during the experiment, so that the process is no longer strictly isothermal. It is this temperature rise which allows the heat released, and hence the calorific value, to be calculated.

The calorific value of gaseous fuels is usually found with the aid of a Junkers or Boys gas calorimeter (*Fig.* 13.9) in which a stream of gas and air is burnt essentially under steady flow, constant pressure conditions. The hot products of combustion, before leaving the calorimeter, are cooled almost to the original temperature of the reactants, in a counterflow heat exchanger, so that the process corresponds closely to the constant temperature, constant pressure reaction forming the basis for the definition of enthalpy of combustion (equation (13.13)), i.e.

$$(\text{C.V.})_{gas\ calor.} \simeq -\Delta H_0 \qquad (13.23)$$

Because of the near equality of ΔU_0 and ΔH_0 already noted for the same reactants at the same temperature, it is unnecessary to

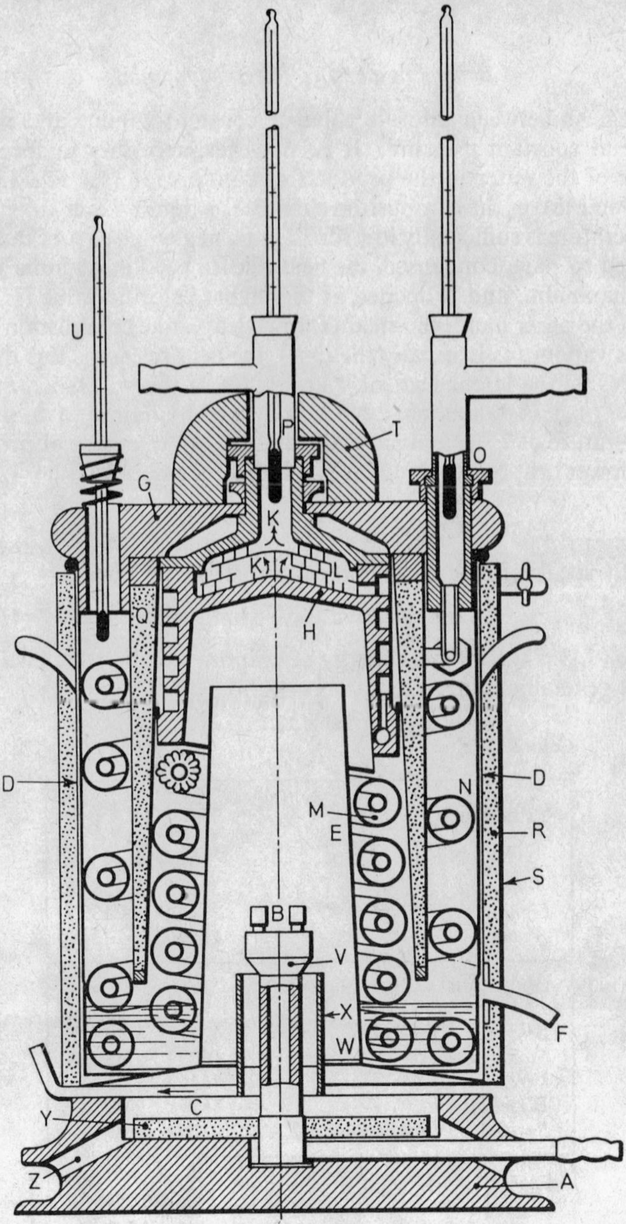

Fig. 13.9 Boys' calorimeter

A. Wood base
B. Gas burners
C. Metal plate
D. Metal vessel
E. Copper chimney
F. Condensed water outlet
G. Wood lid
H. Mixing chamber
K. Dished plates
M. Inner coil
N. Outer coil
O. Inlet water box
Q. Baffle
R. Felt jacket
S. Bright sheet metal
T. Wood shield
W. Ferrule
X. Fibre tube
Y. Felt disc
Z. Holes in base

distinguish between calorific value at constant volume and calorific value at constant pressure. It is, however, necessary to specify the phase of the water in the products of combustion (see *Fig.* 13.7). If the water leaves the combustion chamber as liquid, i.e. if the product temperature is sufficiently low for all but a negligible part of the steam formed to have condensed, the heat release per kilogramme of fuel is a maximum, and is defined as the higher calorific value (H.C.V.). If, on the other hand, the steam formed leaves the combustion chamber as vapour (as is usually the case), the heat release is less than the H.C.V. by the latent heat of evaporation of the water, i.e. $m_m(h_{fg})$ where (h_{fg}) is taken as 2442 kJ/kg corresponding to a product temperature of 25°C and m_m is the mass of water per kg of products. This lower heat release is defined as the lower calorific value (L.C.V.), i.e.

$$\text{L.C.V.} = \text{H.C.V.} - m_m(h_{fg}) \tag{13.24}$$

Strictly speaking, for the bomb calorimeter experiment,

$$\text{L.C.V.} = \text{H.C.V.} - m_m(u_{fg}) \tag{13.24a}$$

where (u_{fg}) is the internal energy of evaporation, but this distinction is not generally made.

Table 13.1

Calorific Values

	HCV	LCV
SOLID FUELS		
	kJ/kg at 25°C	
Anthracite	34583	33913
Lignite	21646	20390
Wood	15826	14319
LIQUID FUELS		
	kJ/kg at 25°C	
Benzene	42217	40519
Motor petrol	46892	43710
Kerosene	46159	43166
Diesel oil	45971	43166
Residual fuel oil (bunker C)	42054	39961
GASEOUS FUELS		
	kJ/m³ at 25°C and 1 bar	
Coal gas (town gas)	19654	17549
Producer gas	5920	5879
Natural gas	35586	31964
Carbon monoxide	11625	11625
Hydrogen	11643	9829

13.5 Efficiency of Combustion Systems

The useful heat release Q per kg of fuel in an actual combustion process, e.g. in a steam boiler, gas turbine combustion chamber or internal combustion engine cylinder, is usually considerably less than the "ideal" value, i.e. $-\Delta U_0$ for closed systems or $-\Delta H_0$ for flow systems. The reasons for this are primarily

(*a*) Incomplete combustion due to
- (*i*) incomplete mixing of fuel and air (insufficient residence time in combustion chamber),
- (*ii*) dissociation and inadequate recombination,
- (*iii*) insufficient air,

(*b*) Heat losses from the combustion chamber,

(*c*) High temperature of the products of combustion when rejected to the surroundings.

The combustion efficiency of "external" combustion devices such as the steam boiler, whose useful output is the heat Q transferred to the working fluid of an adjacent work producing system, in this case the complete sequence of units comprising the steam circuit (see *Fig*. 13.1), may be defined as

$$\eta_c = \frac{Q}{\text{L.C.V.}} \tag{13.25}$$

where $Q = m_w \Delta h_w$,

m_w = kg of working fluid in work producing system/kg of fuel,

Δh_w = increase in sensible* enthalpy per kg of working fluid in work producing system (kJ/kg).

The lower calorific value is taken as the ideal heat release value, since the water contents of the products of combustion will be predominantly in vapour form at the minimum product temperature, i.e. at exit from the combustion chamber.

In the case of "internal" combustion devices such as the reciprocating engine cylinder or the gas turbine combustion chamber (*Fig*. 13.2) in which the reactant and products of combustion are themselves the working fluid of the work producing system, it is the increase in sensible internal energy (in the case of closed systems) or sensible enthalpy (in the case of steady flow systems) of the products over the reactants which constitutes the useful output of the combustion system. Hence, relative to some convenient arbitrary

* See page 354 (footnote).

datum temperature T_d (see equations (13.16) and (13.17)) this increase may be expressed for reactants at state 1 and products at state 2 as follows—

For closed systems

$$(\Delta U)_{sensible} = [(U_P)_2 - (U_P)_d] - [(U'_R)_1 - (U'_R)_d]$$
$$= \sum_P m_i(c_v)_i(T_2 - T_d) - \sum_R m_i(c_v)_i(T_1 - T_d) \qquad \text{(i)}$$

(see equation (13.16)).

For flow systems

$$(\Delta H)_{sensible} = [(H_P)_2 - (H_P)_d] - [(H'_R)_1 - (H'_R)_d]$$
$$= \sum_P m_i(c_p)_i(T_2 - T_d) - \sum_R m_i(c_p)_i(T_1 - T_d) \qquad \text{(ii)}$$

(see equation (13.17)).

$(\Delta U)_{sensible}$ and $(\Delta H)_{sensible}$ may be determined experimentally in terms of the measured temperature rise $(T_2 - T_1)$ and the known composition of the reactants and products. In the limiting case of adiabatic, complete combustion, represented by the horizontal lines $(R_0 - 3)$ in *Figs*. 13.6*a* and *b*, and assuming that the datum temperature T_0 coincides with the datum temperature on which are based the internal energy and enthalpy of combustion, i.e. ΔU_0 and ΔH_0, the temperature rise is a maximum and

$$(\Delta U)_{sensible} = -\Delta U_0 \qquad \text{(iii)}$$

and

$$(\Delta H)_{sensible} = -\Delta H_0 \qquad \text{(iv)}$$

However, in real cases where combustion is incomplete and heat losses from the combustion chamber occur, the temperature rise will be less and

$$(\Delta U)_{sensible} = -\Delta U_0 + \sum_P m_i(\Delta U_0)_i - Q_{loss} \qquad \text{(v)}$$

or

$$(\Delta H)_{sensible} = -\Delta H_0 + \sum_P m_i(\Delta H_0)_i - Q_{loss} \qquad \text{(vi)}$$

where m_i = mass of constituent i,

$(\Delta U_0)_i$ = internal energy of combustion of unburnt or partially burnt constituent i,

$(\Delta H_0)_i$ = enthalpy energy of combustion of unburnt or partially burnt constituent i,

Q_{loss} = heat loss from combustion chamber to surroundings.

Hence we may now define the combustion efficiency for an internal combustion system as follows—

$$\text{Closed system: } \eta_c = \frac{1}{\text{L.C.V.}}\left(-\Delta U_0 + \sum_P m_i(\Delta U_0)_i - Q_{loss}\right) \quad (13.26a)$$

$$\text{Open system: } \eta_c = \frac{1}{\text{L.C.V.}}\left(-\Delta H_0 + \sum_P m_i(\Delta H_0)_i - Q_{loss}\right) \quad (13.26b)$$

EXAMPLE 13.7

A petrol engine burns a stoichiometric benzene air mixture trapped in the cylinder at 100°C and 1 bar. It is subsequently compressed, and due to dissociation the products of combustion at the maximum cylinder temperature and pressure contain unburnt constituents. The composition of the products and their mean molar specific heat at constant volume between 100°C and 2690°C are as follows—

	Mol number m_i *(kg-mol)*	*kJ/kg-mol-K*
CO_2	4·50	49·51
CO	1·50	26·22
H_2O	2·86	37·70
H_2	0·14	24·30
O_2	0·82	28·41
N_2	28·20	25·99
	38·02 kg-mol per kg-mol C_6H_6	

The *molar* internal energies of combustion are as follows—

C_6H_6 : −3163000 kJ/kg-mol
CO : − 282000 kJ/kg-mol
H_2O : − 241500 kJ/kg-mol

Given that the work of adiabatic compression preceding constant volume combustion is 6655 kJ/kg-mol of reactants show that a maximum temperature of 2690°C is compatible with the energy equation. Furthermore show that in the absence of dissociation, the maximum temperature would be approximately 3020°C. It may be assumed that the mean specific heats at constant volume of the constituents in this case are 3 per cent higher than those given in the above table. Find also the combustion efficiency when dissociation occurs.

SOLUTION

Case 1: Combustion with dissociation

Combustion equation

$$C_6H_6 + 7{\cdot}5\,O_2 + 28{\cdot}2\,N_2 = 4{\cdot}5\,CO_2 + 1{\cdot}5\,CO + 2{\cdot}86\,H_2O + 0{\cdot}14\,H_2 + 0{\cdot}82\,O_2 + 28{\cdot}2\,N_2$$

i.e. 36·7 kg-mol reactant → 38·02 kg-mol products

The modified internal energy of combustion is given by equation (13.26a)

$$-\Delta U'_0 = -\Delta U_0 + \sum_P n_i(\Delta U_0)_i \qquad \text{(i)}$$

Since values of $\Delta(U_0)_i$ are based on 1 kg-mol not on unit mass, m_i is replaced by n_i, the mol number.

Substituting for

$$(\Delta U_0)_{CO} = -\ 282000 \text{ kJ/kg-mol}$$

$$(\Delta U_0)_{H_2O} = -\ 241500 \text{ kJ/kg-mol}$$

$$(\Delta U_0)_{C_6H_6} = -\ 3163000 \text{ kJ/kg-mol}$$

$$-\Delta U'_0 = 3163000 - (1{\cdot}5 \times 282000) - (0{\cdot}14 \times 241500)$$

$$= 2706\,200 \text{ kJ/kg-mol}$$

$$= \frac{2706\,200}{36{\cdot}7} = 73920 \text{ kJ/kg-mol reactants} \qquad \text{(ii)}$$

From the non-flow energy equation

$$Q_{12} - W_{12} = \Delta U_{12} \qquad \text{(iii)}$$

where $Q_{12} = -\Delta U'_0 = +\ 73920$ kJ/kg-mol reactants

$$W_{12} = -\ 6655 \text{ kJ/kg-mol reactants}$$

$$Q_{12} - W_{12} = 73920 + 6655 = 80575 \text{ kJ/kg-mol reactants} \qquad \text{(iv)}$$

$$\Delta U_{12} = \sum_P n_i(C_v)_i(T_2 - T_1) - \sum_R n_i(C_v)_i(T_1 - T_1) \qquad \text{(v)}$$

The last term equals 0 relative to datum $T_1 = 100°\text{C}$ and, since values of $(C_v)_i$ are again based on 1 kg-mol and not on 1 kg, m_i is replaced by n_i.

But

$$(C_v)'_i = \frac{\sum_P n_i(C_v)_i}{\sum_R n_i}$$

$$= (4{\cdot}5 \times 49{\cdot}51 + 1{\cdot}5 \times 26{\cdot}22 + 2{\cdot}86 \times 37{\cdot}70 + 0{\cdot}14 \times 24{\cdot}30 + 0{\cdot}82 \times 28{\cdot}41 + 28{\cdot}20 \times 25{\cdot}99)/36{\cdot}7$$

$$= 30{\cdot}91 \text{ kJ/kg-mol reactant-K}$$

Therefore

$$\sum_P m_i(C_v)_i(T_2 - T_1) = (C_v)'_i(T_2 - T_1)$$

$$= 30{\cdot}91(2690 - 100)$$

$$= 80050 \text{ kJ/kg-mol reactant} \qquad \text{(vi)}$$

Since (iv) and (vi) are numerically very nearly equal, the two sides of equation (iii) balance, i.e. the given temperature of 2690°C is compatible with the energy equation.

Case 2: Combustion without dissociation

Combustion equation

$$C_6H_6 + 7{\cdot}5\ O_2 + 28{\cdot}2\ N_2 = 6\ CO_2 + 3\ H_2O + 28{\cdot}2\ N_2$$

i.e. 36·7 kg-mol reactants → 37·2 kg-mol products

Internal energy of combustion

$$\Delta U_0 = -3\,163\,000 \text{ kJ/kg-mol } C_6H_6$$
$$= -86\,250 \text{ kJ/kg-mol reactants} \qquad \text{(i)}$$

Also

$$Q_{12} - W_{12} = \Delta U_{12} \qquad \text{(ii)}$$

where $Q_{12} = -\Delta U_0 = 86250$ kJ/kg-mol reactants
$W_{12} = -6655$ kJ/kg-mol reactants

i.e.

$$Q_{12} - W_{12} = 86250 + 6655 = 92905 \text{ kJ/kg-mol reactants} \qquad \text{(iii)}$$

But

$$\Delta U_{12} = \sum_P n_i(C_v)_i(T_2 - T_1) = (C_v)'_i(T_2 - T_1) \qquad \text{(iv)}$$

where

$$(C_v)'_i = \frac{\sum_P n_i(C_v)_i}{\sum_R n_i}$$

$$= \frac{(6 \times 49{\cdot}51 + 3 \times 37{\cdot}70 + 28{\cdot}2 \times 25{\cdot}99)}{36{\cdot}7} \times 1{\cdot}03$$

$$= 32{\cdot}0 \text{ kJ/kg-mol reactants-K}$$

(cf. 30·91 kJ/kg-mol reactants-K)
Therefore

$$\sum_P n_i(C_v)_i(T_2 - T_1) = (C_v)'_i(T_2 - T_1)$$
$$= 32{\cdot}0(3020 - 100)$$
$$= 93\,450 \text{ kJ/kg-mol reactants} \qquad \text{(v)}$$

Hence (iii) and (v) are again very nearly equal numerically, showing that the given temperature of 3020°C is compatible with the energy equation.

This example shows that dissociation, i.e. incomplete combustion, has a very pronounced effect on the heat release and hence also on the combustion efficiency, and an equally marked effect on the maximum temperature attained by the products of combustion, and thereby on the cycle efficiency.

For a detailed discussion see Ricardo (ref. 13.3).

13.6 Dissociation

As already stated in section 13.3.3, dissociation of the products of combustion, particularly CO_2 and H_2O, according to equations (13.20) and (13.21), becomes an important factor when the products are at a sufficiently high temperature, (> 2000°C), and particularly for near-stoichiometric mixtures.

We may regard the partly dissociated state of the products of combustion as being the result of two reactions proceeding simultaneously in opposite directions and reaching a state of chemical equilibrium determined by the temperature level, i.e.

for CO_2 $$CO + \tfrac{1}{2}O_2 \leftrightharpoons CO_2 \tag{13.20a}$$

for H_2O $$H_2 + \tfrac{1}{2}O_2 \leftrightharpoons H_2O \tag{13.21a}$$

The resultant equilibrium state may be predicted from considerations arising from the second law of thermodynamics, using an artifice known as *van't Hoff's equilibrium box* (*Fig*. 13.10), in which a fully reversible chemical reaction takes place.

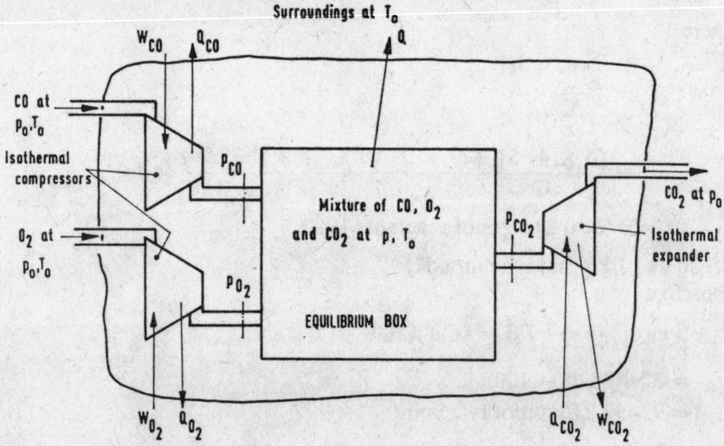

Fig. 13.10 Van't Hoff's equilibrium box

Consider a stoichiometric reaction between CO and O_2, corresponding to the ideal steady flow reaction of *Fig*. 13.5, and defined by equation (13.13).

The reaction is given by equation (13.20), though reversed

$$CO + \tfrac{1}{2}O_2 = CO_2 \tag{13.20b}$$

the CO and O_2 each entering the box at p_0, T_0 and CO_2 leaving at p_0, T_0.

In order to achieve thermodynamic reversibility, it is essential to eliminate finite temperature gradients and all work transfers must take place in accordance with the conditions discussed in Chapter 5. (Basically, heat and work transfers have to be effected infinitely slowly.) As in the case of the discussion of reversible mixing processes in the previous chapter (section 12.3.6) we again make use of the artifice of the semi-permeable membrane permitting gas flow in one direction only.

The two reactants, CO and O_2, enter the system infinitely slowly but in the correct molar proportions (i.e. n kg-mol/sec of CO and $\frac{1}{2}n$ kg-mol/sec of O_2) as given by equation (13.20b), each at pressure p_0 and temperature T_0. They are then compressed isothermally and reversibly in two suitably designed compressors to the respective partial pressures of these constituents in the equilibrium box, i.e. p_{CO} and p_{O_2}. This ensures that the mass transfer process through the semi-permeable membranes dividing the compressors from the equilibrium box take place without pressure change, and hence reversibly (see Chapter 5, section 5.5). The equilibrium box, in addition to CO and O_2, also contains CO_2 in as yet unspecified proportions, the total pressure being given by the partial pressure rule, equation (12.1)

$$p = p_{CO} + p_{O_2}$$

The reaction in accordance with equation (13.20b) then proceeds in the equilibrium box at the temperature T_0 maintained constant by reversible heat transfer Q to the surroundings which are also at T_0. Since the proportions of the constituents of the equilibrium mixture are to remain unchanged, it is essential that the CO_2 formed be removed continuously, i.e. at the rate of n kg-mol/sec. It leaves through a further and semi-permeable membrane against a back pressure, p_{CO_2}, i.e. its own partial pressure in the mixture, and finally through a reversible isothermal expander.

Since all processes involved have been arranged to be reversible, the process as a whole is also reversible, so that an infinitesimal change in external conditions will change the direction of the chemical reaction, which now becomes

$$CO_2 = CO + \tfrac{1}{2}O_2 \qquad (13.20c)$$

as well as of all mass, work and heat transfers.

We can show, by applying the second law of thermodynamics, that in the equilibrium mixture the partial pressures of the constituents, and hence their kg-mol fractions (see Chapter 12, section

12.3.5, equation (12.23b)) are related by a temperature-dependent function K_p, in the form

$$\frac{p_{CO_2}}{p_{CO}(p_{O_2})^{1/2}} = K_p \tag{13.27}$$

K_p is known as the *equilibrium constant.*

Equilibrium constant

Since the various gases taking part in the reaction, i.e. CO, O_2 and CO_2, may each be considered perfect, we may apply the relationships of Chapter 10. Hence in general the work output per unit mass for any reversible, steady flow isothermal process is given by equation (10.49)

$$w = RT \ln\left(\frac{p_1}{p_2}\right) \tag{i}$$

Since we are dealing with n mols having mass M (where M is the molecular weight) and since from equation (3.7), the particular gas constant

$$R = \frac{G}{M} \tag{ii}$$

where G is the universal gas constant, then
for n kg-mol of molecular weight M

$$W = nMRT \ln\left(\frac{p_1}{p_2}\right) = nGT \ln\left(\frac{p_1}{p_2}\right) \tag{iii}$$

Hence the net work performed by the system per unit time becomes

$$W = nGT\left[\ln\left(\frac{p_0}{p_{CO}}\right) + \tfrac{1}{2}\ln\left(\frac{p_0}{p_{O_2}}\right) + \ln\left(\frac{p_{CO_2}}{p_0}\right)\right]$$

$$= nGT\left[\ln\left(\frac{p_{CO_2}}{p_{CO}(p_{O_2})^{1/2}}\right) + \ln(p_0{}^{1/2})\right] \tag{iv}$$

A change in the pressure in the equilibrium box from

$$p = p_{CO} + p_{O_2} + p_{CO_2}$$

to

$$p' = p'_{CO} + p'_{O_2} + p'_{CO_2}$$

would change the net work per unit of time to

$$W' = nGT\left[\ln\left(\frac{p'_{CO_2}}{p'_{CO}(p'_{O_2})^{1/2}}\right) + \ln(p_0{}^{1/2})\right] \tag{v}$$

If it were possible, by changing the system pressure p to change the net work W, i.e.

$$W \neq W' \qquad \text{(vi)}$$

the system producing less work could be reversed and coupled to the remaining system. This would lead to a combined system operating in a cycle and producing no other effect than the extraction of heat from a single heat reservoir and the performance of an equivalent amount of work in direct contradiction to the second law (Chapter 6, section 6.3).

Clearly this cannot be so, and we must therefore conclude that a change in pressure cannot change the net work, i.e.

$$W = W' \qquad \text{(vii)}$$

Hence, from auxiliary equations (iv) and (v)

$$\frac{p_{CO_2}}{p_{CO}(p_{O_2})^{1/2}} = \frac{p'_{CO_2}}{p'_{CO}(p'_{O_2})^{1/2}} \qquad \text{(viii)}$$

so that we may write, at *any given temperature* (inverting the above fractions and subsequently squaring)

$$\frac{(p_{CO})^2}{(p_{CO_2})^2} \cdot p_{O_2} = (K_p)_1 \qquad (13.27)$$

where $(K_p)_1$ is the equilibrium constant for the reaction

$$CO + \tfrac{1}{2}O_2 \rightleftharpoons CO_2 \qquad (13.20a)$$

and is a function of temperature to be experimentally determined.

Similarly, for the reaction

$$H_2 + \tfrac{1}{2}O_2 \rightleftharpoons H_2O \qquad (13.21a)$$

we have a second equilibrium constant

$$\frac{p_{H_2O}}{p_{H_2}(p_{O_2})^{1/2}} = (K_p)_2 \qquad (13.28)$$

Since the equilibrium constants are not dimensionless quantities, it is important to state the unit of pressure.

The constants $(K_p)_1$ and $(K_p)_2$ are of great importance in the combustion of hydrocarbon fuels, in which both the reactions described by equation (13.20a) and (13.21a) occur simultaneously. It is occasionally useful to evolve new forms of the equilibrium constants, e.g. by squaring (13.28)

$$\frac{(p_{H_2O})^2}{(p_{H_2})^2(p_{O_2})} = (K'_p)_2 \tag{13.28a}$$

or combining $(K_p)_1$ and $(K_p)_2$

$$\frac{p_{CO}(p_{O_2})^{1/2}p_{H_2O}}{p_{CO_2}\,p_{H_2}(p_{O_2})^{1/2}} = \frac{p_{CO}p_{H_2O}}{p_{CO_2}\,p_{H_2}} = \sqrt{(K_p)_1(K_p)_2} = (K_p)_3 \tag{13.29}$$

The temperature-dependence of $(K_p)_1$ and $(K_p)_3$, as determined experimentally, may be expressed approximately in the form

$$\log_{10}(K_p)_1 = 8{\cdot}593 - \frac{28\,600}{T} \tag{13.27a}$$

$$\log_{10}(K_p)_3 = 1{\cdot}2065 - \frac{1\,100}{T} \tag{13.29a}$$

where the partial pressures in equations (13.27) and (13.29) are measured in atmospheres and the absolute temperature T in K.

The practical importance of dissociation, particularly in the case of reciprocating internal combustion engines operating with near-stoichiometric mixtures, is very considerable. Due to the very high maximum cylinder temperatures reached (2000–2600°C), dissociation effects are pronounced. The presence of incompletely burnt or unburnt product constituents, i.e. CO and H_2, results in reduced increase of sensible internal energy or enthalpy given by

$$(\Delta U)_{sensible} = -\Delta U_0 + \sum_P [m_i(\Delta U_0)_i] \tag{13.30}$$

or

$$(\Delta H)_{sensible} = -\Delta H_0 + \sum_P [m_i(\Delta H_0)_i] \tag{13.31}$$

(see auxiliary equations (v) and (vi), section 13.5) where $(\Delta U_0)_i$ and $(\Delta H_0)_i$ are the internal energies and enthalpies of combustion of these constituents when their oxidation reactions are taken to completion. Hence the maximum cylinder temperature will be reduced. As the temperature falls during the subsequent expansion stroke, dissociation becomes less pronounced, i.e. partial re-combination occurs with some further release of energy. At the final expansion temperature (usually $\leqslant 1100°C$) re-combination will be virtually complete so that the net energy release is not affected. Nevertheless, since heat release occurs at a lower temperature level than in the absence of dissociation, its availability to do useful work is reduced, and, hence, also the efficiency of the cycle (see Chapter 6, section 6.4 on the effect of temperature on efficiency and Chapter 14 on availability). In conventional steady flow combustion processes the effect of dissociation is much less important due to severe limitations on maximum permissible temperatures. However, recently the introduction of magnetohydrodynamics as a means of producing electricity without the aid of a steam or gas turbine cycle, and without the use of alternators, has led to great interest in the very high speed flow of dissociated gases at very high temperature (ref. 13.4).

EXAMPLE 13.8

A petrol engine burns a stoichiometric benzene-air mixture trapped in the cylinder at 100°C and 1 bar. Find the composition of the products at the maximum temperature and show that this is approximately 2 690°C, given the mean molar specific heats at constant volume as in example 13.7. The combustion may be assumed to be at constant volume and the the volumetric compression ratio is 5:1. The dissociation constants $(K_p)_1$ and $(K_p)_3$ are as follows—

$$\log_{10}(K_p)_1 = 8{\cdot}593 - \frac{28\,600}{T}$$

$$\log_{10}(K_p)_3 = 1{\cdot}2065 - \frac{1\,100}{T}$$

where $(K_p)_1 = \dfrac{(p_{CO})^2}{(p_{CO_2})^2}\,p_{O_2}$ and $(K_p)_3 = \dfrac{p_{CO}\,p_{H_2O}}{p_{CO_2}\,p_{H_2}}$

and all pressures are in bars.

Note: This is the same example as 13.7 with the difference that the product composition is not given but has to be determined from the dissociation constants.

SOLUTION

Combustion equation

$$C_6H_6 + 7{\cdot}5\ O_2 + 28{\cdot}2\ N_2$$
$$= a\,CO_2 + b\,CO + c\,H_2O + d\,H_2 + e\,O_2 + 28{\cdot}2\ N_2 \quad \text{(i)}$$

or $36{\cdot}2$ kg-mol $= (a + b + c + d + e + 28{\cdot}2) = n$ kg-mol

The unknown mol numbers *a, b, c, d, e* may be determined by solving five simultaneous equations, three of which are solved by equating numbers of C, H and O atoms on the two sides of equation (i), and the remaining two by substitution in the two dissociation constants $(K_p)_1$ and $(K_p)_3$, replacing partial pressure fractions by kg-mol fractions.

$$\text{For C} \quad a + b = 6 \tag{ii}$$

$$\text{For H} \quad 2c + 2d = 6 \tag{iii}$$

$$\text{For O} \quad 2a + b + c + 2e = 15 \tag{iv}$$

$$(K_p)_1 = \frac{(p_{CO})^2}{(p_{CO_2})^2} p_{O_2} = \frac{b^2}{a^2} p_{O_2} = \frac{b^2}{a^2} \frac{e}{\sum_P n_i} p_{max}$$

This equation may be expressed more conveniently as follows.

From the generalized perfect gas equation (10.12) in terms of the trapped conditions $p_t = 1$ bar, $T_t = 100°C = 373$ K, but including the change in mol number from 36·2 kg-mol to n_t we have

$$p_{max} = 1 \times \frac{T_{max}}{373} \times 5 \times \frac{\sum_P n_i}{36{\cdot}2}$$

i.e.

$$(K_p)_1 = \frac{b^2}{a^2} \frac{e}{\sum_P n_i} \frac{T_{max}}{373} \times 5 \times \frac{\sum_P n_i}{36{\cdot}2}$$

$$= \frac{b^2}{a^2} e \frac{2690 + 273}{373} \times \frac{5}{36{\cdot}2} = 1{\cdot}095 \frac{b^2}{a^2} e \tag{v}$$

where $\log_{10}(K_p)_1 = 8{\cdot}593 - \dfrac{28\,600}{2690 + 273} = -1{\cdot}057$

i.e. $(K_p)_1 = 0{\cdot}0879$

Finally

$$(K_p)_3 = \frac{p_{CO} p_{H_2O}}{p_{CO_2} p_{H_2}} = \frac{bc}{ad} \tag{vi}$$

where $\log_{10}(K_p)_3 = 1{\cdot}2065 - \dfrac{1100}{2690 + 273} = +\,0{\cdot}8365$

i.e. $(K_p)_3 = 6{\cdot}85$

Equations (ii) to (vi) may now be solved for the unknowns *a, b, c, d, e* by any convenient trial-and-error or other numerical procedure. Usually a value for a is assumed, substituted in equations (ii) to (v) to solve for the remaining unknowns, which in turn are substituted in equation (vi) as a compatibility check.

Thus, assuming $a = 4{\cdot}50$

From (ii) $b = 1{\cdot}50$

From (v) $$e = \frac{0{\cdot}0879}{1{\cdot}095}\frac{a^2}{b^2} = \frac{0{\cdot}0879}{1{\cdot}095}\left(\frac{4{\cdot}5}{1{\cdot}5}\right)^2 = 0{\cdot}82$$

From (iv) $c = 15 - 9 - 1{\cdot}5 - 1{\cdot}64 = 2{\cdot}86$

From (iii) $d = 3 - 2{\cdot}86 = 0{\cdot}14.$

Substituting in equation (vi)

$$6{\cdot}85 = \frac{bc}{ad} = \frac{1{\cdot}5 \times 2{\cdot}86}{4{\cdot}5 \times 0{\cdot}14} = 6{\cdot}85$$

i.e. balance is obtained, so that the solutions are satisfactory, provided the assumed temperature $T_{max} = 2690$°C is itself sufficiently accurate.

This check is obtained by the method already outlined in example 13.7 which is numerically identical with the present example, i.e. by equating the increase in sensible internal energy

$$\Delta U_{12} = \sum_P n_i(C_v)_i(T_2 - T_1)$$

to the combined heat and work input to the system, the former in turn being given by the expression

$$-\Delta U' = -\Delta U_0 + \sum_P n_i(\Delta U_0)_i$$

where $-\Delta U_0$ = internal energy of combustion/kg-mol C_6H_6

$(\Delta U_0)_i$ = internal energy of combustion/kg-mol of unburnt constituents.

If the necessary balance is not obtained the entire procedure has to be repeated, with a new value of the assumed maximum temperature T_{max} and a repetition of the trial and error procedure for the solution of the kg-mol numbers a, b, c, d, e.

It is evident that the methods employed to evaluate energy release and related quantities in combustion reaction involving dissociation are extremely lengthy and tedious. The advent of digital computers has greatly facilitated the solution of these problems, but reference must be made to the pioneer work of Pye (ref. 13.5) before World War II in connexion with performance studies of petrol engines.

Finally, it must be emphasized that, in spite of its length, the treatment outlined above is still not fully comprehensive. At very high temperatures nitrogen ceases to be inert and undergoes an oxidation reaction to form oxide

$$N_2 + O_2 \rightleftharpoons 2\,NO$$

while water vapour undergoes a dissociation reaction into hydrogen and hydroxyl

$$H_2O \rightleftharpoons \tfrac{1}{2}H_2 + OH$$

in addition to the dominant reaction

$$H_2O \rightleftharpoons H_2 + \tfrac{1}{2}O_2$$

These additional reactions involve the introduction of further dissociation constants.

The very important effect of dissociation on the maximum temperature of reciprocating spark ignition (petrol) engines is illustrated by *Fig.* 13.11 (ref. 13.3).

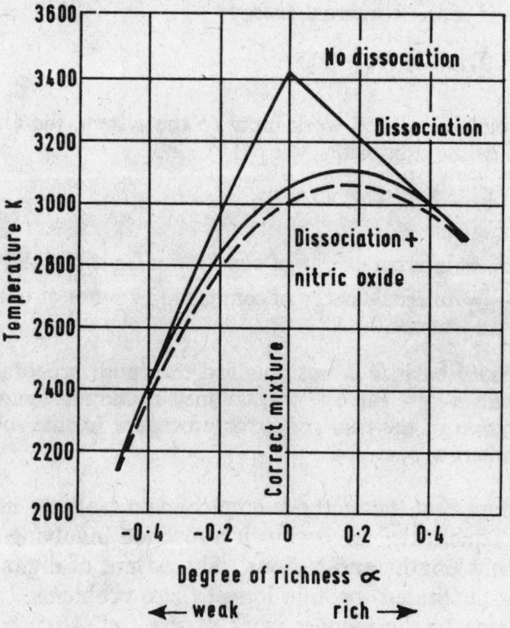

Fig. 13.11 Effect of dissociation on I.C. engine performance (Ricardo)

13.7 Second Law of Thermodynamics Applied to Combustion Processes

In section 13.3 a first law analysis was applied to the combustion process, introducing the concept of internal energy U' and enthalpy H' of a chemical substance to include the energy released by the chemical reaction undergone by the reactants, in addition to the sensible internal energy U and enthalpy H, i.e. energy in virtue of temperature only. Assuming complete combustion of the reactants, i.e. zero chemical energy of the products, we expressed the internal energy and enthalpy changes for an isothermal reactant at temperature T_0 as

$$\Delta U_0 = (U_P)_0 - (U'_R)_0 \qquad \text{(i)}$$

and

$$\Delta H_0 = (H_P)_0 - (H'_R)_0 \qquad \text{(ii)}$$

In Chapter 7, section 7.6, equation (7.23), we showed that for any real process in a non-reacting system

$$dQ \leqslant T\,dS \qquad (13.30)$$

We may extend this important statement to reacting systems by introducing the symbol S' to denote the entropy of a reacting substance, by analogy with U' and H'.

Hence adapting equation (13.30) to a reacting system

$$dQ' \leqslant T\,dS' \qquad (13.30a)$$

Now for the special cases of constant temperature, constant volume closed system combustion, equation (13.11), and constant temperature, constant pressure, steady flow open system combustion, equation (13.13), we may write equation (13.30a) in the form

$$Q_{R_0-P_0} \leqslant T_0[(S_P)_0 - (S'_R)_0] \qquad (13.30b)$$

Comparing equation (13.30b) with equations (13.11) and (13.13) we immediately obtain the following results

$$\Delta U_0 - T_0[(S_P)_0 - (S'_R)_0] \leqslant 0$$

i.e.

$$[(U_P)_0 - T_0(S_P)_0] - [(U'_R)_0 - T_0(S'_R)_0] \leqslant 0 \qquad (13.31)$$

for the closed system constant temperature and volume process, or

$$\Delta H_0 - T_0[(S_P)_0 - (S'_R)_0] \leqslant 0$$

$$[(H_P)_0 - T_0(S_P)_0] - [(H'_R)_0 - T_0(S'_R)_0] \leqslant 0 \qquad (13.32)$$

for the open system constant temperature and pressure process.

Hence an examination of known entropies, internal energies or enthalpies of reacting substances with the aid of the inequalities (13.31) and (13.32) makes it possible to predict the possibility, or otherwise, of certain hypothetical chemical reactions.

Similarly, by combining equation (13.30b) with the general first law statement for a combustion reaction, in which the possibility of useful work was allowed for, viz. equations (13.10) for the closed system and equation (13.12) for the steady flow open system, we obtain the following relationships—

$$W_{R_0-P_0} \leqslant [(U_P)_0 - (U'_R)_0] - T_0[(S_P)_0 - (S'_R)_0] \qquad (13.33)$$

for the closed system, constant temperature process and

$$W_{R_0-P_0} \leqslant (H_P)_0 - (H'_R)_0] - T_0[(S_P)_0 - (S'_R)_0] \qquad (13.34)$$

for the open system, constant temperature process.

Hence it is possible to predict the maximum work which can be extracted from any specified combustion process.

The general problem of the prediction of maximum work or availability for cyclic and non-cyclic processes in closed and open systems, with and without chemical reaction, is discussed in detail in the next chapter (Chapter 14). Combustion processes, in particular, are discussed in section 14.7.

REFERENCES

13.1 Keenan, J. H. and Kaye, J., *Gas Tables* (John Wiley, 1948).

13.2 Geyer, E. W. and Bruges, E. A., *Tables of Properties of Gases with Dissociation Theory and Its Applications* (Longmans, 1948).

13.3 Ricardo, Sir Harry, *The High-speed Internal Combustion Engine* (Blackie).

13.4 Smith, I. E., *The Effect of Dissociation in Rockets, Ramjets and Wind Tunnels*, Thermodynamics and Fluids Mechanics Convention, Cambridge, 1964, Paper 28 (Inst.Mech.Eng., London).

13.5 Pye, Sir David, R., *The Internal Combustion Engine*, Vol. I, Principles (Oxford University Press, 1937).

EXERCISES

13.1 A natural gas having the volumetric composition, 77·5% C_2H_6, 15% CH_4, and 6·5% CO_2, is burned at 15·5°C and 1 bar pressure. Determine

(*a*) the volume of air for complete combustion per cubic metre of fuel,

(*b*) the volumetric analysis of the dry products of combustion.

(*Ans:* (*a*) 14·5 m³/m³ of fuel; (*b*) 13·4% CO_2, 86·6% N_2)

13.2 A high volatility coal having the gravimetric analysis, ash 5·6%, sulphur 1%, hydrogen 5·9%, carbon 66%, nitrogen 1·5%, and oxygen 19·9%, is burned with 30% of excess air. Determine

(*a*) the mass of air required per kg of fuel,
(*b*) the volumetric analysis of the products of combustion.
(*Ans:* (*a*) 11·45 kg/kg of fuel,
(*b*) 13·2% CO_2, 7·1% H_2O, 75% N_2, 4·7% O_2)

13.3 Gas having the volumetric composition: 2% H_2, 2% CO, 92% CH_4, 2% O_2, 2% N_2; is burned with excess air to give the following Orsat analysis of the dry products: 10·5% CO_2, 2% O_2, 87·5% N_2. Determine

(*a*) the mass of dry products per kg of fuel,
(*b*) the mass of air supplied per kg of fuel,
(*c*) the percentage excess air supplied.
(*Ans:* (*a*) 16·1 kg dry products/kg of fuel; (*b*) 17·2 kg air/kg of fuel;
(*c*) 11·7%)

13.4 Determine (*a*) the mass of air required, and (*b*) the volumetric analysis of the *dry* products of combustion when 1 kg of octane (C_8H_{18}) is burned completely in air.
(*Ans:* (*a*) 15·2 kg air/kg of fuel; (*b*) 14·55% CO_2, 85·45% N_2)

13.5 Natural gas of percentage volumetric composition, CH_4 95·0, H_2 1·5, N_2 3·5, is used as fuel for a furnace. What is the stoichiometric air required for combustion and the percentage CO_2 in the dry gas when 40% excess air is supplied?
(*Ans:* 9·08 m^3 air/m^3 gas; %CO_2 = 8·05)

13.6 The analyses of the fuel oil and the exhaust gas derived from an oil-engine test are respectively as follows:

Fuel oil (by weight)		*Exhaust gas* (by volume)	
C	85·8%	CO_2	6·38%
H_2	12·9%	O_2	12·09%
Residue	1·3%	CO	—
		N_2	81·53%

Taking the volumetric analysis of air to be 21% O_2 and 79% N_2, estimate the air consumption of the engine per kilogramme of fuel burnt.
(*Ans:* 33·1 kg air/kg fuel)

13.7 Describe briefly how you would analyse the products of combustion in a boiler flue.

The analysis of a sample of coal taken during a boiler trial is as follows: C 76%, H_2 4%, O_2 9%, the remainder was ash. Analysis of the dry flue gas was as follows: CO_2 13·82%, O_2 6·92%, N_2 79·26% by volume. Calculate the excess air supplied.
(*Ans:* 48·6%)

13.8 The dry exhaust from a spark ignition engine at 1 bar pressure has a volumetric analysis: 12·5% CO_2, 3·1% O_2, 0·3% CO, 84·1% N_2.

(*a*) Assuming the fuel to be a hydrocarbon of the form C_xH_y, find the values of x and y.

Find also

(*b*) the air fuel ratio,
(*c*) the partial pressure of water vapour in the hot exhaust.

(*Ans:* (*a*) $C_{12\cdot8}H_{26\cdot6}$; (*b*) 17·1 kg air/kg fuel; (*c*) 0·119 bar)

13.9 The internal energy of combustion of carbon burning to carbon dioxide at 20°C is −32 650 kJ/kg of carbon. What is the heat transferred when 1 kg of carbon and 4 kg of oxygen at 38°C combine to form CO_2 and O_2 at 205°C, at constant volume?
(*Ans:* −32210 kJ/kg)

13.10 Liquid octane (C_8H_{18}) is burned in a steady flow combustion chamber with 100% excess air, the former entering at 93°C, and the latter at 871°C. Calculate

(*a*) The heat released per kg of fuel in the absence of kinetic energy and work effects,
(*b*) the relationship between heat transferred, work done, and enthalpy change when kinetic energy and work effects are present.

The enthalpy of C_8H_{18} burning to CO_2 and H_2O (vapour) is 5 121 187 kJ/kg-mol and the mean c_p values (kJ/kg-mol-K) for products and reactants are:
between 25°C and 871°C:

CO_2: 44·079 H_2O: 40·662 O_2: 32·988

between 25°C and 93°C:

C_8H_{18}: 213·44 O_2: 29·965

(*Ans:* (*a*) −4 241 368 kJ/kg-mol C_8H_{18})

13.11 Given that the enthalpy of combustion of gaseous benzene at 25°C is −3 175000 kJ/kg-mol, and its latent heat of evaporation at the same temperature is 34050 kJ/kg-mol, determine

(*a*) its enthalpy of combustion at 540°C with benzene and water in vapour form,
(*b*) its internal energy of combustion at 25°C with benzene in vapour form and water in liquid form,
(*c*) its internal energy of combustion at 540°C with benzene and water in vapour form.

Note: the specific heat table given in connection with worked example 13.5 (p. 354) should be used.

(*Ans:* (*a*) −3 209 327 kJ/kg-mol; (*b*) −3296035 kJ/kg-mol:
(*c*) −3230736 kJ/kg-mol)

13.12 In a constant pressure i.c. engine cycle the temperature at the end of compression is 855 K. Fuel is then burnt at constant pressure with 100% excess air. Find the temperature at the end of combustion if the fuel octane (C_8H_{18}) is injected at 27°C.

Lower calorific value of octane at 25°C = 44 450 kJ/kg
Latent heat of evaporation of octane at 25°C = 3470 kJ/kg
(*Ans:* 1925°C)

13.13 Determine the temperature and pressure resulting from constant volume adiabatic combustion of CO with 80% excess air, when the reactants are initially at 298 K and 1·033 bar, neglecting dissociation.
(*Ans:* 2280 K, 7·15 bar)

13.14 A hydrocarbon fuel of composition C_8H_{18} is burned with 20% excess air. Determine the equilibrium composition of the products at 5 bar, on a volumetric basis,

(*a*) at 1670 K,
(*b*) at 2500 K.

(*Ans:* (*a*) 10·52% CO_2; 11·85% H_2O; 3·29% O_2; 74·34% N_2
(*b*) 0·75% CO; 0·15% H_2; 9·74% CO_2; 11·65% H_2O; 3·72% O_2; 74·1% N_2)

13.15 Calculate the internal energy of combustion of octane (C_8H_{18}), the combustion starting at 394°C. Assume gaseous fuel and gaseous products. The internal energy of combustion for this fuel starting and finishing combustion at 25°C as gaseous fuel and gaseous products and at constant volume is −5145000 kJ/kg-mol. The internal energy in kJ/kg-mol is as follows for the gases and vapours listed reckoned from 15°C.

TEMP. °C	C_8H_{18}	O_2	N_2	CO_2	H_2O
394	91120	8450	8055	13360	10120
25	1491	198	192	275	242

If this fuel burns in a 4-stroke engine cylinder together with 25% excess air, and the temperature of 394°C represents conditions at the start of the constant volume combustion, estimate the temperature at the end of the combustion period neglecting heat losses and dissociation. Take the mean volumetric heat of the products 3222 kJ/kg-mol.

What is the percentage volumetric analysis of the products?

(*Ans:* 1517°C CO_2, 11·45%, O_2, 4·47%, N_2, 84·08%)

13.16 The combustion of a chemically correct air-benzene mixture can be represented by the equation

$$C_6H_6 \text{ (vapour)} + 7{\cdot}5O_2 + 28{\cdot}2N_2$$
$$= 6CO_2 + 3H_2O \text{ (vapour)} + 28{\cdot}2N_2.$$

1. If the combustion is at constant pressure and the temperature of the initial mixture and of the products after combustion is 18°C, the enthalpy of combustion is −3168000 kJ/kg-mol of benzene.

Determine the enthalpy of combustion per mol of benzene under the following conditions:

2. At constant pressure at 18°C: C_6H_6 and H_2O as liquids.
3. At constant pressure: C_6H_6 as liquid at 18°C; air and products at 100°C; H_2O as vapour.
4. At constant volume and 18°C: C_6H_6 as vapour; H_2O as vapour.
5. At constant volume and 18°C: C_6H_6 as liquid; H_2O as liquid.
6. At constant volume: C_6H_6 as liquid at 18°C; air and products at 100°C; H_2O as vapour.

Take the mean volumetric heat of N_2 between 18°C and 100°C as 19·880 kJ/kg-mol-K at constant pressure.

(*Ans:* 3269824; 3064277; 3172292; 3266231; 3093292 kJ/kg-mol C_6H_6)

13.17 Show that when iso-octane (C_8H_{18}) is burned adiabatically 1 at bar pressure with 80% of stoichiometric air, with air and fuel supplied at 25°C, the maximum temperature reached is 1930°C.

The lower calorific value of iso-octane may be taken as $5{\cdot}11 \times 10^6$ kJ/kg-mol.

13.18 A compression ignition engine operates on the constant volume cycle with a volumetric compression ratio of 22:1, the trapped conditions being,

$$p_1 = 1{\cdot}015 \text{ bar}$$
$$T_1 = 100°\text{C}.$$

Given the dissociation constants,

$$\left.\begin{aligned} (K_p)_1 &= \frac{(p_{CO})(p_{H_2O})}{(p_{CO_2})(p_{H_2})} = 6{\cdot}45 \\ (K_p)_2 &= \frac{(p_{CO})^2(p_{O_2})}{(p_{CO_2})^2} = 0{\cdot}019 \end{aligned}\right\} \text{ at } 2500°\text{C}$$

and that the hydrocarbon fuel of general form C_6H_6 is burnt with air resulting in an air-fuel "mixture" which is 30% weak, show that the molar composition of the products of combustion given below is consistent with a maximum cycle temperature of 2 500°C.

CO_2: 9·05 mols
CO: 0·35 mols
H_2O: 4·172 mols
H_2: 0·028 mols
O_2: 4·550 mols
N_2: 56·400 mols

You may assume the following:

Index of compression: $n_c = 1{\cdot}35$.

Mean specific heat of products between 100°C and 2 500°C at constant volume $c_v = 29{\cdot}5$ kJ/kg-mol of original mixture.

Heat release, allowing for dissociation:
$Q = 53\,500$ kJ/kg-mol of original mixture.

Universal gas constant $G = 8{\cdot}3143$ kJ/kg-mol-K.

Show further that the temperature at the end of expansion, neglecting recombination, is 950°C, given that the mean specific heat of the products of combustion between 100°C and 950°C is 25·08 kJ/kg-mol of original mixture.

13.19 A mixture, having a volumetric analysis of CO_2 10%, CO 40% and air 50%, is contained in a vessel of fixed capacity at a temperature 0°C. When the mixture is exploded by a spark, estimate

(*a*) the gravimetric composition of the products, and
(*b*) the temperature they would reach

assuming that no dissociation takes place, that the process is adiabatic, and that the specific heats of the gases are constant. Take c_v as 1·050 (CO_2), 1·075 (CO), 0·988 (O_2) and 1·068 (N_2) kJ/kg-K, and the enthalpy of combustion of CO at 25°C as −10 100 kJ/kg.

(*Ans:* CO_2, 45·45%; CO, 17·75%; N_2, 36·80% by weight. 1 860°C)

Chapter 14

Availability

14.1 First and Second Law Analysis of Cycles

The first and second laws of thermodynamics discussed respectively in Chapters 4 and 6 both apply to cyclic processes although important deductions may be made from both laws concerning energy changes in non-cyclic processes. Thus the first law statement (see equation (4.2))

$$\sum \delta W = \sum \delta Q \tag{14.1}$$

was used to establish the important property of energy leading to the non-flow energy equation N.F.E.E.

$$Q_{12} - W_{12} = \Delta E_{12} \tag{14.2}$$

while the second law statement (see equation (7.1a))

$$\oint \frac{dQ}{T} = 0 \tag{14.3}$$

led to the establishment of entropy as a new property and of the entropy equation (7.8a) and (7.8b) governing non-cyclic processes, i.e.

$$T\Delta s_{12} = p\Delta v_{12} + \Delta u_{12} \tag{14.4a}$$

$$= \Delta h_{12} - v\Delta p_{12} \tag{14.4b}$$

In section 4.8 an outline was given of the so-called FIRST LAW ANALYSIS of a complete cycle, in the particular case given a steam cycle incorporating a series of elements each subject to the steady flow energy equation (S.F.E.E.). By summing up the useful work extracted from the cycle, on the one hand, and the gross heat transferred to the cycle on the other, a thermal conversion efficiency was established, equation (4.21), defined as

$$\eta_{th} = \frac{\sum \delta W}{\sum \delta Q_{gross}} \tag{14.5}$$

where $\sum \delta W$ = sum of all positive and negative work transfer terms,

$$\sum \delta Q_{gross} = \text{sum of } \textit{positive} \text{ heat transfer terms only.}$$

[Clearly the expression $\sum \delta W / \sum \delta Q$ would always be unity, as a direct outcome of the first law statement, equation (14.1).]

All assessments of cycle efficiency are based on expressions such as equation (14.5). The information provided by such efficiency expressions is extremely valuable, our efforts in the field of applied thermodynamics being directed continually towards improvements in the processes governing the conversion of heat into work. We know, for instance, that compression ignition or diesel engines operate with very much higher thermal efficiencies than gas turbines, for reasons connected partly with temperature limits, partly with the sequence of processes employed in the cycle, but most importantly with the nature of the various irreversibilities occurring in the cycle.

A first law analysis, by its very nature, cannot provide useful information concerning the latter, while a SECOND LAW, or "AVAILABILITY" ANALYSIS concerns itself particularly with the nature of the various irreversibilities.

14.2 Available Energy

Sources of energy may be divided into two groups, viz. high and low grade energy, depending on whether in theory they may, or may not, be transformed into work without loss. In practice, certain losses are unavoidable even when dealing with conversion of high grade energy, but it is most important to make the above *theoretical* distinction.

Examples of the two kinds of energy are

High grade energy	*Low grade energy*
(1) Mechanical work	(1) Heat or thermal energy
(2) Electrical energy	(2) Heat derived from nuclear fission or fusion
(3) Water power	(3) Heat derived from combustion of fossil fuels
(4) Wind power	
(5) Kinetic energy of a jet	

This distinction may also be expressed by the statement that conversion of high grade energy to mechanical work is exempt from the limitations of the second law, while conversion of low grade energy is subject to them.

In the case of low grade energy, that part which is available for conversion is referred to as AVAILABLE ENERGY, while the part which, according to the second law, must be rejected to a heat sink, is known as UNAVAILABLE ENERGY. Any misuse of available energy constitutes an irreversibility, so that clearly an analysis concerned particularly with availability will necessarily highlight departures from ideal processes, and thereby suggest possible improvements to existing cycles.

Typical examples of irreversibilities in the case of high grade energy conversion are

(1) Mechanical work conversion, i.e. from high speed to low speed shaft or vice versa: frictional losses in gearing.
(2) Electrical energy conversion: copper and iron losses in electric motor.
(3) Water power: frictional losses in ducting, stator and rotor of water turbine.
(4) Wind power: aerodynamic losses in blading of windmill or similar device.
(5) Kinetic energy of jet: frictional losses during nozzle or within blading of impulse turbine.

In the case of low grade energy a similar list may be drawn up—

(1) Heat or thermal energy: irreversibilities arising from heat transfer.
(2) Nuclear fission or fusion: inherent irreversibility of fission or fusion process coupled with those associated with heat transfer.
(3) Combustion of fossil fuels: inherent irreversibilities of combustion process coupled with those associated with heat transfer.

In the following treatment, attention will be confined to low grade energy conversion processes.

14.3 Availability in Cyclic Processes

(*a*) CARNOT CYCLE—INFINITE HEAT RESERVOIR (*Fig.* 14.1)

Since the Carnot cycle constitutes the most efficient cyclic process for the conversion of heat into work, and its efficiency is given by

$$\eta_{th} = 1 - \frac{T_0}{T} \qquad (14.6)$$

(equation (6.5))
where $T =$ temperature of infinite heat reservoir, $T_0 =$ temperature of surroundings, the available part of the heat transferred to the cycle or AVAILABILITY is given by

$$\Delta A = Q_{12}\left(1 - \frac{T_0}{T}\right) = T\Delta S_{12}\left(1 - \frac{T_0}{T}\right) \qquad (14.7)$$

or, in general terms, omitting the suffix 12

$$\Delta A = (T - T_0)\Delta S \qquad (14.7a)$$

$$= (Q_{rev}) - T_0\Delta S \qquad (14.7b)$$

(*b*) REVERSIBLE CYCLE WITH FINITE HEAT RESERVOIR

The significance of the infinite heat reservoir in the case of the Carnot cycle is that no temperature change takes place during the heat transfer process 1–2 (*Fig*. 14.1).

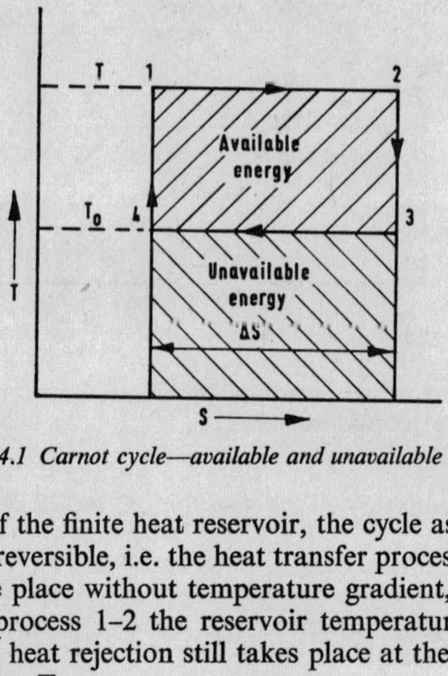

Fig. 14.1 Carnot cycle—available and unavailable energy

In the case of the finite heat reservoir, the cycle as a whole is still assumed to be reversible, i.e. the heat transfer processes 1–2 and 3–4 (*Fig*. 14.2) take place without temperature gradient, but during the heat addition process 1–2 the reservoir temperature falls from T_1 to T_2, although heat rejection still takes place at the temperature of the surroundings, T_0.

The cycle may be treated as a succession of Carnot cycles, the availability following directly from equation (14.7), i.e.

$$\Delta A = \int_1^2 \left(1 - \frac{T_0}{T}\right) dQ = \int_1^2 dQ_{rev} - \int_1^2 \frac{T_0\, T\, dS}{T}$$

$$= \int_1^2 T\, dS - T_0\Delta S_{12} \qquad (14.8)$$

EXAMPLE 14.1

A Carnot engine is supplied by a heat reservoir initially at 538°C and having a water equivalent of 227 kg. Find the loss of availability of the heat source if heat abstraction continues down to 260°C, and compare it with the total heat abstracted.

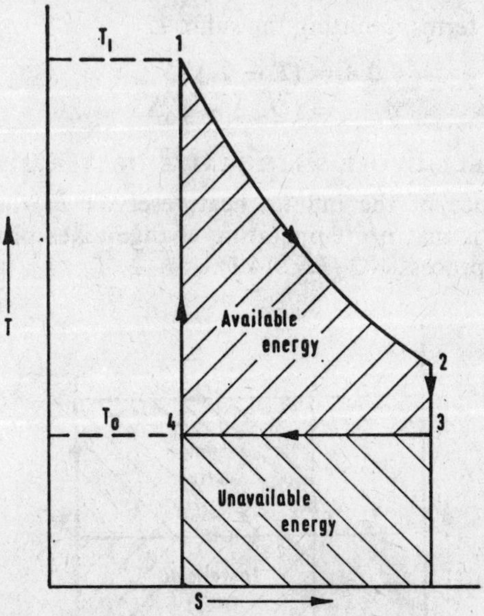

Fig. 14.2 Cycle with finite heat reservoir—available and unavailable energy

The temperature of the surroundings is 15°C and the heat transfer process may be considered reversible.

SOLUTION

Since the water equivalent of the reservoir $M_w = 227$ kg,

$$\frac{dQ}{dT} = 227 \text{ kcal/K} = 227 \times 4{\cdot}1868 \text{ kJ} = 950 \text{ kJ} \qquad \text{(i)}$$

Since the heat transfer process as such is reversible, there is no temperature gradient between the reservoir and the engine, i.e. equation (14.8) applies.

From (i),

$$dS = \frac{dQ}{T} = \frac{950dT}{T} \qquad \text{(ii)}$$

$$\Delta A_{12} = \int_1^2 T\,dS - T_0 \Delta S_{12}$$

$$= \int_{538+273}^{260+273} 950dT - (15 + 273) \int_{538+273}^{260+273} \frac{950}{T}\,dT$$

$$= -950 \times 278 - 288 \times 950 \ln(533/811)$$

$$= -264\,100 - 273\,500(-0{\cdot}420)$$

$$= 149\,000 \text{ kJ} \quad \textit{Ans.}$$

Total quantity of heat abstracted, from (i)

$$Q_{12} = \int_{811}^{533} 950\, dT$$

$$= -950 \times 278 = -264000 \text{ kJ} \quad \textit{Ans.}$$

14.4 Availability in Non-cyclic Closed System Processes

Consider the closed system shown (*Fig.* 14.3). This system can exchange energy with the surroundings by both work and heat transfer until the system has attained the "dead state" of the surroundings, i.e. pressure p_0 and temperature T_0.

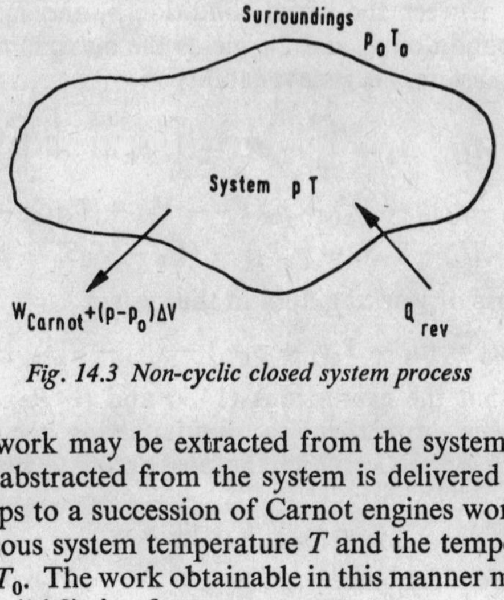

Fig. 14.3 Non-cyclic closed system process

Maximum work may be extracted from the system by assuming that the heat abstracted from the system is delivered in an infinite number of steps to a succession of Carnot engines working between the instantaneous system temperature T and the temperature of the surroundings T_0. The work obtainable in this manner may be derived from equation (14.8), i.e. for any one step

$$dW_{carnot} = \left(1 - \frac{T_0}{T}\right)(-dQ) \qquad \text{(i)}$$

(*Note* that in relation to the system dQ is negative.)

In addition, if the system undergoes a volume change dV the effective work done by the system *on* the surroundings becomes

$$dW_{disp} = (p - p_0)\, dV \qquad \text{(ii)}$$

(See equation (2.8).)

Hence the work transferred *from* the system *to* the surroundings in such a step is given by adding equations (i) and (ii)

$$dW = -\left(1 - \frac{T_0}{T}\right) dQ + p\,dV - p_0\,dV$$

and writing $dS = dQ/T$ for the *reversible* heat transfer process

$$dW = -dQ + T_0\,dS + p\,dV - p_0\,dV \tag{iii}$$

Finally, since for a reversible closed system process the N.F.E.E. assumes the form

$$dQ = dU + p\,dV \tag{iv}$$

(equation (4.4b))
substituting for dQ from equation (iv) in equation (iii) leads to

$$dW = -dU - p_0\,dV + T_0\,dS \tag{v}$$

Integration between the initial condition p_1 and T_1, and the final "dead state" condition p_0 and T_0, yields the maximum work obtainable from the system, i.e. its availability

$$\begin{aligned} W_{max} = A_1 - A_0 &= \int_1^0 -dU - \int_1^0 p_0\,dV + \int_1^0 T_0\,dS \\ &= -(U_0 - U_1) - p_0(V_0 - V_1) + T_0(S_0 - S_1) \\ &= (U_1 - T_0S_1 + p_0V_1) - (U_0 - T_0S_0 + p_0V_0) \end{aligned} \tag{14.9}$$

or for unit mass of working fluid in the system

$$w_{max} = a_1 - a_0 = (u_1 - T_0s_1 + p_0v_1) - (u_0 - T_0s_0 + p_0v_0) \tag{14.9a}$$

It is clear that the expressions (14.9) and (14.9a) represent differences between properties, so that applying the mathematical property check of section 2.13 the availability a itself becomes a property, i.e.

$$a_1 = u_1 - T_0s_1 + p_0v_1 \tag{14.9b}$$

We may therefore conclude

THE AVAILABLE ENERGY OF THE CLOSED SYSTEM IS A PROPERTY KNOWN AS THE AVAILABILITY.

The function

$$a = u - T_0s + p_0v \tag{14.9c}$$

is known as the NON-FLOW AVAILABILITY FUNCTION.

The property function

$$f' = u - T_0s \tag{14.10a}$$

incorporated in equations (14.9) is a special case of the so-called HELMHOLTZ FREE ENERGY FUNCTION

$$f = u - Ts \tag{14.10b}$$

(see equation (14.47)).

Equation (14.9), together with the principle of increase of entropy, i.e. corollary 7 of the second law (section 7.6) lead to a further most important statement:

THE AVAILABILITY DECREASES IN ALL REAL (I.E. IRREVERSIBLE) PROCESSES.

This statement may be proved as follows. From equation (14.9a) the loss of availability of system for any process 1–2 is given by

$$A_2 - A_1 = (U_2 - U_1) - T_0(S_2 - S_1) + p_0(V_2 - V_1) \tag{14.9c}$$

Equation (14.9c) is quite general, but when applied to an *isolated* system, $(S_2 - S_1)$ must represent an increase of entropy due to an INTERNAL IRREVERSIBILITY.

From the N.F.E.E. (4.4b) for a closed, isolated system ($dQ = 0$)

$$-dW = dU \tag{i}$$

where dW is the gross work done by the system.

The *effective* or *useful* work, however, is given by

$$dW_{useful} = dW - p_0\,dV \tag{ii}$$

whence, combining equations (i) and (ii)

$$(dW_{useful})_{1-2} = -(U_2 - U_1) - p_0(V_2 - V_1) \tag{iii}$$

But the combined change in availability of the system and its surroundings is equal to the sum of the loss of availability of the former, given by equation (14.9c) and the gain of availability of the latter in the form of useful work done on it given by equation (iii), so that the net change in availability obtained by adding these two equations

$$\begin{aligned}(\Delta A_{net})_{1-2} &= (U_2 - U_1) - T_0(S_2 - S_1) + p_0(V_2 - V_1)\\ &\quad - (U_2 - U_1) - p_0(V_2 - V_1)\\ &= -T_0(S_2 - S_1)\end{aligned} \tag{14.11}$$

But from corollary 7 of the second law

$$S_2 - S_1 > 0$$

for the isolated system, i.e.

$$(\Delta A_{net})_{1-2} < 0 \tag{14.11a}$$

proving that the availability decreases in all real processes.

The irreversibility of a non-flow process may be obtained as follows—

Expressing the N.F.E.E. (4.4b) in terms of *useful* work on the surroundings where

$$dW_{useful} = dW - p_0\, dV \qquad \text{(i)}$$

we obtain, after differentiation

$$dQ - (dW_{useful} + p_0\, dV) = dU \qquad \text{(ii)}$$

Combining this with the availability equation (14.9b) written in differential form

$$dA = dU - T_0\, dS + p_0\, dV \qquad \text{(iii)}$$

the following equation results

$$dA = dQ - dW_{useful} - T_0\, dS \qquad \text{(iv)}$$

But from the corollary 8 of the second law (section 7.6, equation (7.22b)) i.e. the inequality of Clausius, we have for an infinitesimal irreversible process

$$dS > \frac{dQ}{T} \qquad \text{(v)}$$

so that, substituting from equation (v) in equation (iv) for dQ

$$(dW_{useful})_{actual} < (T - T_0)\, dS - dA \qquad (14.12)$$

the suffix "actual" being used to distinguish the useful work in this case from that in the ideal fully reversible case.

For the limiting case of a fully reversible process, equation (14.12) becomes

$$(dW_{useful})_{ideal} = (T - T_0)\, dS - dA \qquad (14.12a)$$

The irreversibility dI may be expressed quantitatively as the difference between the ideal and actual work as expressed by equations (14.12a) and (14.12) respectively, i.e.

$$dI = (T - T_0)\, dS - dA - (dW_{useful})_{actual} \qquad (14.12b)$$

Finally, substituting for $(dW_{useful})_{actual}$ and dA from equations (ii) and (iii) respectively,

$$dI = (T - T_0)\, dS - (dU - T_0\, dS + p_0\, dV) - (dQ - p_0\, dV - dU)$$
$$= (T\, dS - dQ)$$

or for a finite process 1–2

$$\Delta I_{12} = \int_1^2 T\,dS - Q_{12} \qquad (14.12c)$$

Equation (14.12c) may be regarded as an alternative, quantitative expression for the inequality of Clausius.

EFFECTIVENESS

Just as efficiency is a concept arising from first law analysis of cycles, so effectiveness is a concept appropriate to second law analysis, with the difference that whereas the former applies essentially to cyclic processes only, the latter may be used for both cyclic and non-cyclic processes.

Thus, for an adiabatic non-flow process, the effectiveness may be defined, with reference to equations (14.12) and (14.12a) as

$$\text{Effectiveness } \varepsilon = \frac{\text{actual useful work}}{\text{available useful work}} = \frac{\int_1^2 -(dU + p_0\,dV)}{-\int_1^2 (dA)} \qquad (14.13)$$

where dA is given by equation (14.9c), the entropy gain dS being zero for the ideal process.

Since the actual useful work represents the gain of availability of the surroundings, and the available useful work the loss of availability of the system, we may also write with complete generality

$$\text{Effectiveness } \varepsilon = \left|\frac{\text{gain of availability of surroundings}}{\text{loss of availability of system}}\right| \qquad (14.13a)$$

In this latter form the concept is applicable to processes involving heat transfer, as well as to processes with work transfer or combined work and heat transfer.

AVAILABILITY CHANGES IN SPECIFIC PROCESSES

(see also section 7.4).

Note: All processes are assumed to involve unit mass of working fluid.

(*a*) *Constant volume heating* (*Fig.* 14.4)

For the constant volume heating process 1–2, the change in availability is given by equation (14.9c) putting $v_1 = v_2$, i.e.

$$a_2 - a_1 = (u_2 - u_1) - T_0(s_2 - s_1) \qquad \text{(i)}$$

But for a reversible constant volume process the N.F.E.E. (4.4b) reduces to

$$dq = du = T\,ds \quad \text{or} \quad (u_2 - u_1) = \int_1^2 T\,ds \tag{ii}$$

so that substituting from equation (ii) in equation (i)

$$a_2 - a_1 = \int_1^2 T\,ds - T_0(s_2 - s_1)$$

$$= \int_1^2 (T - T_0)\,ds \tag{14.14}$$

Hence the change in availability is given by the shaded area 1–2–*l*–*m* under the process curve 1–2 in the temperature–entropy

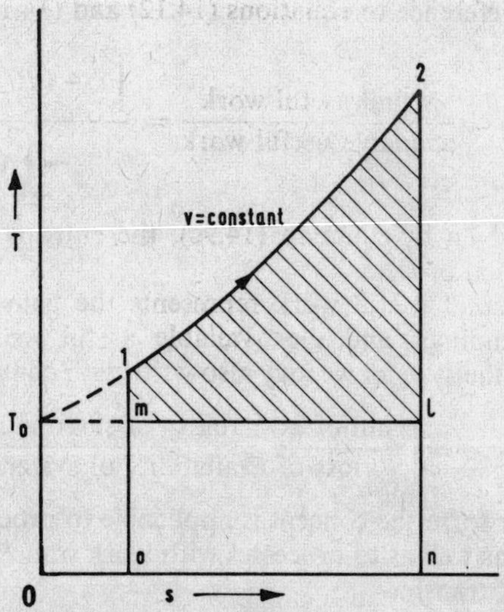

Fig. 14.4 Availability of constant volume heating

diagram, *Fig.* 14.4. Since, evidently, the area 1–2–*n*–*o* under the process curve 1–2 represents the change in internal energy (see also equation (ii)) above,

$$a_2 - a_1 = (u_2 - u_1) - T_0(s_2 - s_1) \tag{14.14a}$$

(*b*) *Constant pressure heating* (*Fig.* 14.5)

Substitution in equation (14.9c) leads to the equation

$$a_2 - a_1 = (u_2 - u_1) - T_0(s_2 - s_1) + p_0(v_2 - v_1) \tag{i}$$

From the definition of enthalpy

$$h = u + pv \qquad \text{(ii)}$$

Substitution from equation (ii) in equation (i) gives

$$a_2 - a_1 = (h_2 - h_1) - T_0(s_2 - s_1) - (p - p_0)(v_2 - v_1) \qquad (14.15)$$

In the special case $p = p_0$ = pressure of surroundings

$$a_2 - a_1 = (h_2 - h_1) - T_0(s_2 - s_1) \qquad (14.15a)$$

So that the change in availability is given by the shaded area 1–2–*l*–*m* under the process curve 1–2 in *Fig.* 14.5, the area 1–2–*n*–*o* under the curve 1–2 representing the change in enthalpy $h_2 - h_1$.

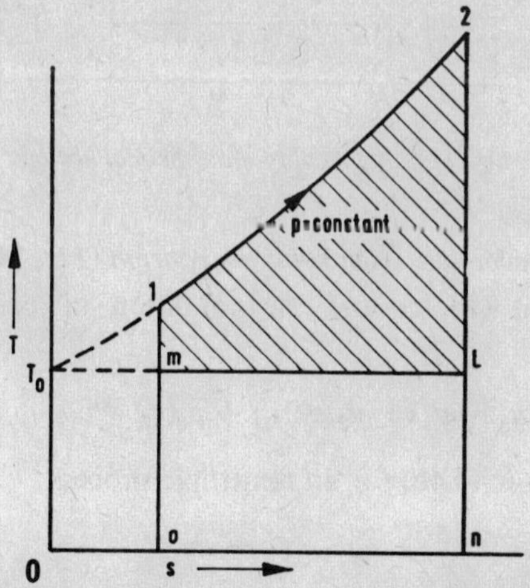

Fig. 14.5 Availability of constant pressure heating

The additional term $(p - p_0)(v_2 - v_1)$ in equation (14.15) represents the work *by* the surroundings *on* the system to maintain the latter in equilibrium.

(c) *Isothermal expansion* (*Fig.* 14.6)
From equation (14.9c)

$$a_2 - a_1 = (u_2 - u_1) - T_0(s_2 - s_1) + p_0(v_2 - v_1) \qquad (14.16)$$

The areas *l*–*m*–*o*–*n* in *Fig.* 14.6*a* and *l*′–*m*′–*o*′–*n*′ in *Fig.* 14.6*b* represent respectively the middle and last terms of equation (14.16).

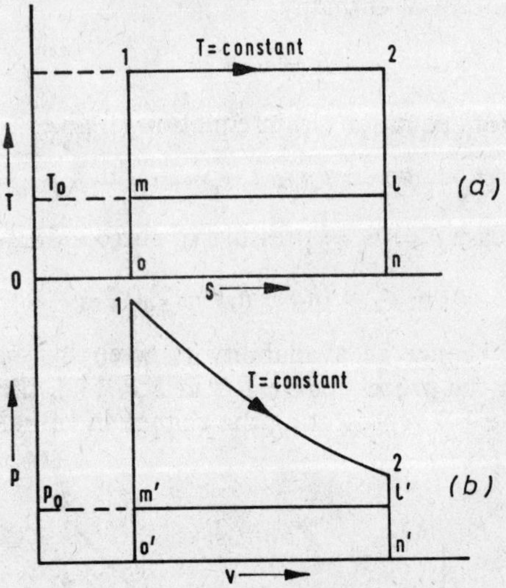

Fig. 14.6 Availability of isothermal heating

(*d*) *Adiabatic reversible* (*isentropic*) *expansion* (*Fig.* 14.7)

From equation (14.9c) and the condition of constant entropy $s_2 = s_1$

$$a_2 - a_1 = (u_2 - u_1) + p_0(v_2 - v_1) \tag{14.17}$$

From the N.F.E.E. (4.4b) for an isentropic process

$$dq - dw = du$$

or

$$0 - \int_1^2 p\,dv = u_2 - u_1 \tag{i}$$

Thus $(u_2 - u_1)$ is represented by the area 1–2–n'–o' in *Fig.* 14.7b, whilst the term $p_0(v_2 - v_1)$ in equation (14.17) is represented by the area l'–m'–o'–n'.

Hence the change in availability for the process 1–2 is obtained from the shaded area 1–2–l'–m' in *Fig.* 14.7*b*.

(*e*) *Adiabatic irreversible expansion* (*Fig.* 14.8)

1–2 represents the actual adiabatic irreversible expansion; 1–2′ represents the ideal adiabatic reversible expansion having the same

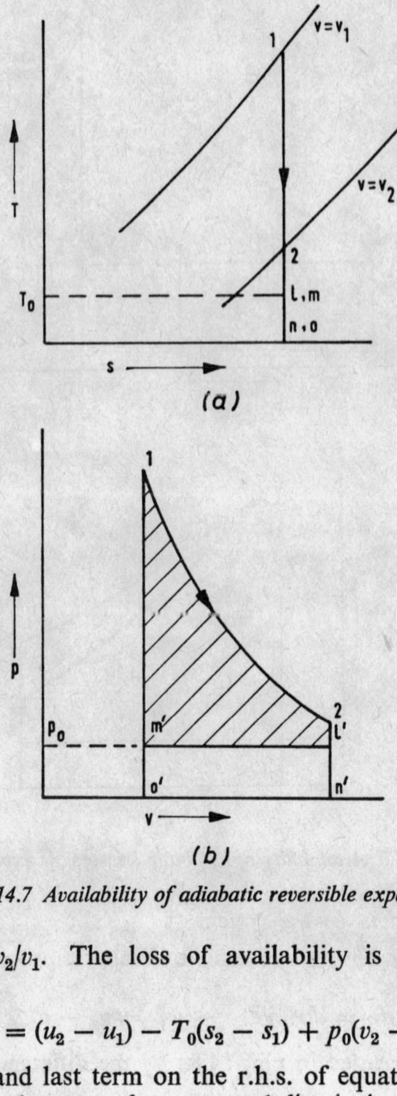

Fig. 14.7 Availability of adiabatic reversible expansion

volume ratio v_2/v_1. The loss of availability is given by equation (14.9c), i.e.

$$a_2 - a_1 = (u_2 - u_1) - T_0(s_2 - s_1) + p_0(v_2 - v_1) \quad (14.18a)$$

The second and last term on the r.h.s. of equation (14.18a) are represented by the areas *l–m–o–n* and *l′–m′–o′–n′* in *Figs.* 14.8*a* and 14.8*b*.

The term $(u_2 - u_1)$ is represented by the difference between the areas under the constant volume curves $r - 1$ and $p - 2$ (T_0 being the datum temperature), i.e. by

$$-(r\text{–}1\text{–}o\text{–}s) + (p\text{–}2\text{–}n\text{–}q)$$

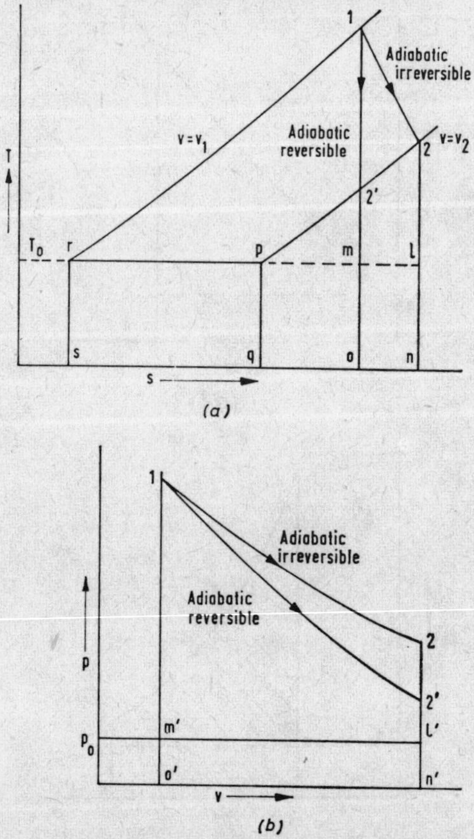

Fig. 14.8 Availability of adiabatic irreversible expansion

The gross work done by the system

From the N.F.E.E. for an irreversible adiabatic process, this is given by

$$dq - dw = du \quad \text{i.e.} \quad -w_{12} = u_2 - u_1 \tag{14.18b}$$

and this is represented in *Fig.* 14.8*a* by the difference in areas

$$-(r\text{–}1\text{–}o\text{–}s) + (p\text{–}2\text{–}n\text{–}q)$$

Loss of useful work is given by

$$\begin{aligned} w_{lost} &= w_{12'} - w_{12} \\ &= (u_1 - u_{2'}) - (u_1 - u_2) \\ &= u_2 - u_{2'} \end{aligned} \tag{14.19}$$

But $(u_2 - u_{2'})$ is represented by the difference between the areas (p–2–n–q) and (p–2′–o–q) i.e. by area (2′–2–n–0) in *Fig.* 14.8*a*.

Since this area is equivalent to the heat transferred during a hypothetical reversible constant volume heating process connecting the "ideal" end state 2′ with the actual end state 2, the term $(u_2 - u_{2'})$ in equation (14.19) is often referred to as the *reheat* effect.

The irreversibility Δi_{12} is clearly represented by equation (14.12c), i.e. by the area

$$1\text{–}2\text{–}n\text{–}o = \int_1^2 T\,ds \quad \text{in } Fig.\ 14.8a$$

(since by definition $q_{12} = 0$).

(In a *non-adiabatic irreversible process*, the irreversibility Δi_{12} is represented by the difference between the area under the process curve 1–2 in the T–s diagram, and the actual heat transferred q_{12}.)

The ratio actual work/ideal work is defined as the isentropic efficiency

$$\eta_{is} = \frac{w_{12}}{w_{12'}} = \frac{u_1 - u_2}{u_1 - u_{2'}} \tag{14.20}$$

while the ratio

$$\left| \frac{\text{gain of availability of surroundings}}{\text{loss of availability of system}} \right|$$

is defined as the effectiveness ε

$$\begin{aligned} \varepsilon &= \left| \frac{[w_{12} - p_0(v_2 - v_1)]}{(u_2 - u_1) - T_0(s_2 - s_1) + p_0(v_2 - v_1)} \right| \\ &= \left| \frac{u_2 - u_1 + p_0(v_2 - v_1)}{u_2 - u_1 - T_0(s_2 - s_1) + p_0(v_2 - v_1)} \right| \end{aligned} \tag{14.21}$$

Summarizing the above—

Loss of availability
of system = − (r–1–o–s − p–2–n–q) − (l–m–o–n)
Fig. 14.8*a* Fig. 14.8*a*
+ (l'–m'–o'–n')
Fig. 14.8*b*

Gross work by system = −(r–1–o–s − p–2–n–q)

Loss of useful work = (2′–2–n–o) = reheat effect

Irreversibility = (1–2–n–o)

EXAMPLE 14.2

1 kg of air, initially occupying 0·01565 m³ and at a temperature of 260°C, is expanded isothermally and reversibly in a cylinder against a piston, to a final volume of 0·0626 m³. Find the loss of availability and compare it with the work done by the gas on the piston.

Air may be treated as a perfect gas and the following values may be assumed—

$p_0 = 1$ bar, $T_0 = 289$ K, $c_v = 0{\cdot}7445$ kJ/kg-K,
$R = 0{\cdot}2871$ kJ/kg-K

SOLUTION

From equation (14.16)

$$a_2 - a_1 = (u_2 - u_1) - T_0(s_2 - s_1) + p_0(v_2 - v_1) \qquad \text{(i)}$$

where

$$u_2 - u_1 = c_v(T_2 - T_1) = 0 \qquad \text{(ii)}$$

$$s_2 - s_1 = R \ln\left(\frac{v_2}{v_1}\right) \quad \text{(from equation (10.51))}$$

$$= 0{\cdot}2871 \ln\left(\frac{0{\cdot}0626}{0{\cdot}01565}\right) = 0{\cdot}398 \text{ kJ/kg-K}$$

i.e.

$$T_0(s_2 - s_1) = 289 \times 0{\cdot}398 = 115 \text{ kJ/kg} \qquad \text{(iii)}$$

$$p_0(v_2 - v_1) = 1 \times 10^5(0{\cdot}0626 - 0{\cdot}01565) \times 10^{-3} \qquad \text{(iv)}$$

$$= 4{\cdot}695 \text{ kJ/kg}$$

Substituting from (ii) to (iv) in (i),
Loss of availability

$$a_2 - a_1 = 0 - 115 + 4{\cdot}695 = 110{\cdot}305 \text{ kJ/kg} \quad \textit{Ans.}$$

Net work done against piston

$$w_{12} = \int_1^2 (p - p_0)\, dv = \int_1^2 (RT \times 10^{-2} - 1)\, dv \times 10^2 \text{ kJ/kg}$$

$$= 0{\cdot}2871 \times (260 \times 273) \ln\left(\frac{0{\cdot}0626}{0{\cdot}01565}\right)$$

$$-10^2(0{\cdot}0626 - 0{\cdot}01565)$$

$$= 222 - 4{\cdot}695 = 217{\cdot}305 \text{ kJ/kg} \quad \textit{Ans.}$$

Note: The work done against the piston exceeds the loss of availability (which is numerically equal to the sum of displacement work and Carnot work) by the Carnot work

$$\int_1^2 \left(1 - \frac{T_0}{T}\right) dQ = \int_1^2 \left(1 - \frac{289}{260 + 273}\right)\frac{RT}{v}\, dv$$

$$= \int_1^2 (1 - 0{\cdot}542) 0{\cdot}2871 \times (260 + 273) \ln\left(\frac{0{\cdot}0626}{0{\cdot}01566}\right)$$

$$= 107 \text{ kJ/kg}$$

14.5 Availability in Non-cyclic Open System Processes (Steady Flow Processes)

The availability of the steady flow open system may be obtained by regarding it as an extension of the closed system considered in section 14.4.

When mass m_{in} enters the system, the availability of the system is increased by the sum of the availability associated with mass m_{in}

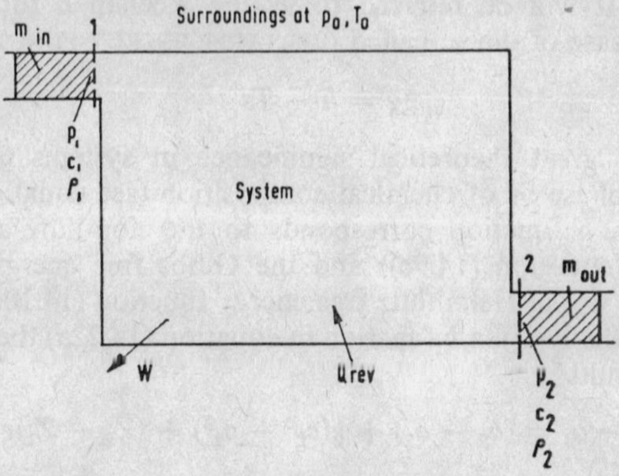

Fig. 14.9 Non-cyclic open system process

given by equation (14.9b) and the net displacement work experienced by this mass element in entering the system, namely $(p_1 - p_0)v_1$, so that, in the absence of kinetic and potential energy effects

$$\Delta A_{in} = m_{in}[(u_1 - T_0 s_1 + p_0 v_1) + (p_1 - p_0)v_1]$$
$$= m_{in}[h_1 - T_0 s_1] \tag{i}$$

Similarly, the loss of availability of the system when mass m_{out} leaves the system is given by

$$\Delta A_{out} = m_{out}(h_2 - T_0 s_2) \tag{ii}$$

Since only steady flow processes are being considered

$$m_{in} = m_{out} \tag{iii}$$

and hence, for unit mass entering and leaving the system simultaneously, the change in availability between sections 1 and 2 (*Fig.* 14.9) becomes, using equations (i) and (ii) above

$$\Delta a = a_2 - a_1 = (h_2 - h_1) - T_0(s_2 - s_1) \tag{14.22}$$

If kinetic and potential energy effects are to be included, equation (14.22) becomes

$$\Delta a = a_2 - a_1 = (h_2 - h_1) - T_0(s_2 - s_1) + \tfrac{1}{2}(c_2^2 - c_1^2) + (Z_2 - Z_1)g \quad (14.22a)$$

The property function

$$b = h - T_0 s \quad (14.23a)$$

has been defined by Keenan as the STEADY FLOW AVAILABILITY FUNCTION. It will be referred to as the Keenan b function and is a special case of the so-called GIBBS FREE ENERGY FUNCTION

$$g = h - Ts \quad (14.23b)$$

which is of great theoretical significance in systems undergoing changes of phase or of chemical composition (see equation (14.5)). The Keenan b function corresponds to the non-flow availability function a (equation (14.9c)) and the Gibbs free energy function corresponds to the Helmholtz free energy function (14.10b).

Inserting the Keenan b function in equation (14.22a) the following equation results

$$a_2 - a_1 = (b_2 - b_1) + \tfrac{1}{2}(c_2^2 - c_1^2) + (Z_2 - Z_1)g \quad (14.22b)$$

Equation (14.22a), for the change of availability of the steady flow open system between given states at the entry and exit sections, corresponds to equation (14.9c) of section 14.4 for the change of availability of the closed system between given initial and final states.

As in the case of closed systems, it may readily be shown that the availability of steady flow open systems decreases in all real processes.

The irreversibility of a steady flow process may be assessed quantitatively in a manner similar to that for the closed system, as shown in section 14.4.

The S.F.E.E. (4.13), neglecting kinetic and potential energy effects and writing in differential form for unit mass, becomes

$$dq - dw_{useful} = dh \quad \text{(i)}$$

But from the availability equation (14.22)

$$da = dh - T_0\,ds \quad \text{(ii)}$$

so that, combining equations (i) and (ii)

$$da = dh - T_0\,ds = dq - dw_{useful} - T_0\,ds \quad \text{(iii)}$$

From Corollary 8 of the second law

$$ds > \frac{dq}{T} \tag{iv}$$

whence, by substituting from equation (iv) in equation (iii),

$$da < [T\,ds - dw_{useful} - T_0\,ds]$$

or
$$(dw_{useful})_{actual} < [(T - T_0)\,ds - da] \tag{14.24}$$

The inequality (14.24) represents the useful work done in any real process. For the limiting case of a fully reversible process

$$(dw_{useful})_{idel} = (T - T_0)\,ds - da \tag{14.24a}$$

so that, by analogy with the closed system process, the irreversibility may be expressed quantitatively as the difference between the ideal and actual work, i.e.

$$\begin{aligned} di &= (T - T_0)\,ds - da - (dw_{useful})_{actual} \\ &= \left[dq\left(\frac{T - T_0}{T}\right) - da\right] - (dw_{useful})_{actual} \end{aligned} \tag{14.24b}$$

where $(dw_{useful})_{actual}$ is obtained from the S.F.E.E. (equation (i)) and da from equation (ii).

Hence, the irreversibility may finally be expressed as

$$\begin{aligned} di &= (T - T_0)\,ds - (dh - T_0\,ds) - (dq - dh) \\ &= (T\,ds - dq) \end{aligned}$$

or for a finite process 1–2

$$\Delta i_{12} = \int_1^2 T\,ds - q_{12} \tag{14.24c}$$

Equation (14.24c) for the steady flow process is identical with equation (14.12c) for the non-flow process, except that the latter applies to unit mass. Both equations are quantitative expressions of the inequality of Clausius.

Equation (14.24c) is still valid when kinetic and potential energy effects are included, since the terms $\frac{1}{2}\,d(c^2)$ and $d(Zg)$ which occur both in the S.F.E.E. and the availability equation (14.22a) cancel out.

EFFECTIVENESS

As in the case of non-flow processes, the effectiveness of an adiabatic process may be defined as

Effectiveness

$$\varepsilon = \left|\frac{\text{actual useful work}}{\text{available useful work}}\right| = \frac{\int_1^2 [dq - dh - d(\text{KE}) - d(\text{PE})]}{\int_1^2 [db + d(\text{KE}) + d(\text{PE})]} \tag{14.25}$$

where db is given by equation (14.23).

As in the case of non-flow processes, we may use the alternative and more general definition

$$\text{Effectiveness } \varepsilon = \left|\frac{\text{gain of availability of surroundings}}{\text{loss of availability of system}}\right| \tag{14.25a}$$

The above definitions apply in cases where the availability of the system is *decreasing*, i.e. when work or heat are being transferred *from* the system *to* the surroundings.

When the opposite is the case, i.e. the availability of the system is increasing due to work or heat being transferred *to* the system *from* the surroundings, the definition is as follows—

$$\text{Effectiveness } \varepsilon = \left|\frac{\text{gain of availability of system}}{\text{loss of availability of surroundings}}\right| \tag{14.25b}$$

AVAILABILITY CHANGES IN SPECIFIC PROCESSES (see also sections 4.7 and 7.5).

(*a*) *Constant pressure heating—single fluid system* (*Fig*. 14.10)

Applying equation (14.22b) to the system shown in *Fig*. 14.10 and neglecting kinetic and potential energy effects, the increase in availability per unit mass of the working fluid becomes

$$a_2 - a_1 = b_2 - b_1 = (h_2 - h_1) - T_0(s_2 - s_1) \tag{14.26}$$

By analogy with the closed system case (*Fig*. 14.5) $(a_2 - a_1)$ is represented by area 1–2–*l*–*m* in *Fig*. 14.10*b*.

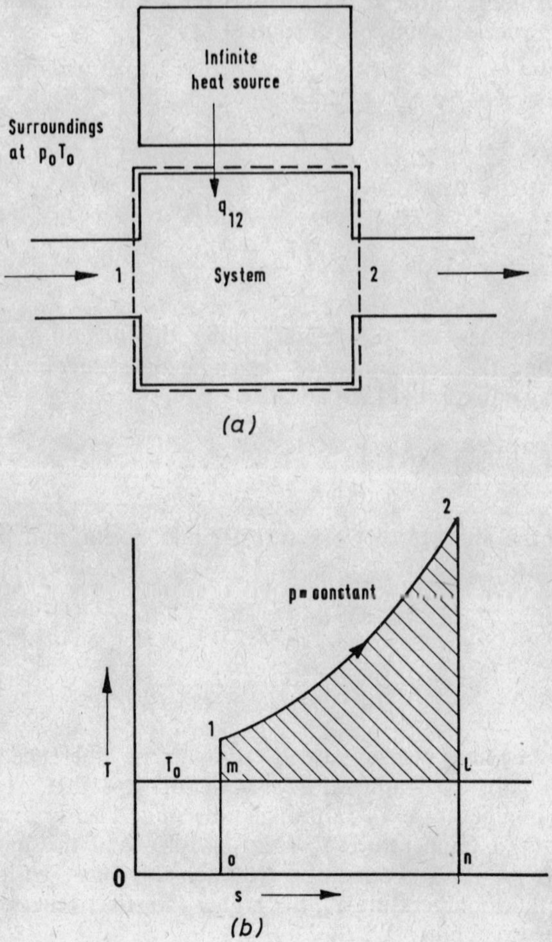

Fig. 14.10 Availability of constant pressure heating—single fluid system

But the loss of availability of the heat source is given by equation (14.7b),

$$\Delta a_{source} = (-q_{12} - T_0(s_2' - s_1')) \tag{i}$$

where $-(s_2' - s_1') \neq (s_2 - s_1)$, unless the heat transfer process is fully reversible, i.e. takes place with zero temperature gradient (see Chapter 5, section 5.3).

Since the heat source is assumed to remain at temperature T_S during the process (infinite heat source)

$$s_2' - s_1' = -\frac{q_{12}}{T_S}$$

i.e.

$$\Delta a_{source} = -q_{12} + T_0\frac{q_{12}}{T_S}$$

$$= q_{12}\left(\frac{T_0 - T_S}{T_S}\right) \qquad \text{(ii)}$$

The effectiveness of the heater, using the definition (14.25b) and regarding the heat source as the immediate surroundings, becomes, using equation (14.26) and (ii)

$$\varepsilon = \left|\frac{b_2 - b_1}{q_{12}(T_0 - T_S)/T_S}\right| \qquad (14.27)$$

Since in the absence of work transfer and kinetic and potential energy effects

$$q_{12} = h_2 - h_1 \qquad \text{(iii)}$$

then

$$\varepsilon = \left|\frac{b_2 - b_1}{(h_2 - h_1)(T_0 - T_S)/T_S}\right| \qquad (14.27a)$$

The irreversibility is given quantitatively as the difference between the denominator and numerator of equation (14.27a). The lower the source temperature T_S, although this must clearly exceed the maximum fluid temperature T_2 (*Fig.* 14.10b), and the higher the temperature of the surroundings, although this must be less than the initial fluid temperature T_1, the higher the effectiveness and the lower the irreversibility ΔI_{12}.

EXAMPLE 14.3

Air enters a heater at 27°C and leaves at 71°C, the pressure béing constant at 3·5 bar. The heat source is at 260°C and the surroundings are at 15°C.

Treating air as a perfect gas find: (*a*) the gain in availability of the air (*b*) the effectiveness of the heater. $c_p = 1{\cdot}005$ kJ/kg-K.

SOLUTION (the notation used is that of *Fig*. (14.10*b*))

(1) *Gain in availability* (from equation (14.26))

$$\Delta a_{12} = b_2 - b_1 = (h_2 - h_1) - T_0(s_2 - s_1)$$

From equation (7.9c)

$$s_2 - s_1 = c_p \ln\left(\frac{T_2}{T_1}\right)$$

also

$$h_2 - h_1 = c_p(T_2 - T_1)$$

whence

$$b_2 - b_1 = c_p\left[(T_2 - T_1) - T_0 \ln\left(\frac{T_2}{T_1}\right)\right]$$

$$= 1{\cdot}005\left[(71 - 27) - 288{\cdot}16 \ln\left(\frac{344{\cdot}16}{300{\cdot}16}\right)\right]$$

$$= 3{\cdot}955 \text{ kJ/kg} \quad \textit{Ans.}$$

(2) *Effectiveness* (from equation (14.27a))

$$\varepsilon = \left|\frac{b_2 - b_1}{(h_2 - h_1)(T_0 - T_S)/T_S}\right| = \left|\frac{3{\cdot}955}{1{\cdot}005 \times 44(15-260)/(1273{\cdot}16+260)}\right|$$

$$= 0{\cdot}194$$

By lowering the heat source temperature from 260°C to 80°C, other conditions remaining unchanged, the effectiveness is increased from the above value to

$$\varepsilon = \left|\frac{3{\cdot}955}{1{\cdot}005 \times 44(15 - 80)/(273{\cdot}16 + 80)}\right| = 0{\cdot}486 \quad \textit{Ans.}$$

Thus the effectiveness provides a quantitative measure of the irreversibility implied by the temperature difference between the heat source and the fluid receiving heat.

(*b*) *Constant pressure heating—two-fluid system* (*Fig.* 14.11)
This is the situation encountered in heat exchangers.

Applying equation (14.23) to the two fluid streams A and B, and assuming that the former is being heated while the latter is being cooled the following relationships hold.

Increase of availability of stream A

$$\Delta A_{12} = m_A(b_2 - b_1) = m_A[(h_2 - h_1) - T_0(s_2 - s_1)] \quad (14.28a)$$

Loss of availability of stream B

$$\Delta A_{34} = m_B(b_4 - b_3) = m_B[(h_4 - h_3) - T_0(s_4 - s_3)] \quad (14.28b)$$

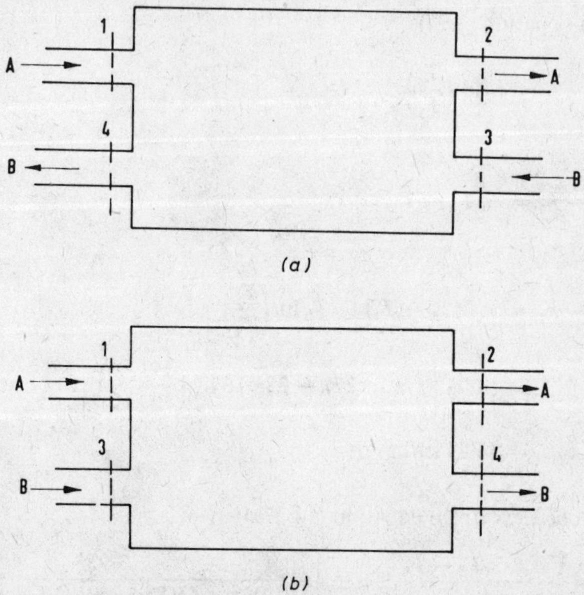

Fig. 14.11 Availability of constant pressure heating—two fluid system
(a) Counter flow heat exchanger
(b) Parallel flow heat exchanger

Applying equation (14.25b)

$$\varepsilon = \left| \frac{\text{gain in availability of cold stream}}{\text{loss in availability of hot stream}} \right|$$

$$= \left| \frac{m_A[(h_2 - h_1) - T_0(s_2 - s_1)]}{m_B[(h_4 - h_3) - T_0(s_4 - s_3)]} \right| \qquad (14.29)$$

Since in heat exchangers of this type, heat transfer is necessarily accompanied by quite large temperature gradients, irreversibilities are unavoidable (see Chapter 5, section 5.3). A first law analysis, in which an efficiency is defined as

$$\eta = \frac{\text{heat transferred to cold fluid}}{\text{heat transferred from hot fluid}} = \frac{m_A(h_2 - h_1)}{m_B(h_4 - h_3)} \qquad (14.30)$$

would necessarily yield an efficiency of unity, and would provide no indication of any irreversibilities present, while the effectiveness given by equation (14.29) is always less than unity, indicating the presence of irreversibilities and underlining the superiority of a second law analysis over a first law analysis.

The usefulness of a second law or availability analysis is demonstrated further by examining more closely the problem of the counter flow and parallel flow heat exchangers illustrated respectively in *Figs.* 14.12*a* and *b*. Typical temperature distribution curves, as functions of the distance ratio x/l, where x is any distance along a concentric tube pair and l is the overall length, show that the effectiveness of the counterflow heat exchanger invariably exceeds that of the parallel flow heat exchanger, the large initial temperature difference $(T_3 - T_1)$ in the latter implying a severe irreversibility.

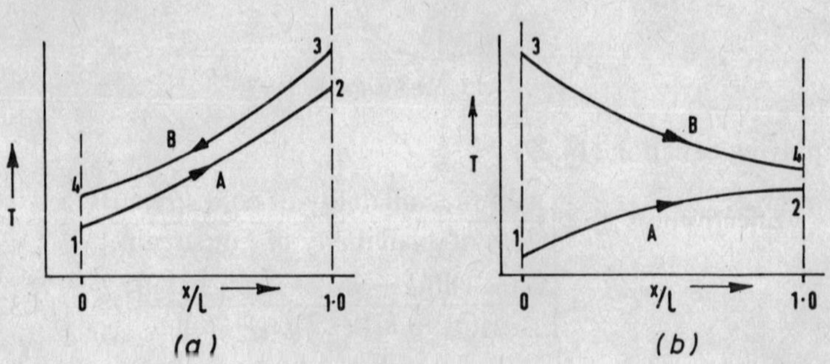

Fig. 14.12
(*a*) *Counter flow heat exchanger*
(*b*) *Parallel flow heat exchanger*

(*c*) *Constant pressure heating with mixing of hot and cold fluids* (*Fig.* 14.13)

This is the situation encountered, for example, in feed water heaters for boiler plant. The irreversibility arises primarily from mixing of the hot and cold fluids, a process not previously discussed as all the systems treated so far have been single-fluid systems.

y kg of heating fluid enter the mixing chamber in state 1, $(1 - y)$ kg of the cold fluid enter in state 2 and 1 kg of mixture leaves in state 3. In the case of feed water heaters, the mixing process is accompanied by condensation of the bleed steam from the turbine, but the arguments used here apply whether a change of phase occurs or not.

An availability analysis gives the following results.

Loss of availability of hot fluid

$$\Delta a_{13} = y(b_3 - b_1) = y[(h_3 - h_1) - T_0(s_3 - s_1)] \qquad (14.31a)$$

Gain of availability of cold fluid

$$\Delta a_{23} = (1 - y)(b_3 - b_2) = (1 - y)[(h_3 - h_2) - T_0(s_3 - s_2)] \qquad (14.31b)$$

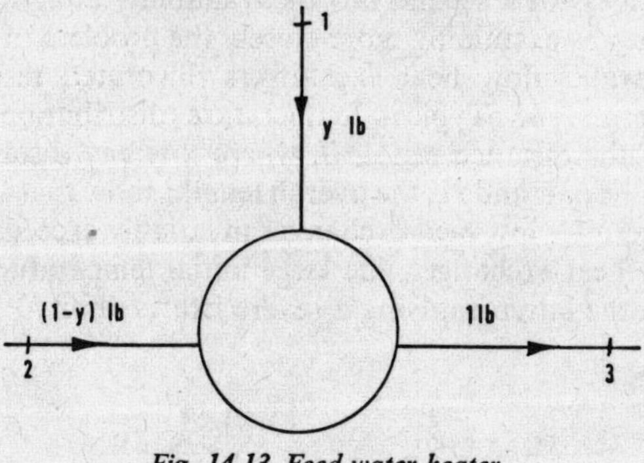

Fig. 14.13 Feed water heater

Applying equation (14.29)

$$\text{Effectiveness } \varepsilon = \left|\frac{\text{gain in availability of cold stream}}{\text{loss of availability of hot stream}}\right|$$

$$= \left|\frac{(1-y)[(h_3-h_2)-T_0(s_3-s_2)]}{y[(h_3-h_1)-T_0(s_3-s_1)]}\right| \qquad (14.32)$$

A first law analysis provides the required solution for y, assuming that the y kg of bleed steam are completely condensed, and that the temperature of the heated feed water at 3 equals that of the bleed steam at 1.

EXAMPLE 14.4

(Using notation of *Fig*. 14.13)

A feed water heater of the mixing type receives bleed steam at 2·5 bar and 0·846 dry, whilst the feed pump return is saturated water at 0·35 bar. Assuming that the mass ratio y is 0·1866 kJ/kg, find the various availability changes as well as the effectiveness of the heater. Assume $T_0 = 288$ K.

SOLUTION

From tables for saturated steam

$$h_1 = 535 + 0{\cdot}846 \times 2182 = 2384 \text{ kJ/kg}$$
$$s_1 = 1{\cdot}607 + 0{\cdot}846 \times 5{\cdot}446 = 6{\cdot}207 \text{ kJ/kg-K}$$
$$h_2 = 112 \text{ kJ/kg}, \; s_2 = 0{\cdot}391 \text{ kJ/kg-K}$$
$$h_3 = yh_1 + (1-y)h_2 = 0{\cdot}1866 \times 2384 + 0{\cdot}8134 \times 112$$
$$= 536 \text{ kJ/kg}$$
$$s_3 = 1{\cdot}609 \text{ kJ/kg-K}$$

Loss of availability of hot fluid

$$\Delta A_{13} = y(b_3 - b_1) = y[(h_3 - h_1) - T_0(s_3 - s_1)]$$
$$= 0{\cdot}1866[(536 - 2384) - 288(1{\cdot}609 - 6{\cdot}207)]$$
$$= -98{\cdot}0 \text{ kJ} \quad \textit{Ans.}$$

Gain of availability of cold fluid

$$\Delta A_{23} = (1 - y)[(h_3 - h_2) - T_0(s_3 - s_2)]$$
$$= 0{\cdot}8134[(536 - 112) - 288(1{\cdot}609 - 0{\cdot}391)]$$
$$= +58{\cdot}5 \text{ kJ} \quad \textit{Ans.}$$

$$\text{Effectiveness } \varepsilon = \left|\frac{\Delta A_{23}}{\Delta A_{13}}\right| = \left|\frac{58{\cdot}5}{-98{\cdot}0}\right| = 0{\cdot}598 \; \textit{Ans.}$$

(*d*) *Adiabatic reversible* (*isentropic*) *expansion* (*Fig.* 14.14)

This represents an ideal, i.e. fully reversible expansion in a turbine, in the absence of heat transfer effects.

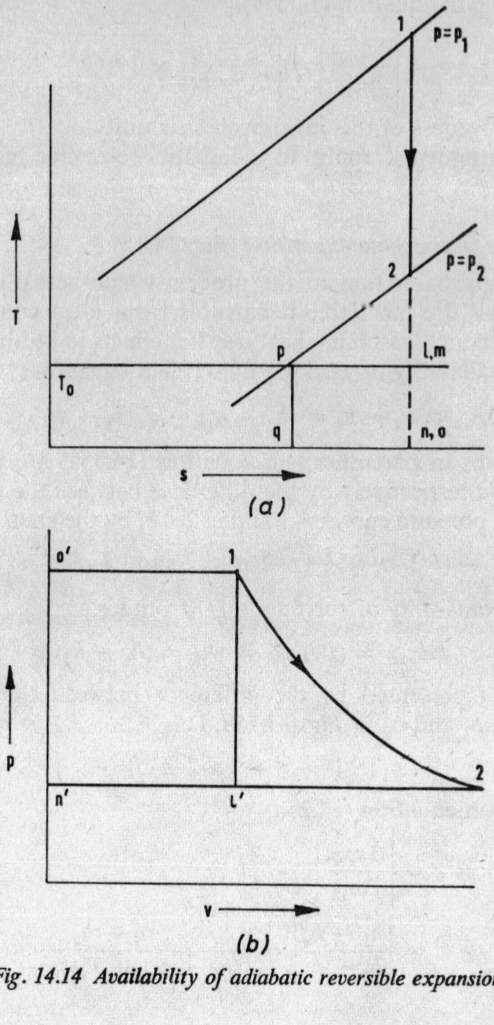

Fig. 14.14 Availability of adiabatic reversible expansion

Loss of availability of working fluid (from equation (14.22))

$$\Delta a_{12} = (h_2 - h_1) - T_0(s_2 - s_1) \qquad (14.33a)$$
$$= h_2 - h_1$$

since the process is fully reversible and adiabatic, i.e. $s_2 - s_1 = 0$. Gain of availability of surroundings (from S.F.E.E.)

$$\Delta a_{surr} = \Delta w_{useful} = (h_1 - h_2) \qquad (14.33b)$$

(Kinetic and potential energy effects are assumed negligible.)

Substituting in equation (14.25a)

$$\varepsilon = \left|\frac{\Delta a_{surr}}{\Delta a_{12}}\right| = \left|\frac{h_2 - h_1}{h_2 - h_1}\right| = 1{\cdot}0 \qquad (14.34)$$

i.e. the effectiveness of this ideal process is unity.

Similar arguments apply to adiabatic reversible compression processes.

(*e*) *Adiabatic irreversible expansion* (*Fig*. 14.15)

This is an approximation to the process which actually occurs in turbines when the inevitable departures from ideal conditions are taken into account, assuming heat transfer effects to be absent.

Loss of availability of working fluid (from equation (14.22))

$$\Delta a_{12} = b_2 - b_1 = (h_2 - h_1) - T_0(s_2 - s_1) \qquad (14.35a)$$

The two terms in parenthesis in equation (14.35a) are represented in *Fig*. 14.15*a* respectively by the difference between the areas under the constant pressure curves *r*–1 and *p*–2, i.e. by the areas

$$-(r\text{–}1\text{–}o\text{–}s) + (p\text{–}2\text{–}n\text{–}q) \quad \text{and} \quad l\text{–}m\text{–}o\text{–}n$$

Gain in availability of surroundings (from S.F.E.E.)

$$\Delta a_{surr} = \Delta w_{useful} = w_{12} = (h_1 - h_2) \qquad (14.35b)$$

and this is represented by the difference between the two areas representing h_1 and h_2 in *Fig*. 14.15*a*, i.e.

$$(r\text{–}1\text{–}o\text{–}s) - (p\text{–}2\text{–}n\text{–}q)$$

Substituting in equation (14.25a)

$$\text{Effectiveness } \varepsilon = \left|\frac{\Delta a_{surr}}{\Delta a_{12}}\right|$$
$$= \left|\frac{h_1 - h_2}{b_2 - b_1}\right| = \left|\frac{h_1 - h_2}{(h_2 - h_1) - T_0(s_2 - s_1)}\right| \qquad (14.36)$$

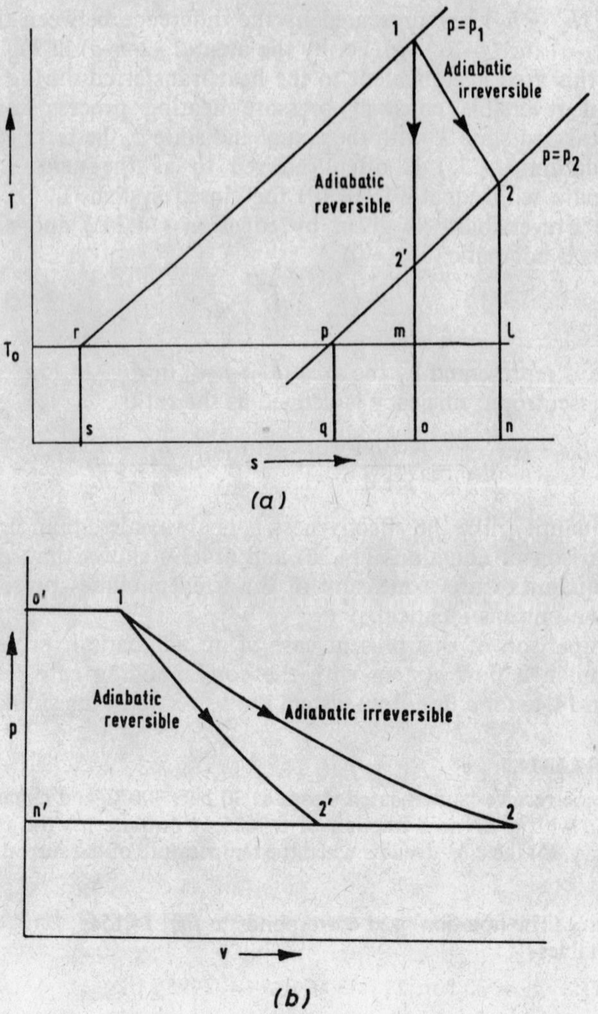

Fig. 14.15 Availability of adiabatic irreversible expansion

Since the irreversibility of the process 1–2 clearly implies an entropy gain, i.e. $(s_2 - s_1) > 0$, the effectiveness ε is less than unity.

The loss of useful work is given by

$$\begin{aligned} w_{lost} &= w_{12'} - w_{12} \\ &= (h_1 - h_{2'}) - (h_1 - h_2) \\ &= (h_2 - h_{2'}) \end{aligned} \tag{14.37}$$

But $(h_2 - h_{2'})$ is represented by the difference between the areas $(p\text{–}2\text{–}n\text{–}q)$ and $(p\text{–}2'\text{–}o\text{–}q)$, i.e. by the area $(2'\text{–}2\text{–}n\text{–}o)$ in *Fig*. 14.15a. Since this area is equivalent to the heat transferred during a hypothetical reversible constant pressure heating process connecting the ideal end state 2′ with the actual end state 2, the term $(h_2 - h_{2'})$ in equation (14.37) is often referred to as the REHEAT EFFECT. (Compare with equation (14.19) for closed systems.)

The irreversibility is given by equation (14.24c) and since the process is adiabatic ($q_{12} = 0$)

$$\Delta i_{12} = \int_1^2 T\,ds \qquad (14.38)$$

This is represented by the area $(1\text{–}2\text{–}n\text{–}o)$ in *Fig*. 14.15*a*.

The isentropic efficiency is defined as the ratio

$$\eta_{is.} = \frac{\text{actual work}}{\text{ideal work}} = \frac{w_{12}}{w_{12'}} = \frac{h_1 - h_2}{h_1 - h_{2'}} \qquad (14.39)$$

This quantity, like the effectiveness ε, is always less than unity, but comparison of equations (14.36) and (14.39) shows that, although both quantities are a measure of the irreversibilities present, they are by no means identical.

Comparison of this present case of an adiabatic irreversible expansion in a flow system with the corresponding case treated in section 14.4e for a non-flow system shows very striking similarities.

EXAMPLE 14.5

A turbine receives superheated steam at 30 bar, 300°C and exhausts it at 3·5 bar with a dryness fraction of 0·9625. Compute (*a*) the isentropic efficiency, (*b*) the effectiveness when the temperature of the surroundings is 15°C.

SOLUTION (The notation used corresponds to *Fig*. 14.15*a*)
From tables—

at (1): $p_1 = 30$ bar, $T_1 = 300$°C, $h_1 = 2995$ kJ/kg
$s_1 = 6{\cdot}541$ kJ/kg-K,
$b_1 = h_1 - T_0 s_1 = 2995 - 288{\cdot}16 \times 6{\cdot}541 = 1107$ kJ/kg

at (2): i.e. after actual expansion
$p_2 = 3{\cdot}5$ bar, $x_2 = 0{\cdot}9625$, $h_2 = 2651$ kJ/kg
$s_2 = 6{\cdot}751$ kJ/kg-K,
$b_2 = h_2 - T_0 s_2 = 2651 - 288{\cdot}16 \times 6{\cdot}751 - 700$ kJ/kg

at (2′): i.e. after ideal reversible expansion
$p_{2'} = 3{\cdot}5$ bar, $x_{2'} = 0{\cdot}924$, $h_{2'} = 2565$ kJ/kg
$s_{2'} = s_1 = 6{\cdot}541$ kJ/kg-K

(*a*) isentropic efficiency from equation (14.39)

$$\eta_{is.} = \frac{w_{12}}{w_{12'}} = \left|\frac{h_1 - h_2}{h_1 - h_{2'}}\right| = \frac{344}{430} = = 0{\cdot}80 \quad \textit{Ans.}$$

(*b*) effectiveness from equation (14.36)

$$\varepsilon = \left|\frac{h_1 - h_2}{b_2 - b_1}\right| = \left|\frac{344}{-407}\right| = 0{\cdot}8455 \quad \textit{Ans.}$$

The effectiveness is thus significantly greater than the efficiency. The reason for this state of affairs is that the loss of availability $(b_1 - b_2)$ which appears in the denominator of equation (14.36) is smaller than the isentropic enthalpy drop $(h_1 - h_{2'})$ which constitutes the denominator of equation (14.39), due to the 'reheat' effect $(h_2 - h_{2'})$ which is itself the result of the irreversible expansion 1–2. This "reheat" effect implies that the availability at entry to the next turbine stage is higher than that in the case of a fully reversible expansion 1–2′, so that indirectly some energy is recovered. While the effectiveness takes this "reheat" effect into account, the isentropic efficiency does not.

(*f*) *Adiabatic irreversible compression* (*Fig.* 14.16)

This is an approximation to the process occurring in aerodynamic or rotary positive displacement compressors in the absence of heat transfer effects, but taking into account internal irreversibilities due to fluid friction and other sources.

Gain in availability of working fluid (equation (14.22a))

$$\Delta a_{12} = b_2 - b_1 = (h_2 - h_1) - T_0(s_2 - s_1) \qquad (14.40a)$$

The two terms in parenthesis in equation (14.40a) are represented respectively by the area difference.

$$+ (p\text{–}2\text{–}n\text{–}q) - (r\text{–}1\text{–}o\text{–}s) \text{ and } (l\text{–}m\text{–}o\text{–}n)$$

in *Fig.* 14.16*a* (cf. *Fig.* 14.15*a*).

Loss of availability of surroundings (from S.F.E.E.)

$$\Delta a_{surr} = w_{12} = h_2 - h_1 \qquad (14.40b)$$

and this is represented by area

$$(p\text{–}2\text{–}n\text{–}q) - (r\text{–}1\text{–}o\text{–}s)$$

The situation now being the reverse of the irreversible expansion process described in the previous sub-section, involving a loss of availability for the surroundings and gain in availability for the working fluid, we have to use the alternative definition of effectiveness as given by equation (14.25b) in place of that given by equation (14.25a) used for the expansion process.

$$\text{Effectiveness } \varepsilon = \left|\frac{\Delta a_{12}}{\Delta a_{surr}}\right| = \left|\frac{b_2 - b_1}{h_2 - h_1}\right| \tag{14.41}$$

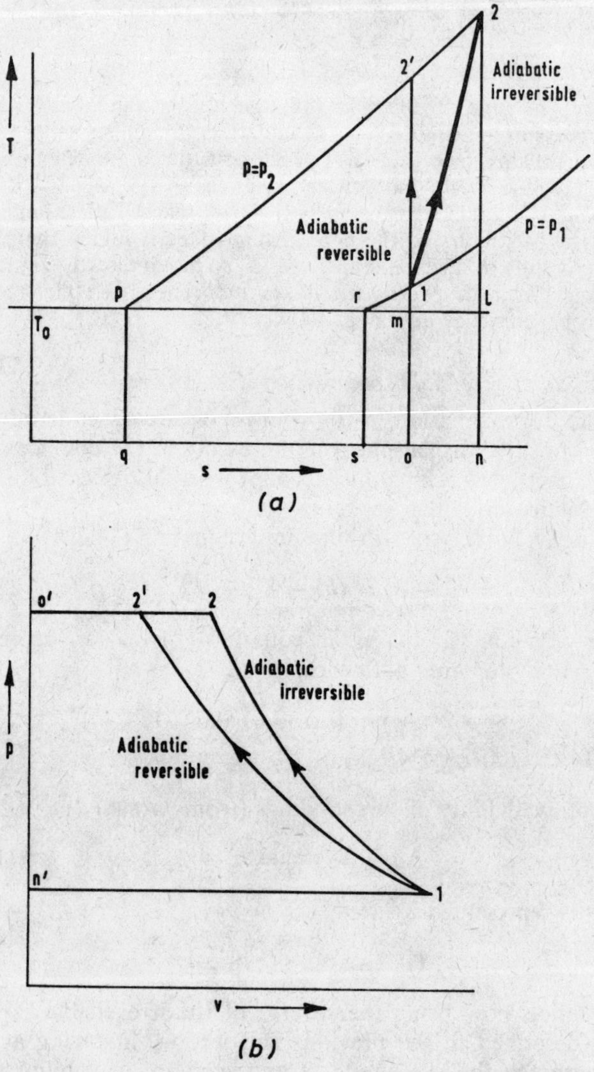

Fig. 14.16 Availability of adiabatic irreversible compression

The excess work (compared with the "ideal" process 1–2′) is given by

$$\begin{aligned} w_{excess} &= w_{12} - w_{12'} \\ &= h_2 - h_{2'} \end{aligned}$$

As before this is represented by the area difference

$$(p\text{–}2\text{–}n\text{–}q) - (p\text{–}2'\text{–}o\text{–}q) = (2'\text{–}2\text{–}n\text{–}o) \text{ in } \textit{Fig. } 14.16a,$$

the term $(h_2 - h_{2'})$ again being referred to as the reheat effect.

The irreversibility is given by equation (14.24c)

$$\Delta i_{12} = \int_1^2 T\,ds \tag{14.43}$$

and is represented by the area (1–2–*n*–*o*) in *Fig.* 14.16*a*.

The isentropic efficiency is defined as the ratio

$$\eta_{is.} = \frac{\text{ideal work}}{\text{actual work}} = \frac{w_{12'}}{w_{12}} = \frac{h_{2'} - h_1}{h_2 - h_1} \tag{14.44}$$

and, like the effectiveness ε, is the reciprocal of the corresponding expression for the expansion process.

EXAMPLE 14.6

Air is compressed in an axial flow compressor from 1·04 bar, 16°C to 4·16 bar, with an isentropic efficiency of 80%. Calculate the effectiveness and compare it with the isentropic efficiency, and comment on the result. Assume air to be a perfect gas (c_p = 1·004 kJ/kg-K) and take the temperature of the surroundings as 16°C.

SOLUTION (The notation used corresponds to *Fig.* 14.16*a*).

From equation (14.41)

$$\text{Effectiveness } \varepsilon = \frac{b_2 - b_1}{h_2 - h_1} \tag{i}$$

where $h_1 = c_pT_1 = 1{\cdot}004 \times 289 = 290$ kJ/kg

and $b_1 = h_1 - T_0s_1 = (1{\cdot}004 \times 289 - 289s_1)$ kJ/kg

Compression temperature

$$T_2 = T_1 + \frac{T_{2'} - T_1}{\eta_{is.}} = T_1 + \frac{T_1\left[\left(\frac{p_2}{p_1}\right)^{(\gamma-1)/\gamma} - 1\right]}{\eta_{is.}}$$

$$= 289\left(1 + \frac{(4)^{0\cdot4/1\cdot4} - 1}{0{\cdot}8}\right) = 465 \text{ K}$$

$$h_2 = c_pT_2 = 1{\cdot}004 \times 465 = 466 \text{ kJ/kg}$$

But from equation (10.33), entropy change

$$\Delta s_{12} = c_p \ln\left(\frac{T_2}{T_1}\right) - \frac{R}{J}\ln\left(\frac{p_2}{p_1}\right)$$

$$= 1{\cdot}004 \ln\left(\frac{465}{289}\right) - 0{\cdot}287 \ln 4$$

$$= 0{\cdot}0807 \text{ kJ/kg-K}$$

i.e.

$$b_2 = h_2 - T_0 s_2 = 466 - 289(0{\cdot}0807 + s_1)$$

Substituting in equation (i),

$$\varepsilon = \frac{(466 - 290) - 289 \times 0{\cdot}0807}{(466 - 290)}$$

$$= \frac{176 - 23{\cdot}3}{176} = 0{\cdot}868 \quad \textit{Ans.}$$

Hence the effectiveness of the compression process, like that for the typical expansion process discussed in example 14.5, considerably exceeds the corresponding isentropic efficiency, and for the same reason, namely the presence of a reheat effect.

14.6 Second Law of Availability Analysis of Complete Cycles

Steam and gas turbine power plant are typical examples of systems operating in a cycle, and employing a series of units in which take place the separate steady flow processes constituting the cycle. By applying the availability analysis of the previous section appropriate to each unit the various irreversibilities may be isolated.

In the following such an analysis will be applied to a simple steam plant, operating on the so-called Rankine or constant pressure cycle. This cycle consists of the following steady flow processes (*Fig.* 14.17*b*)—

1–2: constant pressure heating of water at maximum cycle pressure p_{max} in boiler.
2–3: evaporation of water at maximum cycle pressure p_{max}, also in boiler.
3–4: adiabatic irreversible expansion of initially dry saturated steam in turbine (3–4′ constitutes the equivalent ideal process) to minimum cycle pressure p_{min}.
4–5: condensation of exhaust steam at minimum cycle pressure p_{min}.
5–1: adiabatic irreversible pumping of condensate from minimum pressure p_{min} to maximum pressure p_{max}. (5–1′ constitutes the equivalent ideal process.)

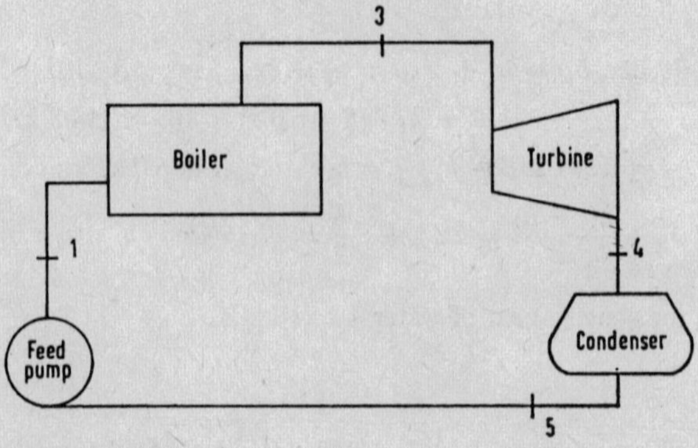

(a)

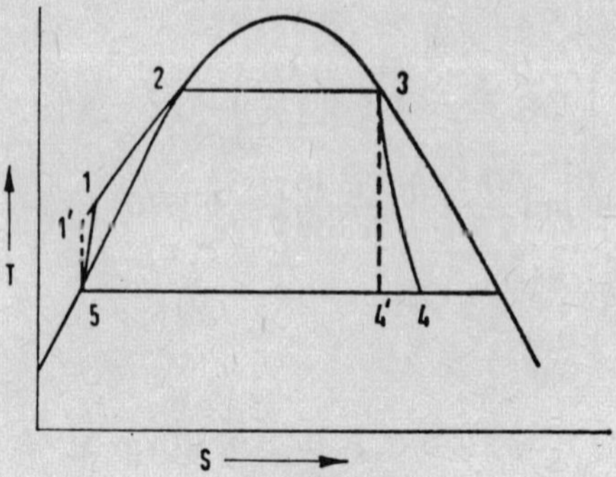

(b)

Fig. 14.17 Complete steam cycle

EXAMPLE 14.7

The analysis will be applied directly to a numerical example, a Rankine cycle having maximum and minimum pressures of 28 bar and 0·035 bar respectively, the isentropic efficiency of both turbine and pump being 80% and the temperature of the surroundings being 5°C. Both a conventional first law and a second law or availability analysis will be carried out.

SOLUTION

Using the notation of *Fig.* 14.17

At (5)

$$p_5 = 0{\cdot}035 \text{ bar}, \; T_5 = 26{\cdot}7°\text{C}, \; b_5 = 112 \text{ kJ/kg}$$
$$s_5 = 0{\cdot}391 \text{ kJ/kg-K}, \; b_5 = h_5 - T_0 s_5 = 3{\cdot}2 \text{ kJ/kg}$$

At (1′)

$$p_{1'} = 28 \text{ bar}, h_{1'} = h_5 + w_{51'} = h_5 + (p_{max} - p_{min})v_1 10^2$$
$$= 112 + 27{\cdot}965 \times 10^{-3} \times 10^2 = 114{\cdot}8 \text{ kJ/kg}$$

$$s_{1'} = s_5 = 0{\cdot}391 \text{ kJ/kg-K}, b_{1'} = h_{1'} - T_0 s_{1'} = 6{\cdot}0 \text{ kJ/kg}$$

$$T_{1'} = T_5 + \frac{h_{1'} - h_5}{c} = 26{\cdot}7 + \frac{2{\cdot}8}{4{\cdot}2} = 27{\cdot}36°\text{C}$$

where c is the specific heat of water 4·2 kJ/kg-K.

At (1)

$$p_1 = 28 \text{ bar}, h_1 = \text{h}_5 + \frac{w_{51'}}{\eta_{51}} = 112 + \frac{2{\cdot}8}{0{\cdot}8} = 115{\cdot}5 \text{ kJ/kg}$$

$$T_1 = T_5 + \frac{h_1 - h_5}{c} = 26{\cdot}7 + \frac{3{\cdot}5}{0{\cdot}8} = 27{\cdot}53°\text{C}$$

$$s_1 = s_{1'} + \int_1^2 c\,\frac{dT}{T} \simeq s_{1'} + c \ln\left(\frac{T_1}{T_{1'}}\right)$$

$$= 0{\cdot}391 + \ln\left(\frac{27{\cdot}53 + 273{\cdot}15}{27{\cdot}36 + 273{\cdot}15}\right) = 0{\cdot}3916 \text{ kJ/kg-K}$$

$$b_1 = h_1 - T_0 s_1 = 6{\cdot}5 \text{ kJ/kg}$$

At (3)

$$p_3 = 28 \text{ bar}, T_3 = 230{\cdot}0°\text{C}, h_3 = 2803 \text{ kJ/kg}$$
$$s_3 = 6{\cdot}213 \text{ kJ/kg-K}, b_3 = 1073 \text{ kJ/kg}$$

At (4′)

$$p_{4'} = 0{\cdot}035 \text{ bar}, T_{4'} = T_5 = 26{\cdot}7°\text{C}$$
$$s_{4'} = s_3 = 6{\cdot}213 \text{ kJ/kg-K}, x_{4'} = 0{\cdot}716,$$
$$h_{4'} = 1853 \text{ kJ/kg}, b_{4'} = 123 \text{ kJ/kg}$$

At (4)

$$p_4 = 0{\cdot}035 \text{ bar}, T_4 = T_5 = 26{\cdot}7°\text{C}$$
$$h_4 = h_3 - (w_{34'} \times \eta_{34}) = 2803 - 950 \times 0{\cdot}8 = 2043 \text{ kJ/kg}$$
$$x_4 = \frac{1931}{2438} = 0{\cdot}793, s_4 = 6{\cdot}841 \text{ kJ/kg-K}$$
$$b_4 = 138 \text{ kJ/kg}$$

(*a*) FIRST LAW ANALYSIS

Energy input	*kJ/kg*	*Percentage of heat input*
Feed pump $h_1 - h_5$	3·5	0·13
Boiler $h_3 - h_1$	2687·5	99·87
	Σ 2691·0	Σ 100·00
Energy output		
Turbine $h_3 - h_4$	760·0	28·30
Condenser $h_4 - h_5$	1931·0	71·70
	Σ 2691·0	Σ 100·00

Net work $= w_{34} - w_{51} = 760{\cdot}0 - 3{\cdot}5 = 756{\cdot}5$ kJ/kg

$$\text{Overall thermal efficiency } \eta_0 = \frac{756{\cdot}5}{2687{\cdot}5} = 0{\cdot}282$$

(*b*) SECOND LAW ANALYSIS

Availability gain	*kJ/kg*	*Percentage of heat input*
Feed pump $b_1 - b_5$	3·3	0·32
Boiler $b_3 - b_1$	1066·5	99·68
	Σ 1069·8	Σ 100·00
Availability loss		
Turbine $b_3 - b_4$	935·0	87·50
Condenser $b_4 - b_5$	134·8	12·50
	Σ 1069·8	Σ 100·00

The irreversibilities themselves may be computed with the aid of equation (14.24c) or alternatively, using the definitions for effectiveness for energy transfer *from* the system (equation (14.25a)) and energy transfer *to* the system (equation (14.25b)), as the difference between the expressions in the denominator and numerator of these equations.

For the *feed pump* (equation (14.41))

$$\text{Effectiveness } \varepsilon = \left|\frac{b_1 - b_5}{h_1 - h_5}\right| = \left|\frac{3{\cdot}3}{3{\cdot}5}\right| = 0{\cdot}944$$

i.e. irreversibility $\Delta i_{51} = 3{\cdot}50 - 3{\cdot}30 = 0{\cdot}2$ kJ/kg

For the *boiler* (equation (14.27))
(assuming a heat source temperature $T_s = 1650°\text{C}$)

$$\text{Effectiveness } \varepsilon = \left|\frac{b_3 - b_1}{(h_3 - h_1)(T_0 - T_s)/T_s}\right| = \left|\frac{1066{\cdot}5}{2687{\cdot}5 \times \left(-\dfrac{1645}{1923{\cdot}15}\right)}\right|$$

$$= 0{\cdot}463$$

i.e. irreversibility $\Delta i_{13} = 2292 - 1066{\cdot}5 = 1225{\cdot}5$ kJ/kg

For the *turbine* (equation (14.36)),

$$\text{Effectiveness } \varepsilon = \left|\frac{h_3 - h_4}{b_3 - b_1}\right| = \left|\frac{760{\cdot}0}{935{\cdot}0}\right| = 0{\cdot}814$$

i.e. irreversibility $\Delta i_{34} = 935{\cdot}0 - 760{\cdot}0 = 175{\cdot}0$ kJ/kg

For the *condenser* (equation (14.27), inverted)

$$\text{Effectiveness } \varepsilon = \left|\frac{(h_4 - h_5)(T_0 - T_4)/T_4)}{b_4 - b_5}\right| = \left|\frac{134{\cdot}8}{134{\cdot}8}\right| = 1{\cdot}0$$

i.e. irreversibility $\Delta i_{45} = 134{\cdot}8 - 134{\cdot}8 = 0$ kJ/kg

Alternatively, if the heat transferred from the condensate is regarded as irretrievably lost, instead of leading to a hypothetical availability gain of the surroundings, the irreversibility is simply the loss of availability of the condensate

$$\Delta i_{45} = b_4 - b_5 = 134{\cdot}8 \text{ kJ/kg}$$

It is evident that the second law analysis provides much more detailed information than the simple first law analysis. In particular, the various irreversibilities may readily be isolated and their importance compared.

It is immediately obvious that the most serious irreversibility occurs in the boiler, due to the very large temperature difference between the heat source, 1650°C, and the steam in the boiler, 230·0°C. In practice, this irreversibility is reduced by incorporating superheaters which abstract some heat from the furnace gases before the boiler, and economizers which pre-heat the feed water and abstract heat from the furnace gases after the boiler.

The usefulness of a second law analysis becomes progressively greater as the complexity of the plant increases, since a quantitative assessment of the various irreversibilities may suggest detailed improvements.

Note: As a check on the above calculations it will be found that the total gain in availability (1069·8 kJ/kg) less the irreversibility in the turbine, condenser and feed pump, regarding the loss of availability of the condensate as a total loss, (175·0 + 134·8 + 0·2) = 310·0 kJ/kg equals the work delivered by the cycle (760 kJ/kg).

14.7 Availability Analysis of Processes involving Chemical Changes

Under this heading two types of processes will be considered—

(*a*) The constant volume combustion process in a closed system, i.e. the bomb calorimeter process (see section 13.4).

(*b*) The constant pressure combustion process in an open or flow system, i.e. the gas calorimeter process (see section 13.4).

(*a*) CONSTANT VOLUME COMBUSTION PROCESS (1 kg of reactants and products) (*Fig.* 14.12)

The isothermal constant volume process 1–2 (*bomb calorimeter*)
Considering the process analysed in section 13.3.1, in which the reactants are initially at pressure p_0 and temperature T_0, and in which the products are returned to the original temperature T_0, though at a different pressure p_0' such as to allow the volume occupied by the products to be equal to that of the reactant, then the internal energy of combustion is defined as

$$\Delta u_0 = (u'_P)_0 - (u'_R)_0 \tag{14.45}$$

where $(u'_P)_0$ = internal energy of products at temperature T_0,
$(u'_R)_0$ = internal energy of reactants at temperature T_0.

The internal energy of combustion, Δu_0, may be represented in the temperature–entropy diagram (*Fig.* 14.18) by resolving the isothermal process 1–2 into an adiabatic combustion process 1–3 during which the temperature rises to T_{max}, followed by reversible constant volume heat abstraction 3–2, until the product temperature is equal to the initial temperature of the reactants, i.e. T_0.

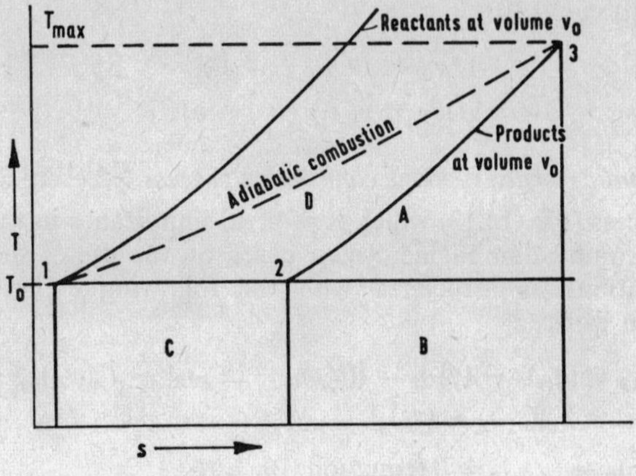

Fig. 14.18 Temperature–entropy diagram for closed system combustion process

The heat abstracted during process 3–2, q_{32}, is given by the negative value of the sum of the areas *A* and *B*; this is numerically equal to the internal energy of combustion, Δu_0.

Thus, for the adiabatic process 1–3, applying the non-flow energy equation, and putting $q_{13} = w_{13} = 0$

$$q_{13} - w_{13} = 0 = (u'_P)_3 - (u'_R)_0 \tag{i}$$

while, for the diabatic reversible process 3–2, putting $w_{32} = 0$

$$-q_{32} = (u'_P)_0 - (u'_P)_3 \qquad \text{(ii)}$$

so that, combining equations (i) and (ii)

$$-q_{32} = (u'_P)_0 - (u'_R)_0 = \Delta u_0 \qquad (14.45a)$$

The maximum work that could be obtained in the change from reactants to products, each at v_0 and T_0, is the change in availability with sign reversed, using equation (14.9a) for the closed systems, i.e.

$$\begin{aligned} w_{max} &= -[(a_P)_0 - (a_R)_0] = -[(u'_P)_0 - T_0(s_P)_0 + p_0v_0] + \\ &\quad + [(u'_R)_0 - T_0(s_R)_0 + p_0v_0] \\ &= -[(u'_P)_0 - (u'_R)_0] + T_0[(s_P)_0 - (s_R)_0] \qquad (14.46) \\ &= -\Delta u_0 + T_0[(s_P)_0 - (s_R)_0] \qquad (14.46a) \end{aligned}$$

This ideal work output is clearly represented in *Fig.* 14.18 by the sum of the areas *A*, *B* and *C*.

It should be noted that equation (14.46) may also be rewritten

$$w_{max} = (f_R)_0 - (f_P)_0 \qquad (14.46b)$$

where f is the Helmholtz free energy function $f = u - Ts$ (14.47) (see equation (14.10))

i.e.

$$(f_R)_0 = (u'_R)_0 - T_0(s_R)_0 \qquad (14.47a)$$

$$(f_P)_0 = (u'_P)_0 - T_0(s_P)_0 \qquad (14.47b)$$

The adiabatic constant volume combustion process 1–3 (1 kg reactants)

This process (*Fig.* 14.18) which is of great importance in theoretical internal combustion engine cycles, based on the use of the perfect gas equations, is associated with the following expression for maximum work

$$\begin{aligned} (w_{max})_{13} &= (f_R)_0 - (f_P)_3 = [(u'_R)_0 - (u'_P)_3] - T_0[(s_R)_0 - (s_P)_3] \\ &= T_0[(s_P)_3 - (s_R)_0] \qquad (14.48) \end{aligned}$$

since $(u'_R)_0 - (u'_P)_3 = 0$ (equation (i)).

Equation (14.48) may also be written with $T_3 = T_{max}$

$$\begin{aligned} w_{max} &= T_0[(s_P)_0 - (s_R)_0] + T_0[(s_P)_3 - (s_P)_0] \\ &= [(f_R)_0 - (f_P)_0] + [(u'_P)_0 - (u'_R)_0] + T_0(c_v)_P \ln\left(\frac{T_3}{T_0}\right) \\ &= -\Delta f_0 + \Delta u_0 + T_0(c_v)_P \ln\left(\frac{T_{max}}{T_1}\right) \qquad (14.48a) \end{aligned}$$

Note on sign convention. The internal energy of combustion, Δu_0, is clearly a negative quantity, since it is numerically equal to the heat *abstracted* during the process 3–2 (*Fig.* 14.18), the latter being, by convention, a negative quantity. The availability clearly decreases during the combustion process, so that the maximum work obtainable is the *negative* of the change in availability, as given by equation (14.46).

(*b*) CONSTANT PRESSURE COMBUSTION PROCESS (1 kg of reactants and products) (*Fig.* 14.19)

The isothermal constant pressure process 1–2 (*gas calorimeter*)

Considering the process analysed in section 13.3.1, in which the reactants are initially at pressure p_0 and temperature T_0, this also

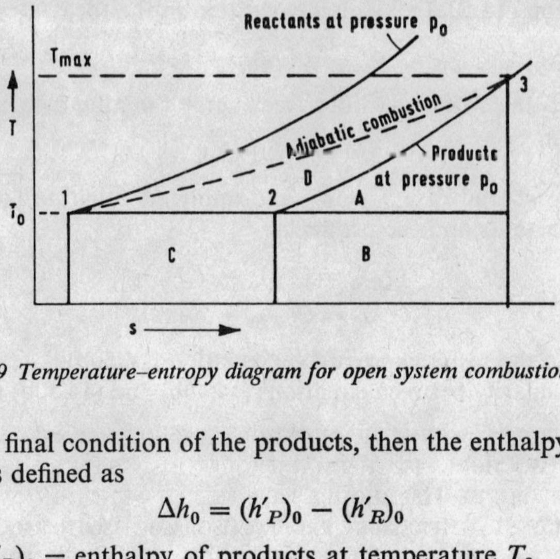

Fig. 14.19 Temperature–entropy diagram for open system combustion process

being the final condition of the products, then the enthalpy of combustion is defined as

$$\Delta h_0 = (h'_P)_0 - (h'_R)_0 \quad (14.49)$$

where $(h'_P)_0$ = enthalpy of products at temperature T_0,

$(h'_R)_0$ = enthalpy of reactants at temperature T_0.

The enthalpy of combustion, Δh_0, may be represented in the temperature–entropy diagram (*Fig.* 14.19) by resolving the isothermal process 1–2 into an adiabatic combustion process 1–3 during which the temperature rises to T_{max} followed by reversible constant pressure heat abstraction 3–2 until the product temperature is equal to the initial temperature of the reactants, i.e. T_0. The heat abstracted during process 3–2, q_{32}, is given by the negative of the sum of the areas A and B, and this is numerically equal to the enthalpy of combustion, Δh_0. The proof is similar to that used for the constant volume process.

Hence $$q_{32} = (h'_P)_0 - (h'_R)_0 = \Delta h_0 \tag{14.49a}$$

The maximum work obtainable is again equal to the negative of the change in availability, i.e.

$$w_{max} = -[(b_P)_0 - (b_R)_0] = [(h'_R)_0 - T_0(s_R)_0] - [(h'_P)_0 - T_0(s_P)_0] \tag{14.50}$$

$$= -\Delta h_0 + T_0[(s_P)_0 - (s_R)_0] \tag{14.50a}$$

where b is the steady flow availability or Keenan b function, equation (14.23).

By analogy with the constant volume combustion process, this ideal work output is the sum of areas A, B and C in *Fig.* 14.19.

Equation (14.50a) may also be written in the form

$$w_{max} = (g_R)_0 - (g_P)_0 \tag{14.50b}$$

where g is the so called Gibbs free energy function defined as

$$g = h - Ts \tag{14.51}$$

or, in the special case $T = T_0$, and applying equation (14.51) both to the reactants and the products

$$(g_R)_0 = (h'_R)_0 - T_0(s_R)_0 \tag{14.51a}$$

$$(g_P)_0 = (h'_P)_0 - T_0(s_P)_0 \tag{14.51b}$$

leading to the result expressed by equation (14.50b).

The similarity between equations (14.46b) and (14.50b) is obvious.

THE MAXIMUM WORK OF REACTION FOR THE CONSTANT VOLUME PROCESS (ISOCHORIC–ISOTHERMAL PROCESS) IS GIVEN BY THE CHANGE IN VALUE OF THE HELMHOLTZ FREE ENERGY FUNCTION, WHILE THAT FOR THE CONSTANT PRESSURE PROCESS (ISOBARIC–ISOTHERMAL PROCESS) IS GIVEN BY THE CHANGE IN VALUE OF THE GIBBS FREE ENERGY FUNCTION.

The adiabatic constant pressure process 1–3

Fig. 14.19 which is of great importance in all steady flow combustion processes, including those in gas turbines and boilers, is associated with the following expressions for maximum work

$$(w_{max})_{13} = (b_R)_0 - (b_P)_3 = [(h'_R)_0 - (h'_P)_3] - T_0[(s_R)_0 - (s_P)_3]$$
$$= T_0[(s_P)_3 - (s_R)_0] \tag{14.52}$$

(by analogy with equation (14.48)).

Equation (14.52) may also be written, with $T_3 = T_{max}$

$$w_{max} = T_0[(s_P)_0 - (s_R)_0] + T_0[(s_P)_3 - (s_P)_0]$$

$$= [(g_R)_0 - (g_P)_0] + [(h'_P)_0 - (h'_R)_0] + T_0(c_p)_P \ln\left(\frac{T_{max}}{T_0}\right)$$

$$= -\Delta g_0 + \Delta h_0 + T_0(c_p)_P \ln\left(\frac{T_{max}}{T_0}\right) \tag{14.52a}$$

(by analogy with equation (14.48a)).

The quantities Δg_0 and Δh_0 may be determined from tables such as those of Geyer and Bruges (ref. 14.1). The temperature after combustion, T_{max}, and the entropy difference between products and reactants may also be determined using gas tables.

In general the quantities g, h, s, etc., have to be computed for the reactants and products in terms of their known volumetric or gravimetric composition, using and rules governing the behaviour of mixtures as outlined in Chapter 12.

(*c*) Evaluation of irreversibility (general case)

The irreversibility may be determined using the approach adopted respectively for closed or open systems, and resulting in equation (14.12c) for the former and the identical equation (14.24c) for the latter. Taking the various cases described above, the following relationships result—

(*i*) *Constant volume combustion, reactants and products at* T_0 (*closed systems*) (*Fig.* 14.18, process 1–2)

Applying equation (14.12c) the irreversibility becomes

$$\Delta i_{12} = \int_1^2 T\, ds - q_{12} \tag{i}$$

In fact it is more convenient to use equation (14.12b), i.e.

$$\Delta i_{12} = \int_1^2 (T - T_0)\, ds - \Delta a_{12} - \underset{=0}{(w_{useful})_{12}}$$

Assuming that the rate of heat abstraction during the process is such as to maintain truly isothermal conditions throughout, i.e. $T = T_0$ along the entire integration path 1–2, $\int_1^2 (T - T_0)\, ds$ vanishes whence

$$\Delta i_{12} = -\Delta a_{12} = -[(a_P)_0 - (a_R)_0]$$

$$= -\Delta u_0 + T_0[(s_P)_0 - (s_R)_0] \tag{14.53}$$

The right-hand side of equation (14.53) is identical with that of equation (14.46a), indicating that for the isochoric–isothermal combustion process the irreversibility is numerically equal to the negative of the availability change, represented by the sum of the areas A, B and C.

(*ii*) *Adiabatic constant volume combustion, reactants at* T_0, *products at* T_{max} (*Fig.* 14.18, process 1–3) (closed systems)

Again applying equation (14.12b)

$$\Delta i_{13} = \int_1^3 (T - T_0)\, ds - \Delta a_{13} - \underset{=0}{(w_{useful})_{13}} \qquad (14.54)$$

But from equation (14.48)

$$w_{max} = -\Delta a_{13} = T_0[(s_P)_3 - (s_R)_0] = \text{area } (B + C)$$

while

$$\int_1^3 (T - T_0)\, ds = \text{area } (D + A)$$

The irreversibility for the adiabatic isochoric combustion process is thus represented by the sum of the areas A, B, C and D.

(*iii*) *Constant pressure combustion, reactants and products at* T_0 (*Fig.* 14.19, process 1–2) (open system)

Applying equation (14.24b)

$$\Delta i_{12} = \int_1^2 (T - T_0)\, ds - \Delta a_{12} - \underset{=0}{(w_{useful})_{12}} \qquad (14.55)$$

Assuming, as in case (i), that the rate of heat abstraction is such as to maintain truly isothermal conditions throughout, expression (14.55) for the irreversibility is once again equal to the negative of the availability change, given by equation (14.50a) which in turn is represented by the sum of the areas A, B and C.

(*iv*) *Adiabatic constant pressure combustion, reactants at* T_0, *products at* T_{max} (*Fig.* 14.19, process 1–3) (open system)

Again applying equation (14.24b)

$$\Delta i_{13} = \int_1^3 (T - T_0)\, ds - \Delta a_{13} - \underset{=0}{(w_{useful})_{13}} \qquad (14.56)$$

But from equation (14.52),

$$w_{max} = -\Delta b_{13} = T_0[(s_P)_3 - (s_R)_0] = \text{area } (B + C)$$

while

$$\int_1^3 (T - T_0)\, ds = \text{area } (D + A)$$

The irreversibility for the adiabatic isobaric combustion process is thus represented by the sum of the areas *A*, *B*, *C* and *D*.

REFERENCE

14.1 GEYER, E. W., and BRUGES, E. A., *Tables of Properties with Dissociation Theory and its Applications* (Longmans, 1948).

14.2 BRUGES, E. A., *Available Energy and the Second Law Analysis* (Academic Press, 1959).

14.3 NASH, L. K., *Elements of Chemical Thermodynamics* (Addison-Wesley, 1962).

14.4 COULE, J. and STUART, B., *Equilibrium Thermodynamics* (John Wiley, 1964).

EXERCISES

Take $T_{surr} = 15°C$, $p_{surr} = 1$ bar unless stated otherwise.

14.1 Find the increase in availability and the irreversibility corresponding to the following processes in each of which the system is 1 kg of H_2O which is initially at 15 bar and 300°C. (Assume perfect gas behaviour.)

(*a*) The system is confined at constant pressure by a piston and is heated until its volume is doubled.

(*b*) The same as (*a*) except that the change results from stirring instead of heating.

(*c*) The system expands reversibly and adiabatically behind a piston until its pressure is 10 bar.

(*d*) It expands reversibly and isothermally behind a piston until its volume is doubled.

(*e*) It expands adiabatically into an adjacent chamber which is initially evacuated. The final pressure is 10 bar.

(*f*) It is in motion with a total kinetic energy of 2·5 kJ/kg. Frictional effects cause it to come to rest. Assume no heat transfer and no change in pressure.

(*Ans:* (*a*) 690, 0 kJ/kg; (*b*) 690, 690 kJ/kg; (*c*) −69·94, 0 kJ/kg; (*d*) −74·35, 0 kJ/kg; (*e*) −27·5, 107·5 kJ/kg; (*f*) −1·51, 2·5 kJ/kg)

14.2 In each of the following processes a steady stream of H_2O enters an apparatus at low velocity and at 15 bar and 300°C. Find for each process the increase in availability of each kilogramme of H_2O as it passes through the apparatus, and their irreversibility.

(*a*) The stream is heated slowly at constant pressure until its temperature is 500°C.

(*b*) It expands reversibly and adiabatically in a nozzle to 10 bar.

(*c*) It expands reversibly and adiabatically to 10 bar in a turbine which it leaves at low velocity.

(*d*) It expands adiabatically to 10 bar through a porous plug which it leaves at low velocity.

(*e*) It expands isothermally to 10 bar through a porous plug which it leaves at low velocity.
(*f*) It expands adiabatically to 10 bar through a turbine of 60% efficiency, leaving at low velocity.
(*g*) It is compressed reversibly and isothermally to 30 bar.
(*Ans:* (*a*) 246·5, 0 kJ/kg; (*b*) 0, 0 kJ/kg; (*c*) −98·0, 0 kJ/kg; (*d*) −52·5, 105·7 kJ/kg; (*e*) as per (*d*); (*f*) −74·3, +30·8 kJ/kg; (*g*) 65, 0 kJ/kg)

14.3 (*a*) 105 kJ of heat is transferred slowly away from a constant temperature reservoir at 65°C which is exposed to a constant pressure of 1 bar. Find the increase in availability.
(*b*) The same as for (*a*) except that −45°C is to be substituted for 65°C.
(*c*) 105 kJ of heat is transferred slowly from the reservoir of (*a*) to that of (*b*). Find the increase in availability of the system comprising both reservoirs and find the total entropy change.
(*Ans:* (*a*) 89·5 kJ/kg; (*b*) −133 kJ/kg; (*c*) −43·5 kJ/kg, 0·150 kJ/kg-K)

14.4 A gaseous system under a pressure of 1 bar and having a constant specific heat at constant pressure of 12·6 kJ/kg-K is used to supply heat for the evaporation of 1 kg of water from the saturated liquid state to the saturated vapour state at a pressure of 1 bar. If the initial temperature of the gaseous system is 1100°C, find values of the increase in availability for the gas and for the water, and find the entropy change of the process as a whole.
(*Ans:* −1728 kJ/kg_{steam}, +510 kJ/kg_{steam}, +4·41 kJ/kg_{steam}-K)

14.5 (*a*) Show that the effectiveness of a turbine of constant stage effectiveness is identical with the effectiveness of each stage.
(*b*) Derive the statement that the effectiveness of a turbine is the weighted mean of the values of the effectiveness for its stages.
(*c*) Show that the condition curve of a turbine having uniform stage effectiveness is a straight line on the *h–s* chart and on the *b–s* chart.

14.6 Saturated steam at 4 bar is fed to a mixing type of feed-water heater. Feed water is pumped from a condenser, which it leaves at a pressure of 0·035 bar and in the saturated state, into the feed-water heater. The combined streams leave the heater as saturated liquid at 4 bar. Assume the compression in the feed pump to be reversible and adiabatic.
(*a*) How much steam enters the heater for each kg of feed water entering?
(*b*) Find the increase in availability for each kg of liquid leaving the heater.
(*c*) Find the effectiveness of the heater.
(*Ans:* (*a*) 0·2312 kJ/kg; (*b*) 93·6 kJ/kg; (*c*) 0·62)

14.7 Steam at 8 bar and 200°C undergoes a non-flow isothermal expansion to 5 bar. Find the change in availability of the cylinder contents and compare it with that for an isentropic process between the same pressure limits, starting from the same initial conditions.
(*Ans:* −38·9 kJ/kg, −57·6 kJ/kg)

14.8 Air at 8 bar and 93°C enters a turbine with a velocity of 100 m/s. It leaves at 1·25 bar, 27°C, and with a velocity of 60 m/s. Find the maximum useful work obtainable per kg of air and compare this with the actual work done. Assume that air is a perfect gas with $c_p = 1{\cdot}005$ kJ/kg-K.
(*Ans:* 151·2 kJ/kg, 69·6 kJ/kg)

14.9 Superheated steam at 30 bar and 300°C enters a turbine and emerges at a pressure of 3 bar, 99·5% dry saturated. Calculate the turbine efficiency, effectiveness and loss of available energy, assuming the process to be adiabatic. Kinetic energy effects may be neglected.
(*Ans:* 59·6%, 67·5%, 400 kJ/kg)

14.10 In a Carnot cycle heat is supplied at 250°C and rejected at 80°C. The working fluid is water which, during heat reception, evaporates completely at 250°C. Assuming 1 kg of working fluid in continuous circulation, calculate

(*a*) the change of entropy during heat reception and rejection,
(*b*) the available energy,
(*c*) the change in available energy when the sink temperature is lowered by 55°C.

(*Ans:* (*a*) ±3·273 kJ/kg-K; (*b*) 557·5 kJ/kg; (*c*) 180 kJ/kg)

14.11 Recalculate Exercise 14.10 on the assumption that, with the same quantity of heat transferred, the temperature of the heat source decreases 60°C, according to the law of the form:

$$\frac{dT}{dQ} = -\text{constant}$$

(*Ans:* (*a*) ±3·475 kJ/kg-K; (*b*) 480 kJ/kg; (*c*) 192 kJ/kg)

14.12 Defining the effectiveness of a cycle as

$$\varepsilon = \frac{\text{net work}}{\text{available energy}}$$

calculate both effectiveness and efficiency for the cycle described in Exercise 14.11.
(*Ans:* 0·678, 0·231)

14.13 Heat is supplied to a Carnot steam cycle by cooling the furnace gases from 1650°C to within 55°C of the maximum temperature, the cycle operating between 140 bar (saturated liquid to saturated vapour), and 0·070 bar. Calculate

(*a*) the stack loss in kJ/kg of flue gas,
(*b*) the loss of availability due to the temperature difference between flue gas and steam, per kg of steam,
(*c*) the effectiveness of the cycle (see Ex. 14.12).

Take c_p for flue gas as 1·05 kJ/kg-K.

(*Ans:* (*a*) 103·1 kJ/kg_{steam}; (*b*) 244 kJ/kg_{steam}; (*c*) 0·621)

14.14 What is the maximum useful work which can be obtained when 100 kJ are abstracted from a heat reservoir at 675 K in an environment at 288 K: What is the loss of useful work if

(*a*) a temperature drop of 50°C is introduced between the heat source and the heat engine, on the one hand, and the engine and the heat sink, on the other,
(*b*) the source temperature drops by 50°C and the sink temperature rises by 50°C during the heat transfer process according to the linear law

$$\frac{dQ}{dT} = \pm \text{constant}$$

(*Ans:* (*a*) 11·20 kJ; (*b*) 5·25 kJ)

14.15 The pressure and temperature of the products of combustion in an engine cylinder at the instant of exhaust opening are 6·28 bar and 2090°C, respectively. The gas is discharged instantaneously, the pressure of the gas remaining in the cylinder falling isentropically to atmospheric (1·013 bar). Assuming the gases to have the properties of air (c_p = 1·05 kJ/kg-K), calculate

(*a*) the loss of availability during the "blowdown" process,
(*b*) the maximum useful work obtainable from the exhausted gases.

Comment on the difference, if any, between (*a*) and (*b*).

(*Ans:* (*a*) 898 kJ/kg; (*b*) 500 kJ/kg)

14.16 Air flows isothermally through a horizontal 0·15 m bore pipe at the rate of 18 kg/s and at 15°C. The initial and final pressure are respectively 7 bar and 5·5 bar. Calculate, per kg of air,

(*a*) the change of entropy
(*b*) the change of availability
(*c*) the irreversibility
(*d*) the external heat addition.

(*Ans:* (*a*) 0·0694 kJ/kg; (*b*) −20 kJ/kg; (*c*) 15·5 kJ/kg;
(*d*) 4·49 kJ/kg)

14.17 Steam at 7 bar, dry saturated, mixes in an open heater with water at the same pressure and having a specific enthalpy of 110 kJ/kg.

The water leaving the heater is saturated, and at 7 bar. Calculate, per kg of water at exit,

(*a*) the mass of entering steam and water,
(*b*) the change of availability,
(*c*) the effectiveness of the heater.

(*Ans:* (*a*) 0·221 kg, 0·779 kg; (*b*) −59·6 kJ/kg; (*c*) 0·617)

14.18 Show that the availability for a steady flow process in the absence of kinetic energy effects, and neglecting gravitational, magnetic or electric field effects, is a property function given by

$$a = h - T_0 s = b$$

where h = specific enthalpy
T_0 = absolute temperature of surroundings
s = specific entropy
b = Keenan function.

Hence deduce expressions for the loss of availability and the irreversibility in the course of adiabatic expansion of the working fluid through a turbine stage. The process should be represented in suitable diagrams.

Steam at 15 bar, 200°C, enters a turbine stage having an isentropic efficiency of 85%, and is expanded to 7 bar. Find the following:

1. the final condition of the steam,
2. the loss of availability,
3. the irreversibility,
4. the effectiveness of the stage.

The surroundings are at a temperature of 21°C.

(*Ans:* $x = 0{\cdot}955$, $\Delta a = -140$ kJ/kg, $\Delta i = 20{\cdot}6$ kJ/kg, $\varepsilon = 0{\cdot}906$)

14.19 The compression process in a gas turbine compressor is adiabatic and irreversible and follows the law

$$pV^n = \text{constant}$$

The working fluid is air initially at pressure p_1 and temperature T_1. If the pressure ratio is r, deduce expressions for the following:

(*a*) the work done per kg of air,
(*b*) the temperature rise,
(*c*) the entropy gain,
(*d*) the excess of work over that which would be done in a perfectly performed process between the same pressure limits,
(*e*) the isentropic efficiency of the compressor, and
(*f*) the effectiveness of the compressor.

Represent the process in a temperature–entropy diagram, and explain the significance of the area under the compression curve.

14.20 A vessel contains air at 13·80 bar and 27°C, part of which is suddenly released until the pressure in the vessel becomes 2·75 bar. After an interval of time the pressure of the air remaining in the vessel stabilizes at 4·36 bar, the temperature then being again 27°C. Stating carefully the assumptions made, calculate the value of the specific heat ratio γ. Calculate also the loss of energy available for doing useful work.

(*Ans:* 1·40; 182 kJ/kg)

14.21 A tube type condenser has a cooling water flow of 4500 kg/h, the water inlet temperature being 15°C, while the steam enters at 0·035 bar with a dryness fraction of 0·8. Assuming that the steam is fully condensed at outlet and that the minimum temperature difference between steam and cooling water is 5·5°C, and that the process as a whole is adiabatic, find the following:

(*a*) the cooling water outlet temperature,
(*b*) the ratio of steam to cooling water flow,
(*c*) entropy changes for cooling water and steam,
(*d*) combined entropy change, and
(*e*) the loss of availability and effectiveness.

Comment on the nature of the irreversibilities present, and suggest means of reducing them.

(*Ans:* (*a*) 21·2°C; (*b*) 0·0133 kg/kg; (*c*) +405·0 kJ/K, – 388·2 kJ/K; (*d*) +16·8 kJ/K; (*e*) −3460 kJ, 0·356)

Chapter 15

Power and Refrigeration Cycles

The concept of a thermodynamic cycle, i.e. of a succession of processes undergone by the working fluid and terminating in a state identical with the initial state, has been referred to repeatedly in this book (see section 4.2). It was first used in the derivation of the First Law of Thermodynamics (section 4.3) and later also in the derivation of the Second Law (section 6.3) and the second, third and fourth corollaries (section 6.4) which made use of the *Carnot cycle* as the ideal reference cycle for all *heat engines*, i.e. of all devices for converting heat into useful work. Again the concept of entropy was introduced in section 7.1 as arising out of a particular integration process $\oint dQ/T$ applied to ideal or reversible cycles, whilst finally in section 14.6 availability or second law analysis was applied to a complete steam cycle (the Rankine cycle) to illustrate the identification of irreversibilities, and thus indirectly point the way to improvements both in the cycle itself and in the components in which the individual processes are performed.

The above considerations have been essentially of a theoretical nature, but in a more practical sense cycles constitute the most important application of engineering thermodynamics. It is fitting, therefore, that the last chapter of this book should be devoted to a brief survey of cycles and of heat engines as the apparatus in which these cycles and the resultant conversion of heat into work are performed.

In Chapters 8 and 10 the properties of VAPOURS, on the one hand, and of GASES, on the other, were discussed in detail. Heat engines may conveniently be divided into those in which the working fluid undergoes a change of phase, usually from liquid to saturated or superheated vapour and vice versa, and those which use gases as the working fluid. Accordingly, this chapter has two major subsections dealing with, respectively, VAPOUR POWER CYCLES and GAS POWER CYCLES. The former include all steam power plant as used in large power stations, irrespective of whether the heat input to the cycle is provided by CHEMICAL ENERGY, i.e. combustion of fossil fuels such as gas, oil or coal, or by ATOMIC ENERGY, i.e. nuclear fission (see section 14.2). Gas power cycles include both GAS TURBINES in which the work transfer processes take place in rotating machinery, i.e. aerodynamic compressors and turbines, and RECIPROCATING ENGINES, particularly internal and external combustion engines. Yet a further subdivision of gas power cycles applicable both to gas turbines and reciprocating engines is into CLOSED and OPEN cycles. In closed cycles the working fluid is used over and over again and, having undergone a series of flow or non-flow processes, with positive or negative heat or work transfer, returns to its original state. In open cycles, on the other hand, the working fluid, usually air, enters the plant to undergo a series of flow or non-flow processes with heat or work transfer and, in addition, takes part in a process involving chemical change, i.e. combustion. Since the latter process is inherently unidirectional, the working fluid cannot be used over and over again, but is rejected having undergone this sequence of processes. Plant operating on open cycles does not conform to the simple definition of a heat engine in section 6.2 as a system operating in a cycle and producing useful work by abstracting heat from a suitable heat source.

Nevertheless, such cycles lend themselves to a simplified form of analysis in which the chemical changes undergone by air and fuel are ignored, and replaced simply by an equivalent heat transfer to a hypothetical closed cycle employing air as the working fluid. Such simplified equivalent cycles are known as AIR STANDARD CYCLES (see *Figs*. 13.2 and 13.3).

The rather complicated subdivisions of gas power cycles are represented in *Fig*. 15.1.

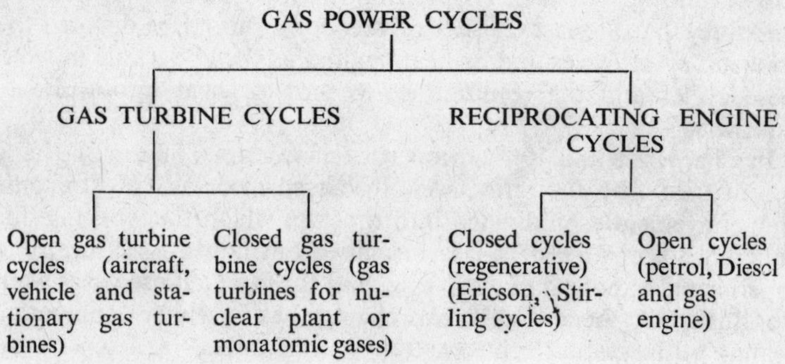

Fig. 15.1

In addition to the WORK-PRODUCING CYCLES outlined above, we have to consider briefly the category of REVERSED HEAT ENGINE

CYCLES (see section 6.2, *Fig.* 6.2) in which a work input results in heat being transferred from a low temperature to a high temperature reservoir. REFRIGERATION CYCLES abstract heat from the system to be maintained at a temperature below that of the surroundings, and reject a larger quantity of heat to the surroundings. (The difference between absorbed and rejected heat quantities being the work input to the cycle.) HEAT PUMP CYCLES abstract heat from the surroundings, and reject it to the system to be maintained at a higher temperature than that of the surroundings. In practice, reversed heat engine cycles, like heat engine cycles, may use gases or vapours as the working fluid. They are however, almost invariably, of the closed type.

15.1 Vapour Power Cycles

The most important parameter associated with any cycle is its OVERALL THERMAL EFFICIENCY (see section 6.2) defined as

$$\text{Efficiency } \eta_{overall} = \frac{\text{Net work output}}{\text{Latent heat in fuel}} \tag{15.1}$$

It has already been shown in Chapter 6 that this efficiency can never exceed unity and that the highest possible efficiency attainable for given upper and lower temperature limits, T_{max} and T_{min}, respectively is that of the Carnot cycles (see section 6.4) which is given by

$$\eta_{carnot} = 1 - \frac{T_{min}}{T_{max}} \tag{15.1a}$$

The Carnot cycle is therefore used as the ultimate standard of reference.

However, thermal efficiency is not the only important criterion. Other important parameters are:

(*a*) The WORK RATIO which is a measure of the susceptibility of the cycle to irreversibilities and is defined as the ratio

$$r_w = \frac{\text{Net cycle work } \oint dw}{\text{Positive work } \int_A^B dw} \tag{15.2}$$

where A and B are the initial and final conditions for that part of the cycle for which work transfer is positive, i.e. from the system to the surroundings. (See *Fig.* 15.2.)

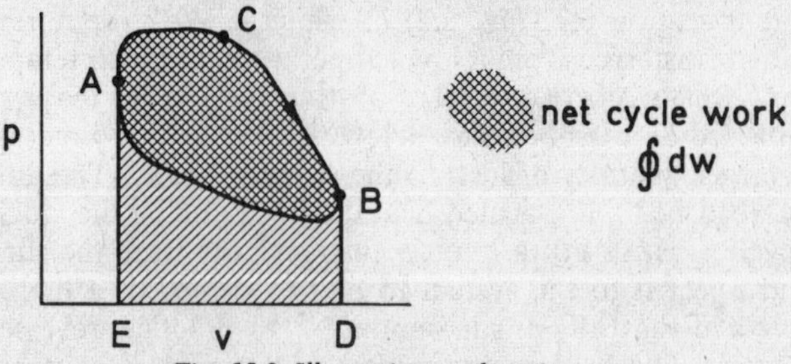

Fig. 15.2 Illustrating work ratio

$$\text{Positive work} = \text{area ACBDE} = \int_A^B (dw)_c$$

A high work ratio is desirable in practice.

(*b*) SPECIFIC CONSUMPTION is defined as the rate of flow of working fluid per unit power, i.e.

$$\text{sc} = \frac{m(\text{kg})}{[\oint dW](\text{kJ})} \tag{15.3}$$

and is a measure of the size of the plant. The higher the useful energy abstracted from unit mass of working fluid, the smaller the mass flow for a given power, and hence the more compact the plant.

The suitability of particular vapour and gas power cycles will therefore be judged with reference to each of the above criteria as expressed by equations (15.1), (15.2) and (15.3).

All vapour power cycles are of the CLOSED type, i.e. they have an external heat source while the working fluid, usually steam or water, circulates continuously through the various components of the plant, in each of which a steady flow process is performed (see sections 4.6, 4.7 and 4.8). The simplest arrangement for such a cycle is shown in *Fig*. 13.1, but is repeated here for convenience (*Fig*. 15.3).

The heat source in such cycles is the combustion chamber or furnace in which the chemical energy of the fuel is converted into heat (Q_{in} in *Fig*. 13.1). This process itself is subject to limitations (see section 13.5) and hence a combustion efficiency η_c should be assigned to this process.

The overall thermal efficiency (equation (15.1)) is thus the product of the combustion efficiency η_c and the cycle efficiency η_{th}, i.e.

$$\eta_{overall} = \eta_c \times \eta_{th} = \left(\frac{\text{Gross heat input to cycle}}{\text{Latent heat in fuel}}\right) \times \left(\frac{\text{Net cycle work}}{\text{Gross heat input to cycle}}\right) \tag{15.4}$$

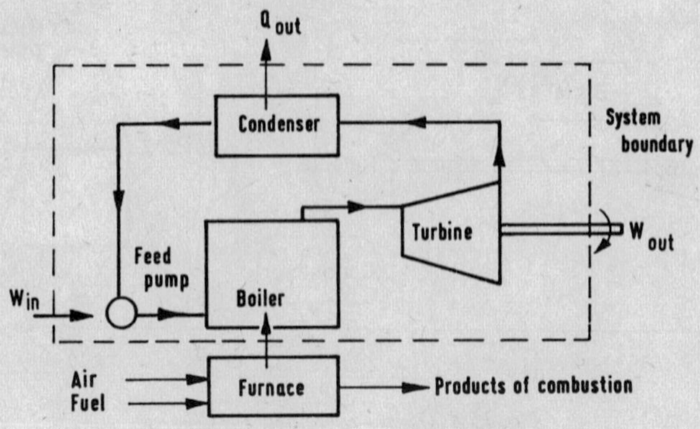

Fig. 15.3 Steam power cycle

Since combustion efficiency has already been dealt with, and may in any case be assumed to be close to unity in a well designed system, we shall confine our attention in the present chapter to cycle efficiency

$$\eta_{th} = \frac{\text{Net cycle work}}{\text{Gross heat input to cycle}} = \frac{\oint dW}{Q_{gross}} \qquad (15.5)$$

which thus replaces the overall thermal efficiency η_{th} (equation (15.1)) as the criterion to be applied.

15.1.1 The Carnot Cycle (*Fig.* 15.4)

This is the fundamental reference cycle for all heat engines, its efficiency being given by equation (15.1a). In section 6.4 it was discussed in its perfect gas form and will now be described as applied to a condensible vapour. *Fig.* 15.4 shows the basic plant and the working cycle in the temperature–entropy diagram. The cycle consists of the following steady flow processes and may be considered as applying, in this instance, to steam:

1–2′: isentropic compression of initially wet steam (state 1) to form saturated water (state 2′) between pressure limits p_1 and $p_{2'}$ and temperature limits $T_1 = T_{min}$ and $T_{2'} = T_{max}$.

2′–3: isothermal heat addition in boiler at $T_{2'} = T_3 = T_{max}$ to convert saturated water (state 2′) into dry saturated steam (state 3). The pressure remains constant, i.e. $p_{2'} = p_3$.

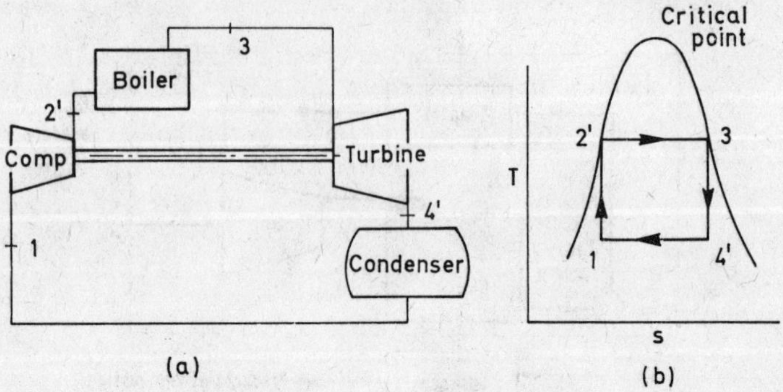

Fig. 15.4 Carnot cycle and T-s diagram

3–4′: isentropic expansion in turbine to pressure $p_{4'} = p_1$ to form wet steam (state 4′) at lower cycle temperature T_{min}.

4′–1: isothermal heat rejection in condenser at $T_{2'} = T_{1'} = T_{min}$ from state 4′ to state 1, i.e. partial condensation. The pressure remains constant, i.e. $p_{4'} = p_1$.

Applying the three criteria of thermal efficiency η_{th} (equation (15.5)), work ratio r_w (equation (15.2)), and specific consumption SC (equation (15.3)) to the cycle, the following relationships result:

$$\text{Cycle efficiency } \eta_{th} = \frac{\oint dW}{Q_{gross}} = \frac{\oint dw}{\oint q_{gross}} \text{ (on unit mass basis)}$$

Now $$\oint dw = w_{34'} + w_{12'} = (h_3 - h_{4'}) - (h_{2'} - h_1)$$

and $$q_{gross} = q_{2'3} = h_3 - h_{2'}$$

whence $$\eta_{th} = \frac{(h_3 - h_{4'}) - (h_{2'} - h_1)}{(h_3 - h_{2'})}$$

$$= 1 - \frac{h_{4'} - h_1}{h_3 - h_{2'}}$$

$$= 1 - \frac{(\Delta s_{14'})T_{min}}{(\Delta s_{2'3})T_{max}}$$

$$= 1 - \frac{T_{min}}{T_{max}} \tag{15.6}$$

since $\Delta s_{14'} = \Delta s_{2'3}$

as expected, since the Carnot efficiency is independent of the working fluid,

$$\text{Work ratio } r_w = \frac{\oint(dw)}{\int dw_{positive}}$$

$$= \frac{(h_3 - h_{4'}) - h_{2'} - h_1)}{(h_3 - h_{4'})}$$

$$= 1 - \frac{h_{2'} - h_1}{h_3 - h_{4'}} \qquad (15.7)$$

Specific steam consumption

$$\text{SSC} = \frac{1}{\oint dw} = \frac{1}{[(h_3 - h_{4'}) - (h_2 - h_{1'})]} \text{ kg/kJ} \qquad (15.8)$$

EXAMPLE 15.1

Evaluate cycle efficiency, work ratio and specific steam consumption for a Carnot cycle (*Fig*. 15.4) operating between 28 bar and 0·025 bar.

SOLUTION

(Using the notation of *Fig*. 15.4b)

At (2′) $p_{2'} = 28$ bar, $T_{2'} = 230{\cdot}0$°C, $h_{2'} = 991$ kJ/kg
$s_{2'} = 2{\cdot}611$ kJ/kg-K

At (3) $p_3 = p_{2'} = 28$ bar, $T_2 = T_3 = 230$°C, $h_3 = 2803$ kJ/kg
$s_3 = 6{\cdot}213$ kJ/kg-K

At (4′) $p_{4'} = 0{\cdot}035$ bar, $T_{4'} = 26{\cdot}7$°C, $x_{4'} = 0{\cdot}716$, $h_{4'} = 1853$ kJ/kg
$s_{4'} = s_3 = 6{\cdot}213$ kJ/kg-K

At (1) $p_1 = p_{4'} = 0{\cdot}035$ bar, $T_1 = T_{4'} = 26{\cdot}7$°C, $x_1 = 0{\cdot}274$
$h_1 = 777$ kJ/kg, $s_1 = s_{2'} = 2{\cdot}611$ kJ/kg-K

$$\eta_{th} = 1 - \frac{h_{4'} - h_1}{h_3 - h_{2'}} = 1 - \frac{1076}{1812} = 0{\cdot}404 \quad \textit{Ans.} \quad \text{(equation (15.7))}$$

cf. $$\eta_{th} = 1 - \frac{T_{min}}{T_{max}} = 1 - \frac{26{\cdot}7 + 273{\cdot}15}{230{\cdot}0 + 273{\cdot}15} = 0{\cdot}404$$

i.e. the efficiency value calculated from steam properties agrees with the general Carnot formulation in terms of temperature (equation (15.6)):

$$r_w = 1 - \frac{h_{2'} - h_1}{h_3 - h_{4'}} = 1 - \frac{214}{950} = 0{\cdot}774 \quad \textit{Ans.} \quad \text{(equation (15.8))}$$

This value deviates substantially from unity and hence indicates susceptibility to irreversibilities.

$$\text{SSC} = \frac{1}{(h_3 - h_{4'}) - (h_{2'} - h_1)} = \frac{1}{(736)} = 0{\cdot}001368 \text{ kg/kJ}$$

$$= 3600 \times 0{\cdot}001368 = 4{\cdot}91 \text{ kg/kWh} \quad \textit{Ans.}$$

The above example illustrates a number of important points.

The efficiency is very sensitive to the upper and lower temperature limits of the cycle. The former is ultimately dictated by metallurgical limitations for such components as boiler drums and turbine nozzles and blades, the latter by the availability of condenser cooling water and the necessity for a temperature gradient within the condenser for rapid heat transfer. This fixes the lower temperature limit between 25°C and 30°C whilst the upper limit is broadly in the region 650–800°C. Hence, the Carnot cycle in the example is very far removed from the upper limit, indeed the upper limit for such a cycle, i.e. the critical temperature $T_c = 374{\cdot}15$°C, is still very much below the metallurgical limit.

The susceptibility to irreversibilities as given by the work ratio r_w is very considerable for the Carnot cycle. This is illustrated by example 15.2 below. Finally, the specific steam consumption is high compared with other cycles. It is mainly on account of its low work ratio and high specific consumption that the Carnot cycle is unsuitable and is replaced in practice by the Rankine cycle (see section 14.6) and its various derivatives.

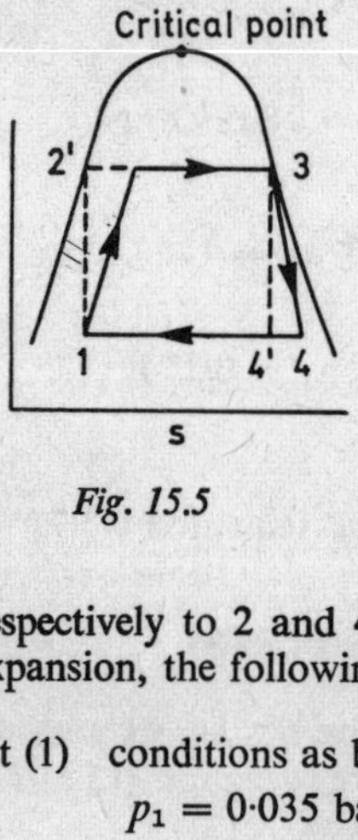

Fig. 15.5

EXAMPLE 15.2

Evaluate the modified efficiency, and specific steam consumption of the Carnot cycle of example 15.1 when the compressor and turbine operate with isentropic efficiencies of 0·8 (See sections 7.6 and 14.5. For definitions of isentropic efficiency see equations (14.39) and (14.44).)

SOLUTION

Assuming that points 1 and 3 remain unchanged, whilst the state points 2′ and 4′ move respectively to 2 and 4 as a result of the irreversible compression and expansion, the following conditions apply:

At (1) conditions as before, i.e.

$p_1 = 0{\cdot}035$ bar, $T_1 = 26{\cdot}7$°C, $x_1 = 0{\cdot}274$,
$h_1 = 777$ kJ/kg, $s_1 = 2{\cdot}611$ kJ/kg

At (2) $w_{12} = w_{12'}/\eta_{is} = (-h_{2'} + h_1)/0{\cdot}8 = -214/0{\cdot}8 = -267{\cdot}5$ kJ/kg

Therefore

$$h_2 = h_1 - w_{12} = 777 + 267{\cdot}5 = 1044{\cdot}5 \text{ kJ/kg}$$

At (3) $p_3 = 28$ bar, $T_3 = 230{\cdot}0$°C, $h_3 = 2803$ kJ/kg
$s_3 = 6{\cdot}213$ kJ/kg

At (4) $w_{34} = (w_{34'})\eta_{ts} = (h_3 - h_{4'})\eta_{ts} = 946 \times 0{\cdot}8 = 746$ kJ/kg

Therefore

$$h_4 = h_3 - w_{34} = 2803 - 760 = 2043 \text{ kJ/kg}$$

$$\eta_{th} = 1 - \frac{h_4 - h_1}{h_3 - h_2} = 1 - \frac{1266}{1758{\cdot}5} = 0{\cdot}279 \quad \textit{Ans.}$$

cf. 0·404 in ideal case

$$\text{SSC} = \frac{1}{(h_3 - h_4) - (h_2 - h_1)} = \frac{1}{478{\cdot}5} = 0{\cdot}002092 \text{kg/kJ}$$

$$= 3600 \times 0{\cdot}002092 = 7{\cdot}53 \text{ kg/kWh} \quad \textit{Ans.}$$

cf. 4·91 kg/kWh in ideal case

15.1.2 The Rankine Cycle (see section 14.6)

(*a*) Simple Rankine cycle (*Fig.* 15.6)

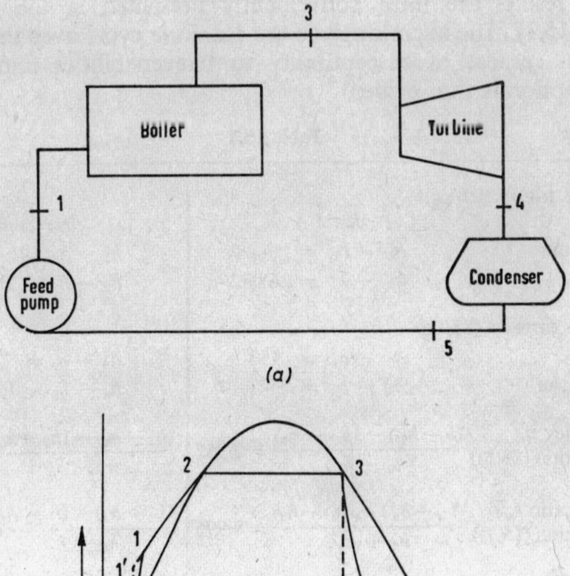

(a)

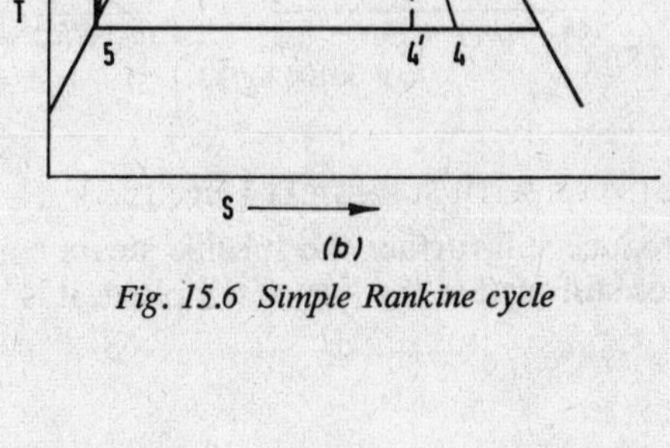

(b)

Fig. 15.6 Simple Rankine cycle

The obvious disadvantages of the Carnot cycle are its very marked susceptibility to irreversibilities, the need for a compressor handling very wet steam, and the relatively high specific consumption.

The Rankine cycle overcomes most of these difficulties. Though its theoretical efficiency, for given temperature limits, is lower than that of the Carnot cycle in practice, when irreversibilities are taken into account, it will actually show better efficiencies.

The compressor in the Carnot cycle is replaced by a feed pump whose work input is very much smaller than that of the compressor in view of the reduced specific volume of water. The condensation process can proceed right up to the saturated liquid line at 5 instead of having to be arrested at some dryness fraction x_1 as in the Carnot cycle (*Fig*. 15.6).

For numerical details the reader is referred to example 14.7 in which the temperature limits and component efficiencies are identical with those of the Carnot cycle of examples 15.1 and 15.2.

The results are most conveniently presented in tabular form (Table 15.1) The superiority of the Rankine cycle over the Carnot cycle in respect of susceptibility to irreversibilities and specific consumption is thus evident.

Table 15.1

	Ideal	Irreversible
ENERGY INPUT (kJ/kg)		
Feed pump	$h_{1'} - h_5 = 2{\cdot}8$	$h_1 - h_5 = 3{\cdot}5$
Boiler	$h_3 - h_{1'} = 2689{\cdot}2$	$h_3 - h_1 = 2687{\cdot}5$
ENERGY OUTPUT (kJ/kg)		
Turbine	$h_3 - h_{4'} = 950$	$h_3 - h_4 = 760$
Condenser	$h_{4'} - h_5 = 1741$	$h_4 - h_5 = 1931$
Efficiency η_{th} (equation (15.5))	$\dfrac{(h - h_{4'}) - (h_{1'} - h_5)}{(h_3 - h_{1'})} = 0{\cdot}352$	$\dfrac{(h_3 - h_4) - (h_1 - h_5)}{h_3 - h_1} = 0{\cdot}282$
Work ratio r_w (equation (15.7))	$\dfrac{(h_3 - h_{4'}) - (h_{1'} - h_5)}{h_3 - h_{4'}} = 0{\cdot}995$	$\dfrac{(h_3 - h_4) - (h_1 - h_5)}{h_3 - h_4} = 0{\cdot}995$
Specific consumption (equation (15.8))	$\dfrac{1}{(h_3 - h_{4'}) - (h_{1'} - h_5)} = 0{\cdot}00106$ kg/kJ	$\dfrac{1}{(h_3 - h_4) - (h_1 - h_5)} = 0{\cdot}001346$ kg/kJ

(*b*) Rankine cycle with superheat (*Fig*. 15.7)

In order to reduce still further the specific steam consumption, and thereby reduce still further the size of the plant, it is usual to employ superheat by passing the saturated steam from the boiler through separate superheater tubes in the furnace, at the saturation pressure $p_1 = p_2 = p_3 = p_4$.

Comparison of *Fig.* 15.7b with *Fig.* 15.6b shows that not only is the work output as represented by the area of the *T–s* diagram greater with superheat than in the simple cycle working between the

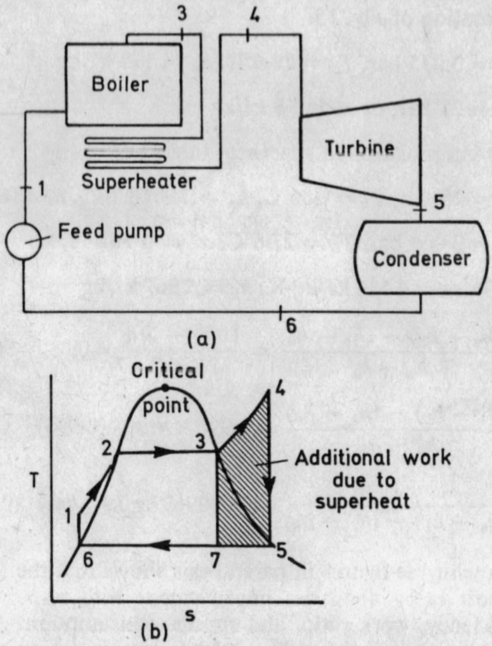

Fig. 15.7 Rankine cycle with superheat

same pressure limits, the gain being represented by area 3–4–5–7 in *Fig.* 15.7b, but the dryness fraction of the steam after expansion in the turbine is considerably higher, i.e. $x_5 > x_7$ with resultant reduction in turbine blade erosion.

In practice, therefore, superheat is always employed. Although upper pressure and temperature limits of approximately 200 bar and 650°C are perfectly feasible in modern power station plant, the advantages of the superheated Rankine cycle will be illustrated by an example in which the upper and lower pressure limits are the same as in the Carnot and simple Rankine cycles of the previous examples, viz. 28 bar and 0·035 bar.

EXAMPLE 15.3

Evaluate the efficiency and specific steam consumption of the Rankine cycle with superheat, operating between 28 bar and 0·035 bar, the maximum superheat temperature being 300°C. The feed pump and turbine isentropic efficiencies may be taken as unity.

SOLUTION

Using the notation of *Fig*. 15.7b

At (6): $p_6 = 0{\cdot}035$ bar, $T_6 = 27{\cdot}6$°C, $h_6 = 112$ kJ/kg

At (1): $p_1 = 28$ bar, $h_1 = 114{\cdot}8$ kJ/kg

At (4): (by interpolation in superheat tables)

$$p_4 = 28 \text{ bar}, T_4 = 300°\text{C}, h_4 = 3001 \text{ kJ/kg}, s_4 = 6{\cdot}586 \text{ kJ/kg-K}$$

At (5): $p_5 = 0{\cdot}035$ bar, $T_5 = 27{\cdot}6$°C, $x_5 = 0{\cdot}761$

$$s_5 = s_4 = 6{\cdot}586 \text{ kJ/kg-K}, h_5 = 1967 \text{ kJ/kg}$$

$$\eta_{th} = \frac{(h_4 - h_5) - (h_1 - h_6)}{h_4 - h_1} = \frac{1034 - 2{\cdot}8}{2886{\cdot}2} = 0{\cdot}357 \quad \textit{Ans.} \quad (0{\cdot}352)$$

$$r_w = \frac{(h_4 - h_5) - (h_1 - h_6)}{h_4 - h_5} = 0{\cdot}996 \quad \textit{Ans.} \quad (0{\cdot}995)$$

$$\text{SSC} = \frac{1}{(h_4 - h_5) - (h_1 - h_6)} = 0{\cdot}000969 \text{ kg/kJ} \quad \textit{Ans.} \quad (0{\cdot}00106 \text{ kJ/kg})$$

Comparison with the figures in parenthesis shows that the Rankine cycle with superheat bring about an improvement with respect to all three criteria—efficiency, work ratio, and specific consumption. Had a higher superheat temperature been chosen, the improvement would have been correspondingly greater.

15.1.3 Reheat Cycle (*Fig*. 15.8)

Yet another method of reducing specific steam consumption, i.e. reducing the size of the plant for a given output, is to expand the steam in two separate turbines, between which the steam passes through a set of "reheating" tubes in which the temperature is restored to its maximum value. A further advantage is that the dryness fraction of the steam after both expansions remains even higher than in the case of the superheat cycle.

It is left as an exercise for the student to perform a typical cycle calculation, again between upper and lower pressure limits of 28 bar

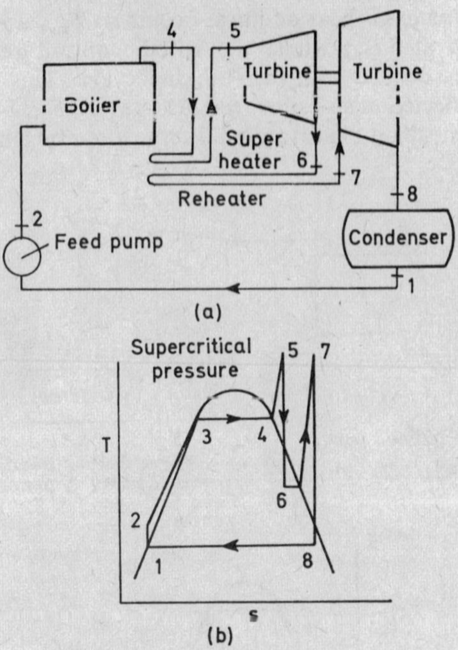

Fig. 15.8 Reheat cycle

and 0·035 bar, and with a maximum temperature $T_5 = T_7 = 400°C$ (*Fig*. 15.8b). The intermediate pressure $p_6 = p_7$ may be taken as 2 bar, although strictly this should be determined by a systematic work or efficiency optimization procedure.

15.1.4 The Regenerative Cycle (*Fig*. 15.9)

This is the cycle which, in conjunction with superheat, is most widely adopted in large steam plant for power generation.

Comparing the simple Rankine cycle with the Carnot cycle (example 15.1 and Table 15.1) it is evident that while the former, for given temperature limits, has the higher efficiency, the latter has by far the higher specific output or lower specific steam consumption, and is therefore the basis of all practical cycles. However, by adopting the principle of *regenerative* heating of the working fluid, i.e. replacing the *external* heat addition to feed water between 1 and 2 (*Fig*. 15.6) by *internal* heat transfer between hot and cold sections of the working fluid, through the medium of a heat exchanger, the efficiency is caused to approach more closely to the Carnot efficiency (in which all *external* heat addition occurs at T_{max} and all *external* heat rejection at T_{min}) whilst the specific output remains at the higher level associated with the Rankine cycle. This internal heat exchange is effected in so-called feed heaters (see *Fig*. 14.13, example 14.4), in which steam bled from an intermediate turbine stage mixes

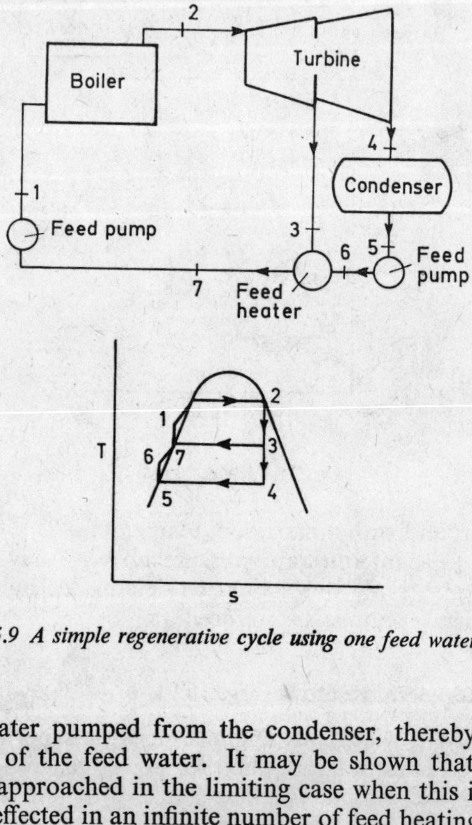

Fig. 15.9 A simple regenerative cycle using one feed water heater

with feed water pumped from the condenser, thereby raising the temperature of the feed water. It may be shown that the Carnot efficiency is approached in the limiting case when this internal heat exchange is effected in an infinite number of feed heating stages each resulting in an infinitesimal temperature rise. In practice the number of feed heating stages is limited to three or four for reasons of cost.

EXAMPLE 15.4

A regenerative cycle operates without superheat and with turbine and feed pump efficiencies of 100% between maximum and minimum pressures of

28 bar and 0·035 bar. A single feed heater of the mixing type is used, with bleed steam at a pressure of 2·5 bar. Find

(*a*) The mass of steam to be bled
(*b*) The cycle efficiency
(*c*) The specific steam consumption.

SOLUTION

(The notation of *Fig*. 15.9 is used.)
Considering first the feed heater:
Bleed steam condition $x_3 = 0{\cdot}849$, $h_3 = 2386$ kJ/kg (by isentropic expansion from state 2, i.e. $s_2 = s_3 = s_4$).

Applying the SFEE to the system shown in *Fig*. 15.10:

$$0 = h_7 - yh_3 - (1 - y)h_6 \qquad \text{(i)}$$

where y kg of steam is bled off at 3 per kg of steam circulating in the boiler.

But

$$h_6 = 112 \text{ kJ/kg } (= h_f \text{ at } 0{\cdot}035 \text{ bar})$$

and

$$h_7 = 535 \text{ kJ/kg } (= h_f \text{ at } 2{\cdot}5 \text{ bar})$$

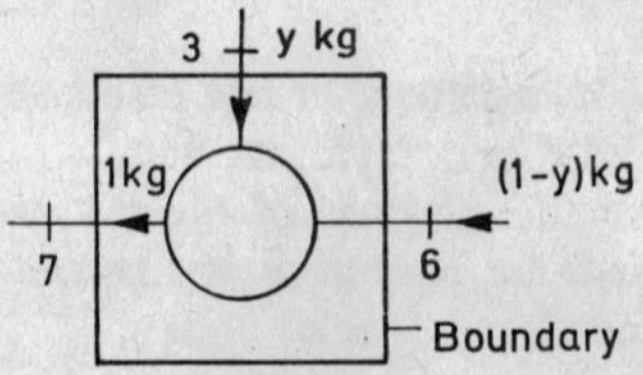

Fig. 15.10

From equation (i)

$$y = \frac{h_7 - h_6}{h_3 - h_6} = \frac{535 - 112}{2386 - 112} = \frac{423}{2274} = 0{\cdot}186 \text{ kJ/kg} \quad \textit{Ans.}$$

A first law analysis can now be applied to the cycle as a whole:

1st stage turbine: $h_2 = 2803$ kJ/kg, $h_3 = 2386$ kJ/kg
Work $w_{23} = (h_2 - h_3) = 2803 - 2386 = 417$ kJ/kg

2nd stage turbine: $h_3 = 2386$ kJ/kg, $h_4 = 1853$ kJ/kg
Work $w_{34} = (1 - y)(h_3 - h_4) = 0{\cdot}814(2386 - 1853) = 434$ kJ/kg
Boiler: Heat addition $q_{12} = h_2 - h_1$ (where $h_1 \simeq h_7 = 535$ kJ/kg)
$= 2803 - 535 = 2268$ kJ/kg

Note: No *external* heat is supplied between 6 and 7, since this process occurs within the feed heater. The small feed pump work terms w_{56} and w_{71} are neglected.

$$\eta_{th} = \frac{w_{23} + w_{34}}{q_{12}} = \frac{417 + 434}{2268} = 0{\cdot}372 \quad \textit{Ans.}$$

$$\text{SSC} = \frac{1}{w_{23} + w_{34}} = 0{\cdot}001175 \text{ kJ/kg} \quad \textit{Ans.}$$

Comparing the results for efficiency and SSC with those for the Carnot cycle (example 15.1) and simple Rankine cycle (Table 15.1) operating between the same pressure limits, the following simple table may be drawn up:

Table 15.2

	Carnot	*Rankine*	*Regenerative*
η_{th}	0·404	0·352	0·372
SSC	0·002092	0·00106	0·001175

It is clear that the efficiency is intermediate between the Rankine and Carnot values, but would be raised to the latter value if an infinite number of bleed stages were provided, thus eliminating all *external* heat addition below the maximum cycle temperature as in the Carnot cycle.

The specific steam consumption is much lower than that of the Carnot cycle, but slightly higher than that of the Rankine cycle, as expected.

In power station practice a high pressure boiler (p_{max} = 75 bar) with a high degree of superheat (T_{max} = 800°C) with a relatively large number of bleed stages and feed heaters would be used to achieve much higher theoretical cycle efficiencies and much lower specific steam consumptions.

For further reading in relation to steam cycles in general, and regenerative cycles in particular, the reader is referred to refs. 15.1 and 15.2, whilst for steam cycles in conjunction with nuclear power stations, ref. 15.3 will be found useful.

15.2 Gas Power Cycles

(see *Fig*. 15.1)

Gas power cycles will be subdivided into gas turbine cycles, on the one hand, and reciprocating engine cycles, on the other. Since the vast majority of gas power cycles are of the OPEN type (see introduction to this chapter) reference to closed cycles will be confined to a single example of such cycles in the field of gas turbines, on the one hand, and reciprocating engines, on the other.

First law analysis of open cycles will be presented using the concept of AIR STANDARD EFFICIENCY, based on the following assumptions:

(*a*) The working fluid is air throughout, with constant specific heat $c_p = 1{\cdot}005$ kJ/kg-K and constant specific heat ratio $\gamma = 1{\cdot}4$.

(*b*) The process of combustion, either in a steady flow combustion chamber as in the case of gas turbines, or in an intermittent combustion system as in the case of reciprocating engines, is replaced by *external heat addition* up to the temperature corresponding to the maximum value reached in the actual cycle.

(*c*) The process of mass rejection of products of combustion, from the tailpipe of the gas turbine or the exhaust pipe of the reciprocating engine, is replaced by a hypothetical *external heat abstraction* process applied to the recirculating air which returns to the first process in the cycle in the initial state.

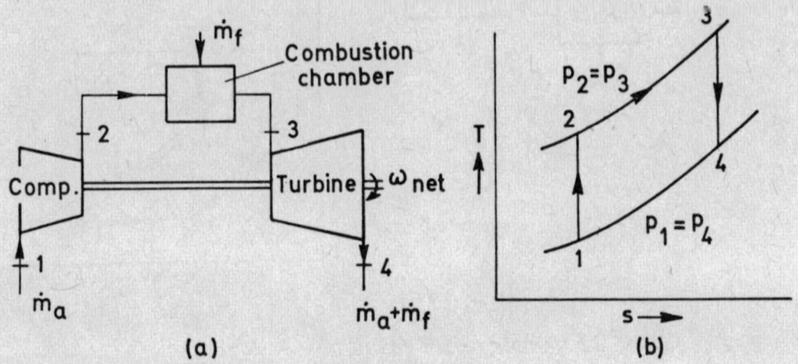

Fig. 15.11 Simple open gas turbine cycle

The principle of the air standard cycle is best illustrated with reference to the *simple open gas turbine cycle.* In this $\dot{m}_a$ kg/s of air are compressed in a compressor of the aerodynamic type (i.e. radial or axial blading, ref. 15.4) from pressure p_1 to pressure p_2. In the combustion chamber $\dot{m}_f$ kg/s of *fuel* burn in $\dot{m}_a$ kg/s of air initially at state 2 producing $(\dot{m}_a + \dot{m}_f)$ kg/s of products of combustion at state 3 which are subsequently expanded in the turbine from state 3 to state 4. The representation of these processes in the temperature–entropy diagram is complicated by the changes in gas properties, compared with those of air treated as a perfect gas, arising from changes in both temperature and chemical composition (ref. 15.5) which precludes the use of simple algebraic expressions for work, efficiency, etc.

If, however, the actual open cycle of *Fig.* 15.11 is replaced by its closed cycle equivalent in *Fig.* 15.12 performance parameters may all be expressed in simple algebraic terms and considerable simplifications result. This type of analysis will be applied first to a true closed cycle and subsequently to a series of open cycles where it leads to useful predictions of general trends, though clearly numerical errors of the order of 5–10% are introduced for the reasons stated.

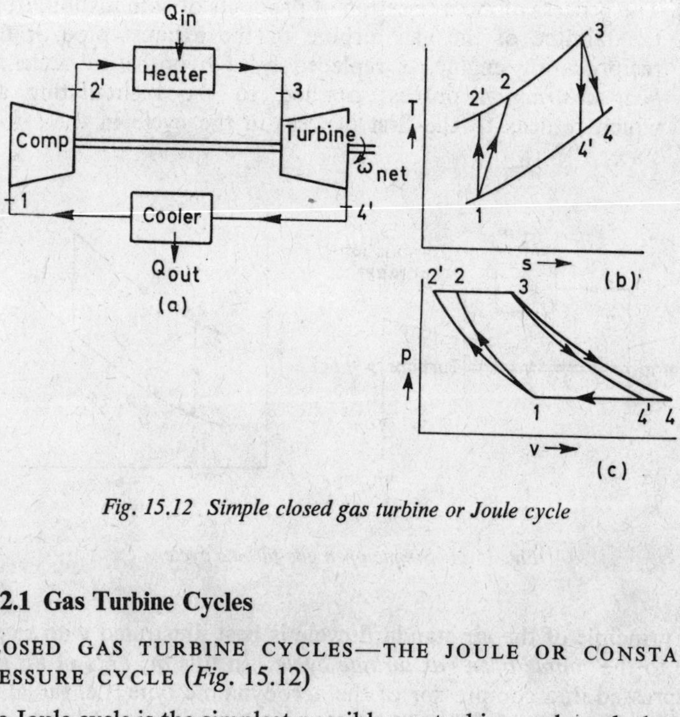

Fig. 15.12 Simple closed gas turbine or Joule cycle

15.2.1 Gas Turbine Cycles

CLOSED GAS TURBINE CYCLES—THE JOULE OR CONSTANT PRESSURE CYCLE (*Fig.* 15.12)

The Joule cycle is the simplest possible gas turbine cycle and, though rarely used in practice as a closed cycle, serves to illustrate the fundamental principles involved. In addition, it also serves as the air standard equivalent of the simple open gas turbine cycle (*Fig.* 15.11). As in the case of vapour power cycles the compression and expansion processes, 1–2 and 3–4 respectively, are assumed in the first instance to be isentropic. The cycle consists of the following steady flow processes applied to air treated as a perfect gas:

1–2′: isentropic compression in a suitable aerodynamic compressor.

2′–3: constant pressure heat addition $q_{2'3}$ in heater with final temperature $T_3 = T_{max}$.

3–4′: isentropic expansion in a suitable turbine from state 3 to state 4′, pressure $p_{4'}$.

4′–1: constant pressure heat rejection $q_{4'3}$ at pressure $p_{4'} = p_1$ in cooler with air returning to its initial state p_1, $T_1 = T_{min}$.

Applying, as in the case of vapour cycles, the three criteria of thermal efficiency η_{th} (equation (15.5)), work ratio r_w (equation (15.2)) and specific consumption SC (equation (15.3)) the following relationships result:

(*a*) *Cycle Efficiency* η_{th} (calculations based on 1 kg of working fluid)

Compression work

$$
\begin{aligned}
w_{12'} &= -(h_{2'} - h_1) \\
&= -c_p(T_{2'} - T_1) \\
&= c_pT_1\left\{\left(\frac{p_{2'}}{p_1}\right)\right\}^{(\gamma-1)/\gamma} - 1
\end{aligned}
$$

Heat addition

$$
\begin{aligned}
q_{2'3} &= h_3 - h_{2'} \\
&= c_p(T_3 - T_{2'}) \\
&= c_p\left\{T_3 - T_1\left(\frac{p_{2'}}{p_1}\right)^{(\gamma-1)/\gamma}\right\}
\end{aligned}
$$

Expansion work

$$
\begin{aligned}
w_{34'} = -(h_{4'} - h_3) &= c_p(T_3 - T_{4'}) \\
&= c_pT_3\left\{1 - \left(\frac{p_{4'}}{p_3}\right)^{(\gamma-1)/\gamma}\right\} \\
&= c_pT_3\left\{1 - \left(\frac{p_1}{p_2}\right)^{(\gamma-1)/\gamma}\right\}
\end{aligned}
$$

But from equation (15.5),

$$\eta_{th} = \frac{\oint dw}{q_{gross}} = \frac{(h_3 - h_{4'}) - (h_{2'} - h_1)}{h_3 - h_{2'}}$$

$$= \frac{c_pT_3\left\{1 - \left(\frac{p_1}{p_{2'}}\right)^{(\gamma-1)/\gamma}\right\} - c_pT_1\left\{\left(\frac{p_{2'}}{p_1}\right)^{(\gamma-1)/\gamma} - 1\right\}}{c_p\left\{T_3 - T_1\left(\frac{p_{2'}}{p_1}\right)^{(\gamma-1)/\gamma}\right\}}$$

$$= \left\{1 - \left(\frac{p_1}{p_{2'}}\right)^{(\gamma-1)/\gamma}\right\} \tag{15.9}$$

Equation (15.9) leads to the interesting and unexpected conclusion that the efficiency of the Joule cycle is a function of pressure ratio $p_{2'}/p_1$ only and increases with the latter. However, when component efficiencies less than unity are introduced for the compressor and turbine (see example 15.6) the cycle efficiency becomes dependent both on the pressure and temperature ratio.

(b) Work Ratio r_w

From equation (15.2),

$$r_w = \frac{\oint dw}{\oint (dw)_{positive}} = \frac{(h_3 - h_{4'}) - (h_{2'} - h_1)}{(h_3 - h_{4'})}$$

$$= 1 - \frac{h_{2'} - h_1}{h_3 - h_{4'}} = 1 - \frac{c_pT_1\left\{\left(\frac{p_{2'}}{p_1}\right)^{(\gamma-1)/\gamma} - 1\right\}}{c_pT_3\left\{1 - \left(\frac{p_1}{p_{2'}}\right)^{(\gamma-1)/\gamma}\right\}}$$

$$= 1 - \left(\frac{T_1}{T_3}\right)\left(\frac{p_{2'}}{p_1}\right)^{(\gamma-1)/\gamma} \tag{15.10}$$

Hence the work ratio, unlike the efficiency is a function both of the temperature ratio

$$\frac{T_1}{T_3} = \frac{T_{min}}{T_{max}}$$

and of the overall pressure ratio $(p_{2'}/p_1)$. Low values of both ratios, i.e. high T_{max}/T_{min} and $p_{2'}/p_1 \rightarrow 1{\cdot}0$, will lead to work ratios close to unity, i.e. minimum susceptibility to irreversibilities.

(*c*) *Specific Consumption* SC (specific work)

From equation (15.3),

$$\text{SC} = \frac{1}{\oint dw} = \frac{1}{(h_3 - h_{4'}) - (h_{2'} - h_1)}$$

$$= \frac{1}{c_p T_3\left\{1 - \left(\frac{p_1}{p_{2'}}\right)^{(\gamma-1)/\gamma}\right\} - c_p T_1\left\{\left(\frac{p_{2'}}{p_1}\right)^{(\gamma-1)/\gamma} - 1\right\}} \quad (15.11a)$$

In gas turbine calculations it is more usual to use specific work SW rather than specific consumption as a criterion of compactness. Clearly the latter is merely the reciprocal of the former, i.e.

$$\text{SW} = \oint dw = c_p T_3\left\{1 - \left(\frac{p_1}{p_{2'}}\right)^{(\gamma-1)/\gamma}\right\} - c_p T_1\left\{\left(\frac{p_{2'}}{p_1}\right)^{(\gamma-1)/\gamma} - 1\right\} \quad (15.11b)$$

It is left as an exercise to the student to evaluate equations (15.9), (15.10), and (15.11) for a range of values of the pressure ratio $p_{2'}/p_1$ and temperature ratio T_3/T_1. The following two examples illustrate two important principles, the first applying to closed cycles only and the second having general validity.

Example 15.5 illustrates the advantage of using monatomic gases (helium, argon, etc.) with their high values of the specific heat ratio ($\gamma = 1{\cdot}667$ — see Chapter 9) rather than air or products of combustion with values of $\gamma = 1{\cdot}4$.

Example 15.6 based on an air standard cycle shows the marked effect of component efficiencies of the compressor and turbine on cycle efficiency.

EXAMPLE 15.5

A simple Joule closed gas turbine cycle operates with a pressure ratio 5:1. Neglecting all internal losses, find the cycle efficiency using

(*a*) helium He, $\gamma = 1{\cdot}667$

(*b*) air, $\gamma = 1{\cdot}4$.

SOLUTION

Using the notation of *Fig.* 15.12 and equation (15.9)

(*a*) For helium

$$\eta_{th} = 1 - \left(\frac{p_1}{p_{2'}}\right)^{(\gamma-1)/\gamma} = 1 - \left(\frac{1}{5}\right)^{0\cdot667/1\cdot1667}$$

$$= 0{\cdot}475 \quad \textit{Ans.}$$

(b) For air

$$\eta_{th} = 1 - \left(\frac{p_1}{p_{2'}}\right)^{(\gamma-1)/\gamma} = 1 - \left(\frac{1}{5}\right)^{0\cdot4/1\cdot4}$$
$$= 0\cdot369 \quad \textit{Ans.}$$

The beneficial effect of using monatomic gases is thus very considerable. A further thermodynamic advantage of closed cycles is that, whereas in open cycles, part load conditions invariably involve a reduction in overall pressure ratio with resultant loss of efficiency, closed cycle plant may be operated at constant overall pressure ratio by altering the density level throughout the system, i.e. reducing the mass flow rate in the main system, by bypassing increasing amounts of gas as the load is reduced.

EXAMPLE 15.6

A simple Joule gas turbine cycle using air operates with a pressure ratio 5:1. Find the cycle efficiency, work ratio and specific work,
(*a*) with compressor and turbine efficiencies of 1·0
(*b*) with compressor and turbine efficiencies of 0·8.
The minimum and maximum temperatures are respectively 300 K and 1000 K.

SOLUTION

(*a*) With $\eta_c = \eta_T = 1\cdot0$
Using equation (15.9),

$$\eta_{th} = 1 - \left(\frac{p_1}{p_{2'}}\right)^{(\gamma-1)/\gamma} = 1 - \left(\frac{1}{5}\right)^{0\cdot4/1\cdot4}$$
$$= 0\cdot369 \quad \textit{Ans.}$$

Using equation (15.10),

$$r_w = 1 - \left(\frac{T_1}{T_3}\right)\left(\frac{p_{2'}}{p_1}\right)^{(\gamma-1)/\gamma} = 1 - \left(\frac{300}{1000}\right)(5)^{0\cdot4/1\cdot4}$$
$$= 0\cdot524 \quad \textit{Ans.}$$

Using equation (15.11b),

$$\text{sw} = c_pT_3\left\{1 - \left(\frac{p_1}{p_{2'}}\right)^{(\gamma-1)/\gamma}\right\} - c_pT_1\left\{\left(\frac{p_{2'}}{p_1}\right)^{(\gamma-1)/\gamma} - 1\right\}$$
$$= 1\cdot005\{1000[1 - (\tfrac{1}{5})^{0\cdot4/1\cdot4}] - 300[(5)^{0\cdot4/1\cdot4} - 1]\}$$
$$= 194\cdot5 \text{ kJ/kg} \quad \textit{Ans.}$$

(*b*) With $\eta_c = \eta_T = 0\cdot8$ (see *Figs.* 15.12b and c), the compression process changes from 1–2′ to 1–2, i.e. adiabatic reversible or isentropic to adiabatic irreversible, and similarly the expansion process changes from 3–4′ to 3–4.

Such irreversibilities have been discussed in detail in sections 14.5e and f.

Compression process 1–2

$$w_{12} = \frac{w_{12'}}{\eta_c} = \frac{-c_pT_1\left\{\left(\dfrac{p_2}{p_1}\right)^{(\gamma-1)/\gamma} - 1\right\}}{\eta_c}$$

$$= \frac{-300 \times 1{\cdot}005[(5)^{0{\cdot}4/1{\cdot}4} - 1]}{0{\cdot}8}$$

$$= -221 \text{ kJ/kg} \qquad (\text{cf. } w_{12'} = 176{\cdot}5 \text{ kJ/kg})$$

Expansion process 3–4

$$w_{34} = \eta_T w_{34'} = \eta_T c_p T_3 \left\{1 - \left(\frac{p_1}{p_2}\right)^{(\gamma-1)/\gamma}\right\}$$

$$= 0{\cdot}8 \times 1{\cdot}005 \times 1000(1 - (\tfrac{1}{5})^{0{\cdot}4/1{\cdot}4})$$

$$= 297 \text{ kJ/kg} \quad (\text{cf. } w_{34'} = 371 \text{ kJ/kg})$$

Heat addition 2–3

$$q_{23} = c_p(T_3 - T_2)$$

where $T_2 = T_1 - \dfrac{w_{12}}{c_p} = 300 + \dfrac{221}{1{\cdot}005} = 520 \text{ K}.$

$$q_{23} = 1{\cdot}005(1000 - 520) = 482 \text{ kJ/kg}$$

(i) *Cycle efficiency* (equation (15.5))

$$\eta_{th} = \frac{\oint dw}{q_{gross}} = \frac{w_{34} + w_{12}}{q_{23}} = \frac{297 - 221}{482}$$

$$= 0{\cdot}158 \quad \textit{Ans.} \qquad (\text{cf. } 0{\cdot}369)$$

Hence the effect of the irreversible compression and expansion on cycle efficiency has been very marked indeed. The example thus illustrates the necessity for keeping component efficiencies as high as possible.

(ii) *Work ratio* (equation (15.2))

$$r_w = \frac{\oint dw}{\int (dw)_{positive}} = 1 + \frac{w_{12}}{w_{34}} = 1 - \frac{221}{297}$$

$$= 0{\cdot}256 \quad \textit{Ans.} \qquad (\text{cf. } 0{\cdot}524)$$

Thus again the effect of reduced component efficiency has been very drastic indeed.

(iii) *Specific work* (equation (15.3))

$$\text{sw} = w_{34} + w_{12} = 297 - 221$$

$$= 76 \text{ kJ/kg} \quad \textit{Ans.} \qquad (\text{cf. } 194{\cdot}5 \text{ kJ/kg})$$

The specific work, like the efficiency, is thus reduced to less than 50% of the corresponding value with ideal component efficiencies $\eta_c = \eta_T = 1{\cdot}0$.

Open gas turbine cycles

The simple open gas turbine cycle, as shown by the air standard analysis of example 15.6, is particularly susceptible to irreversibilities, and hence gives low cycle efficiency and specific work when real component efficiencies are taken into account.

A number of additions, under the headings

(i) Heat exchange
(ii) Intercooling
(iii) Reheating

may be made to the simple cycle in order to improve either its efficiency, or its specific output, or both.

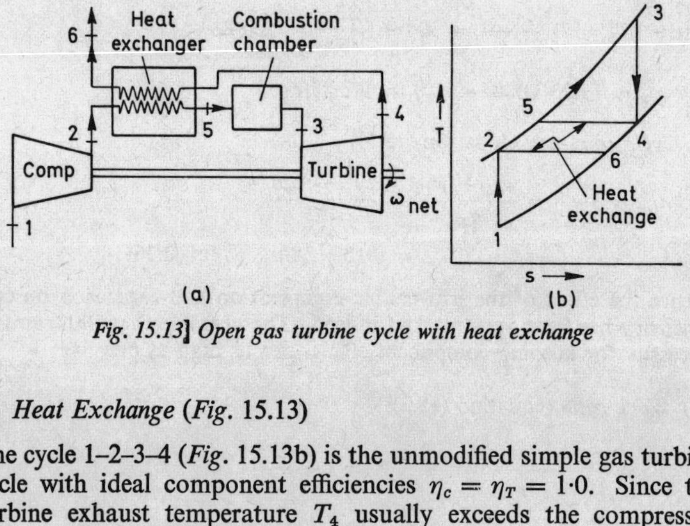

Fig. 15.13 Open gas turbine cycle with heat exchange

(i) *Heat Exchange* (*Fig.* 15.13)

The cycle 1–2–3–4 (*Fig.* 15.13b) is the unmodified simple gas turbine cycle with ideal component efficiencies $\eta_c = \eta_T = 1{\cdot}0$. Since the turbine exhaust temperature T_4 usually exceeds the compressor delivery temperature T_2, a heat exchanger may be inserted between the compressor and combustion chamber, as shown in *Fig.* 15.13a, in which, ideally, the exhaust gas temperature falls from T_4 to T_6 where $T_6 = T_2$ = compressor delivery temperature, whilst at the same time the air temperature rises from T_2 to T_5 where $T_5 = T_4$ = turbine exhaust temperature, thus reducing the temperature rise by external heat addition to $(T_3 - T_5)$.

In practice the heat exchange process is not "perfect", i.e. $T_6 > T_2$ and $T_5 < T_4$ due to

(i) the need for a finite temperature gradient between the two gases (see section 14.5*b*),
(ii) differences in specific heats between the compressed air and turbine exhaust gas.

This may be expressed in the form of a heat exchanger effectiveness (not to be confused with the availability definition, equation (14.29)) where

$$\varepsilon = \frac{(T_5 - T_2)}{(T_4 - T_2)} \tag{15.12}$$

However, assuming "perfect" heat exchange, $\varepsilon = 1{\cdot}0$, and ideal component efficiencies $\eta_c = \eta_T = 1{\cdot}0$ a simple first law analysis leads to the following result:

Specific work

$$\oint dw = c_p T_3 \left\{1 - \left(\frac{p_1}{p_2}\right)^{(\gamma-1)/\gamma}\right\} - c_p T_1 \left\{\left(\frac{p_2}{p_1}\right)^{(\gamma-1)/\gamma} - 1\right\} \quad \text{(i)}$$

i.e. identical with simple gas turbine cycle (equation (15.11b))

External heat addition

$$q_{53} = c_p(T_3 - T_5) = c_p(T_3 - T_4) = c_p T_3 \left\{1 - \left(\frac{p_1}{p_2}\right)^{(\gamma-1)/\gamma}\right\} \quad \text{(ii)}$$

But from equation (15.5), cycle efficiency

$$\eta_{th} = \frac{\oint dw}{q_{gross}} = \frac{\oint dw}{q_{53}}$$

so that, substituting from equations (i) and (ii),

$$\eta_{th} = 1 - \left(\frac{T_1}{T_3}\right)\left(\frac{p_2}{p_1}\right)^{(\gamma-1)/\gamma} \tag{15.13}$$

The expression for work ratio r_w and specific work sw are identical with those for the simple gas turbine cycle, i.e. equations (15.10) and (15.11b) respectively, as will be evident from inspection of the temperature–entropy diagram, *Figs.* 15.11b and 15.13b.

It is again left as an exercise to the student to evaluate equation (15.13) for a range of values of the pressure ratio p_2/p_1 and temperature ratio T_3/T_1. In general, the gain in efficiency compared with the simple cycles is greatest at low pressure ratios. For any one value of T_3/T_1 there is a limiting pressure ratio beyond which heat exchange would lead to a *loss* of efficiency due to a *negative* temperature gradient for heat exchange, i.e. $T_4 < T_2$ or

$$T_3\left(\frac{p_1}{p_2}\right)^{(\gamma-1)/\gamma} < T_1\left(\frac{p_2}{p_1}\right)^{(\gamma-1)/\gamma}$$

The limiting case therefore becomes

$$\left(\frac{p_2}{p_1}\right)^{2(\gamma-1)/\gamma} = \left(\frac{T_3}{T_1}\right) \tag{15.14}$$

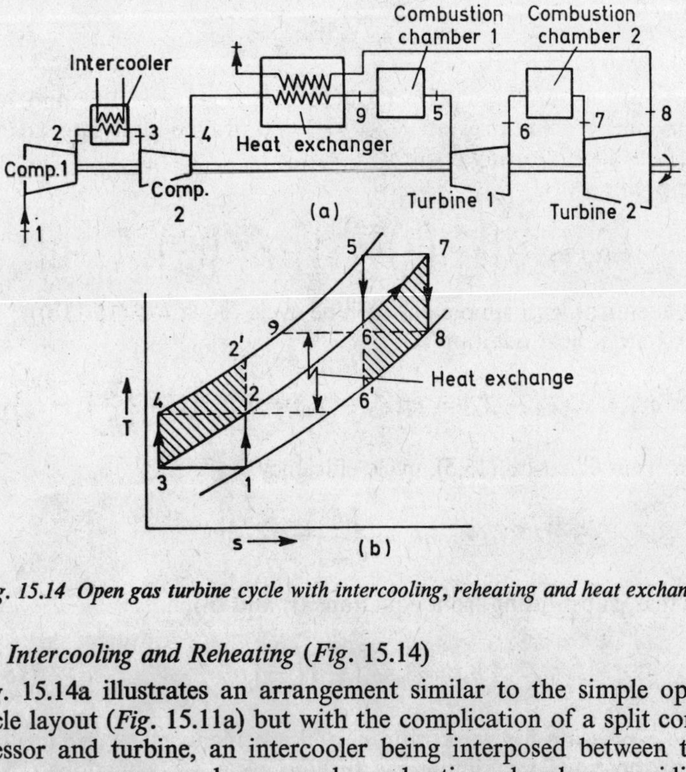

Fig. 15.14 Open gas turbine cycle with intercooling, reheating and heat exchange

(ii) *Intercooling and Reheating* (*Fig.* 15.14)

Fig. 15.14a illustrates an arrangement similar to the simple open cycle layout (*Fig.* 15.11a) but with the complication of a split compressor and turbine, an intercooler being interposed between the two compressors, and a second combustion chamber providing "reheat" between the two turbines. The corresponding temperature–entropy diagram is shown in *Fig.* 15.14b, the intercooling process occurring at the intermediate pressure $p_2 = p_3 (p_1 < p_2 < p_4)$. With maximum cooler effectiveness the temperature after intercooling T_3 may be assumed equal to the initial temperature T_1, i.e.

$$T_3 = T_1 = T_{min}$$

The reheating process similarly occurs at the intermediate pressure $p_6 = p_7$ ($p_5 > p_6 > p_8 = p_1$). The final reheat temperature T_7 may be made equal to the temperature after the main combustion chamber i.e.

$$T_7 = T_5 = T_{max}$$

Compared with the simple cycle 1–2′–5–6′ (*Fig.* 15.14b) the cycle with intercooling and reheat has a considerably higher specific work output, the difference being represented by the two shaded areas 2–3–4–2′ and 6–7–8–6′ respectively. This is due to the fact that intercooling leads to a marked reduction in compression work whilst reheating leads to a marked increase in turbine work.

Thus $w_{comp} = -(h_2 - h_1) - (h_4 - h_3)$ cf. $w_{12'} = -(h_2 - h_1)$

i.e. $$w_{comp} = c_pT_1\left\{\left[\left(\frac{p_2}{p_1}\right)^{(\gamma-1)/\gamma} - 1\right] + \left[\left(\frac{p_4}{p_3}\right)^{(\gamma-1)/\gamma} - 1\right]\right\} \tag{15.15}$$

cf. $$-c_pT_1\left[\left(\frac{p_4}{p_1}\right)^{(\gamma-1)/\gamma} - 1\right]$$

(where, as already stated, $T_3 = T_1 = T_{min}$).

w_{comp} as given by equation (15.15) may be minimized by differentiation, with the intermediate pressure $p_2 = p_3$ as the dependent variable, giving, for minimum work

$$p_2 = \sqrt{(p_1p_4)} \tag{15.16}$$

Similarly

$$w_{turb} = (h_5 - h_6) + (h_7 - h_8) \qquad \text{cf. } w_{56'} = h_5 - h_{6'}$$

i.e. $$w_{turb} = c_pT_5\left\{\left[1 - \left(\frac{p_6}{p_5}\right)^{(\gamma-1)/\gamma}\right] + \left[1 - \left(\frac{p_8}{p_7}\right)^{(\gamma-1)/\gamma}\right]\right\} \tag{15.17}$$

cf. $$c_pT_5\left[1 - \left(\frac{p_8}{p_5}\right)^{(\gamma-1)/\gamma}\right]$$

w_{turb} as given by equation (15.17) may be maximized by differentiation, treating the intermediate pressure $p_6 = p_7$ as the dependent variable and giving, for maximum work

$$p_6 = \sqrt{(p_5p_8)} = \sqrt{(p_1p_4)} \tag{15.18}$$

(since, ideally, $p_1 = p_8$ and $p_4 = p_5$).

Hence the optimum intermediate pressure for both intercooling and reheating has the same value given by equation (15.16).

Whilst both specific work and work ratio are clearly improved by the incorporation of intercooling and reheating it is evident that the cycle efficiency may actually be reduced, as a result of the significant increase in heat addition in the two combustion chambers.

Thus

$$
\begin{aligned}
q_{overall} &= q_{45} + q_{67} \qquad \text{cf. } q_{2'5} \\
&= c_p\{(T_5 - T_4) + (T_7 - T_6)\} \\
&= c_p\,(2T_5 - T_4 - T_6) \qquad \text{cf. } c_p(T_5 - T_{2'}) \quad (15.19)
\end{aligned}
$$

since, as already stated, $T_7 = T_5$.

It is only by the incorporation of heat exchange, as described in the previous section, in addition to intercooling and reheating, that an improvement in efficiency, as well as specific work, becomes possible.

EXAMPLE 15.7

A gas turbine cycle with intercooling and reheating operates with an overall pressure ratio of 5:1 and between temperature limits of 300 K and 1000 K, with intercooling back to the lower limit and reheating to the upper limit. Find the cycle efficiency, work ratio and specific work, assuming $\eta_c = \eta_T = 1{\cdot}0$. Find also the effect on efficiency of incorporating a heat exchanger having an effectiveness of 1·0.

SOLUTION

Using the notation of *Fig*. 15.14b:

From equation (15.18), intermediate pressure ratio is

$$\frac{p_2}{p_1} = \frac{p_6}{p_1} = \sqrt{\frac{p_4}{p_1}} = \sqrt{5} = 2{\cdot}23$$

From equation (15.16),

$$
\begin{aligned}
w_{comp} &= -c_pT_1\left\{\left(\frac{p_2}{p_1}\right)^{(\gamma-1)/\gamma} + \left(\frac{p_4}{p_3}\right)^{(\gamma-1)/\gamma} - 2\right\} \\
&= -c_pT_2\left\{\left(\sqrt{\frac{p_4}{p_1}}\right)^{(\gamma-1)/\gamma} - 2\right\} \\
&= -2 \times 1{\cdot}005 \times 300[(\sqrt{5})^{0{\cdot}4/1{\cdot}4} - 1] \\
&= -155{\cdot}5 \text{ kJ/kg} \qquad \text{(i)}
\end{aligned}
$$

[For the corresponding simple cycle of example 15.6, between the same pressure limits (*Fig*. 15.14b),

$$
\begin{aligned}
w_{comp} &= -c_pT_1\left\{\left(\frac{p_{2'}}{p_1}\right)^{(\gamma-1)/\gamma}\right\} = -1{\cdot}005 \times 300[(5)^{0{\cdot}4/1{\cdot}4} - 1] \\
&= -176{\cdot}5 \text{ kJ/kg}
\end{aligned}
$$

i.e. intercooling gives a reduction in compression work of 21·0 kJ/kg].
From equation (15.17),

$$w_{turb} = c_p T_5 \left\{-2\left(\sqrt{\frac{p_1}{p_4}}\right)^{(\gamma-1)/\gamma} + 2\right\}$$

$$= 2 \times 1{\cdot}005 \times 1000 \left[1 - \left(\frac{1}{\sqrt{5}}\right)^{0{\cdot}4/1{\cdot}4}\right]$$

$$= 412{\cdot}0 \text{ kJ/kg} \qquad \text{(ii)}$$

(For the corresponding simple cycle of example 15.6, *Fig*. 15.14b.)

$$w_{turb} = c_p T_6 \left\{1 - \left(\frac{p_{6'}}{p_5}\right)^{(\gamma-1)/\gamma}\right\}$$

$$= 1{\cdot}005 \times 1000[1 - (\tfrac{1}{5})^{0{\cdot}4/1{\cdot}4}]$$

$$= 371{\cdot}0 \text{ kJ/kg}$$

i.e. reheating gives an increase in expansion work of 41·0 kJ/kg.)
From equation (15.19),

$$q_{overall} = c_p(2T_5 - T_4 - T_6)$$

where $T_5 = 1000$ K

$$T_4 = 300\,(\sqrt{5})^{0{\cdot}4/1{\cdot}4} = 377 \text{ K}$$

$$T_6 = 1000\left(\frac{1}{\sqrt{5}}\right)^{0{\cdot}4/1{\cdot}4} = 631 \text{ K}$$

$$q_{overall} = 1{\cdot}005\,(2000 - 377 - 631)$$

$$= 998{\cdot}0 \text{ kJ/kg} \qquad \text{(iii)}$$

(i) *Cycle efficiency*

Combining (i), (ii) and (iii), thermal efficiency is

$$\eta_{th} = \frac{\oint dw}{q_{gross}} = \frac{\text{(i)} + \text{(ii)}}{\text{(iii)}}$$

$$= -\frac{155{\cdot}5 + 412{\cdot}0}{998{\cdot}0} = 0{\cdot}257 \quad \textit{Ans.}$$

(This may be compared with the corresponding results for the simple cycle, example 15.6, viz.

$$\eta_{th} = 0{\cdot}369$$

indicating that intercooling and reheating actually leads to a considerable *loss* of thermal efficiency.)

(ii) *Work ratio*

$$r_w = \frac{\oint dw}{\int (dw)_{positive}} = \frac{-155{\cdot}5 + 412{\cdot}0}{412{\cdot}0} = 0{\cdot}623 \quad Ans.$$

cf. 0·524 for the simple cycle.

(*iii*) *Specific work*

$$\text{sw} = \oint dw = -155{\cdot}5 + 412{\cdot}0$$
$$= 256{\cdot}5 \text{ kJ/kg} \quad Ans.$$

cf. 194·5 kJ/kg for the simple cycle.

Hence the effect of intercooling and reheating on work ratio and specific work is very beneficial, whilst efficiency is adversely affected.

(*iv*) *Effect of heat exchange*

From *Fig*. 15.14b the maximum temperature which may be reached by the compressor air as a result of heat exchange equals the turbine exhaust temperature T_8 where

$$T_8 = T_6 = 631 \text{ K}$$

i.e. reheating may increase the air temperature from $T_4 = 377$ K to $T_9 = T_8 = 631$ K with corresponding reduction in external heat addition, namely

$$q'_{overall} = c_p[(T_5 - T_9) + (T_7 - T_6)]$$
$$= 1{\cdot}005[(1000 - 631) + (1000 - 631)]$$
$$= 743 \text{ kJ/kg} \qquad \text{cf. } 998 \text{ kJ/kg}$$

giving a modified thermal efficiency

$$\eta'_{th} = \frac{\oint dw}{q'_{overall}} = \frac{-155{\cdot}5 + 412{\cdot}0}{743}$$
$$= 0{\cdot}345 \qquad \text{cf. } 0{\cdot}257$$

15.2.2 Reciprocating Engine Cycles

In reciprocating engine cycles, the steady flow processes in the various components of the gas turbine are replaced by non-flow processes in a single component, namely the engine cylinder sealed by a moving piston.

As in the case of the gas turbine we differentiate between

(i) *Closed cycles* in which the same working fluid undergoes a series of thermodynamic processes without chemical change, with external heat addition and abstraction through suitable heat exchangers and;

(ii) *Open cycles* in which fuel is introduced into the working fluid, usually air, by suitable means, e.g. carburation, i.e. mixing of liquid fuel with air to form a gaseous mixture before entering the cylinder, as in the petrol engine, or fuel injection at high pressure into the air charge in the cylinder following compression, as in the compression ignition or diesel engine. As in the case of the gas turbine, open cycles are conveniently represented by idealized, equivalent closed cycles using pure air with external heat transfer, such cycles again being referred to as *air standard cycles*.

The vast majority of reciprocating engines operate on open cycles, but in recent years and as a result of an increasing awareness of the dangers of *atmospheric pollution*, considerable attention has been given to a number of closed cycle engines, such as the STIRLING and ERICSON engines, the former being dealt with in the following section.

CLOSED RECIPROCATING ENGINE CYCLES:

The STIRLING or Constant Volume Cycle

Historically, the Stirling engine dates back to the first half of the nineteenth century, being the invention of the Rev. Robert Stirling of Dumfries, but in recent years it has undergone intensive development in several countries (refs. 15.6 and 15.7). The working fluid is alternately displaced from a hot space to a cold space by the action of two pistons operating approximately 90° out of phase. During transfer from the cold to the hot cylinder, theoretically at constant volume, the air passes first through a regenerative heat exchanger, and then through an externally supplied heater, whilst during the reverse transfer, again theoretically at constant volume, the air passes first through the heat exchanger and subsequently through a cooler with external heat abstraction (see *Fig*. 15.15a). The actual cycle deviates considerably from the idealized form in *Fig*. 15.15b, which is universally referred to as the Stirling cycle and which has a theoretical efficiency equal to the Carnot efficiency.

The various processes are as follows:

1–2: isothermal compression in *cold space* with *external* heat abstraction q_{12}, at temperature T_{min}.

2–3: constant volume transfer from cold space to hot space with *internal* heat addition q_{23} through regenerative heat exchanger, giving temperature rise

$$(T_3 - T_2) = (T_{max} - T_{min})$$

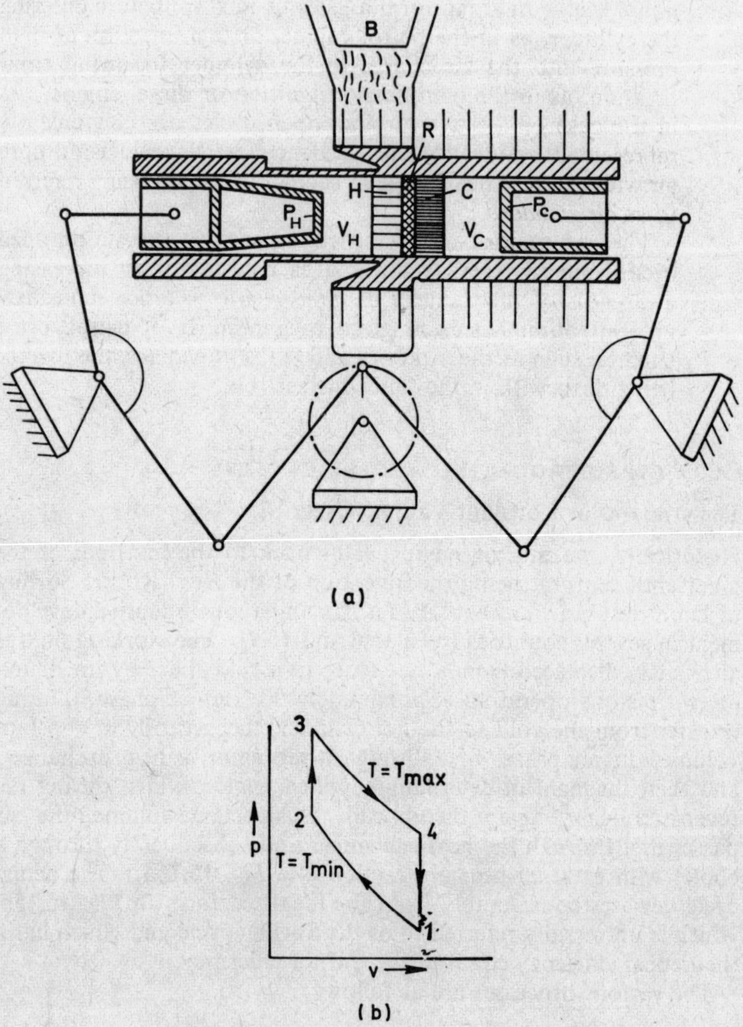

Fig. 15.15 The Stirling or constant volume cycle

In (*a*):	B, burner	P_H, hot piston
	C, cooler	R, regenerator
	H, heater	V_C, cold space
	P_C, cold piston	V_H, hot space

3–4: isothermal expansion in *hot space* with external heat addition q_{34}, at temperature T_{max}.

4–1: constant volume transfer from hot space to cold space with *internal* heat abstraction q_{41} through regenerative heat exchanger, giving temperature drop

$$(T_1 - T_4) = (T_{min} - T_{max})$$

The essential feature of the cycle is that as a result of the equality of the *internal* heat addition and rejection processes

$$|q_{23}| = |c_v(T_{max} - T_{min})| = |q_{41}| = |c_v(T_{min} - T_{max})|$$

the *external* heat transfer is limited to the isothermal processes 3–4 and 1–2, i.e. the cycle conforms to the Carnot condition that the external heat exchange processes should occur respectively at the maximum and minimum temperature of the cycle (see section 6.4). However, it has the great advantage over the Carnot cycle that its work ratio and specific work are considerably higher.

A simple cycle analysis yields the following results:

$$\text{External heat addition } q_{34} = RT_{max} \ln \left(\frac{p_2}{p_4}\right) = w_{34} \qquad \text{(i)}$$

$$\text{External heat abstraction } q_{12} = RT_{min} \ln \left(\frac{p_1}{p_2}\right) = w_{12} \qquad \text{(ii)}$$

where, since $v_2 = v_3 = v_{min}$ $\quad T_3 = T_4 = T_{max}$

and $v_4 = v_1 = v_{max}$ $\quad T_1 = T_2 = T_{min}$

$$\frac{p_3 v_{min}}{T_{max}} = \frac{p_4 v_{max}}{T_{max}} = \frac{p_1 v_{max}}{T_{min}} = \frac{p_2 v_{min}}{T_{min}}$$

i.e.

$$\frac{p_3}{p_4} = \frac{p_2}{p_1} \qquad \text{(iii)}$$

so that (i) and (ii) may be rewritten

$$q_{34} = w_{34} = RT_{max} \ln \left(\frac{p_2}{p_1}\right) \qquad \text{(iv)}$$

$$q_{12} = w_{12} = RT_{min} \ln \left(\frac{p_1}{p_2}\right) \qquad \text{(v)}$$

[N.B. No work is done during the constant volume processes 2–3 and 4–1.]

Thermal efficiency, work ratio and specific work now become:

$$\eta_{th} = \frac{\oint dw}{q_{gross}} = \frac{w_{34} + w_{12}}{q_{34}} = \frac{RT_{max} \ln \left(\frac{p_2}{p_1}\right) - RT_{min} \ln \left(\frac{p_2}{p_1}\right)}{RT_{max} \ln \left(\frac{p_2}{p_1}\right)}$$

$$= \left(1 - \frac{T_{min}}{T_{max}}\right) \quad (15.20)$$

which, as already stated, equals the Carnot efficiency.

$$r_w = \frac{\oint dw}{\int (dw)_{positive}} = \frac{w_{34} + w_{12}}{w_{34}} = 1 - \frac{T_{min}}{T_{max}} \quad (15.21)$$

$$\text{sw} = w_{34} + w_{12} = R(T_{max} - T_{min}) \ln \left(\frac{p_2}{p_1}\right) \quad (15.22)$$

It is left as an exercise for the student to compare the above results for work ratio and specific work with those for a corresponding Carnot cycle.

Open reciprocating engine cycles

This category includes the very important internal combustion (IC) engine in its two main forms employing *either*

> Induction of a pre-mixed charge of air and fuel with subsequent ignition of the mixture by high energy spark, i.e. the spark ignition (SI) engine;
>
> *or*
>
> Induction of pure air with high pressure injection of fuel after compression, and auto ignition as a result of the high pressure and temperature of the air resulting from compression, i.e. the compression ignition (CI) or Diesel engine.

Consideration of such cycles in detail is beyond the scope of this book. Instead a simple treatment based on *air standard cycles* will be given. Traditionally, the SI engine cycle is represented by the constant volume air standard cycle and the CI engine either by the constant pressure cycle or, more approximately, by the dual or mixed cycle.

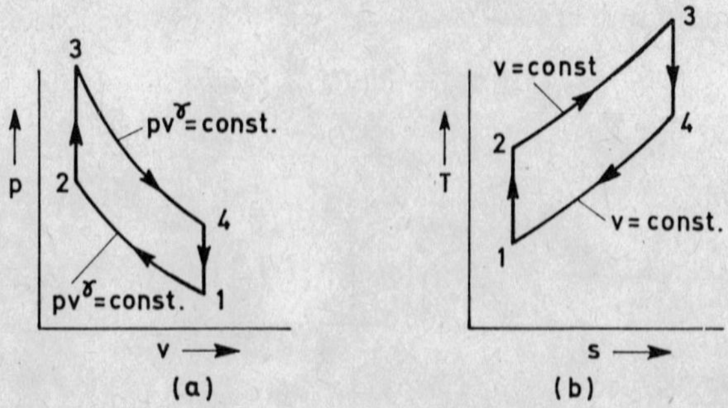

Fig. 15.16 Constant volume cycle

(i) *The Constant Volume or Otto Cycle* (*Fig.* 15.16)

The sequence of processes is as follows:

1–2: isentropic compression, starting at point 1 corresponding to the trapped condition in the cylinder after inlet valve closure, through volumetric compression ratio

$$r_v = v_1/v_2$$

2–3: constant volume heat addition q_{23} to maximum cycle pressure and temperature p_3 and T_3, corresponding to combustion

3–4: isentropic expansion through volumetric expansion ratio

$$r_v = v_4/v_3 = v_1/v_2$$

to condition 4 corresponding to exhaust valve opening.

4–1: constant volume heat abstraction q_{41} to initial state 1, corresponding to complete gas exchange process, i.e. exhaust followed by induction.

Thermodynamic analysis is extremely simple:

Compression work

$$w_{12} = -\,\Delta u_{12} = -\,c_v(T_2 - T_1)$$

$$= -\,c_vT_1\left[\left(\frac{v_1}{v_2}\right)^{\gamma-1} - 1\right]$$

$$= -\,c_vT_1[(r_v)^{\gamma-1} - 1] \qquad \text{(i)}$$

Similarly, expansion work

$$w_{34} = -\,\Delta u_{34} = c_vT_3\left[1 - \left(\frac{1}{r_v}\right)^{\gamma-1}\right] \qquad \text{(ii)}$$

Heat addition

$$q_{23} = \Delta u_{23} = c_v(T_3 - T_2)$$
$$= c_v[T_3 - T_1(r_v)^{\gamma-1}] \qquad \text{(iii)}$$

giving *efficiency*

$$\eta_{th} = \frac{\oint dw}{q_{gross}} = \frac{w_{12} + w_{34}}{q_{23}} = \frac{c_v\left\{T_3\left[1 - \left(\frac{1}{r_v}\right)^{\gamma-1}\right] - T_1[(r_v)^{\gamma-1} - 1]\right\}}{c_v[T_3 - T_1(r_v)^{\gamma-1}]}$$
$$= 1 - \left(\frac{1}{r_v}\right)^{\gamma-1} \qquad (15.23)$$

Equation (15.23) indicates that the cycle efficiency is independent of the temperature limits and a function only of volumetric compression ratio. This is borne out in practice by the great improvement in spark ignition engine efficiencies resulting from increased compression ratios made possible by improved fuel properties (from approximately 6:1 in 1939 to 9:1 in 1969).

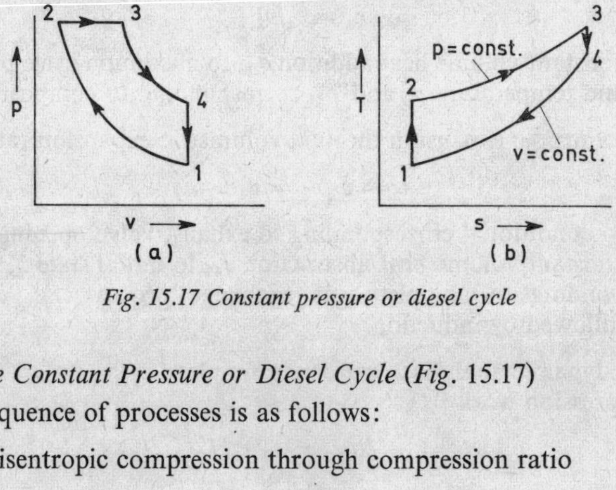

Fig. 15.17 Constant pressure or diesel cycle

(ii) *The Constant Pressure or Diesel Cycle* (*Fig.* 15.17)

The sequence of processes is as follows:

1–2: isentropic compression through compression ratio

$$r_v = v_1/v_2$$

2–3: constant pressure heat addition q_{23}, with "cut-off" ratio

$$r_c = v_3/v_2$$

simulating a moderate rate of fuel injection and heat release, associated with the now outmoded system of air blast fuel injection

3–4: isentropic expansion through expansion ratio

$$r_e = r_v/r_c$$

4–1: constant volume heat rejection corresponding to gas exchange process, as in the case of the Otto cycle.

By analysing the heat and work transfers, using the relationships

$$T_2 = T_1(r_v)^{\gamma-1}$$
$$T_3 = T_2 r_c = T_1(r_v)^{\gamma-1}\, r_c$$
$$T_4 = T_3\left(\frac{r_c}{r_v}\right)^{\gamma-1} = T_1(r_v)^{\gamma-1}\, r_c\left(\frac{r_c}{r_v}\right)^{\gamma-1}$$
$$= T_1(r_c)^{\gamma}$$

we obtain

$$q_{23} = c_p[T_3 - T_2] = c_pT_1[r_c - 1](r_v)^{\gamma-1} \tag{i}$$
$$q_{41} = c_v[T_1 - T_4] = c_vT_1[1 - (r_c)^{\gamma}] \tag{ii}$$

i.e. efficiency

$$\eta_{th} = \frac{\oint dw}{q_{gross}} = \frac{q_{23} + q_{41}}{q_{23}} = \frac{c_pT_1[r_c - 1](r_v)^{\gamma-1} + c_vT_1[1 - (r_c)^{\gamma}]}{c_pT[r_c \quad 1](r_v)^{\gamma-1}}$$
$$= \frac{(r_c)^{\gamma} - 1}{(r_v)^{\gamma-1}\,\gamma[r_c - 1]} \tag{15.24}$$

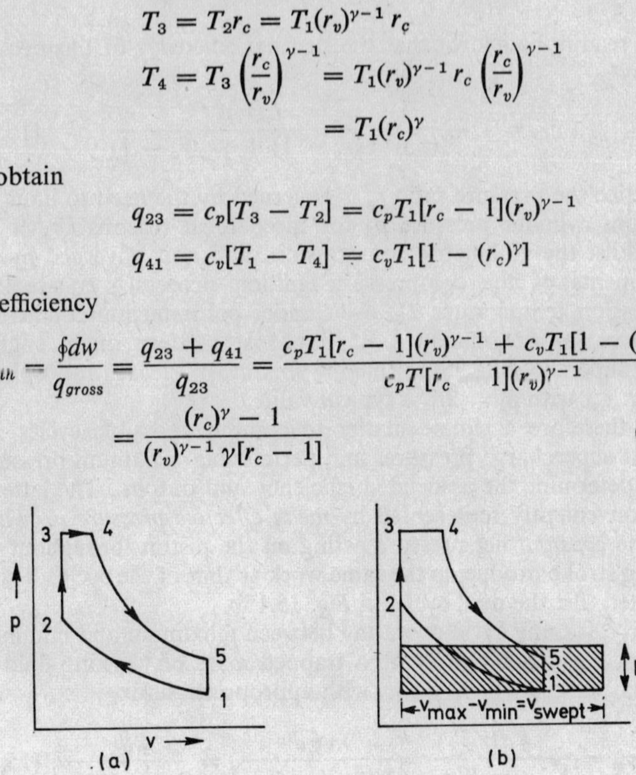

Fig. 15.18 Dual or mixed cycle

(iii) *The Dual or Mixed Cycle* (*Fig.* 15.18)

This cycle corresponds more closely to the modern Diesel engine cycle than the constant pressure cycle of the previous section. The sequence of events is as follows:

1–2: isentropic compression through volumetric compression ratio r_v.

2–3: constant volume heat addition q_{23} through pressure ratio

$$r_p = p_3/p_2$$

3–4: constant pressure heat addition q_{34} with cut-off ratio

$$r_c = v_4/v_3$$

4–5: isentropic expansion through expansion ratio

$$r_e = r_v/r_c$$

It may readily be shown that the thermal efficiency of this cycle is given by

$$\eta_{th} = 1 - \frac{1}{(r_v)^{\gamma-1}} \frac{r_p(r_c)^{\gamma-1}}{(r_p - 1) + \gamma r_p(r_c - 1)} \qquad (15.24)$$

In practice the pressure ratio r_p is governed by the need to limit the maximum cylinder pressure p_3 for mechanical reasons ($p_3 \ngtr 140$ bar), whilst the volumetric compression ratio r_v is governed by the requirements of the compression–ignition process, i.e. adequate compression temperature T_2, even under cold starting conditions. This gives, approximately $r_v \nless 10$. Most modern diesel engines employ supercharging, i.e. the charge pressure p_1 exceeds atmospheric pressure p_a, with $p_1 \simeq 2p_a$ a typical value.

It is therefore a simple matter to compute possible cycles, for different supercharge pressures and permissible maximum pressures and to determine the associated efficiency and output. The latter is most conveniently represented by *mean effective pressure* p_m where p_m is the *constant* net pressure acting on the piston throughout the working stroke producing the same work as that of the cycle. This is illustrated, for the dual cycle, in *Fig*. 15.15b.

Clearly, for any cycle operating between maximum and minimum volumes, V_{max} and V_{min}, with a trapped mass of working fluid m, such that, in general, $V = mv$, with appropriate suffix,

$$10^2\, p_m = \frac{\oint dW}{V_{max} - V_{min}} = \frac{m \oint dw}{m(v_{max} - v_{min})} = \frac{\oint dw}{v_{max} - v_{min}} \qquad (15.25)$$

the constant 10^2 being introduced, as in the case of the perfect gas equation (3.2b), when p_m is measured in bars, where equation (15.25) is quite general, i.e. not limited to the dual cycle.

Some indication of progress over the last twenty years is given by the fact that practical diesel engine mean effective pressures have increased over threefold from approximately 5·5 bar to 18 bar, the former figure being associated with unsupercharged engines, and the latter with a supercharge pressure ratio $p_1 \simeq 3p_a$.

EXAMPLE 15.8 (see *Fig*. 15.8a)

A diesel engine operates on the dual cycle with a volumetric compression ratio $r_v = 10$, trapped pressure 2 bar, maximum pressure $p_3 = 100$ bar and cut-off ratio 1·5.

Find the thermal efficiency and mean effective pressure and comment on the result.

SOLUTION

The pressure ratio $r_p = p_3/p_2$ must first be obtained from

$$p_2 = p_1(r_v)^\gamma$$

i.e. $p_2 = 2(10)^{1\cdot4} = 50$ bar, giving

$$r_p = p_3/p_2 = 100/50 = 2$$

Substituting in equation (15.24) for thermal efficiency

$$\eta_{th} = 1 - \frac{1}{(10)^{0\cdot4}} \times \frac{2\cdot0 \times (1\cdot5)^{1\cdot4} - 1}{(2\cdot0 - 1) + 1\cdot4 \times 2\cdot0(1\cdot5 - 1)}$$

$= 0\cdot580$ *Ans.*

The mean effective pressure p_m is given by equation (15.25)

$$v_{max} - v_{min} = \left(1 - \frac{1}{10}\right) v_1 = 0\cdot9 \left(\frac{RT_1}{p_1}\right) 10^{-2}\ \text{m}^3/\text{kg} \qquad \text{(i)}$$

also

$$\oint dw = \eta_{th} q_{gross} = \eta_{th}(q_{23} + q_{34})$$

where

$$q_{23} = c_v(T_3 - T_2) = c_v T_2(r_p - 1) = c_v T_1(r_v)^{\gamma-1}(r_p - 1) \qquad \text{(ii)}$$

and

$$q_{34} = c_p T_3(r_c - 1)$$
$$= c_p T_1(r_v)^{\gamma-1} r_p(r_c - 1) \qquad \text{(iii)}$$

whence

$$\oint dw = 0\cdot580 T_1(0\cdot718 \times 2\cdot5 \times 1\cdot0 + 1\cdot005 \times 2\cdot5 \times 2\cdot0 \times 0\cdot5)$$
$$= 4\cdot30 T_1\ \text{kJ/kg}$$

$$p_m = \frac{\oint dw}{10^2(v_{max} - v_{min})} = \frac{4\cdot30 T_1}{10^2 \times 0\cdot9 \times \dfrac{0\cdot287 T_1}{2 \times 10^2}}$$

$= 33\cdot3$ bar *Ans.*

It will be noted that this ideal mean effective pressure greatly exceeds the figure quoted for modern diesel engines, viz. 18 bar. The reason is that the latter figure is a practical value, in which both departures of the cycle from the ideal form of *Fig*. 15.18a and frictional losses are included. These two effects are accounted for respectively by the *diagram efficiency* η_d ($\simeq$0·70) and the *mechanical efficiency* η_m ($\simeq$0·8) giving a practical or *brake mean effective pressure*.

$$p_{m'} = p_m \eta_d \, . \, \eta_m \simeq 0{\cdot}56 p_m = 18{\cdot}6 \text{ bar}$$

Likewise the brake thermal efficiency becomes

$$\eta_b = \eta_{th} \eta_d \, . \, \eta_m \simeq 0{\cdot}56 \eta_{th} = 0{\cdot}325$$

These modified figures for mean effective pressure and efficiency are very much in line with modern diesel engine practice.

15.3 Refrigeration and Heat Pump Cycles

Reference was made in section 6.2 to reversed heat engines which in turn may be divided into:

(*a*) *Heat pumps* whose useful "output" is heat transferred to a high temperature sink, e.g. the room to be heated, from a low temperature source, e.g. a supply of river or sea water, for the expenditure of mechanical work. The appropriate thermodynamic criterion is the so-called *coefficient of performance* (equation (6.2)), i.e.

$$(\text{C.O.P.})_{heat\ pump} = \frac{\text{Heat transferred to high temperature sink}}{\text{Work input}} \tag{15.26}$$

(*b*) *Refrigerators* whose useful "output" is heat absorbed from a low temperature source, e.g. a cold chamber to be maintained at a given temperature below that of the surroundings, and transferred to the surroundings, again for the expenditure of mechanical work.

The coefficient of performance in this case is defined by equation (6.3), i.e.

$$(\text{C.O.P.})_{refrig.} = \frac{\text{Heat absorbed from low temperature source}}{\text{Work input}} \tag{15.27}$$

It was also shown that for the same plant treated either as a heat pump or a refrigerator

$$(\text{C.O.P.})_{heat\ pump} = (\text{C.O.P.})_{refrig.} + 1 \tag{15.28}$$

The considerations in Chapter 6 applied essentially to ideal reversed Carnot engines. As in the case of vapour and gas power cycles the

Carnot cycle is impracticable and is replaced by working cycles having lower theoretical coefficients of performance, but realizing satisfactory levels in practice with acceptable mechanical performance.

Only one type of cycle will be considered in any detail in view of its overwhelming practical importance, the *vapour compression cycle*. This is closely related to the vapour power cycles of section 15.1, except for the choice of working fluids other than H_2O for reasons connected with the temperature range over which such machines operate, namely approximately − 30°C to + 30°C for refrigerators and 5°C to 50°C for heat pumps. The main requirement for such fluids are that they should exist both in the liquid and the vapour (*not* solid) phase within these temperature ranges at acceptable pressure levels and that the latent heat of evaporation should be as high as possible in order to reduce the mass of working fluid in circulation. Practical refrigerants are

(1) Ammonia (NH_3).
(2) Sulphur dioxide (SO_2).
(3) Freon 12 (dichloro-difluoro methane) CCl_2F_2.

The properties of (1) and (3) are given in ref. 3.1.

15.3.1 The Simple Vapour Compression Refrigeration or Heat Pump Cycle

Figs. 15.19a and 15.19b show the two main applications of the cycle and emphasize the difference in temperature levels. The *evaporator*, corresponding to T_{min}, absorbs heat either from the cold chamber (refrigerator) or from river or sea water (heat pump). Clearly T_{min} is considerably lower in the former case. Conversely the *condenser*, corresponding to T_{max}, rejects heat either to the atmosphere or the hot chamber, i.e. T_{max} is higher for the heat pump than for the refrigerator. As a very rough approximation, both T_{min} and T_{max} may be considered 20–30°C higher for the heat pump.

The working fluid as in vapour power cycles, undergoes a series of steady flow processes as follows:

1–2: *compression* of wet vapour leaving evaporator to dry saturated vapour condition, with work input w_{12}.

2–3: *condensation* of dry saturated vapour in condenser, the fluid leaving as saturated liquid, with heat rejection q_{23}.

3–4: *throttling* (replacing isentropic expansion) in a specially designed throttle valve to wet vapour condition 4, without work or heat transfer.

4–1: *evaporation* from state 4 to initial state 1 in specially designed evaporator, with heat absorption q_{41}.

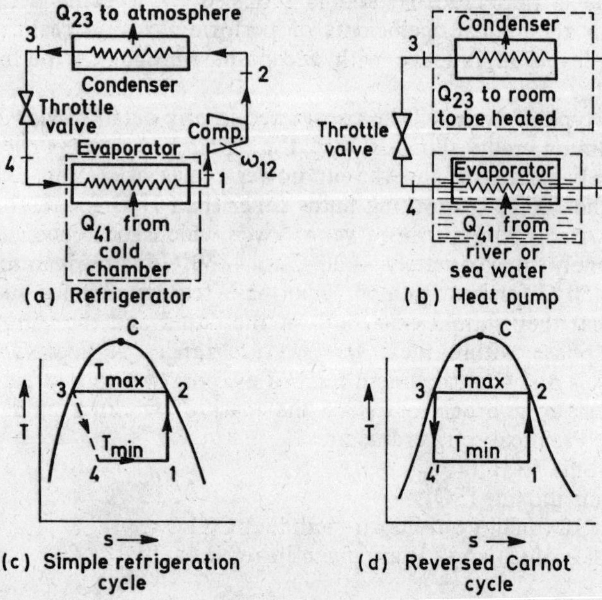

Fig. 15.19 Vapour compression cycle

The coefficient of performance for the two forms of the cycle is given by equations (15.27) and (15.28) respectively, i.e.

$$(\text{C.O.P.})_{refrig.} = \left|\frac{q_{41}}{w_{12}}\right| = \frac{h_1 - h_4}{h_2 - h_1} \tag{15.27a}$$

$$(\text{C.O.P.})_{heat\ pump} = \left|\frac{q_{23}}{w_{12}}\right| = \frac{h_2 - h_3}{h_2 - h_1} \tag{15.28a}$$

where enthalpy values are obtained either from property tables (ref. 3.1) or from suitable phase diagrams, usually in the form of the pressure–enthalpy (p–h) diagram (*Fig.* 8.8). The modulus notation is used in order to avoid confusion arising from the sign convention. Thus q_{41} is a "positive" heat transfer *to* the system, q_{23} is "negative," and w_{12} is work *to* the system, i.e. negative.

Fig. 15.9d shows the reversed Carnot cycle in which the irreversible throttling process 3–4 is replaced by the theoretically reversible expansion process 3–4′ to be performed either in a reciprocating expander or turbine. Theoretically, the expansion work $w_{34'}$ would reduce the net work, i.e.

$$w_{net} = w_{12} + w_{34'}$$

where, clearly, $|w_{12} + w_{34'}| < w_{12}$

whilst, at the same time the heat abstraction would be increased, i.e.

$$|q_{4'1}| > |q_{41}|$$

In practice, however, such gains would be small and the additional complication is not justified.

EXAMPLE 15.9

A vapour compression cycle using freon 12 is used in a refrigerator required to maintain a cold chamber at −20°C. Heat rejection is to the atmosphere at +20°C. Efficient heat transfer in the evaporator and condenser necessitates a temperature difference between working fluid and surroundings of 10°C, in the appropriate direction.

Find the coefficient of performance and the heat absorption per kilogramme of fluid circulating and compare with the corresponding reversed Carnot cycle. Assume compression efficiency $\eta_c = 1{\cdot}0$.

SOLUTION

$t_{min} = (-20 - 10)°\text{C} = -30°\text{C}$ i.e. $t_{surr} > t_{fluid}$ for heat transfer to fluid
$t_{max} = (20 + 10)°\text{C} = +30°\text{C}$ i.e. $t_{surr} < t_{fluid}$ for heat transfer from fluid

Using abridged tables for freon 12, ref. 3.1, and the notation of *Fig*. 15.19c:

At (2): $t_2 = 30°\text{C}, p_2 = 7{\cdot}449$ bar, $h_2 = h_g = 199{\cdot}62$ kJ/kg,
$s_2 = s_g = 0{\cdot}6853$ kJ/kg-K

At (3): $t_3 = 30°\text{C}, p_3 = 7{\cdot}449$ bar, $h_3 = h_f = 64{\cdot}59$ kJ/kg,
$s_3 = s_f = 0{\cdot}2399$ kJ/kg-K

At (4): $t_4 = -30°\text{C}, p_4 = 1{\cdot}004$ bar
After throttling, $h_4 = h_3 = 64{\cdot}59$ kJ/kg

$$x_4 = \frac{h_4 - h_f}{h_{fg}} = \frac{64{\cdot}59 - 8{\cdot}86}{165{\cdot}34} = 0{\cdot}338$$

At (1): $t_4 = -30°\text{C}, p_4 = 1{\cdot}004$ bar

Before isentropic compression, $s_1 = s_2 = 0{\cdot}6853$ kJ/kg-K

$$x_1 = \frac{s_1 - s_f}{s_{fg}} = \frac{0{\cdot}6853 - 0{\cdot}0371}{0{\cdot}6799} = 0{\cdot}995$$

$$h_1 = h_f + x_1 h_{fg} = 8{\cdot}86 + 0{\cdot}955 \times 165{\cdot}34 = 166{\cdot}76 \text{ kJ/kg}$$

Heat abstracted $= q_{41} = h_1 - h_4 = 102{\cdot}17$ kJ/kg *Ans.*
Work input $w_{12} = -(h_2 - h_1) = -32{\cdot}86$ kJ/kg

$$(\text{C.O.P})_{refrig.} = \left|\frac{q_{41}}{w_{12}}\right| = \left|\frac{102{\cdot}17}{-32{\cdot}86}\right| = 0{\cdot}311 \quad \textit{Ans.}$$

For the corresponding Carnot cycle (*Fig*. 15.19d):

At (4′) $t_4 = -30°C, p_4 = 1{\cdot}004$ bar

After isentropic expansion $s_{4'} = s_3 = 0{\cdot}2399$ kJ/kg-K

$$x_{4'} = \frac{s_4 - s_f}{s_{fg}} = \frac{0{\cdot}2399 - 0{\cdot}0371}{0{\cdot}6799} = 0{\cdot}2985$$

$$h_{4'} = h_f + x_{4'}h_{fg} = 8{\cdot}86 + 0{\cdot}2985 \times 165{\cdot}34 = 58{\cdot}26 \text{ kJ/kg}$$

Heat abstracted $= q_{4'1} = h_1 - h_{4'} = 108{\cdot}50$ kJ/kg *Ans.*
cf. 102·17 kJ/kg

$$(\text{C.O.P})_{refrig.} = \left|\frac{q_{4'1}}{w_{12}}\right| = \left|\frac{108{\cdot}50}{-32{\cdot}86}\right| = 0{\cdot}330 \quad \textit{Ans.} \quad \text{cf. } 0{\cdot}311$$

Hence even the theoretical gain is small, and in practice would be *more than* outweighed by additional complications.

It is left as an exercise for the student to perform a similar calculation for the case of a heat pump using low grade heat from a river at 10°C to supply heat at 30°C.

15.3.2 Alternative Vapour Compression Cycles

The simple cycle (*Fig*. 15.19c) is unsuitable in practice since evaporation cannot be arrested at point 1. If evaporation proceeds up to the saturated vapour boundary, or beyond, as shown in *Fig*. 15.20, the subsequent compression results in a superheated state. When the temperature differences between working fluid and surroundings are substantial, they result in subcooling of the liquid leaving the evaporator before entering the compressor. This is shown diagrammatically in *Fig*. 15.20, both in the temperature–entropy and in the pressure–enthalpy diagram. For other modifications of the

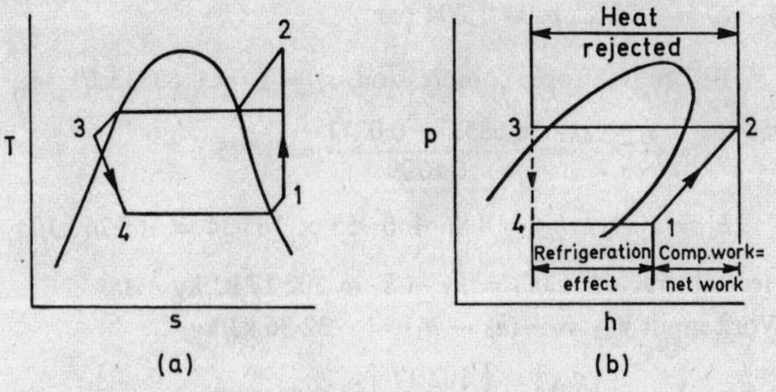

Fig. 15.20 Refrigeration cycle with subcooling and superheating

simple cycle the reader is referred to ref. 15.1, Chapter 13, or more specialized texts.

REFERENCES

15.1 ROGERS, G. F. C. and MAYHEW, R., *Engineering Thermodynamics Work and Heat Transfer*, SI Units (Longmans).

15.2 WEIR, C. D., Optimization of heater enthalpy rises in feed heating trains, *Proc. IMechE*, vol. 174 (1960) pp. 769–96.

15.3 WOOTTON, W. M., *Steam Cycles for Nuclear Power Plant* (Temple Press, 1958).

15.4 COHEN, H. and ROGERS, G. F. C., *Gas Turbine Theory* (Longmans, 1951).

15.5 KEENAN, J. H., and KAYE, J., *Gas Tables* (John Wiley, 1953).

15.6 DE BREY, H., RIMIA, H. and VAN WEENEN, F. L., Fundamentals for the development of the Philips air engine, *Philips Technical Review*, vol. 9, no. 4 (1947).

15.7 VAN WEENEN, F. L., The construction of the Philips air engine, *Philips Technical Review*, vol. 9, no. 5 (1947).

15.8 RICARDO, SIR HENRY, *The High-Speed Internal Combustion Engine* (Blackie, 1955).

EXERCISES

15.1 Draw up a complete energy balance, including the feed pump work term, for a Rankine cycle operating with a boiler pressure of 40 bar and a condenser pressure of 0·16 bar. The steam leaves the boiler dry saturated and the expansion and compression processes are both isentropic. Hence find the cycle efficiency and the work ratio.

(*Ans:* 0·323, 0·995)

15.2 A reheat cycle works between pressures of 110 bar and 0·040 bar. The steam is superheated to 600°C and after expansion to the dry saturated condition is reheated to 500°C. Calculate the ideal thermal efficiency of the cycle taking account of feed pump work.

(*Ans:* 0·450)

15.3 Calculate the efficiency of a regenerative cycle using a single open feed heater. The steam leaves the boiler at 30 bar superheated to 450°C and the condenser pressure is 0·040 bar. The bleed pressure is 2·7 bar.

(*Ans:* 0·357)

15.4 Discuss briefly, but using illustrations, the significance of each of the following terms relating to the performance of thermal power cycles:

(*a*) efficiency
(*b*) work ratio.

In an ideal Rankine cycle steam leaves the boiler and enters the turbine at 40 bar dry saturated. The condenser pressure is 0·075 bar.

Assuming feed pump and turbine isentropic efficiencies to be 100%, evaluate the efficiency and work ratio of the cycle.

Evaluate also the "effectiveness" of the turbine in an actual cycle, assuming the turbine to have an isentropic efficiency of 80% when operating between the above pressure limits. The temperature of the surroundings may be taken as 18°C. Effectiveness is defined as

$$\varepsilon = \left|\frac{\text{Gain in available energy of surroundings}}{\text{Loss of available energy of system}}\right|$$

and available energy in a steady flow system without kinetic and potential energy effects is defined as $b = h - T_0 s$ where T_0 is the temperature of the surroundings.

(*Ans:* 0·347, 0·995, 0·812)

15.5 Briefly discuss the reasons for the choice of the Rankine cycle as a reference cycle for steam plant.

What are the main differences between the operating cycle of a modern power station and the simple Rankine cycle, and what are the reasons for these differences?

In a proposed steam turbine power plant the steam supply is at 90 bar and 500°C. The steam then expands in a multistage turbine having an isentropic efficiency of 78% to an exhaust pressure of 0·035 bar, and is subsequently condensed at that pressure and returned to the boiler by a feed pump having an isentropic efficiency of 75%. Using the *h–s* chart where possible, calculate

(*a*) the thermal efficiency of the cycle,
(*b*) the net output per kg of steam,
(*c*) the work ratio.

Comment briefly on likely differences between the quantities as evaluated above and those for a Carnot cycle operating between the same pressure limits.

(*Ans:* 0·330, 1078 kJ/kg, 0·988)

15.6 A simple gas turbine cycle operates with an overall pressure ratio of 4:1, the compressor inlet conditions being 1·013 bar and 288 K, and the maximum temperature 1100 K. The rate of flow is 9 kg/s. Determine on the basis of air standard cycle assumption:

(*a*) the compressor work, turbine work, heat input and thermal efficiency of the cycle.
(*b*) if the same cycle were employed in a reciprocating engine what would be the mean effective pressure? Why would this cycle be unsuitable?

(*Ans:* (*a*) 139·2 kJ/kg, 362·0 kJ/kg, 676·0 kJ/kg, 0·329
(*b*) 1·225 bar)

15.7 An ideal regenerator is incorporated into the simple gas turbine cycle of Example 15.6. Determine the thermal efficiency of this cycle.

(*Ans:* 0·615)

15.8 A gas turbine cycle has two stages of compression with intercooling and two stages of expansion with reheating. The inlet temperature to each compressor is 288 K and to each turbine 1100 K. The pressure ratio across each compressor and turbine is 2.

Find, on the basis of air standard cycle assumptions:

(*a*) compressor work
(*b*) turbine work
(*c*) heat input
(*d*) overall thermal efficiency.

Compare the above quantities with those of the simple cycle of example 15.6.

If an ideal regenerator is now incorporated in the cycle, what is the effect on:

(*e*) heat input
(*f*) thermal efficiency?

(*Ans:* (*a*) 126·5 kJ/kg, (*b*) 397·0 kJ/kg, (*c*) 952 kJ/kg, (*d*) 0·284, (*e*) 398 kJ/kg, (*f*) 0·680)

15.9 Repeat example 15.8 assuming the following component efficiencies:

$\eta_c = 0{\cdot}80$
$\eta_T = 0{\cdot}85$
$\eta_{\text{regenerator}} = 0{\cdot}60$

(*Ans:* (*a*) 158 kJ/kg, (*b*) 317·5 kJ/kg, (*c*) 895 kJ/kg, (*d*) 0·179, (*e*) 545 kJ/kg, (*f*) 0·293)

15.10 A dual combustion diesel engine cycle (*Fig.* 15.8a) operates with a volumetric compression ratio $r_v = 12$ and with a cut off ratio 1·615. The maximum pressure $p_{max} = p_3$ is given by $p_{max} = 54p_1$, where p_1 is the trapped pressure. Assuming indices of compression and expansion of 1·35, show that the mean effective pressure of the cycle

$$p_i = 10p_1$$

Hence evaluate (*a*) temperature at salient points with $T_1 = 355$ K and (*b*) cycle efficiency, assuming $c_p = 1{\cdot}05$ kJ/kg-K and $\gamma = 1{\cdot}4$.

(*Ans:* (*a*) $T_2 = 805$ K, $p_2 = 29{\cdot}2p_1$, $T_3 = 1490$ K, $T_4 = 2410$ K, $T_5 = 1200$ K
(*b*) $\eta = 0{\cdot}670$)

15.11 Recalculate

(*a*) the conditions at salient cycle points
(*b*) the MEP, and
(*c*) the efficiency when the cycle of example 15.10 is changed to a constant pressure (or diesel) cycle with the same volumetric compression ratio and a cut-off ratio such as to give an expansion curve coincident with the lower part of that of the dual cycle of example 15.10.

(*Ans:* (*a*) $T_2 = 805$ K, $T_3 = 1970$ K, $T_4 = 1142$ K,
(*b*) MEP $= 6{\cdot}82p_1$, (*c*) $\eta = 0{\cdot}513$)

15.12 In recent years there has been a resurgence of interest in engines operating on the regenerative cycle principle, of which the Stirling engine is a particular embodiment. The Stirling cycle consists of two isothermals T_{max} and T_{min} connected by two constant volume processes with complete *internal* heat exchange between the constant volume heating and cooling processes.

Sketch the air standard Stirling cycle with a volumetric expansion ratio of 10 in the pressure–volume and temperature–entropy diagrams, and derive expressions for:

(*a*) the specific output

(*b*) the efficiency

(*c*) the work ratio.

Compare the last two quantities (*b*) and (*c*) with those for a Carnot cycle operating between the same temperature limits, where $T_{max} = 660$ K, $T_{min} = 330$ K, and having the same work output per unit mass of air. Hence suggest reasons for the superiority of the Stirling cycle over the Carnot cycle.

(*Ans:* $\eta_{stirling} = 0{\cdot}5$, $\eta_{carnot} = 0{\cdot}5$

$w_{stirling} = 2200$ kJ/kg, $w_{carnot} = 2200$ kJ/kg

$(r_w)_{stirling} = 0{\cdot}5$ $(r_w)_{carnot} = 0{\cdot}324$)

15.13 The Ericsson regenerative cycle consists of two reversible constant pressure processes and two reversible isothermal processes with internal heat exchange between the two constant pressure processes, and external heat addition and rejection for the two isothermal processes. Show that, with perfect heat exchange, the efficiency of the cycle is equal to that of the Carnot cycle operating between the same temperature limits.

The pressure ratio of an Ericsson cycle using air as the working fluid is 8, and the maximum and minimum temperatures are 1200°C and 200°C respectively. The regenerator allows 75% of the heat rejected at the lower pressure to be absorbed at the higher pressure. Find the efficiency of the cycle, and compare it with that of a cycle operating between the same pressure and temperature limits when the effectiveness of the regenerator is 100%.

Give reasons why the Ericsson cycle with complete regeneration would be preferred to the Carnot cycle as an ideal cycle for a practical engine.

(*Ans:* $\eta = 0{\cdot}529$, cf. $0{\cdot}628$)

15.14 An Ericsson cycle operating with an ideal regenerator operates between 1100 K and 288 K. The pressure at the beginning of isothermal compression is 1·013 bar. Determine (*a*) compressor and turbine work per kg of air and (*b*) the thermal efficiency.

(*Ans:* (*a*) $w_T = 465{\cdot}0$ kJ/kg, $w_c = 121{\cdot}8$ kJ/kg, (*b*) 0·738)

15.15 A vapour compression refrigerator using freon 12 works between temperature limits −10°C and 35°C, the refrigerant leaving the compressor dry saturated. Calculate the refrigeration effect and the coefficient of performance if:

(*a*) the refrigerant leaves the condenser saturated,
(*b*) the refrigerant is subcooled to 15°C before throttling.
(*Ans:* (*a*) 109·2 kJ/kg, 4·82
(*b*) 128·2 kJ/kg, 5·67)

15.16 In the above example assess the effect of substituting ammonia NH_3 for freon 12, the temperature limits remaining unchanged.
(*Ans:* (*a*) 938 kJ/kg, 5·05
(*b*) 1034 kJ/kg, 5·55)

15.17 An ideal heat pump operates on the reversed Carnot cycle, the lower and upper temperature limits of the working fluid (freon 12) being 5°C and 55°C respectively. Assuming compression to end at the saturated vapour condition and expansion to commence at the saturated liquid condition, find the heat output of the unit per kg of freon.

Low-grade heat is obtained from river water, the supply and return temperature of the water being 9°C and 3°C, respectively. The heat pump is to raise the temperature of 11000 m³/h) of air at 1 bar and 10°C by 10°C. Estimate:

(*a*) the rate of circulation of river water,
(*b*) the rate of circulation of freon,
(*c*) the power input to the heat pump.
(*Ans:* 117·65 kJ/kg, (*a*) 4600 kg/h, (*b*) 1158 kg/h, (*c*) 7·76 hp)

Appendix I

Statistical Treatment of Entropy

In Chapter 7 the important thermodynamic property of entropy was derived from considerations arising from the second law of thermodynamics, and more particularly from the Carnot cycle, which for a fully reversible sequence of processes provides a unique relationship between absolute temperature and energy transfer in the form of heat.

From these considerations we deduced that this new property, entropy, could be defined as follows (p. 124)—

Corollary 6. There exists a property of a system, denoted by the symbol S, such that a change in its value is equal to

$$S_2 - S_1 = \int_1^2 \frac{dQ}{T} \qquad \text{(I.1)}$$

for any reversible process undergone by the system between states 1 and 2. This property is called entropy; Q denotes heat transferred and T absolute temperature.

Corollaries 7 and 8, dealing respectively with entropy increases in isolated and non-isolated systems, express the most important fact that, for the former system, gain in entropy is a direct measure of thermodynamic irreversibility, whilst for the latter, the entropy increase under irreversible conditions exceeds that under reversible, but otherwise comparable, conditions (pp. 142 and 144).

However, whilst extensive use was made of this new property in the following chapters, both in connexion with the pure substance in general, and the perfect gas in particular, it proved impossible to establish a satisfactory explanation of the physical nature of entropy in terms of the continuum treatment of matter adopted throughout the book, with the exception of Chapter 9 dealing with the kinetic theory of gases.

It was indicated in the introduction to Chapter 7 and again in Chapter 9, section 9.6, that a satisfactory treatment of entropy must be based on a statistical approach, making use of the concept of thermodynamic probability W. The latter term may be defined in terms of microstate and macrostate which, in turn, are defined as follows—

(*i*) *The microstate* is specified by the position (x, y, z) and energy ε (in the simplest case of a monatomic gas kinetic energy of translation, specified by the velocity components c_x, c_y, c_z, see Chapter 9, of the individual particles in the assembly of particles which is being studied).

(*ii*) *The macrostate* is the observable state arising from the sum of the microstates.

(*iii*) *The thermodynamic probability* W is the number of microstates making up a macrostate.

The above may be illustrated by a simple example.

Let there be two hypothetical cells, i and j, and let four particles, a, b, c, and d, each defined by three space co-ordinates x, y, z and three velocity components c_x, c_y, c_z, be distributed between the two cells. The macrostate may then be indicated diagrammatically by *Fig.* I.1, which merely specifies the number of particles in each cell.

N_i	4	3	2	1	0
N_j	0	1	2	3	4

Fig. I.1

Fig. I.1 indicates that five macrostates are possible. Microstates are indicated diagrammatically in *Fig.* I.2, which applies to the particular macrostate $N_i = 3$, $N_j = 1$.

Cell i	*bcd*	*cda*	*dab*	*abc*
Cell	*a*	*b*	*c*	*d*

Fig. I.2

Hence to the macrostate $N_i = 3$, $N_j = 1$, there correspond four possible microstates. It should be noted that each letter a, b, c or d implies a definite set of space and velocity co-ordinates. For this reason a, b, c, d are also described as PHASE POINTS.

The number of microstates corresponding to a given macrostate is computed by writing down the total number of possible arrangements or permutations of phase points in the macrostate. The number of permutations for N things, i.e. the number of different ways in which they can be arranged, is $N!$ Thus the number of permutations of the four phase points a, b, c and d is $4! = 24$ but this exceeds the number of microstates in the above example since it includes all possible permutations of the three points in cell i, namely $3! = 6$ which, however, do not constitute new microstates since the order of phase points within a particular cell is not considered to change the microstate. Hence the number of microstates becomes $4!/3! = 24/6 = 4$, which agrees with the result derived directly from *Fig.* I.2.

In general, when N phase points are distributed between n cells, the number of microstates for a particular macrostate is given by

$$W = \frac{N!}{N_1!N_2!\dots N_n!} \tag{I.2}$$

where N_1, $N_2 \dots N_n$ denote the number of phase points in cells $1, 2, \dots n$ respectively.

Summarizing the number of microstates for each of the five macrostates represented by *Fig.* I.1 we have ($N = 4$, $n = 2$)—

$$W(N_i = 4, N_j = 0) = \frac{4!}{4!\,0!} = 1$$

$$W(N_i = 3, N_j = 1) = \frac{4!}{3!\,1!} = 4$$

$$W(N_i = 2, N_j = 2) = \frac{4!}{2!\,2!} = 6$$

$$W(N_i = 1, N_j = 3) = \frac{4!}{1!\,3!} = 4$$

$$W(N_i = 0, N_j = 4) = \frac{4!}{0!\,4!} = 1$$

Hence, as expected, the macrostate with maximum probability is that with two points in each cell.

When dealing with a real gas, where the numbers N and N_i are very large, it may be shown, using the so-called Stirling approximation

$$\ln (x!) = x \ln x - x = x (\ln x - 1) \tag{I.3}$$

and taking the logarithm of the thermodynamic probability given by equation (I.2) that

$$\ln W = N \ln N - \Sigma N_i \ln N_i \qquad \text{(I.4)}$$

Now since the number of phase points N_i in individual cells i changes with time, it is clear that the thermodynamic probability given by equation (I.4) will change progressively. Since $N \ln N$ is constant, the probability will be a maximum when $\Sigma N_i \ln N_i$ is a minimum, i.e. when $\delta W = 0$ or, more conveniently $\delta\,(\ln W) = 0$.

Differentiating equation (I.4)

$$\delta\,(\ln W) = -\,[\,\Sigma\, N_i\, \delta\,(\ln N_i) + \Sigma\,(\ln N_i)\,\delta N_i]$$

$$= -\left[\Sigma \frac{N_i}{N_i}\delta N_i + \Sigma\,(\ln N_i)\delta N_i\right]$$

where

$$\Sigma \frac{N_i}{N_i}\delta N_i = \Sigma\,\delta N_i = 0$$

since the total number of particles is constant.

$$\therefore \qquad \delta(\ln W) = -\,\Sigma(\ln N_i)\delta N_i = 0 \qquad \text{(I.5)}$$

where

$$\Sigma(\ln N_i)\delta N_i = \ln N_1 \delta N_1 + \ln N_2 \delta N_2 + \ldots + \ln N_n \delta N_n \qquad \text{(I.6)}$$

where $\delta N_1, \delta N_2 \ldots \delta N_n$ represent the small increases or decreases in numbers $N_1, N_2 \ldots N_n$ resulting from molecular motions or collisions, the total number remaining constant, i.e.

$$\delta N = \Sigma\delta N_i = \delta N_1 + \delta N_2 + \ldots + \delta N_n = 0 \qquad \text{(I.7)}$$

Equations (I.5) and (I.6) constitute the condition of maximum thermodynamic probability, subject to the CONDITION EQUATION (I.7).

A second condition equation is imposed by the fact that the energy of the system as a whole—which is assumed to be isolated—remains constant in accordance with the first law of thermodynamics. Hence we may write, with ε denoting energy (in the case of monatomic gases, kinetic energy of translation)

$$U = \Sigma\, \varepsilon_i N_i = \text{constant}$$

i.e.

$$\delta U = \Sigma \varepsilon_i \delta N_i = \varepsilon_1 \delta N_1 + \varepsilon_2 \delta N_2 \ldots + \varepsilon_n \delta N_n = 0 \qquad \text{(I.8)}$$

Equation (I.8) is the **second condition equation.**

The connexion between entropy and thermodynamic probability may now be established as follows. Equation (I.2) gives the number of microstates for a particular macrostate (see also *Figs*. I.1 and I.2) and is defined as thermodynamic probability. It is evident that if n independent groups, each of two cells, are taken, and each group has a thermodynamic probability given by equation (I.2), i.e. W_1, W_2, ... W_n, the thermodynamic probability of the resultant macrostate becomes

$$W = W_1 . W_2 . . . . W_n \tag{I.9}$$

i.e. a simple product relationship holds.

On the other hand, if the entropy—which is known to be an energy function—of each group is given by S_1, S_2, ... S_n, the entropy of all the groups taken together is clearly the sum of the individual entropies

$$S = S_1 + S_2 + \ldots + S_n \tag{I.10}$$

Hence, if a functional relationship between thermodynamic probability and entropy exists, it must be of the form*

$$S = S(W) = \text{constant} \times \ln W \tag{I.11}$$

in order to satisfy simultaneously equations (I.9) and (I.10).

Hence entropy is a measure of thermodynamic probability. The greatest degree of randomness, i.e. the highest thermodynamic probability given by equation (I.9) is also a state of maximum entropy. We have thus found, in statistical theory, an explanation of the principle of the increase of entropy, expressed in Chapter 7 as the 7th Corollary of the second law, i.e.

THE ENTROPY OF THE ISOLATED SYSTEM EITHER INCREASES OR IN THE LIMIT REMAINS CONSTANT.

which may now be re-expressed in the form—

THE ISOLATED SYSTEM TENDS TOWARDS A STATE OF MAXIMUM THERMODYNAMIC PROBABILITY.

* The constant in equation (I.11) is in fact the Boltzmann constant $k = G/N_0$ where G is the universal gas constant = 8·3143 kJ/kg-mol-K and N_0 is the Avogadro number, $5{\cdot}81 \times 10^{26}$ (see Chapter 9, section 9.2).

Appendix II

Thermodynamic Relationships

The objectives of the following analysis are threefold—

(1) To derive differential equations connecting various thermodynamics properties.
(2) To select, as independent variables, properties which can readily be measured from experimental data, in particular p–v–T data.
(3) To integrate the differential equations by analytical and graphical methods, in order to evaluate new thermodynamic properties from the original experimental data.

II.1 General Thermodynamic Equations

From the first law of thermodynamics applied to closed systems undergoing reversible processes, and from the definition of enthalpy h, and of the Helmholtz and Gibbs free energy functions f (equation (14.47) and g (equation (14.51)) the following differential relationships may be derived—

First law of thermodynamics

$$du = T\,ds - p\,dv \tag{II.1}$$

Substituting in (II.1) for enthalpy $h = u + pv$

$$dh = T\,ds + v\,dp \tag{II.2}$$

Helmholtz free energy function, substituting from (II.1) (equation (14.47))

$$\begin{aligned} df &= du - s\,dT - T\,ds \\ &= -p\,dv - s\,dT \end{aligned} \tag{II.3}$$

Gibbs free energy function, substituting from (II.2) (equation (14.51))

$$\begin{aligned} dg &= dh - s\,dT - T\,ds \\ &= v\,dp - s\,dT \end{aligned} \tag{II.4}$$

Each of these relationships constitutes an exact differential, so that the following further relationships* follow

$$u = \phi_1(s, v) \quad \text{i.e.} \quad du = \left(\frac{\partial u}{\partial s}\right)_v ds + \left(\frac{\partial u}{\partial v}\right)_s dv \qquad \text{(II.1a)}$$

$$h = \phi_2(s, p) \quad \text{i.e.} \quad dh = \left(\frac{\partial h}{\partial s}\right)_p ds + \left(\frac{\partial h}{\partial p}\right)_s dp$$

$$f = \phi_3(v, T) \quad \text{i.e.} \quad df = \left(\frac{\partial f}{\partial v}\right)_T dv + \left(\frac{\partial f}{\partial T}\right)_v dT \qquad \text{(II.3a)}$$

$$g = \phi_4(p, T) \quad \text{i.e.} \quad dg = \left(\frac{\partial g}{\partial p}\right)_T dp + \left(\frac{\partial g}{\partial T}\right)_p dT \qquad \text{(II.4a)}$$

Comparing coefficients in equations (II.1) and (II.1a), (II.2) and (II.2a), etc., we obtain

$$\left(\frac{\partial u}{\partial s}\right)_v = T \qquad \left(\frac{\partial u}{\partial v}\right)_s = -p \qquad \text{(II.1b)}$$

$$\left(\frac{\partial h}{\partial s}\right)_p = T \qquad \left(\frac{\partial h}{\partial p}\right)_s = v \qquad \text{(II.2b)}$$

$$\left(\frac{\partial f}{\partial v}\right)_T = -p \qquad \left(\frac{\partial f}{\partial T}\right)_v = -s \qquad \text{(II.3b)}$$

$$\left(\frac{\partial g}{\partial p}\right)_T = v \qquad \left(\frac{\partial g}{\partial T}\right)_p = -s \qquad \text{(II.4b)}$$

Finally, from equations (II.1b) to (II.4b), the further relationships given below may be derived

$$\left(\frac{\partial u}{\partial s}\right)_v = \left(\frac{\partial h}{\partial s}\right)_p = T \qquad \text{(II.5)}$$

$$\left(\frac{\partial u}{\partial v}\right)_s = \left(\frac{\partial f}{\partial v}\right)_T = -p \qquad \text{(II.6)}$$

$$\left(\frac{\partial h}{\partial p}\right)_s = \left(\frac{\partial g}{\partial p}\right)_T = v \qquad \text{(II.7)}$$

$$\left(\frac{\partial f}{\partial T}\right)_v = \left(\frac{\partial g}{\partial T}\right)_p = -s \qquad \text{(II.8)}$$

* The suffix notation used in conjunction with partial differential coefficients indicates which thermodynamic property is held constant while that in the denominator of the differential coefficient is allowed to change.

Since, for exact differentials, the commutative law may be applied to second differential coefficients, e.g.

$$\frac{\partial^2 u}{\partial s \cdot \partial v} = \frac{\partial^2 u}{\partial v \cdot \partial s} \quad \text{or} \quad \left(\frac{\partial}{\partial v}\right)_s \left(\frac{\partial u}{\partial s}\right)_v = \left(\frac{\partial}{\partial s}\right)_v \left(\frac{\partial u}{\partial v}\right)_s$$

the final set of relationships, known as Maxwell's equations, may now be derived from equations (II.5) to (II.8).

For internal energy u (equations (II.5) and (II.6))

$$\left(\frac{\partial}{\partial v}\right)_s \left(\frac{\partial u}{\partial s}\right)_v = \left(\frac{\partial}{\partial s}\right)_v \left(\frac{\partial u}{\partial v}\right)_s$$

i.e.

$$\left(\frac{\partial}{\partial v}\right)_s (T) = \left(\frac{\partial}{\partial s}\right)_v (-p)$$

or

$$\left(\frac{\partial T}{\partial v}\right)_s = -\left(\frac{\partial p}{\partial s}\right)_v \tag{II.9}$$

For enthalpy h (equations (II.5) and (II.7))

$$\left(\frac{\partial T}{\partial p}\right)_s = \left(\frac{\partial v}{\partial s}\right)_p \tag{II.10}$$

For Helmholtz free energy function (equations (II.6) and (II.8))

$$\left(\frac{\partial p}{\partial T}\right)_v = \left(\frac{\partial s}{\partial v}\right)_T \tag{II.11}$$

For Gibbs free energy function (equations (II.7) and (II.8))

$$\left(\frac{\partial v}{\partial T}\right)_p = -\left(\frac{\partial s}{\partial p}\right)_T \tag{II.12}$$

It will be observed that the partial derivatives on the left-hand side of the Maxwell equations involve p–v–T properties, whilst those on the right-hand side involve, in addition, the entropy s.

The Maxwell equations (II.9) to (II.12) form the basis for further important relationships which may be used for purposes of property evaluation.

II.2 Entropy Equations

Since entropy may be expressed as a function of any other two properties, e.g. temperature T and specific volume v,

$$s = f(T, v)$$

i.e.

$$ds = \left(\frac{\partial s}{\partial T}\right)_v dT + \left(\frac{\partial s}{\partial v}\right)_T dv$$

or

$$T\,ds = T\left(\frac{\partial s}{\partial T}\right)_v dT + T\left(\frac{\partial s}{\partial v}\right)_T dv \qquad \text{(i)}$$

But for a reversible constant volume change

$$dq = c_v\,(dT)_v = T(ds)_v$$

or

$$c_v = T\left(\frac{\partial s}{\partial T}\right)_v \qquad \text{(ii)}$$

whilst from Maxwell's equation (II.11)

$$\left(\frac{\partial s}{\partial v}\right)_T = \left(\frac{\partial p}{\partial T}\right)_v \qquad \text{(iii)}$$

Hence, substituting in equation (i)

$$\boxed{T\,ds = c_v\,dT + T\left(\frac{\partial p}{\partial T}\right)_v dv} \qquad \text{(II.13)}$$

This is known as the first form of the entropy equation.

Similarly, writing

$$s = f(T, p)$$

$$T\,ds = T\left(\frac{\partial s}{\partial T}\right)_p dT + T\left(\frac{\partial s}{\partial p}\right)_T dp \qquad \text{(iv)}$$

where

$$c_p = T\left(\frac{\partial s}{\partial T}\right)_p \qquad \text{(v)}$$

and from Maxwell's equation (II.12)

$$\left(\frac{\partial s}{\partial p}\right)_T = -\left(\frac{\partial v}{\partial T}\right)_p \qquad \text{(vi)}$$

whence, substituting in equation (iv)

$$T\,ds = c_p\,dT - T\left(\frac{\partial v}{\partial T}\right)_p dp \qquad \text{(II.14)}$$

Equation (II.14) is known as the second form of the entropy equation.

Either form of the entropy equation enables entropy to be evaluated from *p–v–T* data and one set of calorimetric data, i.e. either specific heat at constant volume or specific heat at constant pressure,

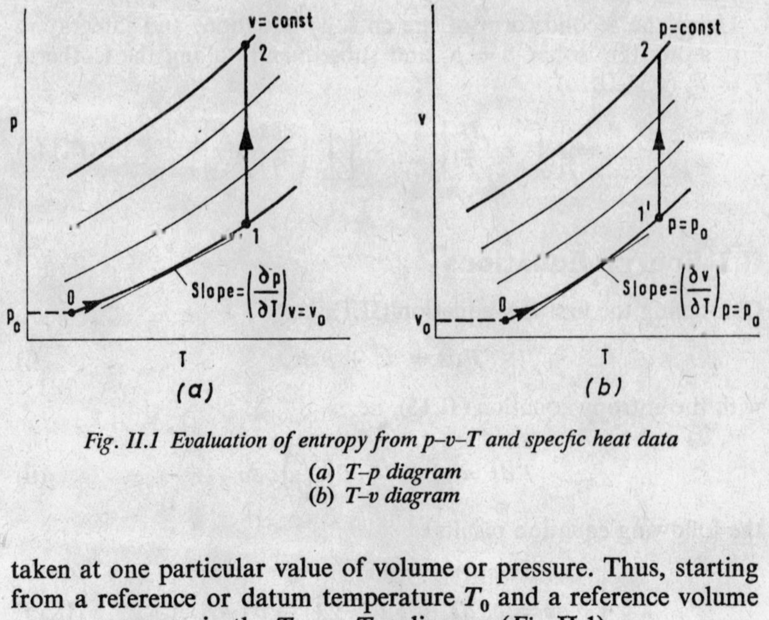

Fig. II.1 Evaluation of entropy from p–v–T and specfic heat data
(a) T–p diagram
(b) T–v diagram

taken at one particular value of volume or pressure. Thus, starting from a reference or datum temperature T_0 and a reference volume v_0 or pressure p_0 in the *T–v* or *T–p* diagram (*Fig.* II.1)—

Integrating equation (II.13) along the isochore $v = v_0$ between 0 and 1 (*Fig.* II.1a)

$$s_1 - s_0 = \int_0^1 \frac{c_v\,dT}{T} + \underbrace{\int_0^1 \left(\frac{\partial p}{\partial T}\right)_v dv}_{=0}$$

$$= \int_0^1 \frac{c_v\,dT}{T} \qquad \text{(i)}$$

and subsequently integrating along the isothermal 1–2

$$s_2 - s_1 = \underbrace{\int_1^2 \frac{c_v\,dT}{T}}_{=0} + \int_1^2 \left(\frac{\partial p}{\partial T}\right)_v dv$$

$$= \int_1^2 \left(\frac{\partial p}{\partial T}\right)_v dv$$

so that the entropy at any state 2 (p_2, v_2, T_2) relative to the datum state p_0, v_0, T_0 becomes

$$s_2 = \left[\int_0^1 c_v \frac{dT}{T}\right]_{v=v_0} + \left[\int_1^2 \left(\frac{\partial p}{\partial T}\right)_v dv\right]_{T=T_2} \qquad \text{(II.13a)}$$

Using the second form of the entropy equation, and integrating first along the isobar $p = p_0$ and subsequently along the isotherm $T = T_2$ (*Fig.* II.1*b*)

$$s_2 = \left[\int_0^{1'} c_p \frac{dT}{T}\right]_{p=p_0} - \left[\int_{1'}^2 \left(\frac{\partial v}{\partial T}\right)_p dp\right]_{T=T} \qquad \text{(II.14a)}$$

II.3 Energy Equations

Combining the first law equation (II.1), i.e.

$$T\,ds = du + p\,dv \qquad \text{(i)}$$

with the entropy equation (II.13), i.e.

$$T\,ds = c_v\,dT + T\left(\frac{\partial p}{\partial T}\right)_v dv \qquad \text{(ii)}$$

the following equation results

$$\boxed{du = c_v\,dT + \left[T\left(\frac{\partial p}{\partial T}\right)_v - p\right] dv} \qquad \text{(II.15)}$$

Similarly, combining the alternative first law equation (II.2) with the second entropy equation (II.14), i.e.

$$T\,ds = dh - v\,dp \qquad \text{(iii)}$$

with

$$T\,ds = c_p\,dT - T\left(\frac{\partial v}{\partial T}\right)_p dp \qquad \text{(iv)}$$

there results the alternative energy equation

$$\boxed{dh = c_p\,dT + \left[v - T\left(\frac{\partial v}{\partial T}\right)_p\right] dp} \qquad \text{(II.16)}$$

Equations (II.15) and (II.16) are known as the first and second form of the energy equation and point to the existence of a dependence of internal energy and enthalpy on properties other than temperature. The perfect gas is thus a very special case in which the p–v–T relationship is such as to make the terms

$$\left[T\left(\frac{\partial p}{\partial T}\right)_v - p\right] \quad \text{and} \quad \left[v - T\left(\frac{\partial v}{\partial T}\right)_p\right]$$

in the two forms of the energy equation, vanish.

Equations (II.15) and (II.16), like the two forms of the entropy equation, i.e. (II.13) and (II.14), may be used to evaluate internal energy and enthalpy by the use of one set of calorimetric data and the p–v–T relations, using the T–v diagram for internal energy and the T–p diagram for enthalpy.

Referring to *Fig.* II.1a and using the same notation as for the entropy determination

$$u_2 = \left[\int_0^1 c_v\,dT\right]_{v=v_0} + \left[\int_1^2 \left[T\left(\frac{\partial p}{\partial T}\right)_v - p\right] dv\right]_{T=T_2} \qquad \text{(II.15a)}$$

and

$$h_2 = \left[\int_2^1 c_p\,dT\right]_{p=p_0} - \left[\int_1^2 \left[v - T\left(\frac{\partial v}{\partial T}\right)_p\right] dp\right]_{T=T_2} \qquad \text{(II.16a)}$$

II.4 Difference in Specific Heats

Equating the entropy equations (II.13) and (II.14)

$$c_p\,dT - T\left(\frac{\partial v}{\partial T}\right)_p dp = c_v\,dT + T\left(\frac{\partial p}{\partial T}\right)_v dv$$

i.e.

$$dT = \frac{T\left(\frac{\partial p}{\partial T}\right)_v dv + T\left(\frac{\partial v}{\partial T}\right)_p dp}{c_p - c_v} \qquad \text{(i)}$$

But

$$dT = \left(\frac{\partial T}{\partial v}\right)_p dv + \left(\frac{\partial T}{\partial p}\right)_v dv \qquad \text{(ii)}$$

whence, comparing (i) and (ii),

$$\left(\frac{\partial T}{\partial v}\right)_p = \frac{T\left(\frac{\partial p}{\partial T}\right)_v}{c_p - c_v} \tag{iii}$$

and

$$\left(\frac{\partial T}{\partial p}\right)_v = \frac{T\left(\frac{\partial v}{\partial T}\right)_p}{c_p - c_v} \tag{iv}$$

whence, from either equation (iii) or (iv),

$$c_p - c_v = T\left(\frac{\partial v}{\partial T}\right)_p\left(\frac{\partial p}{\partial T}\right)_v \tag{II.17}$$

Hence the difference between the specific heats, constant for the perfect gas, is again expressed in terms of p–v–T data, so that one set of calorimetric data, either for c_v or c_p, suffices.

An alternative form of equation (II.17), making use of the self-evident relationship

$$\left(\frac{\partial p}{\partial T}\right)_v = -\left(\frac{\partial v}{\partial T}\right)_p\left(\frac{\partial p}{\partial v}\right)_T$$

results in

$$c_p - c_v = -T\left(\frac{\partial v}{\partial T}\right)_p^2\left(\frac{\partial p}{\partial v}\right)_T \tag{II.17a}$$

Making use of the *definitions* of the coefficient of cubical expansion, i.e.

$$\beta = \frac{1}{v}\left(\frac{\partial v}{\partial T}\right)_p \tag{v}$$

and of the compressibility K, i.e.

$$K = -\frac{1}{v}\left(\frac{\partial v}{\partial p}\right)_T \tag{vi}$$

substitution in equation (II.17a) results in

$$c_p - c_v = \frac{\beta^2 T v}{K} \tag{II.18}$$

This is a particularly useful form of the equation giving the difference between the specific heats at constant pressure and constant volume.

II.5 Ratio of Specific Heats

From the entropy equation (II.13), under conditions of constant entropy,

$$c_v(dT)_s = -T\left(\frac{\partial p}{\partial T}\right)_v (dv)_s$$

or

$$c_v\left(\frac{\partial T}{\partial v}\right)_s = -T\left(\frac{\partial p}{\partial T}\right)_v \tag{i}$$

whilst, from the second entropy equation (II.14), by similar reasoning

$$c_p\left(\frac{\partial T}{\partial p}\right)_s = T\left(\frac{\partial v}{\partial T}\right)_p \tag{ii}$$

Dividing (ii) by (i)

$$\frac{c_p}{c_v} = \frac{-\left(\dfrac{\partial v}{\partial T}\right)_p}{\left(\dfrac{\partial p}{\partial T}\right)_v}\left(\frac{\partial p}{\partial T}\right)_s\left(\frac{\partial T}{\partial v}\right)_s$$

$$= -\left[\left(\frac{\partial v}{\partial T}\right)_p \div \left(\frac{\partial p}{\partial T}\right)_v\right]\left(\frac{\partial p}{\partial v}\right)_s \tag{iii}$$

From the self-evident relationship

$$\left(\frac{\partial v}{\partial T}\right)_p\left(\frac{\partial T}{\partial p}\right)_v\left(\frac{\partial p}{\partial v}\right)_T = -1 \tag{iv}$$

the quantity in square brackets may be simplified to become

$$-\left(\frac{\partial v}{\partial p}\right)_T$$

i.e.

$$\frac{c_p}{c_v} = \left(\frac{\partial p}{\partial v}\right)_s \div \left(\frac{\partial p}{\partial v}\right)_T \tag{II.19}$$

Hence the ratio of specific heats may again be determined from *p–v–T* data, although one set must be determined under conditions of constant entropy.

II.6 Joule–Thomson Coefficient

The classical throttling experiment which led to the conclusion that the constant enthalpy process represented by *Fig.* 10.2 could be accompanied by changes in temperature, thus indicating the dependence of enthalpy not only on temperature, but also on pressure, led to the formulation of the Joule–Thomson coefficient (equation (10.9)) defined as

$$\mu = \left(\frac{\partial T}{\partial p}\right)_h \tag{II.20}$$

(see also *Fig*. 10.3)

Using the self-evident relationship

$$\left(\frac{\partial T}{\partial p}\right)_h \left(\frac{\partial p}{\partial h}\right)_T \left(\frac{\partial h}{\partial T}\right)_p = -1 \tag{i}$$

where by definition

$$c_p = \left(\frac{\partial h}{\partial T}\right)_p \tag{ii}$$

equation (II.20) becomes

$$\mu = -\frac{1}{c_p}\left(\frac{\partial h}{\partial p}\right)_T \tag{iii}$$

But from the second energy equation (II.16)

$$\left(\frac{\partial h}{\partial p}\right)_T = +v - T\left(\frac{\partial v}{\partial T}\right)_p$$

whence

$$\mu = \frac{1}{c_p}\left[T\left(\frac{\partial v}{\partial T}\right)_p - v\right] \tag{II.21}$$

Thus the Joule–Thomson coefficient may again be derived from p–v–T data in conjunction with specific heat values. Hence the behaviour indicated in *Fig*. 10.3 for gases may be predicted from property and specific heat data.

The Joule coefficient η, corresponding to the Joule–Thomson coefficient but applicable to non-flow adiabatic irreversible processes in which the internal energy u remains constant, may be defined as

$$\eta = \left(\frac{\partial T}{\partial v}\right)_u$$

and may be shown to be given by the following relationship, corresponding to equation (II.21)—

$$\eta = \frac{1}{c_v}\left[p - T\left(\frac{\partial p}{\partial T}\right)_v\right] \tag{II.22}$$

Appendix III

p-v-T Relationships

III.1 General

The perfect gas equation [equation (3.2) or (10.1)] and the van der Waals' equation (9.8) are two examples of equations of state for gases, connecting pressure, volume and temperature.

They have a limited range of application, and consequently many other equations of state have been devised to represent the relationship between pressure, volume and temperature.

Such equations of state have two main purposes—

(1) They provide a simple means of calculating any one of the three p–v–T properties in terms of the other two.
(2) Partial derivatives from p–v–T relationships are used to evaluate other property data, in terms of the thermodynamic relationships of Appendix II.

p–v–T data are represented in their most general form on a three-dimensional surface (*Figs.* 2.11 or 8.1) or, more conveniently, as a graph of the compressibility function pv/RT (derived from the perfect gas equation and equal to unity in that particular case), against pressure.

A typical graph of this compressibility function, for different values of absolute temperature, is shown in *Fig.* III.1. Temperature and pressure are usually expressed as ratios of these properties and their values at the critical point, and are referred to as reduced properties, i.e.

$$\frac{T}{T_c} = T_r \qquad \text{(III.1)}$$

$$\frac{p}{p_c} = p_r \qquad \text{(III.2)}$$

$$\frac{v}{v_c} = v_r \qquad \text{(III.3)}$$

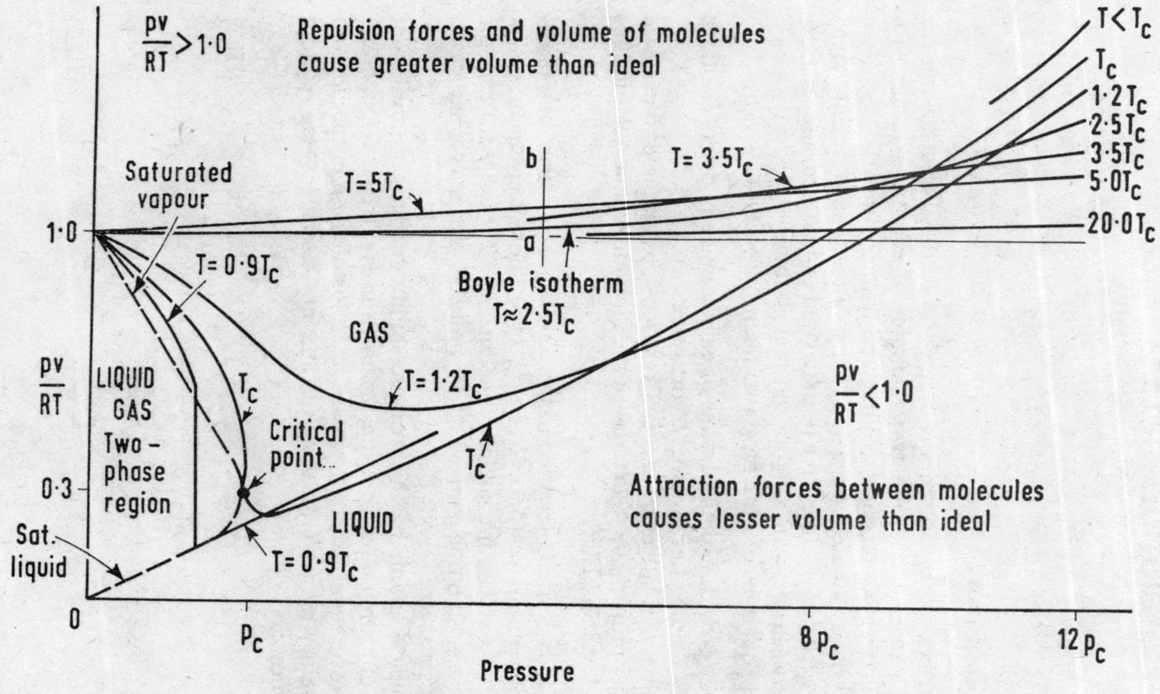

Fig. III.1 Typical compressibility chart for real gas
(From E. F. Obert, *Concepts of Thermodynamics* (McGraw-Hill, 1960))

The compressibility plot for the critical isothermal $T/T_c = T_r = 1$ shows a distinct minimum near the critical pressure, i.e. near $p/p_c = p_r = 1$.

The Boyle isothermal occurs at approximately $T/T_c = T_r = 2{\cdot}5$ and is characterized by zero slope at zero pressure.

It should be noted that at very low pressures all isothermals tend towards a value of the compressibility function equal to unity, i.e. towards perfect gas behaviour, as already stated in the introduction to Chapter 10, and as predicted by the kinetic theory of gases (Chapter 9).

This important fact may be expressed in the form—

$$\left(\frac{pv}{RT}\right)_{p\to 0} \to 1 \text{ for real gases} \qquad \text{(III.4)}$$

The general approach to the derivation of state equations is illustrated in the following section, dealing with van der Waals' equation (9.8) which was one of the earliest and is still one of the most useful of these equations.

III.2 Van der Waals' Equation, equation (9.8)

The equation may be stated in the form—

$$\left(p + \frac{a}{v^2}\right)(v - b) = RT \qquad \text{(III.5)}$$

the significance of the terms a/v^2 and $-b$ having been discussed in section 9.5.

The constants a and b may be found either from a systematic curve fitting procedure in terms of measured p–v–T property values, or more readily from certain conditions known to be valid at the critical point, viz.—

(1) The equation itself is valid

$$\left(p_c + \frac{a}{{v_c}^2}\right)(v_c - b) = RT_c \qquad \text{(i)}$$

(2) The critical isothermal has zero slope (see *Fig.* 9.5)

$$\left(\frac{\partial p}{\partial v}\right)_{T=T_c} = 0 \text{ at } p = p_c \qquad \text{(ii)}$$

(3) The critical isothermal has a point of inflexion

$$\left(\frac{\partial^2 p}{\partial v^2}\right)_{T=T_c} = 0 \text{ at } p = p_c \qquad \text{(iii)}$$

It is thus evident that three separate conditions have to be satisfied, while only two constants have to be evaluated, namely a and b. Hence van der Waals' equation may be expressed in two forms, namely

(*i*) *The first form* in which the gas constant R is allowed to assume a value R' which allows conditions (1) to (3), above, to be satisfied at the critical point. This form necessarily results in large errors under those conditions where perfect gas behaviour is normally approached, i.e. at very low pressures (see equation III.4) due to the substitution of a value of R' differing greatly from the theoretically correct one given by equation (3.7), i.e.

$$R = \frac{G}{M} \tag{III.6}$$

(*ii*) *The second form* in which the correct gas constant given by equation (III.6) is retained, but in which the critical volume v_c is adjusted to a new value v_c'. In this form van der Waals' equation tends to be accurate under conditions remote from the critical point.

III.3 Derivation of First and Second Form of van der Waals' Equation

(*a*) *First form*

From conditions (i), (ii) and (iii) above,

$$p_c = \frac{R'T_c}{v_c - b} - \frac{a}{v_c^2} \tag{iv}$$

$$\left(\frac{\partial p}{\partial v}\right)_{T=T_c} = \frac{-R'T_c}{(v_c - b)^2} + \frac{2a}{v_c^3} = 0 \tag{v}$$

$$\left(\frac{\partial^2 p}{\partial v^2}\right)_{T=T_c} = \frac{2R'T_c}{(v_c - b)^3} - \frac{6a}{v_c^4} = 0 \tag{vi}$$

Dividing (v) by (vi)

$$v_c - b = \tfrac{2}{3}v_c \qquad \text{i.e.} \quad b = \tfrac{1}{3}v_c$$

Substituting in (v)

$$a = \tfrac{9}{8}R'T_c v_c$$

Substituting for a and b in (iv)

$$p_c = \frac{R'T_c}{\frac{2}{3}v_c} - \frac{\frac{9}{8}R'T_c}{v_c}$$

$$= \frac{3}{8}\frac{R'T_c}{v_c}$$

i.e.

$$R' = \frac{8}{3}\frac{p_c v_c}{T_c} \qquad \text{(vii)}$$

resulting in the following form of van der Waals' equation—

$$p = \left(\frac{\frac{8}{3}p_c v_c}{T_c}\right)\left(\frac{T}{v - \frac{1}{3}v_c}\right) - \frac{3p_c v_c^{\,2}}{v^2} \qquad \text{(III.7)}$$

(*b*) *Second form*

Introducing an adjusted critical volume v_c', and retaining the gas constant R from equation (III.6), conditions (i), (ii) and (iii), above reduce to

$$p_c = \frac{RT_c}{v_c' - b} - \frac{a}{v_c'^2} \qquad \text{(viii)}$$

$$\left(\frac{\partial p}{\partial v}\right)_{T=T_c} = \frac{-RT_c}{(v_c' - b)^2} + \frac{2a}{v_c'^3} = 0 \qquad \text{(ix)}$$

$$\left(\frac{\partial^2 p}{\partial v^2}\right)_{T=T} = \frac{2RT_c}{(v_c' - b)^3} - \frac{6a}{v_c'^4} = 0 \qquad \text{(x)}$$

whence $b = \frac{1}{3}v_c' \qquad a = \frac{9}{8}RT_c v_c' \qquad v_c' = \frac{3}{8}\frac{RT_c}{p_c}$ (xi)

i.e.

$$a = \frac{27}{64}\frac{R^2T_c^{\,2}}{p_c} \qquad b = \frac{1}{8}\frac{RT_c}{p_c}$$

resulting in the second form of van der Waals' equation—

$$p = \left(\frac{RT}{v - \frac{1}{8}RT_c/p_c}\right) - \left(\frac{(27/64)(R^2T_c^{\,2})/p_c}{v^2}\right) \qquad \text{(III.8)}$$

III.4 Reduced Form of van der Waals' Equation

As already stated in section III.1, the use of the reduced properties T_r, p_r and v_r results in considerable simplification and generalization of the compressibility plot, *Fig.* III.1. These reduced properties may readily be incorporated in the two forms of van der Waals' equation, i.e. (III.7) and (III.8) and lead to the following forms—

(*a*) *First reduced form* (from equation (III.7))

$$\frac{p}{p_c} = \frac{\frac{8}{3}(T/T_c)}{(v/v_c) - \frac{1}{3}} - \frac{3}{(v/v_c)^2}$$

i.e.

$$p_r = \frac{8}{3}\frac{T_r}{(v_r - \frac{1}{3})} - \left(\frac{3}{v_r^2}\right) \qquad \text{(III.9)}$$

(*b*) *Second reduced form* (from equation (III.8))

After transposition, equation (III.8) becomes

$$\frac{p}{p_c} = p_r = \frac{T_r}{v_r\left(\frac{v_c p_c}{RT_c}\right) - \frac{1}{8}} - \frac{27/64}{v_r^2\left(\frac{v_c p_c}{RT_c}\right)^2} \qquad \text{(xii)}$$

and introducing the critical point compressibility function

$$z_c = \frac{v_c p_c}{RT_c}$$

then

$$p_r = \frac{T_r}{v_r z_c - \frac{1}{8}} - \frac{27/64}{v_r^2 z_c^2} \qquad \text{(III.10)}$$

These two forms of the equation may now be used to produce compressibility plots of the form of *Fig*. III.1.

Introducing the compressibility function

$$z = \frac{pv}{RT}$$

(see equation (III.4)), we have

$$z = \frac{pv}{RT} = \frac{\left(\frac{p}{p_c}\right)\left(\frac{v}{v_c}\right)}{\left(\frac{T}{T_c}\right)}\left(\frac{p_c v_c}{RT_c}\right)$$

$$= \frac{p_r v_r}{T_r} z_c \qquad \text{(III.11)}$$

where p_r is given either by equation (III.9) or (III.10).

Hence, substituting in equation (III.11)

(*a*) *First form*

$$z = z_c \left[\frac{8v_r}{3v_r - 1} - \frac{3}{v_r T_r} \right] \qquad \text{(III.12)}$$

(*b*) *Second form*

Introducing the so-called *ideal reduced volume*

$$v_r' = v_r z_c \qquad \text{(xiii)}$$

and expressing equation (III.10) in terms of the compressibility function z we have

$$z = \frac{v_r'}{v_r' - \frac{1}{8}} - \frac{27/64}{T_r v_r'} \qquad \text{(III.13)}$$

Since van der Waals' equation in its first form is highly inaccurate under those conditions where perfect gas behaviour is approached, it is clear that for general purposes the compressibility plot should be based on the second form of the compressibility factor, given by equation (III.13). For the critical isothermal $T = T_c$ this gives remarkably good accuracy over the range $0 < p < p_c$, though a considerable deviation occurs beyond $p = p_c$, but for higher isothermals agreement between calculated and measured compressibility factors is very close (see *Fig*. III.2).

III.5 Principle of Corresponding States

Van der Waals concluded from the existence of certain common characteristics for all gases, e.g. the critical point, the common form of the critical isothermal, the Boyle isotherm (see *Fig*. III.1) that a common representation must be possible for all gases. This led to the formulation of the reduced properties p_r, v_r, T_r, already defined in section III.1, and given by equations (III.1) to (III.3), in which actual property values are expressed as ratios of those values and the corresponding critical values. According to the principle of corresponding states, representation of those reduced properties on a three-dimensional surface in the form

$$f(p_r, v_r, T_r)$$

leads to a common representation for all gases. In fact, the principle is obeyed much more closely when the ideal reduced volume v_r', equation (xiii), section III.4, is used instead of the actual reduced volume v_r. This leads to some discrepancies in the immediate vicinity of the critical point, but otherwise generally good agreement.

Whilst the principle of corresponding states may be demonstrated using a generalized three-dimensional plot in the form of a p_r, v_r, T_r surface corresponding to p–v–T surfaces for particular gases

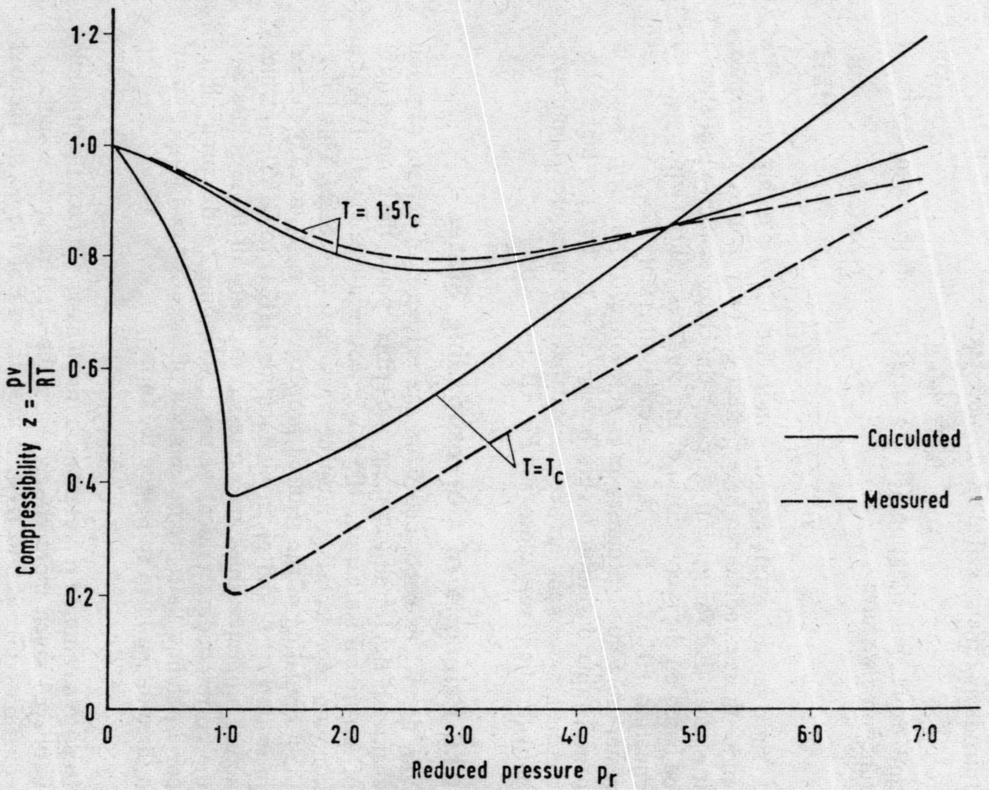

Fig. III.2 Comparison of calculated and measured compressibility factors

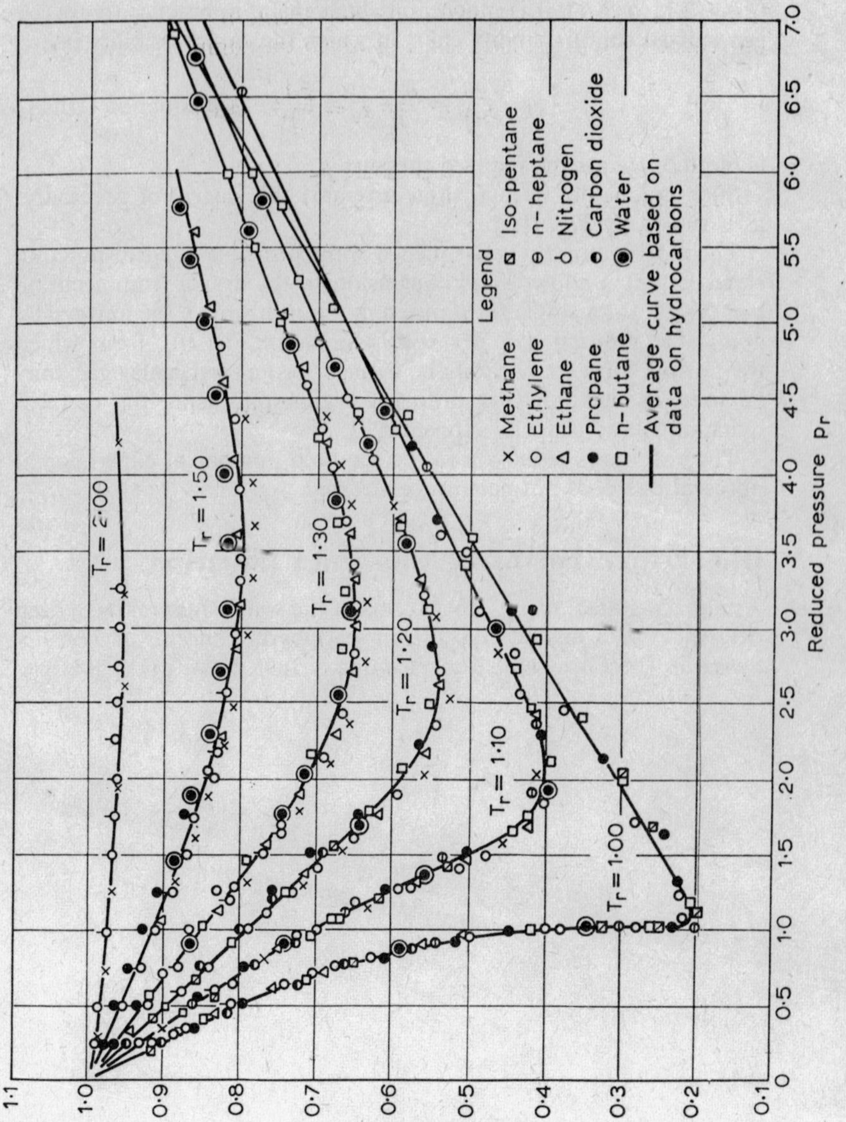

Fig. III.3 Generalized compressibility chart for various gases

(*Figs*. 2.11 or 8.1), it is much more convenient in practice to use the generalized compressibility chart in which the compressibility factor

$$z = \frac{pv}{RT} = \frac{p_r v_r}{T_r} z_c = \frac{p_r v_r'}{T_r} \quad \text{see equation (III.11)}$$

is plotted against the reduced pressure p_r.

Such graphs do, in fact, show a remarkable degree of generality, as is shown by *Fig*. III.3.

The most important consequence of the principle of corresponding states is that it allows valid conclusions to be drawn from accurate p–v–T data for a single substance only. For example, the universally applicable reduced van der Waals' equation (III.10), from which the normal form may readily be derived for any particular gas, may be used as the basis for property evaluations using the thermodynamic relationships of Appendix II.

The enormous reduction in the required number of experimental observations does not need to be stressed.

III.6 Other Forms of the State Equation

As already stated, many other forms of the state equation have been devised. For a detailed discussion see Obert, *Concepts of Thermodynamics* (McGraw-Hill Book Company Inc., New York, 1960).

Index

Basic Engineering Thermodvnamics

1985년 3월 1일 인 쇄
1985년 3월 15일 발 행

著 者 Wallace
發行者 金 炳 喜

發行處 塔 出 版 社
서울특별시 종로구 신문로 1가 135-2
전 화 : 722 - 0981 ~ 3 725 - 8400
등 록 : 1975. 5. 21. 제1-428

값 원